Lecture Notes in Civil Engineering

Volume 141

Lecture Notes in Civil Engineering (LNCE) publishes the latest developments in Civil Engineering - quickly, informally and in top quality. Though original research reported in proceedings and post-proceedings represents the core of LNCE, edited volumes of exceptionally high quality and interest may also be considered for publication. Volumes published in LNCE embrace all aspects and subfields of, as well as new challenges in, Civil Engineering. Topics in the series include:

- Construction and Structural Mechanics
- Building Materials
- Concrete, Steel and Timber Structures
- Geotechnical Engineering
- Earthquake Engineering
- Coastal Engineering
- Ocean and Offshore Engineering; Ships and Floating Structures
- Hydraulics, Hydrology and Water Resources Engineering
- Environmental Engineering and Sustainability
- Structural Health and Monitoring
- Surveying and Geographical Information Systems
- Indoor Environments
- Transportation and Traffic
- Risk Analysis
- Safety and Security

To submit a proposal or request further information, please contact the appropriate Springer Editor:

– Mr. Pierpaolo Riva at pierpaolo.riva@springer.com (Europe and Americas);
– Ms. Swati Meherishi at swati.meherishi@springer.com (Asia - except China, and Australia, New Zealand);
– Dr. Mengchu Huang at mengchu.huang@springer.com (China).

All books in the series now indexed by Scopus and EI Compendex database!

More information about this series at http://www.springer.com/series/15087

Anton Sinitsyn
Editor

Sustainable Energy Systems: Innovative Perspectives

SES 2020

Springer

Editor
Anton Sinitsyn
Vologda State University
Vologda, Russia

ISSN 2366-2557 ISSN 2366-2565 (electronic)
Lecture Notes in Civil Engineering
ISBN 978-3-030-67653-7 ISBN 978-3-030-67654-4 (eBook)
https://doi.org/10.1007/978-3-030-67654-4

This Springer imprint is published by the registered company Springer Nature Switzerland AG
The registered company address is: Gewerbestrasse 11, 6330 Cham, Switzerland

Contents

Optimization Model for Determination of Characteristics for the Development of Construction Project Involving Labor Forces

Anastasia Kiseleva[1], Anton Radaev[1(✉)], and Marlena Rajczyk[2]

[1] Peter the Great St. Petersburg Polytechnic University, St. Petersburg, Russian Federation
radaev_ae@spbstu.ru
[2] Częstochowa University of Technology, Czestochowa, Poland

Abstract. The paper considers the issues of determination of the characteristics of development of a construction project in the field of road construction involving labor forces. The relevance of the research is determined both by the existing features of development of the road construction industry and by the limited scientific developments, which fully take into account the particular circumstances of work distribution among labor force units. During the research, an optimization model has been developed to provide calculation of unknown variables – the amount of work assigned to each labor force unit as well as the duration of individual stages of a road construction project. The calculations are based on the source data, including indicators of sourcing efficiency and unit costs of individual labor force units, as well as the estimated value of the total amount of work under the project. It is important to note that the developed model is linear and therefore can be effectively implemented with application of simplex method that is available in a large number of software environments for optimization modeling.

The developed model has been implemented on a practical example connected to distribution (among units – workers of design organization) the work carried out as part of a particular construction project. On the basis of the results it has been concluded that the developed model is highly adequate and, as a consequence, has a high practical significance.

Keywords: Construction project · Labor forces · Optimization model

1 Introduction

The current conditions of the construction industry's development in relation to road construction are characterized by reduced timing of construction projects executed by specialized organizations (in order to achieve competitive advantages on regional and international markets). As a result, there is a growth of a need for rational organization of work processes at various stages of a construction project life cycle, including the stages of the facility design and construction. In this context, the issues of rational

A. Sinitsyn (Ed.): SES 2020, LNCE 141, pp. 1–13, 2021.
https://doi.org/10.1007/978-3-030-67654-4_1

organization of design development of a road construction project, including distribution of work among the organizational and technological resources involved in this process, are of particular importance. That circumstance is also confirmed by growing interest of the design organizations in the tools for determination of characteristics of the construction project development process involving labor forces. That can be explained, among other reasons, by insufficiency of relevant scientific and methodological developments. The mentioned-above features have determined the feasibility of the research, which purpose is to develop a tool for determination of the characteristics of project development process in the field of road construction involving labor forces. The following issues have been identified as main objectives of the research:

1. Review and analysis of scientific works connected to the research area.
2. Creation of tool for modeling of the process of development a road construction project involving labor forces.
3. Implementation of the developed tool on a practical example.

Thus, the research object is the process of development of a road construction project involving labor forces, and the research subject is defined by the characteristics of the process in terms of the volume of work assigned to individual units, as well as terms of execution for the main stages of the process.

At the initial stages of the research, a review and analysis of scientific developments [1–22] in the research area has been carried out. Based on the results of mentioned-above procedure, the following partial conclusions were made:

1. All scientific developments regarding determination of the characteristics for the process of a construction project's development can be classified into the following main categories:
   1.1. Basic category of tasks: constant work-in-process (CONWIP) in multi-class dynamic Project Evaluation and Review Technique (PERT) networks [1]; discrete time – cost tradeoff problem – DTCTP [2]; construction project cost estimation [21]; multi-objective decision-making – MODM [3, 16]; resource constrained project scheduling problem – RCPSP [2, 3, 14, 19]; resource constrained multi-project scheduling problem – RCMPSP [15, 22]; multi-mode resource-constrained project scheduling problem – MRCPSP [4–9, 13, 17, 20]; resource-constrained project scheduling problem with flexible resource profiles – FRCPSP [10–12].
   1.2. Object of study: a set of projects [1, 5, 15, 17, 22]; a single project [2–4, 6–14, 19–21]; production or service process [16, 18].
   1.3. Validated parameter(s): the amount of resources assigned to carry out each individual task within the project [1, 3, 11, 18, 19]; deviation of the actual duration value of each individual task within the project from the planned value [2]; starting point and end point of each individual task within the project [5, 22]; execution mode of each individual task within the project [5–7, 9, 10, 12, 13, 16, 17, 20]; execution mode of each individual task within the project at a given time [8]; projected cost of the project [21].

1.4. Task solving tool: Markov chain calculation algorithm [1]; genetic algorithms [2, 6, 9, 12, 21]; Monte-Carlo method [3]; lexicographic optimization method [3]; ant colony optimization method [4]; simulated annealing method [7, 13]; firefly algorithm [14]; cuckoo search algorithm [19]; linear optimization method [16]; mixed-integer linear optimization method [8, 11, 17]; heuristic methods [10, 18]; direct calculation methods [18]; simulation modeling methods [22, 24]; critical path formation method [5]; method of priority rules comparative analysis [5, 15]; training methods for artificial neural networks [21].

2. The following issues have been identified as key disadvantages of the studied research: lack of consideration of particular characteristics of individual resource units [1–15, 17, 20–22]; impossibility of determining the optimal distribution of resources (the results determine only the requirements for the forces involved) [2, 5]; lack of consideration of the work aspects of design organizations in the field of construction [16, 18]; the tool is applicable only for planning of project implementation process before its execution [1, 3–15, 17–22]; the tool is applicable only for adapting an ongoing project to the external environment changes [2]; the tool is difficult in application to real workflow conditions in terms of stationarity of the process [1]; the process of determining an optimal solution is approximate [8, 10, 11, 17, 20–22]; the process of finding the optimal solution is largely complex [1, 3, 8, 11, 15, 17, 20]; lack of mechanism identifying the optimal solution [2, 4, 6, 7, 9, 12–14, 18, 19]; dependence of optimization results on source data values [4, 7, 9, 12–14, 19, 21, 22].

Based on the above, it follows that the vast majority of research and development related to design of construction projects do not take into account the particular characteristics of individual labor force units (including their productivity and working time fund). At best, the complexity of the workforce structure is determined by certain corresponding classes, which, however, does not take into account the features of the issue in question.

Research works connected to the direct evaluations of the individual workforce characteristics consider the production or service process as object of study and therefore do not take into account the actual work aspects of design organizations in relation to design of construction projects. Thus, on the basis of the results received at the initial stage of the study, it has been concluded that there is no relevant research of construction project involving labor forces and characteristics of its development process, which once again confirmed the high relevance of the research. A description of the proposed mathematical model for solving the considered problem is presented in the following section of the paper.

2 Methods

2.1 Development of Optimization Model

The main provisions of the developed optimization model are as follows:

1. The object of the study is the process of development of a project in the field of road construction, involving relevant work in the context of individual stages using specific labor force units.
2. A labor force unit is an individual employee of a design organization or a separate unit of technological equipment performing manual and mechanized or automated labor operations (the term "labor force unit" was introduced to generalize the initial problem and the possibility of using the developed tool for the widest possible range of homogeneous technological process organization tasks) [23].
3. The stages of project's development are performed continuously and sequentially: the end point of current project stage corresponds to the starting point of the next one.
4. Performing of each project stage is ensured by involving labor force units (for example, employees of a design organization), each assigned a certain amount of work, performed independently of other units.
5. The work within each individual project stage is carried out by the labor force units involved simultaneously. Thus, the nominal duration of each individual project stage is determined by the highest value of the work duration for each individual labor force unit involved.
6. The process of performing the work under a separate project stage is generally described by a limited set of labor force units involved as both mandatory and optionally.
7. Effectiveness of involving labor force units is described by a specific indicator that determines the amount of work performed per unit of time (for project organization employees it is the qualification coefficient); this specific indicator is, to some extent, similar to productivity.
8. Additional characteristics of labor force units are the unit costs of work (for employees of the project organization it is remuneration), as well as the maximum amount of work performed within the project in question [24, 25].
9. It's necessary to determine the amount of work performed by labor force unit at each stage of the project, that provides the minimum total work costs of labor force units and the project stage duration within acceptable values.

The source data and unknown variables of the developed model are presented in Table 1. The calculated characteristics calculated in the framework of the model are shown in Table 2.

Table 1. Source data and unknown variables of the generated optimization model

No.	Name of source data element/unknown variable	Meas. units[(a)]	Designation
1	**Source data**		
1.1	**General source data**		
1.1.1	Number of labor force units	unit	m
1.1.2	Number of work stages	unit	n
1.1.3	Estimated value of the total amount of work	WVU	$[t_{\Sigma}]$
1.1.4	Minimum project duration	hour	$T^{\min}$
1.1.5	Maximum project duration	hour	$T^{\max}$
1.2	**Indexes**		
1.2.1	Labor force unit index	–	$i = 1, 2, \ldots m$
1.2.2	Work stage index	–	$j = 1, 2, \ldots n$
1.3	**Parameters of individual labor force unit** i ($i = 1, 2, \ldots, m$)		
1.3.1	Name	–	–
1.3.2	Involvement efficiency indicator[(b)]	WVU/hour	k_i
1.3.3	Unit costs of work	CU/hour	c_i
1.3.4	Maximum permitted amount of work[(c)]	WVU	$t_i^{\max}$
1.4	**Parameters of individual work stage** j ($j = 1, 2, \ldots, n$)		
1.4.1	Name	–	–
1.4.2	Proportion of the stage duration j ($j = 1, 2, \ldots, n$) of the general project duration	–	γ_j
1.5	**Work parameters of individual labor force unit** i ($i = 1, 2, \ldots, m$) **within work stage** j ($j = 1, 2, \ldots, n$)		
1.5.1	Index of possibility of involving	–	y_{ij}
1.5.2	Minimum permitted amount of work[(d)]	WVU	$t_{ij}^{\min}$
2	**Unknown variables**		
2.1	Work performed by labor force unit i ($i = 1, 2, \ldots, m$) within project stage j ($j = 1, 2, \ldots, n$)	WVU	t_{ij}
2.2	Actual duration of project stage j ($j = 1, 2, \ldots, n$)	hour	T_j

Note:
[(a)]WVU – work volume unit – is a generalized unit of measurement describing the quantitative equivalent of elementary volume of work performed by one or a group of labor force units (in the particular case it corresponds to a work hour of an organization employee or equipment unit, a page of project documentation, etc.);
[(b)]the source data element is used as an equivalent when matching WVU and time units; when labor force units relate to employees of a design organization, the indicator value for an individual unit is inversely proportional to the skill level of the corresponding employee;
[(c)]the source data element is used for taking into account the limited ability of involving labor force units into the project;
[(d)]the source data element is used for provision of the necessary involvement of individual labor force units into the project.

Table 2. Calculated characteristics of the generated optimization model

№	Name of calculated characteristic	Units	Expression
1	**Estimated characteristics calculated for each individual labor force unit** i $(i = 1, 2, \ldots, m)$		
1.1	Total workload	WVU	$t_i = \sum_{j=1}^{n} t_{ij}$
1.2	Actual value of the work duration[(a)]	hour	$T_i = \sum_{j=1}^{n} T_{ij} = \frac{\sum_{j=1}^{n} t_{ij}}{k_i}$
1.3	Total work costs	CU	$C_i = c_i \cdot T_i = \frac{c_i}{k_i} \cdot \sum_{j=1}^{n} t_{ij}$
2	**Estimated characteristics calculated for each individual work stage** j $(j = 1, 2, \ldots, n)$		
2.1	Minimum stage duration[(b)]	hour	$T_j^{\min} = \gamma_j \cdot [T^{\min}]$
2.2	Maximum stage duration[(b)]	hour	$T_j^{\max} = \gamma_j \cdot [T^{\max}]$
3	**Estimated characteristics calculated for each individual labor force unit** i $(i = 1, 2, \ldots, m)$ **within the work stage** j $(j = 1, 2, \ldots, n)$		
3.1	Maximum permitted workload	WVU	$t_{ij}^{\max} = k_i \cdot T_j \cdot y_{ij}$
3.2	Work duration	hour	$T_{ij} = \frac{t_{ij}}{k_i}$
4	**Aggregated design characteristics**		
4.1	Actual value of the total workload	WVU	$t_\Sigma = \sum_{i=1}^{m} t_i = \sum_{i=1}^{m}\sum_{j=1}^{n} t_{ij}$
4.2	Actual value of the total work duration	hour	$T_\Sigma = \sum_{j=1}^{n} T_j$
4.3	Total cost of operation of all labor force units	CU	$C_\Sigma = \sum_{i=1}^{m} C_i = \sum_{i=1}^{m}\left(\frac{c_i}{k_i} \cdot \sum_{j=1}^{n} t_{ij}\right)$

Note:
[(a)]the calculated characteristic in the central part of the expression is determined by paragraph 3.2 of the table;
[(b)]the calculated characteristic value does not depend on the values of unknown variables.

The optimization model in generalized form is as follows:

$$\begin{cases} C_\Sigma(\{t_{ij}\}) \to \min; & (1) \\ t_{ij} \geq t_{ij}^{\min},\ i = 1, 2, \ldots, m;\ j = 1, 2, \ldots, n; & (2) \\ t_{ij} \leq t_{ij}^{\max}(T_j),\ i = 1, 2, \ldots, m;\ j = 1, 2, \ldots, n; & (3) \\ T_j^{\min} \leq T_j \leq T_j^{\max},\ j = 1, 2, \ldots, n; & (4) \\ t_i(\{t_{ij}\}) \leq t_i^{\max},\ i = 1, 2, \ldots, m; & (5) \\ t_\Sigma(\{t_{ij}\}) = [t_\Sigma]. & (6) \end{cases}$$

The optimization model in expanded form is as follows:

$$\begin{cases} \sum_{i=1}^{m}\left(\frac{c_i}{k_i}\cdot\sum_{j=1}^{n}t_{ij}\right)\to \min; & (7)\\ t_{ij}\geq t_{ij}^{\min},\ i=1,2,\ldots,m;\ j=1,2,\ldots,n; & (8)\\ t_{ij}\leq k_i\cdot T_j\cdot y_{ij},\ i=1,2,\ldots,m;\ j=1,2,\ldots,n; & (9)\\ T_j^{\min}\leq T_j\leq T_j^{\max},\ j=1,2,\ldots,n; & (10)\\ \sum_{j=1}^{n}t_{ij}\leq t_i^{\max},\ i=1,2,\ldots,m; & (11)\\ \sum_{i=1}^{m}\sum_{j=1}^{n}t_{ij}=[t_{\Sigma}] & (12)\end{cases}$$

As can be seen from the above expressions, the developed optimization model is linear regarding to unknown variables and therefore can be successfully implemented with application of simplex method in modern software for optimization modeling, in particular, the “Solver” add-in for “Microsoft Excel” software, the “Optimization Toolbox” add-in for “Matlab”, and others [26, 27].

In order to validate the developed model, it was applied to a practical case. A more detailed description of the test is given in the next subsection of the work.

3 Results and Discussion

At the final stage of the research, the developed optimization model has been implemented on the practical example for distribution of the work related to implementation of a construction project with 6 key stages to 11 labor force units – employees (contractors) of a design organization [28, 29]. Normalized temporary units were considered as units of workload, while the lower deadline limit for the entire project was not set. The optimization model has been created in “Microsoft Excel” software and implemented with application of the “Solver” add-in. A general view of the Microsoft Excel worksheet with the values of the source data elements, the model implementation results — the optimal values of the variables, and the values of the calculated characteristics is shown in Fig. 1. The guidelines for filling out the worksheet are presented in Table 3, the settings for the “Solver” add-in are given in Table 4.

Table 3. The guidelines for filling out the cells of the "Microsoft Excel" worksheet for implementation of the developed model on a practical example

Worksheet cell addresses[(a)]	Content/Formula	Structural element of mathematical description
Table 1		
E4:J4	values of source data elements	Table 1, no 1.4.1
N4(:S4)	=E4	–
B8:B18	values of source data elements	Table 1, no 1.3.1
E8:J18	values of source data elements	Table 1, no 1.5.1
K8:K18	values of source data elements	Table 1, no 1.3.2
L8:L18	values of source data elements	Table 1, no 1.3.4
M8:M18	values of source data elements	Table 1, no 1.3.3
N8:S18	values of source data elements	Table 1, no 1.5.2
E19:J19	values of source data elements	Table 1, no 1.4.2
E22(:J22)	=E19*I54	Table 2, no 2.1
E25(:J25)	=E19*I55	Table 2, no 2.2
Table 2		
E31(:J31)	=E4	–
N31(:S31)	=E4	–
B35(:B45)	=B8	–
E35:J45	values of unknown variables	Table 1, no 2.1
K35(:K45)	=SUM(E35:J35)	Table 2, no 1.1
L35(:L45)	=SUM(E35:J35)/K8	Table 2, no 1.2
M35(:M45)	=L35*M8	Table 2, no 1.3
N35(:S45)	=E8*E$47*$K8	Table 2, no 3.1
E47:J47	values of unknown variables	Table 1, no 2.2
Table 3		
I54	values of source data elements	Table 1, no 1.1.3
I55	values of source data elements	Table 1, no 1.1.4
I56	values of source data elements	Table 1, no 1.1.5
I58	=SUM(K35:K45)	Table 2, no 4.1
I59	=SUM(E47:J47)	Table 2, no 4.2
I60	=SUM(M35:M45)	Table 2, no 4.3

Note:
[(a)]the abstract designation A1 (: B10) means that in the cell A1 you need to enter the formula indicated in the corresponding column of the table, after which "stretch" (copy) the result to cell B10.

Table 4. Settings for the "Solver: add-in of "Microsoft Excel" software used for implementation of the model on the practical example

Setting item	Value	Structural element of mathematical description
Target cell	I60	Expressions (1), (7)
Optimization type	Minimization	
Variables cells	E35:J45;E47:J47	Table 1, nos 2.1, 2.2
Constraints	E35:J45 >= N8: S18	Expressions (2), (8)
	E35:J45 <= N35: S45	Expressions (3), (9)
	E47:J47 >= E22: J22	Expressions (4), (10), left side
	E47:J47 <= E25: J25	Expressions (4), (10), right side
	K35:K45 <= L8: L18	Expressions (5), (11)
	I58 = I54	Expressions (6), (12)
Solution search method	Simplex LP	–

Based on the results, the following partial conclusions have been made:

1. Given the maximum amount of the performers' work ensuring minimum project implementation costs, it is necessary to attract the most qualified employees, according to the appropriate ratios of the performance indicators of attracting performers and the work costs.
2. For the given boundary values of the total project duration, the actual value of the specified characteristic corresponding to the optimal solution is $209.5/450 \approx 46.55\%$ of the initial value of characteristic.

Thus, the values obtained during the implementation of the developed optimization model on the practical example correspond to the actual activity conditions of design organizations, and that, in turn, indicates the adequacy and high practical significance of the developed tool.

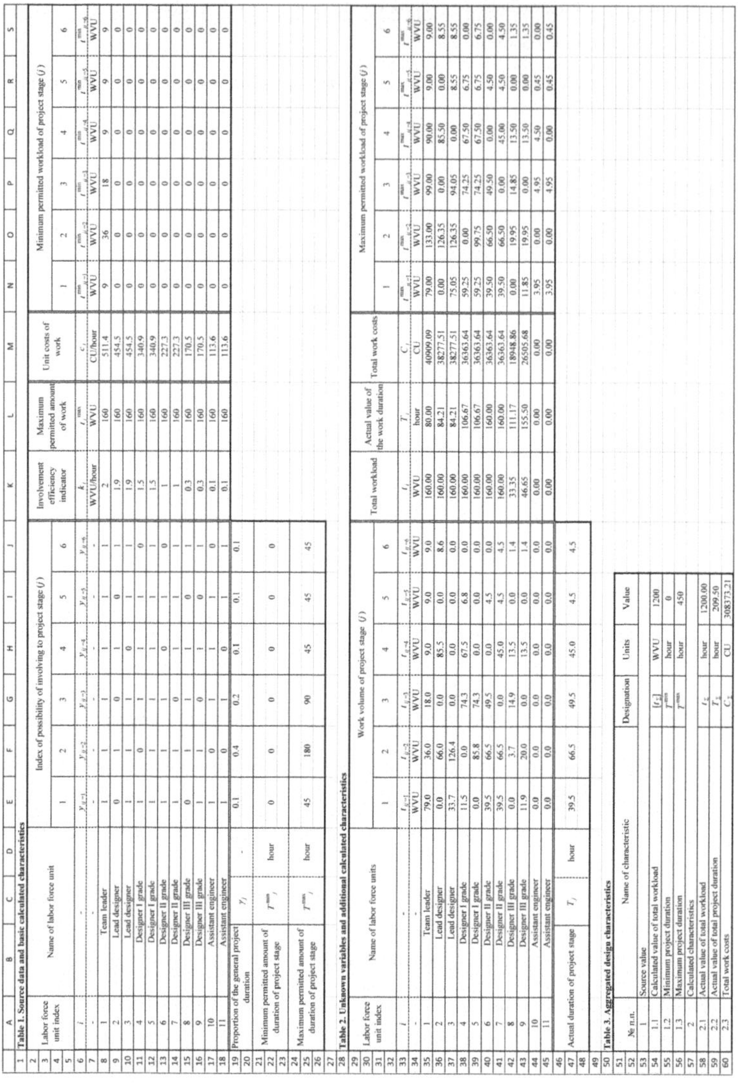

Fig. 1. General view of the "Microsoft Excel" worksheet as part of the developed model on a practical example

4 Conclusions

Based on the research, the following main conclusions were made:

1. Results of the analysis of scientific developments in the research area have showed that the studies connected to the determination of the characteristics for the process of a construction project's development involving labor forces are of high necessity

due to lack of mathematical modeling tools taking into account the actual distribution of workloads to labor forces in designing a construction project.

2. Optimization model for determination of characteristics of development process of a construction project involving labor forces has been developed and implemented on a practical case. It has been concluded, based on the relevant results, that the developed tool has a high practical value.

Nevertheless, it is important to note certain features of the model that limit the scope of tasks for which the tool can be considered effective. The features are the following:

1. The adequacy of results of the optimization model depend on the adequacy of setting the estimated value of the total work amount during the project (as an element of the initial data). Complexity of its forecast directly depends on the complexity and scale of the project structure, as well as on the number and scale of labor forces involved in the project. In this regard, taking into account the negative impact of various factors on the estimated value of the total workload, it is advisable to preliminarily determine the range of the total workload calculated values. On the basis of that, through a series of launches of the optimization model, alternative solutions (combinations of optimal variable values) should be determined with the subsequent selection of the preferred one based on the specified probability of forecast value deviation of the total workload from the one actually received.
2. The difficulty of taking into account of the involvement of labor force units in other projects that implemented simultaneously with the project under consideration, regarding the maximum permitted work amount for each individual performer. To eliminate this drawback, it is advisable to modify the initial optimization model to be able to take into account several projects to be performed in the context of individual stages.

During the next stages of the research, it is planned to develop the aforementioned modified optimization model, as well as to create the procedure for implementation of computational experiments with taking into account the possible deviations of the predicted values of the total work volume under a projects from the actual values.

References

1. Yaghoubi, S., Noori, S., Mazdeh, M.M.: A heuristic method for consumable resource allocation in multi-class dynamic PERT networks. J. Ind. Eng. Int. **9**(1) (2013)
2. Shahriari, M.: Multi-objective optimization of discrete time–cost tradeoff problem in project networks using non-dominated sorting genetic algorithm. J. Ind. Eng. Int. **12**(2), 159–169 (2016). https://doi.org/10.1007/s40092-016-0148-8
3. Hashemin, S.S., Fatemi Ghomi, S.M.T.: Constrained consumable resource allocation in alternative stochastic networks via multi-objective decision making. J. Ind. Eng. Int. **8**(1), 18 (2012)
4. Garmsiri, M., Abassi, M.R.: Resource leveling scheduling by an ant colony-based model. J. Ind. Eng. Int. **8**(1), 7 (2012)

5. Browning, T.R., Yassine, A.A.: Resource-constrained multi-project scheduling: priority rule performance revisited. Int. J. Prod. Econ. **126**(2), 212–228 (2010)
6. Alcaraz, J., Maroto, C., Ruiz, R.: Solving the multi-mode resource-constrained project scheduling problem with genetic algorithms. J. Oper. Res. Soc. **54**(6), 614–626 (2003)
7. Józefowska, J., Mika, M., Rózycki, R., Waligóra, G., Węglarz, J.: Simulated annealing for multi-mode resource-constrained project scheduling. Ann. Oper. Res. **102**(1–4), 137–155 (2001)
8. Sabzehparvar, M., Seyed-Hosseini, S.M.: A mathematical model for the multi-mode resource-constrained project scheduling problem with mode dependent time lags. J. Supercomputing **44**(3), 257–273 (2008)
9. Peteghem, V.V., Vanhoucke, M.: A genetic algorithm for the preemptive and non-preemptive multi-mode resource-constrained project scheduling problem. Eur. J. Oper. Res. **201**(2), 409–418 (2010)
10. Naber, A., Kolisch, R.: MIP models for resource-constrained project scheduling with flexible resource profiles. Eur. J. Oper. Res. **239**(2), 335–348 (2014)
11. Baumann, P., Fündeling, C.U., Trautmann, N.: The resource-constrained project scheduling problem with work-content constraints. In: Handbook on Project Management and Scheduling, vol. 1, pp. 533–544 (2015)
12. Ranjbar, M., Kianfar, F.: Resource-constrained project scheduling problem with flexible work profiles: a genetic algorithm approach. Scientia Iranica **17**(1 E), 25–35 (2010)
13. Tao, S., Dong, Z.S.: Multi-mode resource-constrained project scheduling problem with alternative project structures. Comput. Ind. Eng. **125**, 333–347 (2018)
14. Kassandra, T., Rojali, Suhartono, D.: Resource-constrained project scheduling problem using firefly algorithm. Procedia Comput. Sci. **135**, 534–543 (2018)
15. Wang, Y., He, Z., Kerkhove, L.-P., Vanhoucke, M.: On the performance of priority rules for the stochastic resource constrained multi-project scheduling problem. Comput. Ind. Eng. **114**, 223–234 (2017)
16. Shahnazari-Shahrezaei, P., Tavakkoli-Moghaddam, R., Kazemipoor, H.: Solving a multi-objective multi-skilled manpower scheduling model by a fuzzy goal programming approach. Appl. Math. Model. **37**(7), 5424–5443 (2013)
17. Chakrabortty, R.K., Sarker, R.A., Essam, D.L.: Multi-mode resource constrained project scheduling under resource disruptions. Comput. Chem. Eng. **88**, 13–29 (2016)
18. Mourtzis, D.: An integrated system for managing ship repair operations. Int. J. Comput. Integr. Manuf. **18**(8), 721–733 (2005)
19. Maghsoudlou, H., Afshar-Nadjafi, B., Akhavan Niaki, S.T.: Multi-skilled project scheduling with level-dependent rework risk; three multi-objective mechanisms based on cuckoo search. Appl. Soft Comput. J. **54**, 46–61 (2017)
20. Pinha, D.C., Ahluwalia, R.S.: Flexible resource management and its effect on project cost and duration. J. Ind. Eng. Int. **15**(1), 119–133 (2018). https://doi.org/10.1007/s40092-018-0277-3
21. Arabzadeh, V., Niaki, S.T.A., Arabzadeh, V.: Construction cost estimation of spherical storage tanks: artificial neural networks and hybrid regression – GA algorithms. J. Ind. Eng. Int. **14**(4), 747–756 (2018)
22. Araúzo, J.A., Pajares, J., Lopez-Paredes, A.: Simulating the dynamic scheduling of project portfolios. Simul. Model. Pract. Theory **18**(10), 1428–1441 (2010)
23. Skrypnik, I.L., Ksenofontov, Yu.G., Kaverzneva, T.T., Rumyantseva, N.V., Kiss, V.V.: Assessment of technical level of new, promising models of equipment at the stage of their development in modern engineering practice. In: IOP Conference Series: Materials Science and Engineering, vol. 862, no. 4, p. 042031 (2020)

24. Bukunova, O., Bukunov, A.: Information modelling as a tool to manage construction project information flows. In: IOP Conference Series: Materials Science and Engineering, vol. 753, no. 4, p. 042030 (2020)
25. Makovetskaya-Abramova, O., Ivanov, A., Lazarev, Y., Shakhova, M., Rozov, A.: Economic assessment of construction of the roadside service facilities. In: E3S Web of Conferences, vol. 157, p. 04035 (2020)
26. Radaev, A., Korneeva, E.: Method for forecasting pollution of urban areas. In: E3S Web of Conferences, vol. 140, p. 09005 (2019)
27. Radaev, A., Leventsov, V.: The methodology for substantiating the materials handling equipment of a unit load warehousing system. Int. J. Qual. Res. **12**(4), 989–1016 (2018)
28. Gamayunova, O.: The labor market in the construction industry: demand, supply, professional training. Procedia Eng. **117**(1), 1058–1064 (2015)
29. Bolotin, S., Dadar, A., Ptuhina, I.: Construction work tasks duration: pessimistic scenarios based on PERT method. Adv. Mater. Res. **945–949**, 3026–3031 (2014)

Development of a Waterproof Composition Based on Organic Polymers, Which Increases the Durability of Building Materials and Structures and Increases Their Resistance to Aggressive Factors

A. Y. Velsovskij[1(✉)], A. A. Sinitsyn[1], N. N. Gabibov[1], and L. R. Mukhametova[2]

[1] Department of Highways, Vologda State University, Lenina ul. 15, 160000 Vologda, Russia
v2u@yandex.ru
[2] Kazan State Power Engineering University, Str. Krasnoselskaya, 51, 420066 Kazan, Russia

Abstract. The article describes exploratory research and the creation of a new waterproof composition based on organic polymers. The most common measures to reduce concrete corrosion are described. The feature of the new waterproof coating is that, being polyalkene, it has increased resistance to aggressive influences, is inert and environmentally friendly. The developed composition increases the durability of building materials and structures, and increases their resistance to aggressive natural and man-made factors.

Keywords: Aggressive factors · Waterproof composition · Polymers

1 Introduction

Concrete and reinforced concrete are the most widespread building materials. Its application in construction still faces the problem of ensuring durability under the influence of aggressive media, both natural and anthropogenic [1–8].

The most common measures to reduce concrete corrosion are as follows:

1. Coating the concrete surface with waterproof materials of containing remineralizing additives capable of reversing the already ongoing corrosion processes.
2. Coating the concrete surface with polymerizing waterproof materials that form a tight-fitting waterproof film on the surface.
3. Coating the concrete surface with waterproof agents (with or without metal corrosion inhibitors), which form a tight-fitting waterproof membrane with good adhesion.
4. Irrigation of concrete surface with solutions containing migrating inhibitors of metal corrosion with high penetrating ability.

A. Sinitsyn (Ed.): SES 2020, LNCE 141, pp. 14–19, 2021.
https://doi.org/10.1007/978-3-030-67654-4_2

5. Treatment of the concrete surface with bridging mortars and coating agents containing remineralizing additives with high penetrating ability.
6. Sealing of seams and joints of reinforced concrete structures.

The authors of this article, with the financial support of the Government of the Vologda Region, within the framework of the state scientific grant (contract No. 8 of December 25, 2019), developed a new waterproof composition. It combines several measures to reduce concrete corrosion by reducing the material water resistance and imparts hydrophobic properties to the surface of structures.

The scientific novelty of this work consists in the study of possibility of using organic polymers to protect building materials from aggressive media and factors of a natural and man-made nature. These measures are aimed at giving the surface and structure waterproof and hydrophobic properties simultaneously.

To implement the work, the following priority tasks were formulated:

- To theoretically substantiate the possibility of increasing the strength and durability of building materials and structures by applying a protective composition with organic polymers.
- To investigate the impact of organic polymers on the performance characteristics of building materials and structures (water resistance, frost resistance, resistance to aggressive environments, etc.).
- To develop a preliminary model of the technological scheme for the industrial production of a waterproof mixture on the basis of an industrial enterprise.

2 Materials and Methods

To achieve the set tasks we analysed the scientific and technical literature, regulatory and technical documentation and other materials related to the considered topic. A patent research was conducted, the models of the research object were constructed, a list of experiments necessary to confirm certain provisions of theoretical research and obtain specific values was established. A matrix of options for the composition of a waterproof mixture based on selected groups of chemicals was developed. The prototypes of a waterproof mixture of various chemical composition were obtained and studied in laboratory. A preliminary model of a technological scheme for the industrial production of a waterproof mixture on the basis of an industrial enterprise was developed. Pilot batches of waterproof mixture samples with the best performance (based on the results of laboratory studies) were obtained and their impact on building materials and structures was tested, with subsequent optimization the waterproof composition according to the test results [9–12].

To determine the declared parameters, the purchased laboratory equipment for sample preparation was used, as well as the equipment belonging to the industrial enterprise. In addition to the joint laboratory research, a pilot batch of the waterproof mixture was manufactured on the basis of the enterprise (the production site is located in Vologda).

There are several methods for impregnating a porous material. The first method is as follows. The material being impregnated is placed in a chamber and evacuated, after which it is immersed in an impregnating liquid under vacuum and maintained. After that the pressure is instantly increased to atmospheric by air input into the chamber. Evacuation and air supply are repeated several times until the desired material impregnation is achieved. After impregnation, the material is removed from the chamber. Bitumen, petrolatum are used as the impregnating liquid, which are heated to a liquid state before impregnation [13].

The disadvantages of the first method are low bitumen adhesion and high impregnation temperature (up to 160 °C), as well as the need to use a special chamber, which can't be achieved when coating very large structures and products.

The second method consists in applying a thin-layer waterproof coating based on polymer sulfur on porous building materials and is as follows. The polymer sulfur is heated, the products being treated are dried, heated, and placed in an impregnating bath. The drying and heating of products is carried out at a temperature of 120–125 °C for 1–5 min in vacuum. benzoyl peroxide, corresponding out at a temperature of 130–140 °C for 0.5–10 min, after which the impregnation is repeated, having previously rotated the product, and the amount of impregnation depends on the number of edges of the product [14].

The disadvantages of the second method are the need to use high temperatures, as well as low chemical resistance and environmental safety of sulfur.

The authors developed a solution which is akin to the third way to increase the chemical resistance and strength of concrete products and structures. It consists in treating their surfaces with a composition based on monomers with an initiating system consisting of benzoyl peroxide and dimethylaniline taken in a ratio of 1:0.5. The initiating system is 2% by weight of monomers, the composition for processing consists of two monomers - styrene and methyl methacrylate, taken in a molar ratio of 2:1, which are copolymers. Surface treatment of concrete products and structures is carried out in two stages with an exposure time of 30 min between stages with preliminary heating of the composition with monomers at a temperature of 80 °C and a subsequent cooling at room temperature for 15–20 min. At the first stage the composition with monomers are heated for 30 min, and at the second stage - for 60 min [15].

The disadvantages of the third method include low chemical resistance of the resulting polymer and insufficient environmental safety.

The technical result achieved by using a new waterproof coating is increasing the resistance of the protected structures to aggressive impacts in combination with the environmental safety of the most protective waterproof coating. The essence of the invention is in the method of obtaining a waterproof coating of porous surfaces. A waterproof coating containing pre-activated and polymerizable liquid unsaturated aliphatic hydrocarbons with a boiling point above 100 °C, which are alkenes or a mixture thereof, is applied to the porous the surface with a brush, spray, etc., and alkenes or their mixture are initiated immediately before application to the surface by adding 1–2% benzoyl peroxide, previously dissolved in a solvent.

To obtain a waterproof coating, the following components can be used: alkene or a mixture of C10 - C17 alkenes, in purity corresponding to the content of the basic substance ≥95%, benzoyl peroxide, corresponding to the Russian state Standard GOST 14888 or with purity characteristics ≥98% of the basic substance. The method is as follows. The waterproof coating is prepared. Pre-activated and polymerizable liquid unsaturated aliphatic hydrocarbons (alone or as a mixture) with a boiling point above 100 °C are used. The waterproof coating is applied to the porous surface with a brush or spray. Immediately after the application of a mixture of aliphatic alkenes activated with the help of polymerization catalysts, the latter begin to polymerize, forming a continuous waterproof coating.

The matrix of options for the composition of the waterproof mixture was compiled basing on 4 groups of substances:

1. Monomer(s) or mixture of monomers (1 readily available monomer with relatively low volatility is selected);
2. Auxiliary low molecular weight polymer (LMPE), capable of forming an emulsion in organic solvents and slowing down the process of the monomer evaporation;
3. High molecular weight alkane with proven ecological purity which changes the rheological properties of the system;
4. Catalyst(s) or their mixture (benzoyl peroxide, ChKhZ-57), designed to start the polycondensation reaction.

The best results are obtained with a combination of components (option 3).

During implementation of the work, prototypes of the two-component composition of the waterproof mixture were obtained.

The first and main component A is an emulsion of LMPE in a paraffin solution in decene-1. The second component B is a solution of benzoyl peroxide in an alkane (hexane, heptane or their mixture) and is added to the main component just before use.

On the industrial site, one pilot batch of samples was produced in the form of 10 sets, each of which consists of components A and B in volumes of 1 l and 100 ml, respectively, with the best indicators according to the laboratory studies. The results of the laboratory tests are shown below.

The list of used measuring and testing instruments is presented in Table 1.

Table 1. The used measuring and testing instruments.

Test means, type, serial number	Measurement accuracy	Measurement range
Electronic scales AJ-32OOCE	±10 mg	0–23.2 kg
Electronic scales HR-150AG	±0.10 mg	0–150 g
Installation for testing concrete samples for water resistance UVB-MG4.01	±3%	0–2 MPa

The composition of the dry mix mortar (DMM) and the parameters of samples for testing are presented in Table 2.

Table 2. The composition of the DMM and the parameters for testing.

Name	Component of DMM	Moisture, %
Control sample	Quartz sand mixture	83.080
	Polydecylene (1 type)	16.670
	Plasticizer (polycarboxylate)	0.250
Test sample	Quartz sand mixture	83.080
	Polydecylene (1 type)	16.670
	Plasticizer (polycarboxylate)	0.250
	Waterproof coating of porous surfaces	–
Cylinder dimensions	h = 5 cm; d = 15 cm	
Age of samples, days	28	
Moisture, %	0.070	

The appearance of the waterproof compound can be described as a colorless opalescent liquid. When applied to the sample surface, it is well absorbed. The waterproof compound reacts with calcium ions, during the reaction, new chemical products are formed, closing the pore space of the cement stone. As a result, the water permeability and chemical resistance of the concrete surface are significantly increased. Tests for water permeability showed that the brand for water resistance corresponds to a practically non-permeable sample.

Concrete samples were tested for resistance to the action of a 2% hydrochloric acid solution for 24 h. The samples had the form of cubes with an edge of 5 cm, covered with a protective waterproof coating. In particular, the weight reduction associated with the possible release of carbon dioxide and the appearance of the samples were investigated.

The research results showed an excellent result - the reduction in the mass of the samples does not exceed 0.24% of the mass of the cement included in their composition. The test results are shown in the Table 3 below.

Table 3. Test results on the chemical resistance.

Type of coating	Exposure time, h	Average weight loss, % of cement mass
Without coating	24	3.12
Polydecylene	24	0.24

The laboratory tests show that the waterproof coating for porous surfaces significantly increases the water resistance of fine-grained concrete samples, and effectively protects concrete from the aggressive effects of natural and man-made factors.

3 Conclusions

The feature of the new waterproof coating is that, being polyalkene, it has increased resistance to aggressive influences, is inert and environmentally friendly. Based on the data obtained, not only the composition of the waterproof coating has been developed, but the developed matrix of the composition has been adapted to the conditions of industrial production, which made it possible to obtain an effective protective coating for building materials and structures. The use of a new waterproof mixture allows slowing down the process of corrosion of building materials and structures. It prolongs their service life, reduces the risks of unplanned repairs and the costs of their implementation. The developed waterproof mixture is environmentally friendly and provides high chemical resistance due to the high inertness of the material. The cost of the developed material is twice as economical as compared to materials of a similar purpose.

References

1. Bazhenov, Y., Kozlova, I., Nechaev, K., Kryuchkova, A.: E3S Web of Conferences, vol. 91, p. 02044 (2019)
2. Ustinov, A., Zybina, O., Tanklevsky, L., Lebedev, V., Andreev, A.: E3S Web of Conferences, vol. 33 (2018)
3. Ahmad, S.: Cem. Concr. Compos. (2003)
4. Bertolini, L., Elsener, B., Pedeferri, P., Polder, R.: Corrosion of Steel in Concrete: Prevention, Diagnosis, Repair (2005)
5. Nilsson, L.O.: Dev. Formul. Reinf. Concr. (2019)
6. Cabrera, J.G.: Cem. Concr. Compos. (1996)
7. Potekhin, I., Mischenko, V., Mottaeva, A., Zheltenkov, A.: E3S Web of Conferences, vol. 33, p. 03020 (2018)
8. Usman, A., Sutanto, M.H., Napiah, M.: E3S Web of Conferences, vol. 65, 05027 (2018)
9. Kagan, G.L., Shorin, V.A., Velsovskij, A.Y.: IOP Conference Series: Materials Science and Engineering, vol. 463, p. 22073 (2018)
10. Kagan, G.L., Shorin, V.A., Velsovskij, A.Y.: E3S Web of Conferences, vol. 161, p. 1038 (2020)
11. Vel'sovskij, A., Karpov, B., Smirnova, E.: Proc. Inst. Civ. Eng. Civ. Eng. **168**, 49 (2015)
12. Shorin, V.A., Kagan, G., Velesovskii, A.Y.: Soil Mech. Found. Eng. **45**, 144 (2008)
13. Korenkova, S.F., Rudakova, E.M.: Patent 2581067 Russian Federation, IPC C04B 41/48. Method of Increasing the Chemical Resistance and Strength of Concrete Products and Structures. Applicant and Patentee SGASU. - No. 2015100946/03; Declared 01/12/2015; Publ. 10.04.2016, Bul. No. 10 (2016)
14. Busygin, I.G., Busygina, N.V., Yurenkova, E.O., Beregovaya, N.G., Kuznetsov, O.A., Mezhueva, L.V.: Patent 2562636 Russian Federation, IPC C04B 41/45. A Method of Applying a Thin-Layer Waterproofing Coating Based on Polymer Sulfur on Porous Building Materials. No. 2013135930/03 (2015)
15. Krasnov, A.M., Budanov, N.P., Chetverikova, I.V.: Patent 1463731 DMMP, IPC C04B 41/62. Method of Impregnation of Porous Material. Applicant and Patent Holder Mari Polytechnic Institute Named after A.M. Gorky. - No. 3978228; Declared 18.11.1985; Publ. 03.07.1989 (1989)

Road Cement-Mineral Materials with Granulometric and Hydrophobic Additives

Lenar Mavliev, Evgenii Vdovin(✉), Victor Stroganov, and Nikita Konovalov

Kazan State University of Architecture and Engineering, Kazan, Russia
lenarmavliev@yandex.ru, vdovin007@mail.ru

Abstract. Studies of the modification of limestone rock crushing waste treated by Portland cement were carried out. Its consist in structuring the clay-carbonate fine component of mineral material by introducing granulometric additives (natural sand) and mineral mixture hydrophobization by organosilicon additives (potassium methylsilicate). The influence of natural sand and potassium methylsilicate on physical and mechanical properties of cement-mineral material is determined. It is shown that the complex modification ensured the achievement of the required level of physical and mechanical parameters of cement-mineral material for road surfaces. Dependences of influence of Portland cement, sand, and potassium methylsilicate on physical and mechanical properties of treated rock crushing waste are obtained. The optimal compositions of modified cement-mineral materials for road pavement with the strength grades M40, M60, M75 and frost resistance F15 were established.

Keywords: Stone crushing waste · Structuring and optimization of composition · Physical and mechanical properties · Hydrophobization · Potassium methylsilicate · Frost resistance

1 Introduction

When economic activities carrying out in the field of mining, there is a need of rational use of waste and local raw materials [1–3]. From the variety of non-metallic minerals, it is possible to distinguish the waste of stone crushing of building stone, which in rare cases are used as a building material [4–6].

In areas where there are no reserves of durable rock stone, the using of local mineral materials, in particular limestone rock crushing waste treated by binding material, becomes one of the opportunities to reduce the construction cost, save energy, resources and time [7–10].

This problem can be solved by modification the composition and properties of resulting material by purposeful additives [11–14].

For example, it is known that the granulometric additives introduction in building materials composition leads to an increase of physical and mechanical properties [15].

A. Sinitsyn (Ed.): SES 2020, LNCE 141, pp. 20–28, 2021.
https://doi.org/10.1007/978-3-030-67654-4_3

Today organosilicon compounds are widely used to modify and increase the durability of building materials, which are fixed on material surface, hydrophobize the walls of pores and capillaries [16–22]. This fact has a positive effect on frost resistance improving, since the adhesion of salt and ice crystals to the hydrophobic pore surface was reduced.

The purpose of this work is to modify by granulometric and hydrophobic additives the limestone rock crushing waste treated by cement, to produce cement-mineral materials for road surfaces structural layers.

2 Materials and Methods

In study the limestone rock crushing waste composited sample was used. For the initial mineral materials, the granulometric composition and basic physical properties were determined according to EN 1260 (Tables 1, 2, 3 and 4).

Table 1. Granulometric composition of stone crushing waste

Screen size, mm	8,0	4,0	2,0	1,0	0,5	0,25	0,125	<0,125
Private balances, %	9,78	14,99	22,36	18,84	8,02	7,17	5,68	13,17
Full balances, %	9,78	24,77	47,13	65,97	73,99	81,16	86,83	100

Table 2. Physical properties of stone crushing waste

Indicator name	Actual indicator value
Content of powdery and clay particles, in % by weight	6,96
Content of clay particles, in % by weight	0,73
Content of flake particles, in % by weight	23,89
The fineness modulus	3,07

Table 3. Sand granulometric composition

Screen size, mm	8,0	4,0	2,0	1,0	0,5	0,25	0,125	<0,125
Private balances, %	0,25	2,60	14,98	11,28	18,00	24,76	13,39	14,74
Full balances, %	0,25	2,85	17,83	29,11	47,11	71,86	85,26	100,00

Table 4. Sand physical properties

Indicator name	Actual indicator value
Content of powdery and clay particles, in % by weight	0,98
Content of clay particles, in % by weight	–
The fineness modulus	2,44 (medium size)

The natural quartz sand was used as a granulometric additive. It was introduced into the stone crushing waste as a percentage 10/90, 20/80, 30/70, 40/60, 50/50, 60/40, 70/30.

For mineral mixtures processing by a binder, Portland cement (PC) CEM I 42.5 N was used in 6%, 8% and 10% quantities (by mixture weight). The mineral and chemical compositions are shown in Table 5 and 6.

Table 5. Mineral composition of cement

Name	C_3S	C_2S	C_3A	C_4AF
Content, %	58	17	8	13

Table 6. Chemical composition of cement

Name	SiO_2	Al_2O_3	Fe_2O_3	CaO	MgO	SO_3	R_2O (Na_2O + 0,65 K_2O)	CaO_{cb}
Content, %	21,1	5,85	4,2	65,4	1,13	1,03	1,07	0,16

The 40% aqueous solution of potassium methylsilicate $(CH\text{-}Si\text{-}OK)_n$, n = 1 − 2, was used as a hydrophobizing additive (MSK). It was administered in 0.05; 0.07; 0.09; 0.11; 0.13% of the mixture mass.

The cement-mineral mixture preparation was carried out by sequential stone crushing waste mixing with sand, Portland cement and water solutions of hydrophobic additives in a laboratory mixer with a forced vertical discharge (10 L) at a temperature of (20 ± 5) °C.

The modification effect on quality road construction materials indicators was evaluated according to standard methods (EN 12390-2) after samples hardening during 28 days at a temperature of (20 ± 2) °C and humidity at least 90%:

- the compressive strength was determined on water-saturated samples (during 2 days) with a size of 10 × 10 × 10 cm (EN 12390-3);
- the tensile strength was determined on samples with dimension 10 × 10 × 40 cm (EN 12390-6).
- frost resistance - on samples-cubes with an edge of 10 cm (CEN/TS 12390-9). The frost resistance coefficient was defined as the ratio of the sample strength after 15 freeze-thaw cycles to the sample strength before the test.

3 Results

It is shown that the sand introduction in ratios from 10/90 to 70/30 at a PC consumption of 6%, 8%, 10%, causes the increase of material compressive strength by 1.7–2.7 times, and tensile strength by 1.8–1.9 times and frost resistance by 3.7–14.6 times (Figs. 1, 2 and 3).

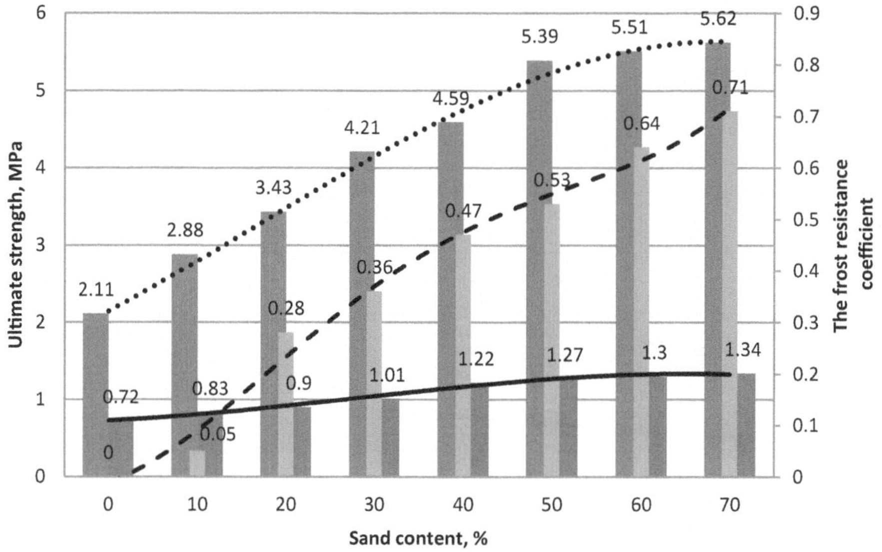

Fig. 1. Dependence of the physical and mechanical properties of the stone crushing waste treated with 6% PC on the sand content in the mixture: ——— - tensile strength during bending; ••••• - compressive strength; - - - - - - frost resistance coefficient

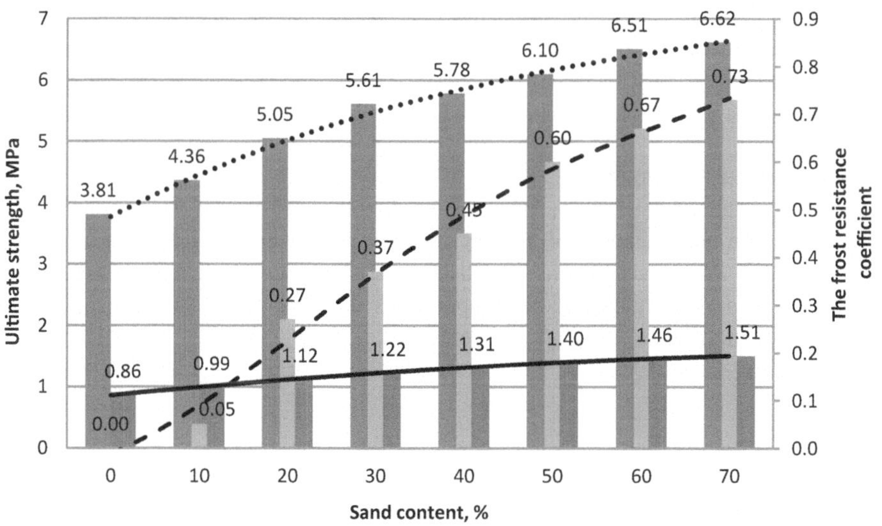

Fig. 2. Dependence the physical and mechanical properties of stone crushing waste treated by 8% PC on sand content in mixture: ——— - tensile strength during bending; ••••• - compressive strength; - - - - - - frost resistance coefficient

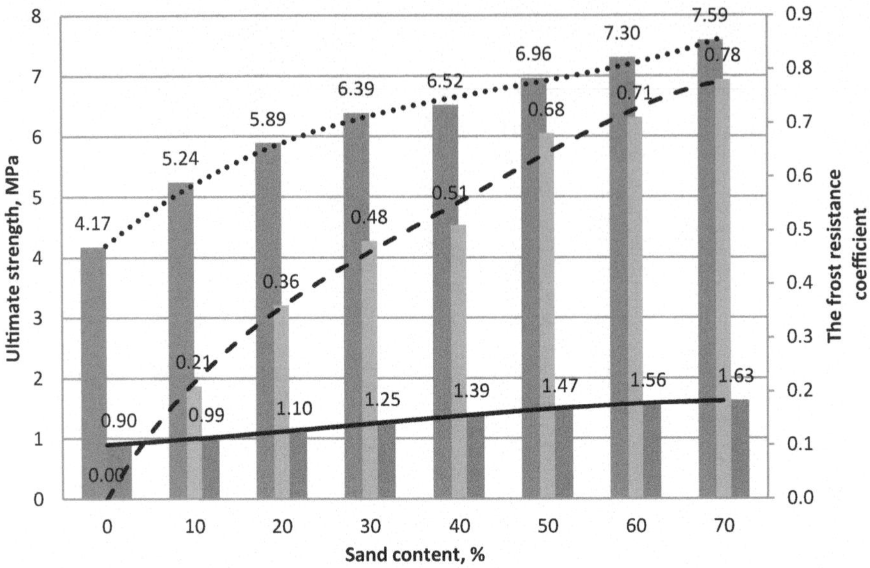

Fig. 3. Dependence the physical and mechanical properties of stone crushing waste treated by 10% PC on sand content in mixture: ——— - tensile strength during bending; ••••• - compressive strength; - - - - - - frost resistance coefficient

The obtained data confirmed that the required indicators level for providing the required frost resistance grade of cement-mineral material from treated rock crushing waste (for example, F15 for the bearing base or road surface in climatic conditions of the Central part of Russia) is achieved when the maximum amounts of PC (at least 10%) and sand (at least 70%) are introduced, which reduces the technical and economic efficiency of used materials.

It is known that one of the most promising directions in solving the frost resistance problem is the modification of treated material by hydrophobizing organosilicon additives [23]. Previously, it was established [12], that the potassium methylsilicate (PMS) usage as an organosilicon compound is effective, the feasibility of it was confirmed by studies results (Figs. 4, 5 and 6).

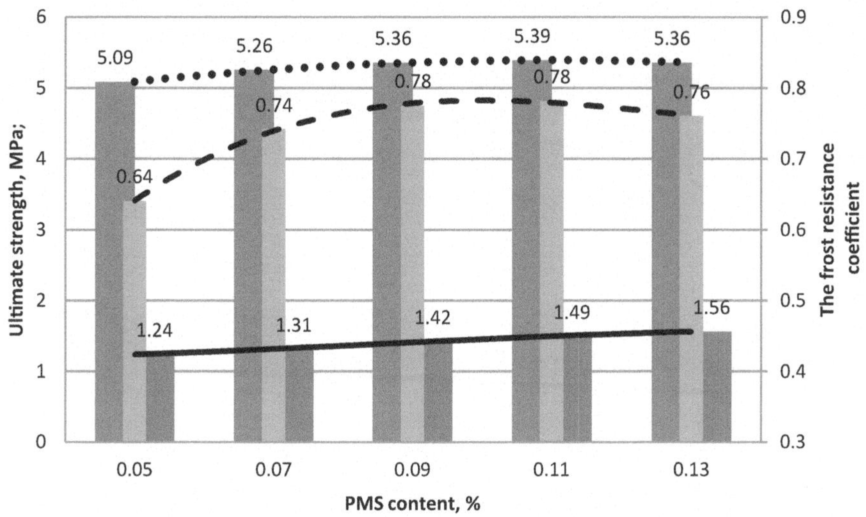

Fig. 4. Dependence the physical and mechanical properties of stone crushing waste with sand (in the ratio 50/50), treated by 6% PC, on PMS content in mixture: ——— - tensile strength during bending; •••••• - compressive strength; - - - - - - frost resistance coefficient

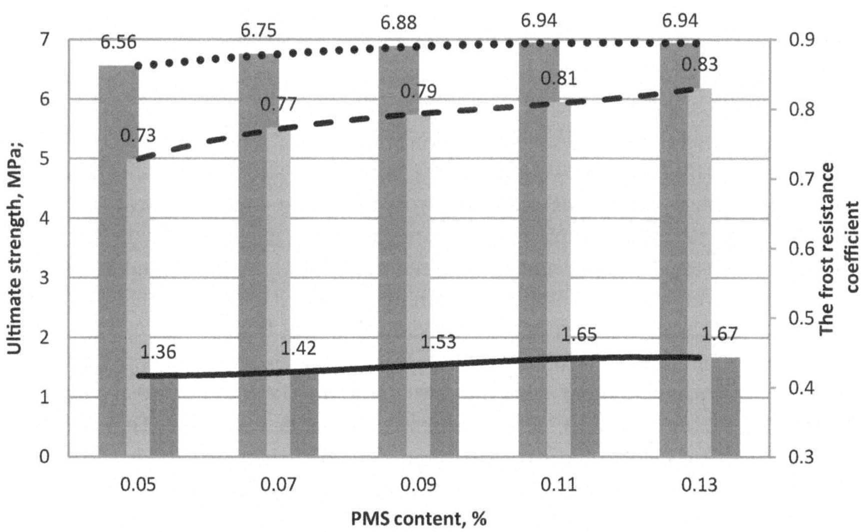

Fig. 5. Dependence the physical and mechanical properties of stone crushing waste with sand (in the ratio 50/50), treated by 8% PC, on PMS content in mixture: ——— - tensile strength during bending; •••••• - compressive strength; - - - - - - frost resistance coefficient

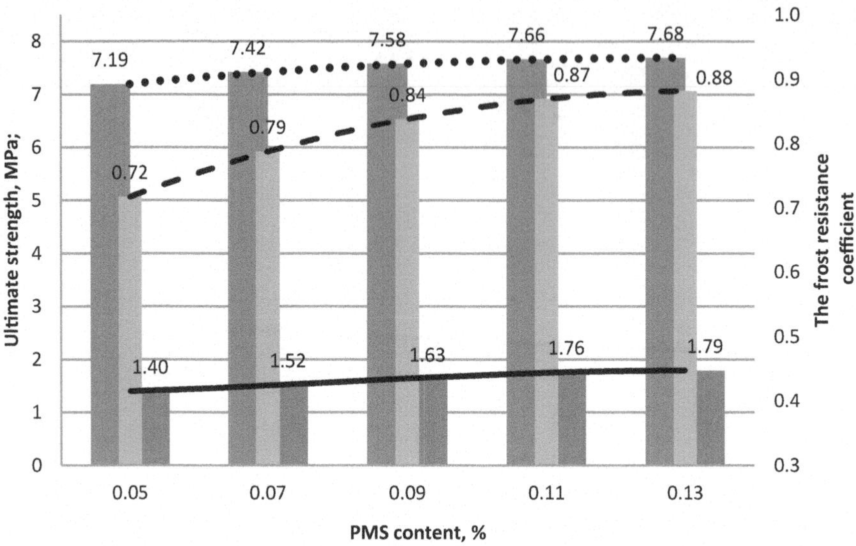

Fig. 6. Dependence the physical and mechanical properties of stone crushing waste with sand (in the ratio 50/50), treated by 10% PC, on PMS content in mixture: ——— - tensile strength during bending; ••••• - compressive strength; - - - - - - frost resistance coefficient

Results show that:

- achieving the strength of M40 and frost resistance of F15 is provided with the optimal composition: sand/crushing screening in the ratio-50/50, PC-6.0% and PMS-0.075% (by mixture weight). The compressive strength was 5.31 MPa, the tensile strength was 1.37 MPa, and the frost resistance coefficient was 0.76.
- achieving the strength of M60 and frost resistance F15 is provided with the optimal composition: sand/crushing screening in the ratio-50/50, PC-8.0% and PMS-0.08% (by mixture weight). The compressive strength was 6.81 MPa, the tensile strength was 1.48 MPa, and the frost resistance coefficient was 0.78.
- achieving the strength of M60 and frost resistance F15 is provided with the optimal composition: sand/crushing screening in the ratio-50/50, PC-10.0% and PMS-0.09% (by mixture weight). The compressive strength was 7.58 MPa, the tensile strength was 1.63 MPa, and the frost resistance coefficient was 0.84.

4 Discussion

In performed studies, the modification of treated cement-mineral road construction materials was carried out, which consists in structuring of the clay-carbonate fine component of mineral material by introducing granulometric additives (natural sand) and mineral mixture hydrophobization by organosilicon additives (PMS).

The sand introduction into the treated rock crushing waste contributed to increase of strength and frost resistance, which is consistent with previous studies. However, the required grade of frost resistance is achieved only with increased dosages of sand (more than 70%) and cement (at least 10%).

The volumetric mixture hydrophobization with the PMS introduction provided a positive effect on frost resistance improving by creating hydrophobic layers on the mineral particles surface, in pores and capillaries of material.

The complex modification ensured the achievement of required level of physical and mechanical parameters of cement-mineral material for road surfaces.

5 Conclusions

The complex modification effect of treated carbonate stone crushing waste by introducing structuring (natural sand) and hydrophobizing (potassium methylsilicate) additives on physical and mechanical properties of cement-mineral material was studied. The dependences of effect of Portland cement consumption, sand and potassium methylsilicate dosages on complex properties of the treated material are established. Optimal treated rock crushing waste compositions with strength grades M40, M60, M75 and frost resistance F15 for road construction were obtained.

References

1. Gálvez-Martos, J.L., Styles, D., Schoenberger, H., Zeschmar-Lahl, B.: Construction and demolition waste best management practice in Europe. Resour. Conserv. Recycl. **136**, 166–178 (2018). https://doi.org/10.1016/j.resconrec.2018.04.016
2. Rakhimova, N.R., Rakhimov, R.Z.: Reaction products, structure and properties of alkali-activated metakaolin cements incorporated with supplementary materials - a review. J. Mater. Res. Technol. **8**(1), 1522–1531 (2019)
3. Park, S.S., Kim, S.J., Chen, K.Q., Lee, Y.J., Lee, S.B.: Crushing characteristics of a recycled aggregate from waste concrete. Constr. Build. Mater. **160**, 100–105 (2018). https://doi.org/10.1016/j.conbuildmat.2017.11.036
4. Gautam, P.K., Kalla, P., Nagar, R., Jethoo, A.S.: Laboratory investigation on use of quarry waste in open graded friction course. Resour. Policy **59**, 62–67 (2018). https://doi.org/10.1016/j.resourpol.2018.02.009
5. Khudyakova, L.I., Timofeeva, S.S.: Practical use of nonmetalliferous raw materials copper-nickel deposits. South Russ. Ecol. Dev. **13**, 157–165 (2018). https://doi.org/10.18470/1992-1098-2018-4-157-165
6. Singh, S., Choudhary, J., Kumar, B., Gupta, A.: Effects of using kota stone as filler on mechanical properties of asphalt concrete mixes. In: LNCE, pp. 249–259 (2020)
7. Rana, A., Kalla, P., Csetenyi, L.J.: Recycling of dimension limestone industry waste in concrete. Int. J. Mining, Reclam. Environ. **31**, 231–250 (2017). https://doi.org/10.1080/17480930.2016.1138571
8. Felekoglu, B.: Utilisation of high volumes of limestone quarry wastes in concrete industry (self-compacting concrete case). Resour. Conserv. Recycl. **51**, 770–791 (2007). https://doi.org/10.1016/j.resconrec.2006.12.004

9. Mo, K.H., Alengaram, U.J., Jumaat, M.Z., Lee, S.C., Goh, W.I., Yuen, C.W.: Recycling of seashell waste in concrete: a review. Constr. Build. Mater. **162**, 751–764 (2018)
10. Shafigh, P., Nomeli, M.A., Alengaram, U.J., Mahmud, H.B., Jumaat, M.Z.: Engineering properties of lightweight aggregate concrete containing limestone powder and high volume fly ash. J. Clean. Prod. **135**, 148–157 (2016). https://doi.org/10.1016/j.jclepro.2016.06.082
11. Polymer modified concrete. Int. J. Cem. Compos. Light. Concr. **10**, 267 (1988). https://doi.org/10.1016/0262-5075(88)90058-9
12. Mavliev, L., Bulanov, P., Vdovin, E., Zaharov, V., Gimazov, A.: Road soil cement with complex additives based on organosilicon compounds and electrolytes. ZKG Int. **69**, 49–54 (2016)
13. Wu, Z., Deng, Y., Liu, S., Liu, Q., Chen, Y., Zha, F.: Strength and micro-structure evolution of compacted soils modified by admixtures of cement and metakaolin. Appl. Clay Sci. **127–128**, 44–51 (2016). https://doi.org/10.1016/j.clay.2016.03.040
14. Vdovin, E.A., Stroganov, V.F.: Properties of cement-bound mixes depending on technological factors. Mag. Civ. Eng. **93**, 147–155 (2020). https://doi.org/10.18720/MCE.93.12
15. Popov, L.N.: Effect on micro-additives on properties of concretes. In: 7th International Congress on the Chemistry of Cement, vol. 3, no. VI, pp. 192–194 (1980)
16. Wojciechowski, M., Baryła, P., Lefik, M.: Main properties of sands hydrophobized by alkoxysilane emulsions. Acta Geophysica **62**(5), 1192–1201 (2014)
17. Daniels, J.L., Hourani, M.S., Harper, L.S.: Organo-silane chemistry: a water repellent technology for coal ash and soils. In: 3rd World of Coal Ash, WOCA Conference - Proceedings (2009)
18. Dai, J.G., Akira, Y., Wittmann, F.H., Yokota, H., Zhang, P.: Water repellent surface impregnation for extension of service life of reinforced concrete structures in marine environments: the role of cracks. Cem. Concr. Compos. **32**, 101–109 (2010). https://doi.org/10.1016/j.cemconcomp.2009.11.001
19. Daniels, J.L., Mehta, P., Vaden, M., Sweem, D., Mason, M.D., Zavareh, M., Ogunro, V.: Nano-scale organo-silane applications in geotechnical and geoenvironmental engineering. Int. J. Terrasp. Sci. Eng. **1**, 19–27 (2009)
20. Li, S., Zhang, W., Liu, J., Hou, D., Geng, Y., Chen, X., Gao, Y., Jin, Z., Yin, B.: Protective mechanism of silane on concrete upon marine exposure. Coatings **9** (2019). https://doi.org/10.3390/coatings9090558
21. Czachor, H., Doerr, S.H., Lichner, L.: Water retention of repellent and subcritical repellent soils: new insights from model and experimental investigations. J. Hydrol. **380**, 104–111 (2010). https://doi.org/10.1016/j.jhydrol.2009.10.027
22. Zhang, X., Li, S., Zhao, T., Jin, Z.: Effect of octyl-triethoxysilane emulsion on protection of concrete. In: Key Engineering Materials, pp. 504–509 (2015)
23. Grabowska, K., Koniorczyk, M.: The effect of hydrophobic treatment by organosilicon admixtures of cement mortar [Skuteczność hydrofobizacji poprzez zastosowanie domieszek krzemoorganicznych w zaprawie cementowej]. Cem. Wapno Bet. (2019). https://doi.org/10.32047/CWB.2019.24.4.7

Impact of Climatic Factors on Electricity Losses on Aerial Power Lines

Muhayo Islomovna Toshkhodzhaeva[1(✉)], Elena Ivanovna Gracheva[2], Shakhboz Tolibjonovich Dadabaev[1], and Umeda Bobojonovna Odirmatova[3]

[1] Department of Power Supply and Automation, Khujand Polytechnic Institute of the Tajik Technical University named after Academician M.S. Osimi, Khujand 735700, Tajikistan
shukrona14_01_2011@mail.ru

[2] Department Power Supply of Industrial Enterprises, Kazan State Energy University, Kazan 420066, Russian Federation

[3] Department of Power Supply, Mining-Metallurgical Institute of Tajikistan, Buston 735730, Tajikistan

Abstract. This article discusses the theoretical solution of problems to determine losses in the process of transmission of electricity through overhead power lines, the fluctuation of resistivity is determined depending on the change in ambient temperature, the structure of losses on overhead power lines is given. As an example, for solving these problems, the data of power transmission lines with a voltage of 35 kV in the territory of the Sughd region of Republic of the Tajikistan were applied, the change in losses between the two regions was calculated and compared. Recommendations for reducing losses on overhead power lines are offered.

Keywords: Electricity losses · Power lines

1 Introduction

Recently, the attention of researchers has been drawn to the problem of loss of electricity in power lines. This issue is relevant for the transmission of high voltages on aerial power lines (APL), which has its own characteristic features. The existing literature data is mainly devoted to the variety of power loss in electricity, taking into account local climatic conditions and the influence of external destabilizing factors.

However, these works mainly analyze the estimation of losses with the use and transformation of energy from renewable sources. The issue of calculating voltage losses at 35 kV overhead transmission lines remains open, taking into account their application in the power supply system of the Sughd region. The results interpreted in this work are devoted solving this problem. At the same time, in the above works, no attention was paid to the losses of APL with a voltage of 35 kV, typical for the local area of the Sughd region of Tajikistan [1, 2].

In this regard, this work considers the main characteristics and definitions of losses of APL with a voltage of 35 kV in the territory of the Sughd region, taking into account

A. Sinitsyn (Ed.): SES 2020, LNCE 141, pp. 29–34, 2021.
https://doi.org/10.1007/978-3-030-67654-4_4

climatic features. The solution to this problem is conditioned by the implementation, taking into account the law of the Government of the Republic of Tajikistan about energy saving and energy efficiency.

2 Methodology

As an analysis, the data on losses in two systems are compared: with minimum consumers - the city of Khujand and with maximum consumers - B. Ghafurov district. The calculation of electricity losses due to resistivity was made according to the following formulas [3, 4]:

$$\Delta P_a = \Delta U_a \sqrt{3 r_0 * \cos\varphi * I * L} \tag{1}$$

where $\boldsymbol{\Delta U_a}$ - active component of voltage loss, $\boldsymbol{r_0}$ - specific active resistance for AC wires - 35,50,70,95,120,150, Ohm/km, $\mathbf{cos}\,\varphi$ - power factor, $\boldsymbol{I}$ - rated current A, $\boldsymbol{L}$ - length of power lines, km.

$$\Delta P_p = \Delta U_{\mathrm{p}} \sqrt{3 x_0 * \sin\varphi * I * L} \tag{2}$$

where $\boldsymbol{\Delta U_p}$ - reactive component of voltage loss, $\boldsymbol{x_0}$ - specific reactance Ohm/km.

The results of the above formulas and informational definitions are shown in Tables 1 and 2.

In the Sughd region, from 1946 to 1992, more than 67 APL were built, with a voltage of 35 kV. The characteristic of resistivity depending on weather conditions was compiled according to the types of wires of these lines. According to the climate forecast, until 2050, the average air temperature in Asian countries will rise to +60 °C, which gives an idea that the resistivity in power lines will grow, which will not leave the quality and reliability of electricity transmission unaffected [5].

Comparison of the tables gives definitions that each territory has its own technical, load and climatic features. From 2007 to January 1, 2017, one of the main problems of the state was the provision of electricity to the country's inhabitants, especially in the autumn-winter period [6, 7]. Gradually considering and applying certain measures, the Republic of Tajikistan was able to solve this problem to the maximum. The diagram

Table 1. Specific voltage losses, taking into account the length and transmitted power on the 35 kV APL on the territory of Khujand city.

№	Linenames	Lengthoflines, km	Inom, A	P_{nom}, MWt	$\sum \Delta U$, V
1.	L-L_Hp	8,3	40	2,0	205,47155
2.	L-N-Em-1	5	50	2,5	127,78213
3.	L-N-Em-2	3	50	2,5	92,83353
4.	L-A-Ts-1	2,16	60	3,0	66,242254
5.	L-A-Ts-2	2,15	120	6,0	159,67367

Table 2. Specific voltage losses, taking into account the length and transmitted power on the 35 kV APL in the territory of B. Ghafurov district.

№	Linenames	Lengthoflines, km	Inom. A	P_{nom}, MWt	$\sum \Delta U$, V
1.	L-P-K	4,4	40	2,0	90,537128
2.	L-Hb-K	15,61	120	6,0	1178,6193
3.	L-Kostak	3,88	80	4,0	193,78967
4.	L-D2,3	7,62	160	1,0	350,2274
5.	L-Shorkul	5,5	20	1,0	81,10586
6.	L-D-K	18,5	160	3,0	1707,992
7.	L-DC12	5,42	10	0,5	31,82981
8.	L-A-P2	21,2	dismantled		
9.	L-K-Uldu	1,6	30	2,0	36,5102
10.	L-Karaj	0,3	20	1,0	5,48064
11.	L-L-U	5,22	190	11	581,2835
12.	L-SD -U	5	60	3,0	176,1797
13.	L-SD-V	8,9	100	5,0	425,1112
14.	L-H-D	13,9	100	5,0	814,6643
15.	L-K-D	12,35	60	3,0	354,3255
16.	L-D-P	5,7	80	4,0	180,5983
17.	L-N-A	16,5	12	8,0	116,2786
18.	L-S-V	5,6	100	5,0	323,7923
19.	L-S-3	2,65	60	3,0	92,46753
20.	L-L-R	5,3	30	2,0	93,93641
21.	L-VODOZABOR	3,7	60	3,0	171,7824

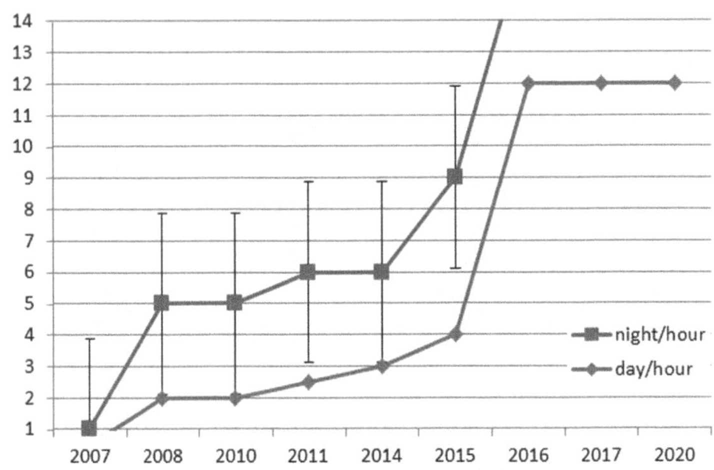

Fig. 1. Diagram of electricity supply in the country.

characterizes the supply of electricity per day in the autumn - winter period [10] (Fig. 1).

The problem with the supply of electricity found its definite solution with the construction of the Roghun Hydroelectric Power Plant (HPP) and a small number of small and large HPP on the territory of the Republic of Tajikistan. This objects has introduced a new challenge to improve and improve the quality of the power supply to the researchers in the field [8, 9].

In the power supply system, losses are one of the main problems of this structure, according to Tables 1 and 2, it is clear that a percentage of the total power is lost depending on the resistivity, in which climatic factors have an important influence. According to the characteristics of the wires, it has already been determined that at a temperature of −237 °C, the specific resistance of the wire gradually disappears, but with an increase in temperature, it begins to grow. According to this formula, the changes in resistivity were theoretically determined:

$$r_0 = r_{20}[1 + \alpha(t - 20\,^{\circ}\mathrm{C})] \tag{3}$$

where $\boldsymbol{r}_0$ - specific resistance of the wire depending on the change in ambient temperature, $\boldsymbol{r}_{20}$ - resistivity at temperature +20 °C, $\boldsymbol{\alpha}$ - temperature coefficient of the wire depending on the material, $\boldsymbol{t}$ - ambient air temperature.

Table 3. Specific resistance of wires depending on temperature +5 °C.

№	Wiretype	SpecificResistance	Temperature coefficient 1 °C
1.	AC 35	0,864	0,00403
2.	AC 50	0,60	0,00403
3.	AC 70	0,43	0,00403
4.	AC 95	0,32	0,00403
5.	AC 120	0,25	0,00403
6.	AC 150	0,197	0,00403
At a temperature of +25 ^{0}C			
7.	AC 35	0,938	0,00403
8.	AC 50	0,65	0,00403
9.	AC 70	0,469	0,00403
10.	AC 95	0,346	0,00403
11.	AC 120	0,275	0,00403
12.	AC 150	0,214	0,00403

Table 3 shows the data based on the calculation results.

The distribution sector is considered the weakest link in the entire energy structure. The transmission loss is approximately 17% and the distribution is 50%. There are two main types of losses.

1. Technological losses.
2. Non-technical losses (commercial losses).

Technical loss – is a phenomenon that occurs due to energy dissipated in conductors and equipment used in power lines, transformers, auxiliary lines, distribution lines, and magnetic losses in transformers. Technical losses usually amount to 22.5% and directly depend on the characteristics and operating mode of the network.

Commercial electricity losses – these are losses caused by theft, inconsistency of meter readings with payment for electricity by household consumers and other reasons in the field of organizing control and energy consumption.

Structural losses – are losses for own needs of stations and distribution networks.

In addition to these, there are still a number of losses, such as: Instrument errors, Energy underestimation factors, Load losses, Weather losses, No-load losses, Energy transportation losses.

The above listed losses are always acceptable and are formed in the energy structure. But this also has its own principles of solution.

3 Conclusion

The data given in the tables prove that weather conditions affect power line losses. Since the lines were installed, during the Soviet Union they have their age disadvantages. This is the deterioration of the lines, on average, the age of each of these lines is about 40 to 57 years old, or even more. Supports of the type are wooden, reinforced concrete. The method for solving this problem is to reduce the resistance, for this we increase the cross-section of the wires. But the operational age of the lines shows that the pylons of these power lines are not able to withstand a certain load passing through these wires.

The solution to this problem can be found by increasing the voltage on the power lines. A more acceptable and alternative solution method is the reconstruction of these APL using wires of the 35 kV SIP type.

References

1. Chorshanbiev, S.R., et al.: Modeling of high-voltage 110–35 kV dushanbe urban electric networks of Republic of Tajikistan for analysis and assessment of power losses. In: 2020 IEEE Conference of Russian Young Researchers in Electrical and Electronic Engineering (EIConRus), pp. 1184–1187. IEEE (2020)
2. Latypov, I.S., Sushkov, V.V.: Reduction of active power losses in wires of overhead power transmission lines with voltage of 6–35 kV. Dyn. Syst. Mech. Mach. **3**, 1 (2016)
3. Shvedov, G.V., Chorshanbiev, S.R., Nazirov, Kh.B.: Structural analysis of electricity losses in 35–500 kV electrical networks of the Republic of Tajikistan. Polytech. Bull. Ser. Eng. Res. **1**(1), 74–86 (2018)
4. Shvedov, G.V., Schepotin, A.S.: On the clarification of calcuzlations of load losses of electricity in the wires of overhead power lines, **6**, 52–57 (2020)

5. Tolibjonovich, D.S., Islomovna, T.M., Saidulloevna, M.D.: Modeling of starting transition processes of asynchronous motors with reduced voltage of the supply network. Eur. J. Electric. Eng. **22**(1), 23–28 (2020). https://doi.org/10.18280/ejee.220103
6. Dadabaev, S.T.: Optimization of starting modes of high-voltage electric drives of irrigation pumping station in hot climate. Russ. Electromech. **2**, 86–91 (2018). (In Russian), https://doi.org/10.17213/0136-3360-2018-2-86-91
7. Gracheva, E.I., Naumov, O.V.: Vliyanie rezhima raboty na veroyatnostnye kharakteristikielektricheskikh ustroystv (Operating mode influence on probability characteristics of electric devices). J. Eng. Appl. Sci. **11**(1), 2934–2938 (2016)
8. Gracheva, E.I., Naumov, O.V.: Estimation of power losses in electric devices of the electrotechnical complex. In: International Conference on Industrial Engineering, Applications and Manufacturing (ICIEAM), vol. 6 (2019)
9. Khasanov, S.R., Gracheva, E.I., Toshkhodzhaeva, M.I., Dadabaev, S.T., Mirkhalikova, D.S.: Reliability modeling of high-voltage power lines in a sharply continental climate. In: E3S Web Conference, vol. 178, p. 01051 (2020). https://doi.org/10.1051/e3sconf/202017801051
10. Law on Energy Saving [Electronic resource]. https://energo-cis.ru/wyswyg/file/Zakon/Nacional/Tadghikistan, Accessed 25 Sept 2020

Influence of the Roughness and Shape of the Canal of Trapezoidal Canals on the Pressure Loss of Hydropower Structures

Bakhtiyor Uralov, Sanatjon Khidirov(✉), Fotima Artykbekova, and Bobur Shodiev

Tashkent Institute of Irrigation and Agricultural Mechanization Engineers, Tashkent, Uzbekistan
Vohidov.oybek@bk.ru

Abstract. It is known that the problem of determining the hydraulic resistance in free-flow channels (including machine and derivation channels) until recently was solved by using the concept of "maximum permissible speed" V_{max}, (referring to the uniform movement of water). The value of this speed was assigned (and is currently assigned) based on reference data from the type of soil (and in some cases, depending on the depth of water in the canal). Knowing V_{max} and the flow rate, it is easy to find the cross-sectional area, as well as the slope of the channel (using formulas to determine the Shezy coefficient "*C*" or the coefficient of hydraulic friction λ and in accordance with the accepted value of the roughness coefficient). In engineering practice, in hydraulic calculations of the channels we are considering, the Shezy coefficient "*C*" is usually used. Meanwhile, there is an opinion shared by us that when performing the above calculations it is more expedient to use the coefficient of hydraulic friction λ. As you know, the value of "*C*" depends mainly on the coefficient of roughness "*n*", the average value of which is the so-called "group roughness", which gives only a descriptive characteristic of the state of the wetted surface. The absolute roughness, expressed by the value of Δ, is a much more sensitive (in comparison with "*n*") estimate of the roughness of the wetted channel surfaces, and therefore its introduction into the practice of hydraulic calculations increases their accuracy. In connection with this formulation of the question, in this work, research was carried out on escaping; the influence of the shape of the open section of the channel on the amount of pressure loss, in the case of a "smooth" wetted surface and in the case of a channel with a rough wetted surface, as well as the influence of the "degree of roughness" of the wetted surface on the amount of pressure loss. It is this kind of research that makes it possible to put into the hands of the designer, as it seems to us, a complete method for solving one of the most important and urgent problems of the practice of hydraulic engineering.

Keywords: Trapezoidal canal · Hydropower · Hydraulic engineering · Hydrodynamic

A. Sinitsyn (Ed.): SES 2020, LNCE 141, pp. 35–46, 2021.
https://doi.org/10.1007/978-3-030-67654-4_5

1 Introduction

The problem of determining the head loss along the length with uniform steady motion in cylindrical channels has more than a century of history. The currently widely used design relationships for determining the Chezy coefficient are based on the assumption that the longitudinal shear stresses acting from the flow on the channel walls are uniformly distributed over the wetted perimeter. Evaluating the acceptability of this assumption for flows with different cross sections, one main group of channels can be distinguished. In these groups of channels, the assumption does not accurately correspond to reality, and the error in performing practical calculations is not permissible, however, for a wide range of Reynolds numbers, an adjustment can be made taking into account the geometric features of the shape of the free section, and thereby ensure the required accuracy of calculations. We will say that the channels belonging to the main group have a "correct" shape. As shown by the preliminary analysis, this group for the case of free-flow movement can include, in particular, rectangular, trapezoidal, parabolic, semicircular, which are often found in the practice of hydraulic engineering and hydropower construction, including machine and derivation channels of hydropower structures. In the present work, it is planned to study the head losses along the length during the free-flow movement of water in the channels of the "correct" shape, i.e. in the channels of the main group.

Until recently, it was believed that in free-flow channels with a "correct" cross-sectional shape, the effect of the cross-sectional shape on the head loss value can be reasonably accurately estimated using the hydraulic radius. However, this position is not always ensured in the case of channels having a "regular" cross-sectional shape. For free-flow channels with a "correct" cross-sectional shape, the hydraulic radius as a parameter that must take into account its influence on the value of pressure losses in these channels is insufficient [1–3]. The dependences obtained for calculating the hydraulic resistance in round pressure pipes cannot be extended without appropriate adjustments to the free-flow channels (provided that the pipe diameter is replaced in the corresponding calculations D - value $4R$, where R- hydraulic radius) [3, 4]. This position is justified by the presence of a number of factors that distinguish the pressure movement of liquid in pipes from its free-flow movement in the channels, where there is a free flow surface, a wider range of roughness of the channel bottom and walls, a different (than in pipes) distribution of tangential stresses along the wetted perimeter, the possibility of two different flow states (depending on the slope of the channel bottom). From this, in particular, it follows (and this is confirmed by the data of the corresponding experimental studies published in the literature) that in the case of free-flow channels, the pressure loss coefficient, and, consequently, the pressure loss, depends not only on the relative roughness and Reynolds number, but also on the shape of the transverse section of the channel. Moreover, if for hydraulically smooth free-flow channels this dependence in the first approximation can be considered clarified, then for channels, the wetted surface of which is characterized by the presence of one or another roughness, the question of its influence, as well as the shape of the channel cross-section on the pressure loss value, is far from even roughly approximate solution. Since until now there are no analytical dependencies that fully describe the turbulence mechanism and are suitable for practical calculations related to the free-flow movement of a liquid, one has to resort to experimental studies [5–15].

The object of research in this work is the Amu – Bukhara machine canal (ABMC), in the area of a damless water intake from the Amudarya riverbed of the Bukhara region of Uzbekistan. The sections of the ABMC machine canals, from the head structure to the AB – I and AB – II pumping stations, were taken as the object of hydraulic research.

Research Subject. The Amudarya river bed in the area of the damless water intake into the water supply machine channel of the ABMC; hydraulic and channel processes occurring in the area of the water intake and the supply machine channel of the pumping station; turbidity of the stream; computer programs, basic equations of hydrodynamics.

Research Goal. The purpose of the work is in general terms arranged from the above. Specifying this, it can be emphasized that, in the final analysis, we propose, when designing machine channels of pumping stations (PS), large earthen ("non-fixed" and "fixed") channels, the following "design provisions" justified by us: 1) dependence of hydraulic resistance on the Reynolds number and the shape of the channel cross-section in channels with a "smooth" or close to it wetted surface; 2) dependences that take into account the influence of the shape of the free cross-section of the channel on the value of the pressure loss in the channels with different degrees of roughness of the wetted surface.

Research Objectives. In connection with the above, the following main research tasks can be formulated: 1) to collect experimental data on the pressure losses along the length at steady-state free-flow motion; 2) highlight the materials related to the channels of the "correct" form and analyze them; 3) to clarify the influence of the shape of the free cross-section of the channel on the value of the pressure loss; in the case of a "smooth" wetted surface and in the case of a channel with a rough wetted surface; 4) to generalize the issue of head losses along the length, taking into account the shape of the cross-section of the unpressurized flow using a special geometric parameter.

2 Methods

In the process of research, experimental methods, methods of field observation, as well as generally accepted methods in hydraulics, methods of compiling mathematical models based on the laws of hydromechanics and their numerical calculations were used.

3 Results and Discussion

The experimental data obtained showed that the intensity of the influence of the degree of roughness and the shape of the open section of the trapezoidal machine and derivation channels on the hydraulic resistance of hydropower structures is in direct proportion to the operating mode of pumping stations and hydroelectric power plants. Experimental studies published in the literature, carried out in order to clarify the above assumption and elucidate the above-mentioned regularities in free-flow channels, as well as for machine and derivation channels, were carried out at different times and in different conditions, some of their results are not always consistent with each other, but recommended for them On the basis of the calculated hydraulic dependences are very contradictory. In particular, the question of the influence of the morphometric elements of the channel and its dimensions on the regularities of the hydraulic resistance of the machine and derivation channels of hydropower structures is not completely clear [16–19].

When liquid flows in free – flow channels, machine and derivation channels of hydropower structures, a number of factors are added that usually do not occur during pressure flow of liquid in pipelines (where their entire living section is filled with liquid); the presence of a free surface, the existence of suspended materials in the flow, the difference in the shape of the cross-section of the channels from the circular cross-section, the existence of two different states of the flow depending on the slope of the channel, the presence of hydropower structures in free-flow channels, machine and derivation channels, a wider range of roughness. If the average velocity in a channel with a different correct cross-section is calculated by the usual equation of the average velocity, and in this case will have almost the same form, then it can be found that the expressions for the average velocity in this case will have almost the same form as the expressions obtained for average speed in the channel for a trapezoidal section (Eqs. (1) and (2));

$$v/v_* = a_{gl} - b + b\ln(Rv_*/v) + b\Phi - \overline{\kappa}v/v_* \tag{1}$$

$$v/v_* = a_{sh} - b + b\ln(Rv_*/\Delta) + b\Phi - \overline{\kappa}v/v_* \tag{2}$$

Only ϕ and $\overline{\kappa}$ depending on the geometry of the channel cross-section will change (from section to section). In view of the above, Eqs. (1) and (2) can be considered rational equations for determining the average flow velocity in channels with a constant section and slope.

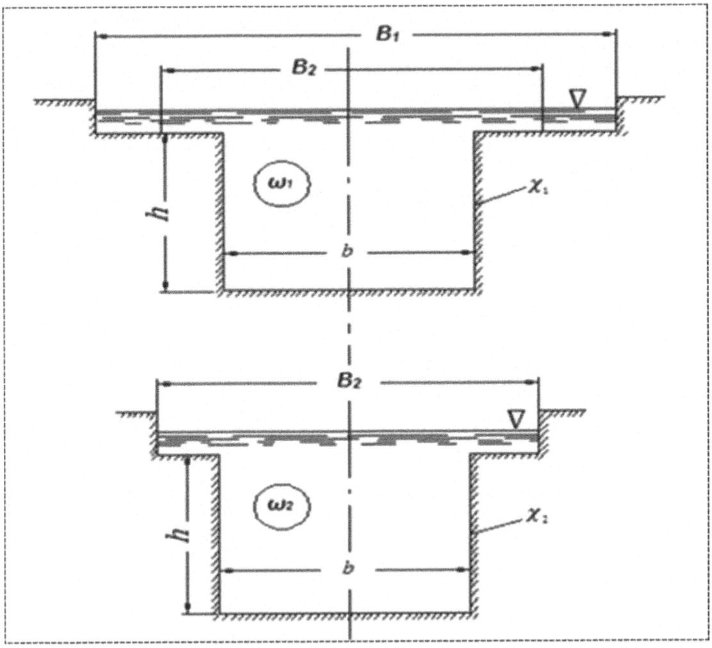

Fig. 1. Cross-sections of two channels with different hydraulic radii and approximately the same flow capacity

If these general equations are compared with the corresponding equation for a canal of infinite width (Figs. 1 and 2), then one can see that they differ in the presence of terms « $b\phi$» and $\overline{\kappa}\,\overline{u}/\overline{u}_*$, from the formula (1) and (2). These terms can be interpreted as reflecting the joint effect on the pressure loss of the presence of a free surface and the inhomogeneous distribution of shear stresses at the bottom and walls of the channel.

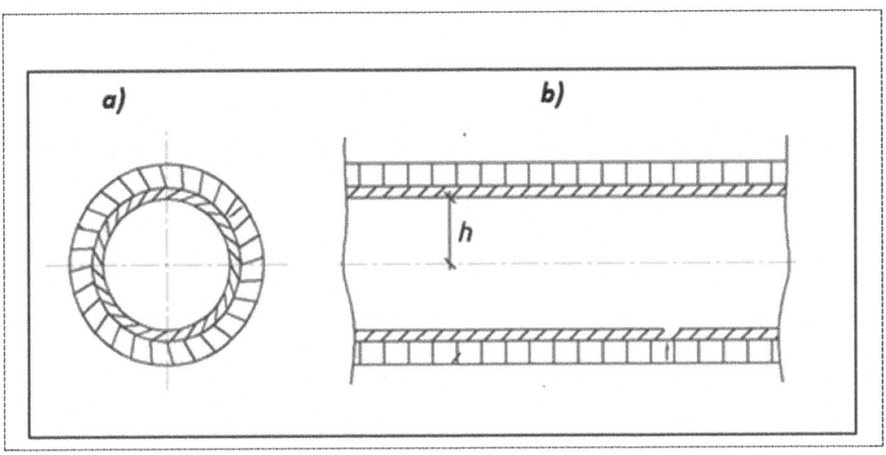

Fig. 2. Shear stress distribution: a) in round; b) in wide rectangular pipes and channels

Below, an expression for the average velocity in a channel with a trapezoidal section will be derived and it will be shown that neglecting the correction terms due to the change in shear stresses on the wall (at the solid boundary) and the apparent shear stresses on the free surface can lead to an error amounting to a significant percentage. In other words, it is not enough to take the conventional hydraulic radius as the characteristic size for such a channel.

Here it is necessary to consider two cases: 1) when the bisectors of the inner angles of a given shape intersect over the living section of the channel and 2) when the bisectors of the inner angles of a given shape intersect in the living section of the channel. Let us first consider when the bisectors of the inner angles of a given shape intersect over the living section of the channel.

In Fig. 3, the internal angles of the channel section are denoted by Ψ_1 and Ψ_2, base length through B_0, and depth through h. We divide the free area of the channel into zones of infinitely small width dy in such a way that all parts of a given area are at the same minimum distance y from the wall.

The speed at point P in one of the zones can be expressed as:

$$\frac{u}{\overline{u}_*} = a + b\ln\left(\frac{yu_*}{\nu}\right) - K_f\frac{\overline{u}}{\overline{u}_*} \tag{3}$$

where, y is the distance to the point P along the normal from the wall, a u_* - dynamic velocity corresponding to the local shear stress at the base of the normal from point P.

Term $K_f\frac{\overline{u}}{\overline{u}_*}$ - is a correction that takes into account the effect of the free surface, at that $\overline{u}$ - average speed in the free area, a $\overline{u}_*$ - average dynamic speed along a hard boundary. Obviously, it depends on the position of point P. The ratio of the local dynamic speed u_* to the average dynamic speed can be written as follows:

$$\frac{u_*}{\overline{u}_*} = 1 + K_* \tag{4}$$

Replacing u_* in dependency (3) with its value from dependency (4)

$$\frac{u_*}{\overline{u}_*} = a + b\ln\frac{\overline{u}_*y}{\nu} + b\ln\frac{\overline{u}_*}{\overline{u}_*} - k_f\frac{u}{\overline{u}_*} \tag{5}$$

k_* and discarding small quantities in $b\ln\frac{\overline{u}_*}{\overline{u}_*} = k_*\frac{u}{\overline{u}_*}$ containing k_* we get

$$\frac{u_*}{\overline{u}_*} = a + b\ln\frac{\overline{u}_*y}{\nu} - \left(k_f - k_*\right)\frac{u}{u_*} \tag{6}$$

Dependence (6) can be called the exact law of distribution of velocities in a channel with smooth surfaces. If we neglect the flow in the laminar sublayer, then the total flow rate through the open area will be expressed as follows:

$$Q = \upsilon\omega = \int_0^h ub(y)dy = \int_0^h ud\omega \tag{7}$$

Moreover: $d\omega = b(y)dy$, where, Q – canal water flow, $\upsilon = u$ – average flow rate. $b(y)$ – channel free area, length $b(y)$ any zone is expressed by the ratio:

$$b(y) = \chi - \varphi y \tag{8}$$

where, χ - wetted perimeter, φ – point-dependent angle function A, where their bisectors intersect, and in this case, when the intersection point of the bisectors of the internal angles lies above the living section of the channel, φ has the form:

$$\varphi = ctg\Psi_1 + ctg\Psi_2 + 2\cos ec\,\Psi_1 + 2\cos ec\,\Psi_2 \tag{9}$$

The free area ω is defined as:

$$\omega = \int_0^\omega d\omega = \int_0^h b(y)dy \tag{10}$$

Substituting the value $b(y)$ from dependence (10) into dependence (11) and integrating, we obtain:

$$\omega = \int_0^h (\chi - \phi y)dy = \chi h - \frac{\phi h^2}{2} \tag{11}$$

The average flow rate is determined by the expression:

$$\upsilon = \frac{1}{w}\int_0^w ud\omega = \frac{1}{w}\int_0^h ub(y)dy \tag{12}$$

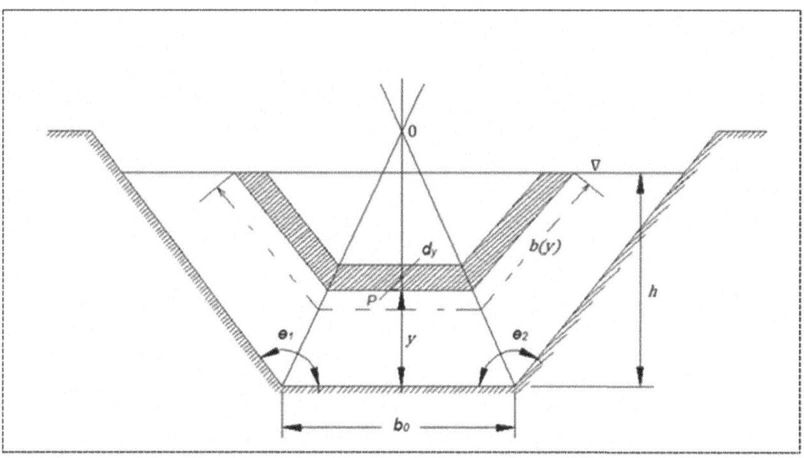

Fig. 3. The case when the bisectors intersect over a live section of a trapezoidal channel

and denoting equality through Φ:

$$\ln\frac{h}{R}-\frac{\varphi h^2}{4\omega}=\Phi \quad (13)$$

From the last expression (13) and Eq. (6), the expression for the average velocity of currents in a channel with a trapezoidal cross – section with a smooth bottom surface and slopes with the adopted $u=\upsilon$ and $u_*=\upsilon_*$ follows, and for Fig. 3, will have the form:

$$\frac{\upsilon}{\upsilon*}=a_{gl}-b+b\ln\left(\frac{RV_*}{\nu}\right)+b\varphi-K\frac{\upsilon}{\upsilon_*} \quad (14)$$

If the bottom and slopes of the channel are rough, a_{gl} in Eq. (14) should be replaced by a_{sh}, then the equation for the average velocity in such a channel will take the form:

$$\frac{\upsilon}{\upsilon_*}=a_{\omega}-b+b\ln\left(\frac{R}{\Delta}\right)+b\varphi-K\frac{\upsilon}{\upsilon_*} \quad (15)$$

When the bisectors of the inner angles of a given shape intersect in the living section of the channel, the free surface is above the point of intersection of the bisectors of the inner angles, then Φ is expressed differently (see Fig. 4). Let the bisectors of the interior angles intersect at point A^1 at a distance h_0 along the vertical of the wall given. A vertical line through A^1 intersects the free surface at point A. The two normals drawn from A to the side walls have lengths l_1 and l_2. The vertical line passing through A and A^1 divides the area ω of the trapezoid into two parts ω_1 and ω_2. Then Φ is determined by the ratio:

$$\begin{gathered}2\omega\Phi=2ctg\psi_1/2+tg\psi_1/h_0^2\ln(h_0/l_1)+ctg\psi_2/2\\+tg\psi_2/h_0^2\ln(h_0/l_2)+2\omega.\ln(l./R)+2\omega_2h(l_2/R)-\omega\end{gathered} \quad (16)$$

where, R is the hydraulic radius. Substituting the value of Φ, you can determine the average flow rate for this case as well.

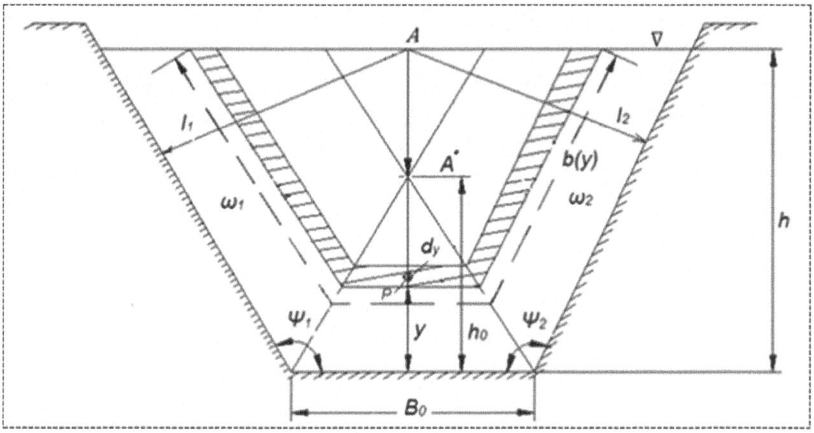

Fig. 4. Cases when bisectors intersect in live section of the trapezoidal channel

On the other hand, the indicated general Eqs. (14), (15) and (16) make it possible to find the magnitude of the error in determining the pressure loss that would occur if the terms are odd « $b\phi$» and $\overline{\kappa}v/v_*$. The "$b\phi$" term can be calculated for any given channel cross-sectional shape, since it is determined only by its geometry. Calculation according to our method and calculation according to Kelegan [20], show that in the channels of triangular cross-section the value "ϕ" does not depend on the water depth, and in this case ϕ = *0.19*. For channels of rectangular cross-section, the expression for "ϕ" takes the form:

$$\phi = \lambda n(1 + 2h/B_0) - h/B_0 \tag{17}$$

For ducts with a semicircular cross – section:

$$\varphi = \int_0^h \left[\ln\left(\frac{y}{R}\right)\right]\frac{B_0}{R}\frac{dy}{\chi} + 1.0 \tag{18}$$

Table 1. Several calculated values of Φ for channels of rectangular and semicircular cross – section

Rectangular section		Semicircular section	
R/BO	Φ	*R/B*$_0$	Φ
0.05	0.049	0.05	0.178
0.10	0.098	0.15	0.200
0.15	0.143	0.20	0.210
0.20	0.175	0.30	0.227
0.25	0.193	0.40	0.244
		0.50	0.259

To find the value of $\overline{K}$, it will probably be necessary to introduce some parameter expressing the ratio of the transverse dimension of the free flow surface in the channel to the wetted perimeter. It is quite possible that the best way $\overline{K}$ can be found from experiments. However, as follows from Eqs. (14), (15) and (16), before carrying out these experiments, the characteristics of the bottom and walls of the channel must be determined in advance (also from experiments – preferably with very wide channels of rectangular cross-section, as well as the trapezoidal shape of a living sections (Fig. 3 and Fig. 4)).

According to our method and according to the method of G. Kelegan [20], the formulas of hydraulic resistance for machine and diversion channels of hydropower structures, trapezoidal shape and other forms of correct cross-section can be represented as:

$$\frac{1}{\sqrt{\lambda}} = \frac{1}{\overline{\chi}\sqrt{2}}\left(\ln\frac{\eta_\Lambda R}{\delta_\Lambda} - 1 + \ln\frac{h}{\eta_\Lambda R} - \frac{\xi h^2}{4\omega}\right) \tag{19}$$

The same ratio is obtained according to V.T. Chow [] for curved cross-section channels. In relation (19) it is accepted: $\overline{\chi}$ - Karman's constant [21]; $\overline{\chi}$ = 0.4, η_Λ- Reynolds number.

For a viscous sublayer, $\eta_\Lambda = \delta_\Lambda v_* / v$; δ_Λ; δ_Λ- thickness of the viscous sublayer; h – filling the channel; ξ – function of the shape of the free section of the channel in the ratio $b(y) = \chi - \xi y$; χ – wetted perimeter; ω – the area of the free section of the channel (Fig. 3 and Fig. 4). Theoretical formula (19) for determining the coefficient of hydraulic friction λ_R, is valid as for the movement of fluid in smooth (η_Λ=1/9), and in rough canals ($\eta_\Lambda = 1/30$, $\eta_\Lambda = \delta_\Lambda / \Delta E$). The third and last terms in this formula take into account the effect of the shape of the free cross-section of the channel on its hydraulic resistance. However, formula (19) does not fully take into account the influence of the free surface on the distribution of velocities and head losses. Bearing this in mind and some other assumptions made when deriving formula (19), it should be assumed that formula (18) only allows us to outline the general form of the terms that determine the dependence of the hydraulic resistance of the channel on the shape of its open section. The specific form of the corresponding dependence can be established only from consideration of the corresponding experimental data for free-flow machine and derivation channels of hydropower structures, with turbulent fluid movement.

4 Conclusions

Summarizing (briefly) the above, the following can be noted:

1. The function Φ depends on the shape of the free section of the channel, being a characteristic of this shape.
2. As can be seen from Table 1, this function can be significant in magnitude (which is important to keep in mind when determining for specific conditions).

3. Little is known about the K coefficient. Nevertheless, it can be assumed that it will also depend on the shape of the free cross-section of the channel, but its value will be insignificant, especially in those cases when the suppression of fluctuations of the maximum velocity is small.
4. To find the "K" value, it will probably be necessary to introduce some parameter expressing the ratio of the transverse dimension of the free flow surface in the channel to the wetted perimeter. It is quite possible that K can best be found from experiments. However, as it follows from Eqs. (14) and (15), before carrying out these experiments, the characteristics of the bottom and walls of the channel must be determined in advance (also from experiments - preferably with very wide channels of rectangular cross-section (Fig. 1)).
5. The theoretical formula (19) for determining the coefficient of hydraulic friction $\omega = \int_0^{\omega} d\omega = \int_0^h b(y)dy$ is valid both for the movement of fluid in smooth (η_Λ= 1/9) and rough channels.
6. With pressure flow in a round pipe ($R = D/4$) and infinitely wide rectangular channels (at$b \gg h;\ R = h$), as well as in the machine and diversion channels of hydropower facilities, where a uniform distribution of shear stresses is ensured (τ_0) around the wetted perimeter $\tau_0 \approx \tau_{osr}$ (Fig. 2), the geometric interpretation of the hydraulic radius is justified, in other cases (where $\tau_0 \neq \tau_{osr}$) – makes no sense.
7. The unpressurized machine and derivation channel of the correct cross-section corresponds to the law of hydraulic resistance determined by the shape of the free cross-section ϕ and K - taking into account the effect of the free flow surface during the free flow of water in the machine and diversion channels of hydropower facilities.

References

1. Troitsky V.P., Uralov B.R.: The influence of the shape of the pressureless cylindrical channel and roughness on the pressure loss. In: Environmental Protection from Industrial Emissions from Pulp and Paper Industry, Interuniversity Collection of Scientific Papers (1981)
2. Mamajonov, M., Bazarov, D.R., Uralov, B.R., Djumabaeva, G.U., Rahmatov, N.: The impact of hydro-wear parts of pumps for operational efficiency of the pumping station. J. Phys. Conf. Ser. **1425**, 012123 (2019). https://doi.org/10.1088/1742-6596/1425/1/012123
3. Uralov, B., Saidkhodjaeva, D., Kurbonova, U., Baymanov, R.: Influence of the shape of the pressureless trapezoidal channel and roughness on the pressure loss of the machine channels of the pumping stations. IOP Conf. Ser. Mater. Sci. Eng. **883**, 012012 (2020). https://doi.org/10.1088/1757-899x/883/1/012012
4. Troitsky V.P.: The main provisions of the design and hydraulic calculation of large earthen uncoated channels. In: Proceedings of the LPI Named After M.N. Kalinin, pp. 38–42 (1976)
5. Bazarov, D.R., Vokhidov, O.F., Lutsenko, L.A., Sultanov, S.: Restrictions applied when solving one-dimensional hydrodynamic equations. In: Proceedings of EECE 2019, Lecture Notes in Civil Engineering, vol. 70. pp. 299–305 (2019). https://doi.org/10.1007/978-3-030-42351-3_26.

6. Bazarov, D., Shaazizov, F., Erjigitov, S.: Transfer of Amudarya flowing part to increase the supportability of the Uzbekistan southern regions. IOP Conf. Ser. Mater. Sci. Eng. **883**, 012068 (2020). https://doi.org/10.1088/1757-899x/883/1/012068
7. Bazarov, D., Markova, I., Norkulov, B., Isabaev, K.: Operational efficiency of water damless intake (2020). https://doi.org/10.1088/1757-899X/869/7/072051
8. Bazarov, D., Uralov, B., Matyakubov, B., Vokhidov, O.: The effects of morphometric elements of the channel on hydraulic resistance of machine channels of pumping stations. Mater. Sci. Eng. 869, (2020). https://doi.org/https://doi.org/10.1088/1757-899X/869/7/072015.
9. Ergashev, R., Azizov, O., Dehkanova, N., Bozorov, A.: Development of energy-saving modes of irrigation pump stations. IOP Conf. Ser. Mater. Sci. Eng. **883**, 012017 (2020). https://doi.org/10.1088/1757-899x/883/1/012017
10. Kan, E., Mukhammadiev, M., Ikramov, N., Majidov, T.: Full-scale testing of the pump unit with a frequency converter. IOP Conf. Ser. Mater. Sci. Eng. **883**, 012112 (2020). https://doi.org/10.1088/1757-899x/883/1/012112
11. Shaazizov, F., Shukurov, D.: Physical modeling of the filtration process through the dam (2020). https://doi.org/10.1088/1757-899X/869/7/072037
12. Uralov, B., Isabaev, K., Jamolov, F., Akhmadi, M., Mirzaev, M.: The influence of the shape the living section of the pressureless machine channel and the roughness of its wetted surface on the hydraulic resistance. IOP Conf. Ser. Mater. Sci. Eng. **883**, 012006 (2020). https://doi.org/10.1088/1757-899x/883/1/012006
13. Uralov, B., Xidirov, S., Matyakubov, B.: River channel deformations in the area of damless water intake River. Mater. Sci. Eng. **869** (2020). https://doi.org/10.1088/1757-899X/869/7/072014.
14. Obidov, B., Choriev, R., Vokhidov, O., Rajabov, M.: Experimental studies of horizontal flow effects in the presence of cavitation on erosion – free dampers. IOP Conf. Ser. Mater. Sci. Eng. **883**, 012051 (2020). https://doi.org/10.1088/1757-899x/883/1/012051
15. Obidov, B., Vokhidov, O., Shodiev, B., Ashirov, B., Sapaeva, M.: Hydrodynamic loads on a water drain with cavitation quenchers. IOP Conf. Ser. Mater. Sci. Eng. **883**, 012011 (2020). https://doi.org/10.1088/1757-899x/883/1/012011
16. Antropovskii, V.I.: hydraulic resistance of different-type river channels with manifestations of karst and suffosion processes. Water Resour. **30**, 650–652 (2003)
17. Artikbekova, F.K.: estimating the processes in inlet canal with the consideration of the operational features of pump stations, Tashkent (2020)
18. Karaushev A.V.: Problems of the dynamics of natural water flows. Hydrometeoizdat (1960)
19. Militeev A.N.: Solving the problems of hydraulics in shallow reservoirs and hydroelectric water reservoirs using numerical methods (1982)
20. Keulegan, G.H.: Laws of turbulent flow in open channels. Natl. Bur. Stand. **21**, 707–741 (1938)
21. Karman T.: Mechanical similarity and turbulence (1936)

Technical Condition of 35 kv Overhead Lines in "Orelenergo", a Branch of the PJSC "IDGC of the Center"

Maksim Vladimirovich Borodin[1(✉)], Tatyana Anatolyevna Kudinova[2], Alexander Ivanovich Psarev[3], and Nailya Kamilevna Miftakhova[4]

[1] Department "Power Supply", Orel State Agrarian University Named After N.V. Parakhin, Orel, Russia
maksimka-borodin@yandex.ru
[2] Department "Linguistics and Humanities", Orel State Agrarian Universitynamed After N.V. Parakhin, Orel, Russia
[3] Department "Power Supply", Orel State Agrarian University Named After N.V. Parakhin, Orel, Russia
[4] Department "Power Supply of Industrial Enterprises", Kazan State Power Engineering University, Kazan, Russia

Abstract. The article discusses the main violations and malfunctioning of 35 kV overhead lines. The analysis of the technical condition of 35 kV overhead lines of "Orelenergo", a branch of the PJSC "IDGC of Center" is presented. The results of the given analysis show that about 60% of elements of 35 kV overhead lines are in good technical condition, about 39% of elements of 35 kV overhead lines are in satisfactory condition. The relevance of the development of a program aimed at improving the efficiency of the operation of 35 kV overhead lines of "Orelenergo", a branch of the PJSC "IDGC of Center" is indicated. The analysis of the technical condition of 35 kV overhead lines presented in the article will allow the personnel of power grid companies and other specialists to propose measures aimed at increasing the efficiency of 35 kV overhead lines of the company under consideration.

Keywords: Power lines · Technical condition · High voltage

1 Introduction

The authors [1–9] point out that improving the reliability of power supply to consumers is a priority area for the development of the modern electric power system in Russia as a whole. Power supply safety and reliable operation of the electrical network, capable of providing continuous power supply of the required quality to consumers, are highly dependent on the solution of issues of preventing failures of elements of overhead lines (OHL) operating in various conditions.

In accordance with modern requirements for the reliability of the network, the disconnection of any network element should not lead to power limitation of consumers.The authors [10–16] indicate that the reliability of power supply in 35 kV

A. Sinitsyn (Ed.): SES 2020, LNCE 141, pp. 47–54, 2021.
https://doi.org/10.1007/978-3-030-67654-4_6

electrical networks depends on the network configuration, its elements, circuit and operating parameters, the state of electrical equipment, etc.Therefore, maintaining the elements of the electrical network in good technical condition directly determines not only the quality, but also the minimum level of electricity losses in this network.

The task of increasing the reliability of 35 kV electrical networks and optimizing the modes has always been given close attention. In this regard, the issues of developing multi-criteria approaches to optimizing 35 kV networks in order to improve the reliability and efficiency of their functioning are relevant.

2 Research Methodology

"Orelenergo", a branch of the PJSC "IDGC of Center" pays great attention to the issue of increasing the reliability of power supply to consumers. The company is developing various plans for reconstruction, scheduled repairs, etc., but there is no specific, clear program aimed at solving the above mentioned issue. At the same time, "Comprehensive Program for the Development of Electric Grids with a Voltage of 35 kV and Higher in the Orel Region for the Period of 2019–2023", developed at "Orelenergo", does not take into account a full analysis of indicators affecting the efficiency of the operation of 35 kV overhead lines and existing long-term faults. To develop an appropriate program, it is necessary to analyze the technical condition of 35 kV overhead lines according to the following criteria:

- servicetime of 35 kV OHL;
- existing violations and malfunctioning of 35 kV OHL;
- "Good technical condition of 35 kV OHL";
- "Satisfactory technical condition of 35 kV OHL";
- "Unsatisfactory technical condition of 35 kV OHL";
- "Unsuitable technical condition of 35 kV OHL".

3 Results and Their Discussion

"Orelenergo", a branch of the PJSC "IDGC of Center" has 117 35 kV overhead lines. The results of the analysis of the service life of 35 kV overhead lines are shown in Fig. 1.

Analyzing Fig. 1, it is obvious that the service life of many 35 kV overhead lines is longer than the standard one. Moreover, only four 35 kV overhead lines have a service life of less than 20 years. Besides, more than 50% of all 35 kV overhead lines operate longer than the service time established by regulatory documents. The 35 kV Kulikovskaya - Kombikormovaya overhead line has the shortest service time equal to 12 years.

At the moment, the following violations and malfunctions are observed on the 35 kV overhead lines of "Orelenergo":

presence of individual trees on the edge of the clearing, threatening to fall on the OHL or grow towards the OHL at unacceptable distances;

- presence of individual trees on the edge of the clearing, threatening to fall on the OHL or grow towards the OHL at unacceptable distances;
- in sufficient width of the clearing along the OHL;
- presence of trees and shrubs with a height of 4 m and higher under the wires;
- presence of vegetation on the land allotted for support;
- performance of various works in security zones without written approval from the enterprise operating the OHL;
- lack of symbols, numbering of supports, warning posters;
- rake of the supports along or across the line in excess of the permissible norms, deformation of individual parts of the support, the lack of alignment of the racks and footrests of the stayed towers;
- deepening of foundations of supports, racks of reinforced concrete supports or attachments of wooden supports less than provided for by the project;
- subsidence or swelling of the soil around the foundation, subsidence or extrusion of the foundation;
- cracks and damage to attachments, foundations, supports;
- loose fit of the heel of the support to the surface of the foundation, inconsistency of the diameters of the nuts with the diameters of the anchor bolts, welding of the anchor bolts to the heel of the support instead of fastening with nuts, the absence of nuts on the anchor bolts; lack of parts on metal supports;
- corrosion of support parts and metal foot plates, defects in riveted and bolted joints;
- deformation of support elements and defects of welded seams;
- malfunctions of fastening parts of wooden supports;
- lack of bolts, nuts, keys and wedges, insufficient length of bolt cutting, breakage or loosening of wire ties, bolted connections, poor-quality fastening of brackets;
- rotting of support parts;
- presence of bird nests and other foreign objects on the supports;
- presence of corrosion of wires and cables;
- mechanical damage to porcelain or glass of insulators (chipping of a part of the insulator plates, the appearance of cracks);
- yytraces of overlapping garlands and individual insulators (damage to the glaze, destruction of porcelain, glass, traces of melting on the reinforcement of insulators and reinforcement of garlands);
- presence of defective (unusable) insulators;
- dirtiness of insulators, causing strong corona in wet weather;
- deviation of insulating supporting suspensions from the design position in excess of the permissible value;
- corrosion of fittings and insulator caps;
- cracks in reinforcement, rubbing or deformation of individual reinforcement parts;
- damage or breaks of grounding slopes on the support and near the ground;
- unsatisfactory contact in the bolted joints of the lightning protection cable with the grounding slopes or the support body;
- unsatisfactory contact of the ground electrode connection with the support body (reinforcement of a reinforced concrete support);

- exceeding the permissible value of the support grounding resistance;
- lack of brackets attaching grounding slopes to the support;
- destruction of the grounding device circuit by corrosion.

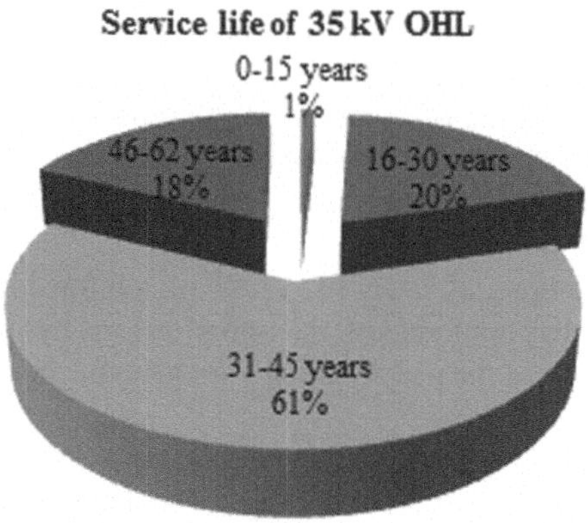

Fig. 1. Service life of 35 kV overhead lines of "Orelenergo", a branch of the PJSC "IDGC of Center" in 2019.

The results of the analysis of the technical condition of 35 kV overhead lines of "Orelenergo", a branch of the PJSC "IDGC of Center" are presented in Table 1.

According to the results, presented in Table 1, it can be concluded that in the northern section of the power transmission line 58% (3900 elements) of 35 kV OHL are in good technical condition, 41% (2795 elements) have a satisfactory technical condition and 1% (41 elements) have an unsatisfactory technical condition. In the southern section of the power transmission line, 51% (5368 elements) of 35 kV OHL are in good technical condition, 48% (2795 elements) have a satisfactory technical condition and 1% (107 elements) have an unsatisfactory technical condition. At the same time, in the eastern section of the power transmission line 67% (6216 elements) of 35 kV OHL are in good technical condition, 33% (3045 elements) have a satisfactory technical condition. It is also necessary to note that in the eastern section of the power transmission line there is a 35 kV line with wooden poles that were put into operation 56 years ago, while the standard service life of wooden poles of 35 kV OHL is 25 years [17].

Table 1. Technical condition of 35 kV overhead line of "Orelenergo", a branch of the PJSC "IDGC of Center"in 2019.

Name	Technical condition			
	Good	Satisfactory	Unsatisfactory	Unsuitable
Northern section of power transmission line				
Metal supports 35kV, pcs	89	37	2	0
Reinforced concrete poles 35 kV, pcs	2577	420	39	0
Span of 35 kV overhead lines, pcs	1234	2338	0	0
Southern section of power transmission line				
Surge arrester 35 kV, pcs	6	0	0	0
Metal supports 35 kV, pcs	212	240	14	0
Reinforced concrete poles 35 kV, pcs	3214	1098	74	0
Span of overhead lines, pcs	1936	3804	19	0
Eastern section of power transmission line				
Metal supports 35 kV, pcs	167	26	0	0
Reinforced concrete poles 35 kV, pcs	3488	358	0	0
Span of 35 kV overhead lines, pcs	2534	2414	0	0
Wooden poles 35kV, pcs	27	247	0	0

The technical condition of 35 kV OHL in percentage terms for each element is shown in Figs. 2, 3, 4 and 5.

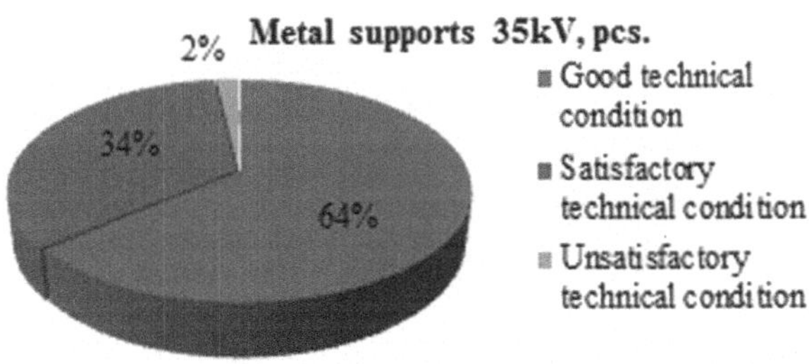

Fig. 2. Technical condition of metal supports of 35 kV OHLof "Orelenergo", a branch of the PJSC "IDGC of Center".

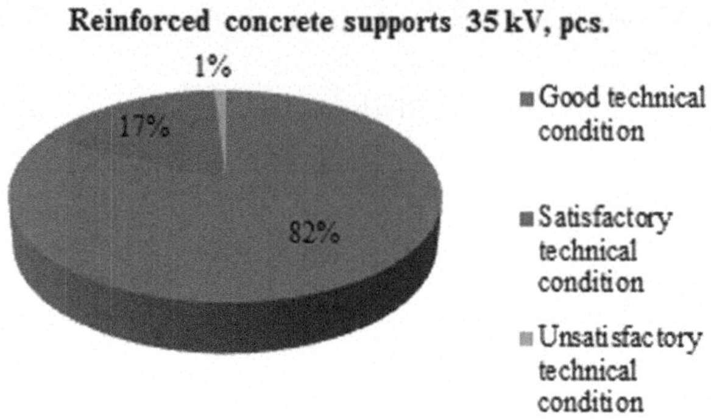

Fig. 3. Technical condition of reinforced concrete supports of 35 kV OHL of "Orelenergo", a branch of the PJSC "IDGC of Center".

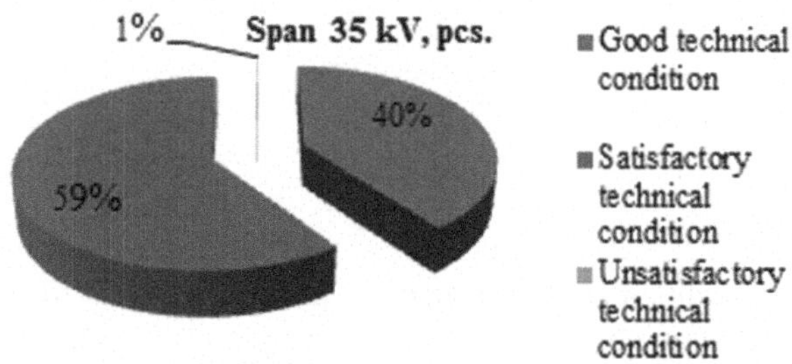

Fig. 4. Technical condition of spans of 35 kV OHLof "Orelenergo", a branch of the PJSC "IDGC of Center".

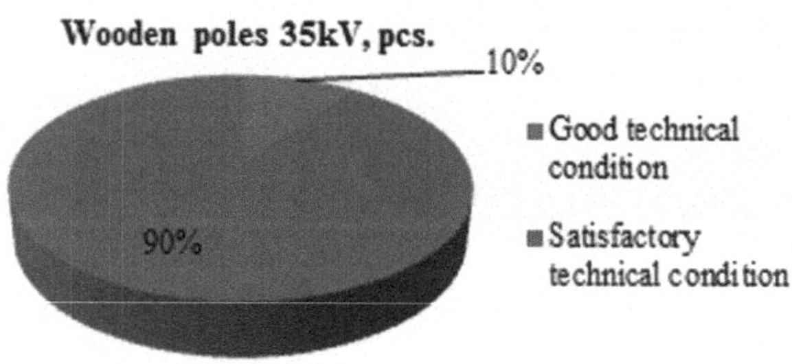

Fig. 5. Technical condition of wooden poles of 35 kV OHL of "Orelenergo", a branch of the PJSC "IDGC of Center".

The results, presented in Figs. 2, 3, 4 and 5 allow us to conclude that about 60% of the elements of 35 kV OHL are in good technical condition, about 39% of the elements of 35 kV OHL are in satisfactory condition; here with, a satisfactory condition at any time can become unsatisfactory. Consequently, to improve the reliability of power supply, it is necessary to reconstruct many overhead lines now.

4 Conclusions

The results of the analysis of the technical condition of the 35 kV OHL of "Orelenergo", a branch of the PJSC "IDGC of Center" show that today it is necessary to develop a program aimed at improving the efficiency of the 35 kV OHL.

The analysis of the technical condition of 35 kV OHL, presented in the article will allow the personnel of power grid companies and other specialists to propose measures aimed at increasing the efficiency of 35 kV OHL of "Orelenergo". In turn, the implementation of the proposed measures will ensure:

- an increase in the availability and development of the power grid infrastructure for sustainable socio-economic growth in the region, growth of competition in the electricity (power) market, which will lead to the integrated development of transmission and distribution networks;
- creation of favorable conditions for technological connection of electrical installations of consumers to the electric networks of "Orelenergo", a branch of the PJSC "IDGC of Center";
- elimination of installed capacity deficit in power centers;
- improving the reliability of power supply to consumers;
- reduction of energy losses;
- yreduction ofcash costsfor the operation of 35 kV OHL.

References

1. Soloviev, Yu.V., Samarin, G.N.: Analysis of the Conditions for the Development of Defects in Polymer Structures of Equipment for 6–35 kV Overhead Lines with Insulated Wires Under Ice and Wind Loads. Scientific and Technical Bulletin of the St. Petersburg State Polytechnic University 4, 135, pp. 63–67 (2011)
2. Borodin, M.V., Uryupin, N.S.: Justification of energy-efficient cross-section of power lines with voltage deviations of ±5%, ±10%. Bull. Nizhny Novgorod Inst. Eng. Econ. **2**(81), 20–29 (2018)
3. Glybina, Yu.N., Belikov, R.P., Fomin, I.N.: Analysis of the types and amount of faults in electric networks of voltage class 6–10KV. Agrotechnics Power Supply **3**(16), 43–49 (2017)
4. Vinogradov, A.V., Chernyshov, V.A., Semenov, A.E., Moroz, A.O.: Method for monitoring the technical condition of overhead power transmission line elements. Agrotechnics Power Supply **3**(3), 64–68 (2014)
5. Borodin, M., Psarev, A., Kudinova, T., Mukhametzhanov, R.: Improving power quality by calculating voltage losses. In: E3S Web of Conferences, vol. 1041 (2019)
6. Fomin, I., Belikov, R., Zelyukin, V., Mikhailova, E.V.: Improvement of power supply reliability by means of remote control of the automatic repeated switching-on of sectionalizing circuit-breakers. In: E3S Web of Conferences, 01042 (2019)
7. Vinogradov, A., Bolshev, V., Vinogradova, A.,Kudinova, T., Borodin, M., Selesneva, A., Sorokin, N.: A system for monitoring the number and duration of power outages and power quality in 0.38 kV electrical networks. In: Vasant, P., Zelinka, I., Weber, G.W. (eds.) Intelligent Computing & Optimization, ICO 2018. Advances in Intelligent Systems and Computing, vol. 866, pp. 1–10. Springer, Cham (2019)

8. Gimadiev, R.A., Ivshin, I.V., Mukhortov, I.S., Bilalov, F.F., Valiullin, R.R., Miftakhova, N.K.: The effectiveness of the Methodology for Assessing the Results of Monitoring the Technical Condition of Medium Voltage Cable Lines: Power Engineering of Tatarstan **2**(42), 68–71 (2016)
9. Gracheva, E.I., Fedorov, O.V.: Forecasting Reliability Electrotechnical Complexes of In-Plant Electric Power Supply Taking Into Account Low-Voltage Electrical Apparatuses, 2019 International Conference on Industrial Engineering. Applications and Manufacturing, ICIEAM 2019, 8743057 (2019)
10. Borodin, M.V., Dorozhkina, A.A., Semenov, A.E.: Methodology for the selection and determination of the most effective measuring transformer for a substation. News High. Educ. Inst. Chernozem Region **4**(58), 58–68 (2019)
11. Borodin, M.V., Pustovaya , O.A. (Ed.), Topical Issues of Power Engineering, Materials of the 7th All-Russian Scientific Conference with International Participation, dedicated to the professional holiday Power Engineer Day (2020)
12. Borodin, M.V., Belikov, R.P., Makhiyanova, N.V.: Improving the power quality by calculating voltage losses. Bull. Don Agrarian Sci. **3**(47), 35–40 (2019)
13. Olshansky, V.G., Stolpovskaya, E.V., Lazhintseva, N.V.: Experience of modernization of lattice supports of 35–110 kV overhead lines used in the Northern Regions of Western Siberia. Electric Stations **5**(1030), 40–46 (2017)
14. Sorokin, N.S.: Improving the efficiency of functioning of electric networks with voltage of 6–35 kV by Monitoring the Parameters of the Electric Network. Scientific Bulletin of the National University of Bioresources and Environmental Management Ukraine, Series: Technique and Power Engineering of Agro-Industrial Complex **209**(2), 92–95 (2015)
15. Vinogradov, A.V.: To determine the efficiency of power supply systems. Bull. Nizhny Novgorod Inst. Eng. Econ. **7**(74), 26–35 (2017)
16. Astakhov, S.M., Belikov, R.P.: State and ways to improve the efficiency of distribution networks in the agro-industrial complex. Bull. OrelSAU **29**(2), 106–108 (2011)
17. Technical Documentation 34.20.504–94, Typical Operating Instructions for Overhead Power Transmission Lines With Voltage of 35–800 kV, approved by RAO UES of Russia, 19 September 1994. https://www.consultant.ru/cons/cgi/online. cgi?req=doc&base=EXP&n=378706#06693717546709732. Accessed 22 June 2020

Algorithm for Eliminating "Noise" When Measuring Voltages and Currents to Deter-Mine Damage in Power Systems

Nadezhda Buryanina[1(✉)], Yurii Korolyuk[1], Evdokiia Maleeva[1], Elena Lesnykh[2], and Konstantin Suslov[3]

[1] Chukotka Branch "M. K. Ammosov North-Eastern Federal University", Studencheskaya Street, 3, Chukotka Anadyr 689000, Russia
bns2005_56@mail.ru

[2] Dusi Kovalchuk Street, Siberian Transport University University, 191, Novosibirsk 630049, Russia

[3] Irkutsk National Research Technical University, Lermontov Street, 83, Irkutsk 664074, Russia

Abstract. The article discusses an algorithm for determining the fundamental harmonics of currents and voltages in the measuring devices of relay protection at saturation of the magnetic circuits of current transformers. The aperiodic component arises due to the presence of complex resistances of generators and power lines in the parameters of the power system. Periodic components appear in microprocessor technology due to imperfection of the measuring devices. The exclusion from the short-circuit current entering the measuring devices of relay protection, non-periodic components and harmonics, except for the first one, makes it possible to increase the reliability of relay protection.

Keywords: Power systems · Fundamental harmonics of currents and voltages

1 Introduction

The introduction of digital control systems in the control of power systems allows performing the functions of relay protection and determining the locations of damage both during transient processes and in normal modes. Unlike analogue protection relays and fault locating devices, digital protection relays process information discretely. This allows you to monitor the change in the parameters of the power system mode instantly and also instantly make a decision on control during transient processes, and analog relay protection makes a decision later, when the mode change is confidently recorded, i.e. after the exclusion of free components arising from transient processes. That is, digital relay protection is faster than analog. This is due to the negative influence of the presence of aperiodic components and non-industrial frequency parameters in the measuring devices of relay protection.

The difference between digital relay protections and analog ones is that analog protections receive mode information continuously, and digital ones - discretely at certain intervals. This allows you to fix the instantaneous values of the mode

A. Sinitsyn (Ed.): SES 2020, LNCE 141, pp. 55–61, 2021.
https://doi.org/10.1007/978-3-030-67654-4_7

parameters. Not fast enough identification of an emergency can lead to the development of an emergency, which is often inherent in power systems controlled by analog relay protection. The progression of an accident can lead to a loss of synchronism and destruction of equipment. In both cases, to economic damage.

High voltage power lines are one of the most frequently damaged elements of an electrical power system. Damage to the power line leads to a drop in voltages and possible damage to industrial facilities. Therefore, quick determination of the location of damage by relay protection is the most important problem.

Both foreign and domestic scientists are engaged in the analysis of digital relay protection and automation. Of greatest interest, from the point of view of research, are protections that use information about changes in currents [1–11].

Below is the statistics of damage on 330–500 kV lines (Table 1).

Table 1. The main reasons for technological disturbances in the operation of 330–500 kV overhead lines.

The reasons	%, of the total number of violations
Loads from the effects of ice, snow, wind loads, atmospheric overvoltages, other climatic influences	30
Extraneous influences, vandalism	25
Disadvantages of operation, defects of repair	20
Fires on the highway and near overhead lines	12
Wear, change of materials during operation, other reasons	8
Falling trees on wires	5

Improving the reliability of power lines is the main aim of researchers in the world. One of the tasks is to improve relay protection and automation, in particular when using current transformers from magnetic circuits without a magnetic gap and with it. The objective of this work is to separate the power frequency current from the short-circuit current in the measuring devices of digital relay protections containing aperiodic components [12].

2 Materials and Methods

With the sampling frequency of analog information recommended in [1] and equal to 1600 Hz, the sampling interval Δt is 0.625 ms. Let's select the current values at the input of the measuring device of digital relay protection in the time interval from 0 to 2.5 ms. With an angular frequency equal to $\omega t = 2\pi f$, the instantaneous value of the sinusoidal current is

$$i(\mathrm{t}) = i[\sin(\omega \cdot t)] \tag{1}$$

With a sampling interval equal to a quarter of the period of the fundamental frequency, the current in the measuring organ of digital relay protection doesn't contain harmonic components [13].

This article proposes algorithms for eliminating the third, fifth and seventh harmonics when using magnetic circuits with a non-magnetic gap, as well as elimination with an aperiodic component when using current transformers without a non-magnetic gap.

The third harmonic component of the current in the measuring device of the relay protection and automation system is excluded when using the equation:

$$i3(t) = \frac{i\left(t + \frac{0{,}02}{3}\right) - i(t)}{2 \cdot \sin\left(\frac{2\pi}{3}\right)},$$

$$i5(t) = \frac{i\left(t + \frac{0{,}04}{5}\right) - i(t)}{2 \cdot \sin\left(\frac{2\pi}{5}\right)}, \tag{2}$$

$$i7(t) = \frac{i\left(t + \frac{0{,}04}{7}\right) - i(t)}{2 \cdot \sin\left(\frac{2\pi}{7}\right)}$$

The amplitude of the third harmonic can be - 30–35% of the fundamental, the fifth - up to 10%, the seventh - up to 5%. Subsequent harmonics have amplitudes from fractions of a percent to 3%. The following harmonics can be eliminated with a low pass filter with a cut off frequency corresponding to the seventh harmonic. In the absence of aperiodic components in the measured currents, and this is possible when using current transformers with a non-magnetic gap in the magnetic circuit, the proposed algorithm excludes the harmonic components of the minor frequency. The running time of the algorithm doesn't exceed three quarters of the period of the fundamental frequency.

The number of measured current samples for each phase of the protected object should be four, separated by three sampling intervals. At a sampling rate of 1.6 kHz, the three sampling intervals are 1.75 ms. This makes it possible to ensure a quick response of the relay protection when the magnetic cores of the current transformers are saturated.

3 Results

Let us consider how to determine the currents and voltages of the forward and reverse sequences through the instantaneous values of the parameters [].

Consider how to determine the positive- and negative-sequence currents and voltages through the instantaneous values of the parameters. At time t, the phasor of the current is obtained as follows:

$$\dot{I}(t) = i(t) + j \cdot i^{'(t)},$$

$$I(t) = \sqrt{i(t)^2 + i^{'}(t)^2}, \varphi(t) = arctg[\frac{i^{'}t}{i(t)}] \tag{3}$$

Many scientists are invited to determine the currents of the direct and reverse sequences in a special phase A through the complexes:

$$\dot{I}_1 = \frac{\dot{I}_A + \dot{I}_B \cdot e^{j\frac{2\pi}{3}} + \dot{I}_C \cdot e^{-j\frac{2\pi}{3}}}{3}$$

$$\dot{I}_2 = \frac{\dot{I}_A + \dot{I}_B \cdot e^{-j\frac{2\pi}{3}} + \dot{I}_C \cdot e^{j\frac{2\pi}{3}}}{3} \tag{4}$$

Since the tg(φ) in the first and third quadrants is positive, and in the second and fourth quadrants it is negative, the angles should be specified: *i(t) > 0, i'(t) < 0, then* $\varphi = -\varphi$; *i(t) < 0, i'(t) > 0, then* $\varphi = 180 - \varphi$; *i(t) < 0, i'(t) < 0, then* $\varphi = 180 + \varphi$.

The disadvantages of calculating the direct and negative sequence currents through complex quantities are the duration of the time spent and the error. The time for calculating the currents of direct and reverse subordination in the presence of aperiodic and harmonic components is at least 1.33 periods of industrial frequencies.

We propose calculating the instantaneous values of positive- and negative-sequence currents using two formula

$$\dot{I}_1(t) = \frac{\dot{I}_A(t) + \dot{I}_B(t) \cdot e^{j\frac{2\pi}{3}} + \dot{I}_C(t) \cdot e^{-j\frac{2\pi}{3}}}{3}$$

$$\dot{I}^2(t) = \frac{\dot{I}_A(t) + \dot{I}_B(t) \cdot e^{-j\frac{2\pi}{3}} + \dot{I}_C(t) \cdot e^{j\frac{2\pi}{3}}}{3} \tag{5}$$

This Eqs. (5) suggest the calculation of phases B and C in time intervals that are distant from the time of phase A current measurement by a third of the period before and after the measurement when we use (5), i.e. the processing time current is less 11 ms. By this time, the magnetic cores of current transformers in some cases may reach saturation, and the calculation results will prove incorrect [14, 15].

Consider the effect produced by the non-utility frequency components in currents on the accuracy of determining an emergency situation during short circuits. The short circuit current in time is determined by the equation.

$$i(t) = \frac{E_{eq}}{\sqrt{R_{eq}^2 + X_{eq}^2}} \cdot \left(\sin(\omega \cdot t + \psi - \varphi) - \sin(\psi - \varphi) \cdot e^{\frac{-t}{T_c}}\right) \tag{6}$$

where E_{eq} – equivalent EMF; R_{eq}, X_{eq} – equivalent active and reactive resistances from the equivalent EMF to the short-circuit point; ω – angular frequency equal to 2πf;

f – frequency; $T_C = \frac{L}{R}$ – time constant; ψ – voltage angle at which the short circuit occurred, expressed in radians; φ – impedance angle of a short-circuit branch.

The calculated current value of the current at times (nΔt) is proposed to be determined by:

$$i(n \cdot \Delta t) = \frac{2 \cdot i(n \cdot \Delta t) - i[(n-1) \cdot \Delta t] - i[(n+1) \cdot \Delta t]}{4 \cdot sin^2\left(\omega \cdot \frac{\Delta t}{2}\right)} \tag{7}$$

where n – is the sequence number of the sample. The time constant and the angular frequency is defined as:

$$T_c = \frac{-\Delta t}{\ln\left(\frac{i_{ev}(2\Delta t) - i(3\Delta t)}{i_{ev}(\Delta t) - i(2\Delta t)}\right)}, \varphi = arctg(T_c \cdot \omega) \tag{8}$$

The aperiodic component is calculated as:

$$i_A(t) = \frac{E}{\sqrt{R^2 + X^2}} \cdot [-sin(\psi - \varphi)] \cdot e^{-\frac{t}{T}}, i_{sin}(t) = i_t - i_A(t) \tag{9}$$

The instantaneous complex value of the current in the measuring device of the relay protection does not contain an aperiodic component. This means that it is not contained in the instantaneous values of the currents calculated in subsequent intervals. Thus, by measuring currents at certain points in time, it is possible to determine the currents at any moment in the "future" time after the occurrence of a short circuit. This is relevant for emergency automation, the onset of which is much longer than the response time of relay protection [15–17].

When analyzing the algorithm for separating the first harmonic components of currents in the measuring elements of relay protection, harmonic components were investigated up to the thirteenth with amplitudes equal to the amplitude of the fundamental harmonic, which is much less than the real amplitudes. The proposed algorithm selects sinusoidal current components in the following sequence: four instantaneous current values are measured according to Eq. (6) at equal time intervals, separated by three sampling intervals. Further, the time constant of the electrical network from the point of short circuit to the equivalent EMF is determined. Then the aperiodic component of the current is determined and excluded from the current in the measuring device of the relay protection and, as a result, the complex value of the current is determined.

The calculations were carried out shows the dependences of the calculated voltages of the positive and negative sequences, which are changed on both sides of the line in case of a short circuit at a distance of 90 km from the beginning of a 150 km long line (Fig. 1).

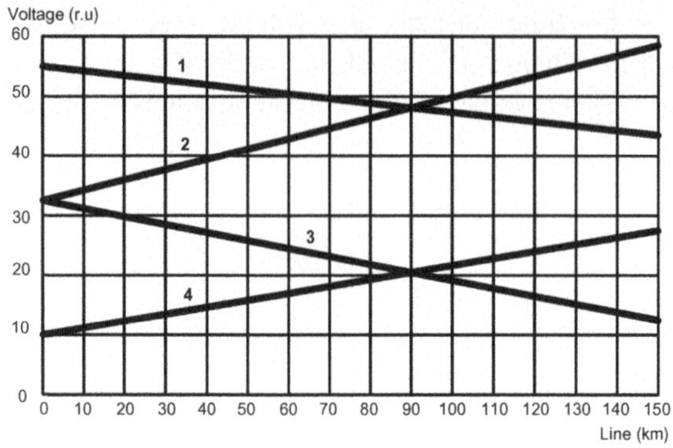

Fig. 1. Changing the voltages when determining a single-phase short circuit at a distance of 90 km from the beginning of a 150 km long line. 1 and 2 - positive sequence voltages, 3 and 4 - negative sequence voltages.

4 Conclusion

An algorithm has been developed that selects the fundamental current harmonic in the measuring device of relay protection in the presence of aperiodic and harmonic components of non-fundamental frequencies in short-circuit currents. The algorithm filters the harmonic component in the presence of aperiodic and harmonic components of non-fundamental frequencies in the measuring organ.

Acknowledgements. The study was carried out with the financial support of the Russian Foundation for Basic Research and the Subject of the Russian Federation - the Republic of Sakha (Yakutia) № 18–48-140 010. Research and the Subject of the Russian Federation - the Republic of Sakha (Yakutia) № 18–48-140 010.

References

1. Mecke, A., Lee, I., Baker jr., J.R., Banaszak Holl, M.M., Orr, B.G.: Eur. Phys. J. E **14**, 7 (2004)
2. Kiliçkiran, H.C., Şengör, İ, Akdemir, H., Kekezoğlu, B., Erdinç, O., Paterakis, N.G.: El. Power Sys. Res. **164**, 89–102 (2018)
3. Hasanpour, M., Babaei, M., Tabrizian, M.: En. Procedia **100**, 280–286 (2016)
4. Akke, M., Thorp, J.: IEEE Transl. Power Delivery **13**, 66–72 (1998)
5. Hosemann, G., Steigerwald, H.: IEEE Transl. on Power Delivery **8**, 933–940 (1993)
6. Ajaei, F.B., Sanaye-Pasand, M., Sanaye, M., Homaei, H., Iravani, R.: IEEE Trans. Power Delivery **26**, 2531–2540 (2011)
7. Gangadharan, P.K., Sidhu, T.S., Finlayson, G.J.: IEEE Transl. Power Delivery **22**, 108–115 (2007)
8. P.A. Zvadal, D.S. Tuchinal, T.V. Strebkoval, A.V.: Mironov 2nd International Youth Scientific and Technical Conference on Relay Protection and Automation Conf., pp. 1–15 (2019)

9. Saldarriaga-Zuluaga, S., López-Lezama, J., Muñoz-Galeano, N.: Electronics **9**, 1740 (2020)
10. Evans, F.J., Wells, G.: IEEE Trans. Instrum. Meas. **19**, 144–147 (1970)
11. Andreev, M.V., Rudnik, V.E., Bay, Yu.D.: Thermophysical Basis of Energy Technologies Conf., 1–10 (2020)
12. E.V. Shneerson, Digital relay protection (2007). (in Russian)
13. Greshnov, E.B., Korolyuk, Y.F.: Electronic modeling **5**, 71–74 (1982). (in Russian)
14. Buryanina, N.S., Korolyuk, Yu.F., Lesnykh, E.V.: Method for instantaneous voltages and currents values sampling. Rus. Patent, 07 December 2017
15. Buryanina, N.S., Korolyuk, Y.F., Koryakina, M.L., Lesnykh, E.V., Suslov, K.V.: IOP Conference Series: Materials Science and Engineering 860, 012035 (2020)
16. Kuzhekov, S.L., Nudel'man, G.S.: Izvestiya Vuzov "Elektromekhanika" 4, 12–19 (2009)
17. Kulikov, A.L.: Vestnik NGIEI **9**, 29–38 (2017)

Sewage Sludge Treatment Using the “freezing-Thawing” Method

Leonid Sokolov[1(✉)], Aleksandr Tyanin[1], and Liliya Mukhametova[2]

[1] Vologda State University, Lenina Street 15, Vologda 160000, Russia
sokolovli@mail.ru

[2] Kazan State Power Engineering University, Street Krasnoselskaya, 51, 420066 Kazan, Russia

Abstract. Dewatering of wastewater treatment sludge by cyclic freezing-thawing method was investigated. The parameters determining the moisture yield of the compacted excess activated sludge were specified. Three cycles of sludge freezing result in a decrease in the specific resistance of filtration and an improvement in the dynamics of sewage sludge filtration. Relationship between the sludge specific resistance and the number of freezing-thawing cycles and the freezing temperature were established.

Keywords: Waste treatment · Sewage sludge · Cyclic freezing-thawing method

1 Introduction

Sludge drying beds with an artificial waterproof base and drainage are widely used for the gravitational dewatering of sludge formed at biological wastewater treatment plants of low productivity in the North-West region of Russia [1, 2]. This method of sludge dewatering is also widely used in Finland, Norway, Sweden [3–8]. The proper operation of such sites strongly depends on climatic factors, which change the physical properties of sludge and, accordingly, affect its moisture yield.

It is known from literature [2, 9, 10] that freezing changes the sediment structure, contributes to its rapid dewatering. Thus, freezing can be considered as a process of preparing sediments for dewatering at sludge sites [11]. All sediments differ in composition, properties, structure. It is known that thin suspensions and colloidal fractions of sediments have a strong impact on their moisture yield and filterability of untreated sediments [12]. Sludge properties depend on the type of oxidized organic matter, the age of the sludge, the ratio of the amount of substrate to the mass of bacteria, temperature and the content of dissolved oxygen. Under endogenous conditions, the amount of extracellular substance on the surface of bacteria increases [13]. Biological sediments are diverse in their composition, structure, which is determined by the place of their formation and accumulation at different stages of wastewater purification [14–16]. Taking into account their high humidity, significant areas are required for their placement and storage, therefore, the main task of their conditioning is to increase moisture yield by reducing the specific resistance of the sediment during filtration and other methods of dewatering. The average specific resistance, determined from the

A. Sinitsyn (Ed.): SES 2020, LNCE 141, pp. 62–67, 2021.
https://doi.org/10.1007/978-3-030-67654-4_8

Kozeny-Carman equation [17, 18], characterizes the porosity of the dewatered sludge and its ability for filtering. It depends on several properties, i.e. particle size and shape, dry matter concentration, compressibility, moisture. There are many combinations of physical properties that can result in the same average specific resistance. The Kozeny-Carman equation cannot adequately describe the influence of the dry matter concentration [12] and assumes that the particles are incompressible. For these reasons, one should beware of comparing the resistance of sediments of different nature, because they are formed at specific treatment facilities with different quality of wastewater, in special natural and climatic conditions.

The aim of this paper is to study the influence of repeated freezing and thawing of the sediment of compacted excess activated sludge (CEAS) on decrease in the specific resistance of filtration, decrease in the time of sludge compaction and increase in the rate of gravitational dewatering with the use of filter material.

2 Materials and Methods

At the first stage, a series of experiments was carried out to determine the concentration, moisture content, ash content and specific filtration resistance of the initial CEAS sludge. For the considered samples the obtained concentration was 32.7–33.3 g/l, humidity was 96–97%, ash content was 30.4–31.6% and specific filtration resistance was within the range $(196.7–230.2)\cdot 10^{-10}$ cm/g. The optimal range of sediment freezing temperatures was identified, which for CEAS was from −5 to −10 °C. So, the experiments on triple (cyclic) freezing of the sediment were carried out in this temperature range.

Laboratory assessments of conditioning cryogenic impact on biological sediments included gravity compaction and filtering experiments using paper filters and capillary suction experiments. A cascade of experiments was set up on cyclic freezing of the sediment, followed by the determination of the filtration sediment specific resistance after each cycle. The duration of each freezing cycle was 24 h. The duration of the cascade experiment was 72 h. The excess activated sludge was collected at the sewage treatment plant after secondary sedimentation tanks, CEAS was collected after the sludge compactor before being treated with any chemical reagents. Sludge samples were delivered to the laboratory within an hour. Experiments began at the time of delivery of samples to the laboratory, thus excluding the sludge aging. The experiments were carried out immediately after stirring each sample of the precipitate at a rate of 500 rpm for 20 s in the analytical beaker using a three-blade stirrer 50 mm in diameter. For filtration, a 130 mm funnel, 125 mm white ribbon paper filter was used.

The research methodology included the simultaneous placement of six 500 ml samples of CEAS sediment in a freezer at a constant temperature. Each of the three cycles of temperature exposure included: a 18-h freezing and subsequent defrosting for 6 h at a temperature of +20 °C. After each cycle of freezing and thawing in five repetitions, the filtration specific resistance of sediment (R) was determined according to the standard methods.

3 Results

The specific resistance of filtration of CEAS sediment after each of the three freezing cycles are shown in Fig. 1.

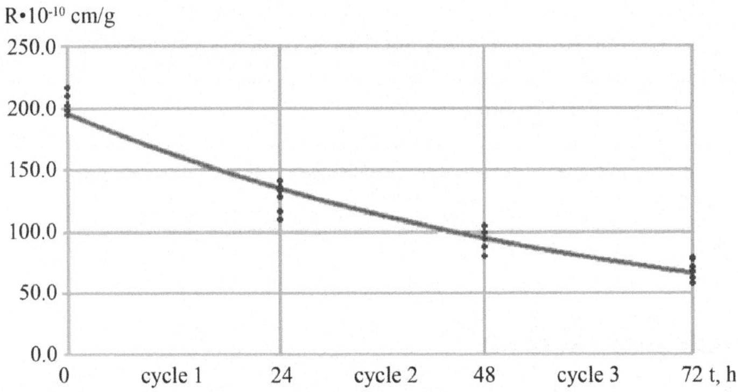

Fig. 1. Change in specific resistance of filtration of CEAS sediment by freezing cycles: 1st cycle (0–24 h); 2nd cycle (24–48 h); 3rd cycle (48–72 h).

Cyclic freezing followed by thawing made it possible to reduce the specific resistance of the sediment filtration after the first phase of freezing by 1.4 times relative to the initial value, and by 2.3 and 3.0 times after the second and third phases, respectively. The obtained effect confirms a significant change in the structure of the sediment, contributing to an increase in its moisture yield.

The gravity compaction examination of the freezed CEAS was performed by simultaneously placing six samples in 1000 ml graduated cylinders and recording the resulting sediment volume for one hour at ten minute intervals.

The results of the CEAS gravity compaction study are shown in Fig. 2.

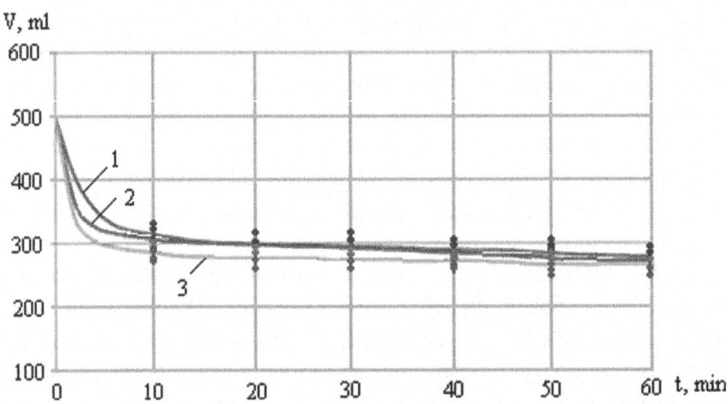

Fig. 2. Change in the CEAS volume during gravitational compaction of the sediment after three cycles of its freezing: 1-1st cycle; 2-2nd cycle; 3-3rd cycle.

The gravitation filtration of CEAS was studied by passing six samples of 100 ml volume through white ribbon paper filters, and measuring the filtrate volume every ten minutes.

The obtained results on gravity filtration of CEAS sediment are presented in Fig. 3.

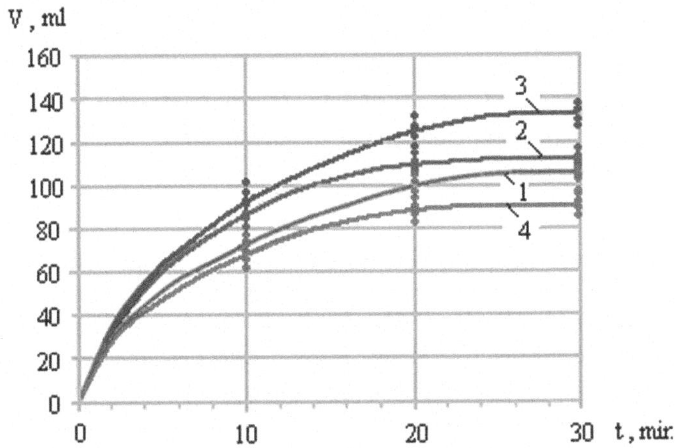

Fig. 3. Change in the volume of the filtrate (V) as a function of duration of filtering the sediment after three cycles of its freezing: 1-1st cycle; 2-2nd cycle; 3-3rd cycle; 4-initial sediment.

4 Discussion

The obtained results show that an increase in the number of freezing cycles from one to three has a positive effect on the gravitational compaction and filtration of the sediment in relation to samples that were not treated by temperature. Intense moisture yield (MY) was recorded during the first 20 min of filtration for all sediment samples (Fig. 3). Further filtration takes place in a quiet phase, without a significant change in speed. The investigated sediment is more difficult to dehydrate than just excess activated sludge, which is consistent with the results of the studies [2, 17].

The average suspended matter (SM) of the initial test sediment was 32.5 g/l and practically did not change during the freezing cycles. At the same time, the moisture content of the sediment obtained during the first freezing cycle was 97.5%, and decreased by 1% on average after the next two cycles (Table 1).

Table 1. The results of CEAS sediment examination.

No. of experiment	Freezing cycle								
	1st cycle			2nd cycle			3rd cycle		
	$R \cdot 10^{-10}$, cm/g	SM, mg/l	MY, %	$R \cdot 10^{-10}$, cm/g	SM, mg/l	MY, %	$R \cdot 10^{-10}$, cm/g	SM, mg/l	MY, %
1	110.9	31965	97.6	81.7	31880	97.0	58.4	30735	96.6
2	142.6	32315	97.4	100.4	32015	96.8	67.8	33210	96.3
3	140.3	33060	97.5	105.6	34270	96.4	75.9	32450	96.1
4	133.2	32565	97.3	98.5	32410	96.9	71.2	33515	96.3
5	115.8	32465	97.5	90.2	31640	97.0	61.2	32980	96.6
Av	128.6	32474	97.5	95.3	32443	96.8	66.9	32578	96.4

The presented research results confirm the change in the properties of the CEAS sludge. Therefore, it shows that the repeated freezing under the operating conditions of sludge sites in winter can be used to intensify the dewatering process. In this case, the technology of sludge treatment at sludge beds in winter includes sludge spilling, its gravitational compaction, draining of the upper layer of clarified moisture, freezing of the entire sediment layer, removal of released free moisture by filtration through the loading layer after complete thawing of the sediment.

The conducted studies show that during the sediment freezing on sludge areas, at the first stage free moisture crystallization takes place, forming a surface layer of ice. Solid particles are displaced out of the crystallizing liquid layer into the sediment layer. The water molecules are oriented in such a way to ensure the further formation of ice crystals. It was also established that sediment frozen at low negative temperatures at low speeds of the front of forming ice has a higher moisture yield, since during slow freezing process, bound moisture has time to spread into the intercellular space, where it crystallizes. The resulting internal compressive stress contributes to the enlargement and coagulation of dewatered sludge particles and the accumulation of a layer of compacted sludge at the bottom of the site, i.e. from the opposite side relative to the heat exchange surface. Under these conditions, freezing of solid and colloidal particles occurs last, and this ensures the maximum formation of free moisture when the structure of the sediment changes.

5 Conclusion

It was found that the specific resistance is a characteristic of the substance of the biological sludge and does not depend on its concentration and on most of the dewatering conditions. Studies of cryogenic treatment of biological sludge have shown an effective decrease in filtration resistance for excess activated sludge at a freezing temperature of −50 °C, and at −10 °C for CEAS. Intense moisture yield was recorded during the first 20 min of filtration for all sediment samples. The results of this study of biological sediments using cryogenic treatment confirm the change in the structure and properties of sediments, which makes it possible to consider the use of multiple freezing under the operating conditions of sludge beds in winter to intensify the dewatering process. It was established that the process of freezing sediment in the studied temperature range (from 0 to −20 °C) dominates in relation to mainly gravitational moisture, but if necessary, conditions can be created (low temperature, cyclic and cascade freezing, pressure, incl. Vacuum) to remove film, capillary and hygroscopic moisture. Thus, the freezing temperature determines the kinetics of the process of changing the physicochemical structure of the sediment and has a positive effect on its water-yielding capacity. The change in the sediment structure during freezing results in a decrease in its specific resistance to filtration. Accordingly, it leads to an increase in the productivity of sludge sites and a decrease in the suspended matter of drainage water.

References

1. Politaeva, N., Prokhorov, V., Lezhnev, E.I., Polyanskii, V.A., Matisov, B.G., Mukhametova, L.R., Mukhin, I.A.: IOP Conference Series: Earth Environment Sciences, vol. 337 (2019)
2. Turovsky, I.S.: Wastewater Sludge Treatment, 2nd edn. Stroyizdat, Moscow (1982)
3. Surikova, A., Sytsyanko, E., Solov'Eva, E., Mukhametova, L.: E3S Web Conference, vol. 178 (2020)
4. Wójcik, M., Stachowicz, F., Masłoń, A.: E3S Web Conference, vol. 17, p. 00100 (2017)
5. Swanwick, J.O., Marian, F.M.: Water Waste Treat. **8**, 386 (1961)
6. Fytili, D., Zabaniotou, A.: Renew. Sustain. Energy Rev. (2008)
7. Andrews, R.J., Lloyd, J.W., Lerner, D.N.: Geol. Soc. Spec. Publ. (1998)
8. Madaeni, S.S., Yeganeh, J. M.K.: Porous Mater (2003)
9. Tyanin, A.N., Kuznetsov, N.A.: Hous. Communal Infrastruct. **3**, 42 (2019)
10. Sokolov, L.I., Lebedeva, E.A., Pavlikov, D.A.: Ecol. Ind. Russ. **6**, 24 (2010)
11. Technological Control Technique for the Operation of Urban Sewage Treatment Plants. Stroyizdat, Moscow (1977)
12. Coackley, P., Jones, B.R.S.: Sew. Ind. Wastes **28**, 963 (1966)
13. Paroni, J.L., Tenney, M.W., Eschelberger, W.F.J.: J. Water Pollut. Control Fed. **361** (1972)
14. Kasprzyk, M.: E3S Web Conference (2019)
15. Physico-chemical studies of sorption materials based on biomass waste. In: E3S Web of Conferences, vol. 161, p. 02003 (2020)
16. Politaeva, N., Smyatskaya, Y., Fedyukhin, A.: Desalin. Water Treat. **174**, 116 (2020)
17. Kozeny, K.: Über Kapillare Leitung Des Wässerigen Bodes. Aufstieg, Versicherung und Anwendung auf die Bewässerigen (1927)
18. Carman, P.C.: Chem. Eng. Res. Des. **16a**, 168 (1938)

Steel and Composite Tapes in Timber to Concrete Joint

Arturs Skincs[1], Dmitrijs Serdjuks[1(✉)], Karina Buka-Vaivade[1], Vadims Goremikins[1], and Ali Yusuf Mohamed[2]

[1] Riga Technical University, Kipsalas Street 6B/6A, Riga, Latvia
Dmitrijs.Serdjuks@rtu.lv
[2] Peter the Great St. Petersburg Polytechnic University, Politekhnicheskaya Street 29, St. Petersburg, Russia

Abstract. Timber-concrete composite panels cause certain interest now due to possibility to combine advantages of pure timber and pure concrete panels in one structural member. The structures have to be designed, considering behaviour of such composite panels in order to benefit from these advantages. In the analyse of load-carrying capacity for timber-concrete composite panels, subjected to flexure, the important component is connection system between concrete layer and timber, which affects the stress distribution and the deformations of the structure. Behaviour of timber-concrete composite panels, with the rigid timber to concrete joint, were evaluated in this research. The rigid timber to concrete joint was developed by the using of the crushed granite pieces as the keys in combination with the perforated steel and carbon fibre reinforced plastic composite tapes to increase effectiveness of the structural materials use. Behaviour of the timber-concrete composite panels were investigated by the transformed section method, finite element method and by experiment for the purpose of this study. Four timber-concrete composite panels with the span in 1.8 m were statically loaded till the failure by the scheme of three-point bending. Perforated tapes made of S355 steel grade with the dimensions 2000X45X3 mm and total area of perforation in 10% were used in timber to concrete joint. Possibility to increase by 19% load-carrying capacity of the timber-concrete composite panels by the adding of perforated steel tapes was stated. Maximum vertical displacements of the panels decrease by 25% at the same time.

Keywords: Cross-laminated timber · Perforated steel tape · Rigid timber to concrete joint · Bending test · Timber-concrete composite panels · Structural analysis · Carbon fibre reinforced plastic

1 Introduction

Hybrid timber-concrete composite structural members subjected to flexure, such as panels and floor systems, are enough popular especially during the last years [1–8]. Effectiveness of hybrid timber-concrete members strongly dependent from the type of timber to concrete joint. Compliant and rigid once are two types of timber to concrete joints in hybrid timber-concrete composite structural members, which are used now in

A. Sinitsyn (Ed.): SES 2020, LNCE 141, pp. 68–79, 2021.
https://doi.org/10.1007/978-3-030-67654-4_9

the practice. Investigations of behaviour of timber to concrete joints indicates, that the timber-concrete composite structural members realized by the compliant and rigid timber to concrete joints, are characterized by the comparable load-carrying capacity [9, 10]. Studies shows that the adhesive composite connection of the timber-concrete structural members is more effective in comparison with the compliant one so as provides higher bending stiffness and leads to smaller deflections and a better structural members behaviour [9, 10]. Theoretically timber-concrete composite structural members realized by the rigid timber to concrete joints, possess a potential for increase of it load-carrying capacity in comparison with the analogous members realized by the compliant joints due to the lack of the bonds compliance in the first case [9]. More rational materials use and probable increase of the load-carrying capacity in timber-concrete composite structural members is joined with the normal stresses distribution between the joined layers caused by the different compliances of the joints in the both mentioned above cases [10]. However, experimental testing of hybrid timber-concrete composite structural members, with the rigid timber to concrete joint, subjected to flexure, indicate, that the brittle failure can took place, especially in the case, when the tensioned part of the structural member is strengthened by the FRP tapes [10]. Perforated steel tapes as shear connector used for the development of rigid timber to concrete joint in combination with the adhesives [11]. Perforation of the steel tapes usually is necessary to provide placement of steel reinforcement of the concrete layer better interaction between concrete and the tapes in case of absence of reinforcement. Adding of the steel or composite tape as a shear connector to the rigid timber to concrete joint probably enables to obtain hybrid timber-concrete composite structural member with increased stiffness and reliability due to the brittle failure probable prevention in the case.

Therefore, the objective of the study is investigation of influence of the steel or composite tape applied as a shear connector in the rigid timber to concrete joint of the behaviour of hybrid timber-concrete composite structural members subjected to flexure. Laboratorian, analytical and FEM analyses of the hybrid timber-concrete composite structural members subjected to flexure should be conducted for the purpose. Influence of perforation of the steel tapes on the behaviour of hybrid timber-concrete composite structural members subjected to flexure also should be evaluated.

2 Materials and Methods

2.1 General Approach

Influence of the steel and composite tape applied as a shear connector in the rigid timber to concrete joint of the behaviour of hybrid timber-concrete composite structural members subjected to flexure evaluated based on the laboratorian, analytical and FEM analyses.

Two laboratorian experiments were carried out. The first one was necessary to compare influence of structural material and perforation areas of the tapes, applied as a shear connector in the rigid timber to concrete joint, on the load-carrying capacity and deflections of the hybrid timber-concrete composite structural members subjected to flexure. It was realised with sixteen small-scale hybrid timber-concrete specimens, which are differed by the tape's materials and perforation area. The small-scale specimens have the dimensions in 400 x 90 x 47 mm. The boards of C18 strength class timber with dimensions in 400 x 90 x 17 mm were used as a base of small-scale specimens (see Fig. 1).

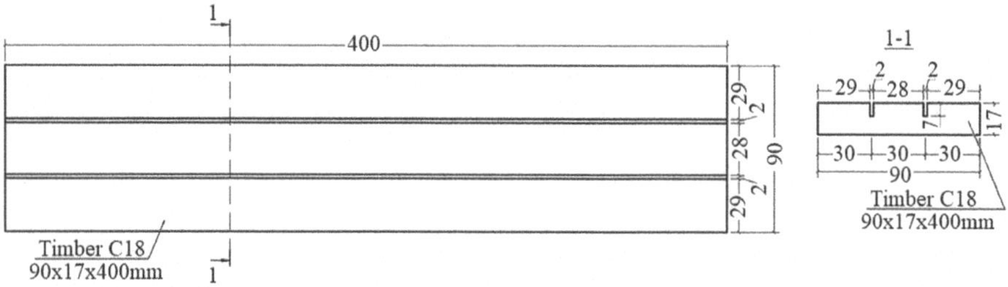

Fig. 1. Dimensions of the small-scale specimen timber board.

The layers of cement base finishing mass Sacret BAM with thickness equal to 30 mm were joined with the timber base by the using as dowels of pieces of crushed granite with dimensions changing within the limits from 16 to 18 mm (see Fig. 2). The pieces of crushed granite were strengthened on the surface of the timber boards by epoxy glue Sica Dur 330. The epoxy glue Sica Dur 330 has modules of elasticity and shear strength equal to 12800 and 15 MPa, correspondingly. Three sub-groups of small-scale specimens with four specimens in each were realised with the adding as a shear connector in the rigid timber to concrete joint the steel tapes made of S355 grade steel. Perforation areas were equal to 0, 10.6 and 55.3%, correspondingly. One small-scale sub-group was created with the adding as a shear connector in the rigid timber to concrete joint the carbon fibre reinforced plastic tape Mapei Carboplate E 250. The carbon fibre reinforced tape Mapei Carboplate E250 has the modulus of elasticity, tensile strength and thickness equal to 250000 MPa, 2500 MPa and 1.4 mm, correspondingly [12]. All specimens were supplied by the two tapes, which were added in the gropes with depth and width in 7 and 2 mm, correspondingly. The tapes were strengthened with the timber by the epoxy glue Sica Dur 330. Dimensions of steel and carbon fibre reinforced tapes were equal to 400 x 25 x 1.27 mm.

The second experiment was realised with four hybrid timber-concrete composite panels realised with the steel tapes applied as a shear connector in the rigid timber to concrete joint with perforation areas equal to 0 for the first two panels and 10.6% for other two panels.

The second experiment is necessary for investigation of influence of the steel tape applied as a shear connector in the rigid timber to concrete joint of the behaviour of hybrid timber-concrete composite panel. Timber to concrete joint in timber-concrete composite panels was provided by the granite pieces with dimensions changing within the limits from 16 to 25 mm only (see Fig. 3 b)).

The granite pieces were joined to the surface of the timber by the epoxy glue Sica Dur 330. Timber-concrete composite panels consist from the cross-laminated timber

Fig. 2. Strengthening of crushed granite pieces and steel tapes on the surface of timber board in the mould of the small-scale specimen.

a) b)

Fig. 3. Placement of a) steel tapes; b) and crushed granite pieces on the surface of the CLT panels.

(CLT) components, which are strengthened by the carbon fibre reinforced plastic tape Mapei Carboplate E250 in the tensioned zone and with the layer of finishing mass Sacret BAM with thickness in 40 mm in compressed zone.

The cross-laminated timber base was formed by the three layers of strength class C24 timber boards with thickness in 20 mm. Fibres of outer layers are oriented parallel to the longitudinal axis of the panels. The panels have length, width and thickness equal to 2000, 350 and 60 mm, correspondingly (see Fig. 4). The panels were produced by the Sconto enterprise, Ltd, Jelgava, Latvia.

The load-carrying capacity and maximum vertical displacements of the panels were determined analytically by the transformed section method and FEM, realized by the software RFEM Dlubal 5.17.02. Analytical analyse of the behaviour of hybrid timber-concrete composite structural members enables to carry under the control results obtained in course of laboratorian experiments [9, 10, 14].

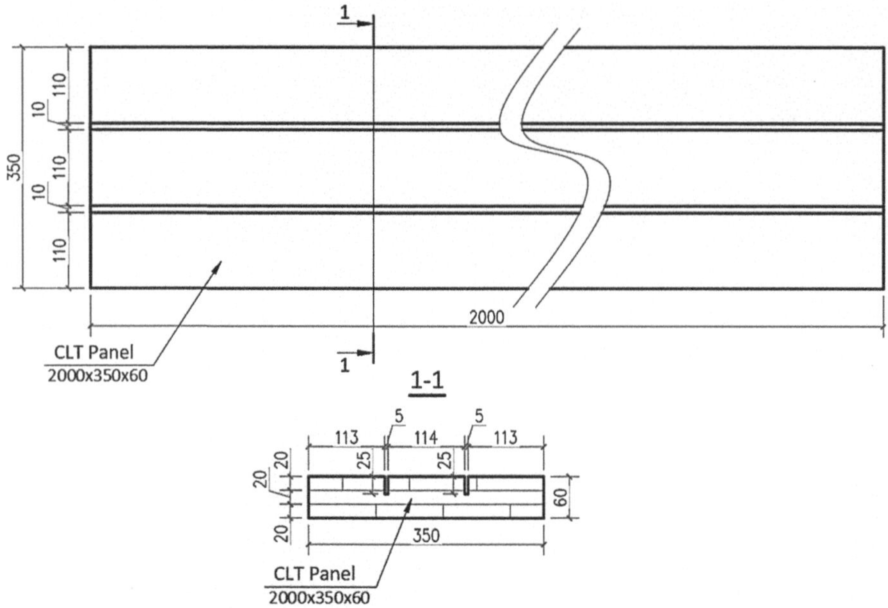

Fig. 4. CLT panels with the gropes on the surface for the steel tapes placement.

2.2 Analytical Analyse of the Behaviour of Hybrid Timber-Concrete Composite Structural Members

Transformed section method was chosen for determination of load-carrying capacity and maximum vertical displacements for timber-concrete composite panels so as it is characterized by the simplified design procedure and comparable precision in comparison with the k-method, gamma method and shear analogy method. Transformed cross-section method is joined with the replacement of the real cross-section of element by the equivalent transformed cross-section [13–16]. Transformation of cross – section is based on the ratios of modulus of elasticity of the layers to the modulus of elasticity of the outer layers of CLT panels, which were oriented in the longitudinal direction. The widths of the internal layer of the CLT panels, layer of finishing mass Sacret BAM and carbon fibre reinforced tape Mapei Carboplate E250 must be multiplied by the reduction factor which is equal to relations of modulus of elasticities. Obtained transformed cross-section then is considered as glued homogenous cross-section. Checks of ultimate limit state (ULS) and serviceability limit state (SLS) must be conducted based on the recommendations of [17].

Load-carrying capacity of timber-concrete composite panels was carried out analytically by the transformed section method and by the FEM [18], which was realized by the software RFEM Dlubal 5.17.02. The plate finite elements were used to develop the FEM model of the timber-concrete composite panels (see Fig. 5).

The model was divided into the finite elements by the rectangular mesh. Timber-concrete panels analyse by the software RFEM Dlubal 5.17.02 include following

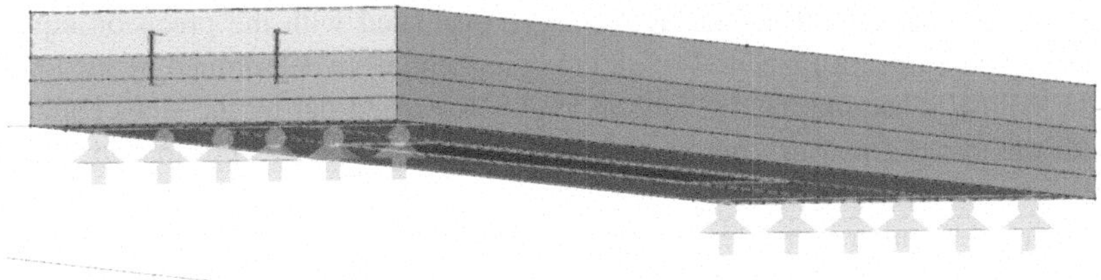

Fig. 5. The FEM model of the timber-concrete composite panel developed by the software RFEM Dlubal 5.17.02 [16].

stages: development of the model, applying of loads, solution of the task and analyse of obtained results.

2.3 Description of Laboratorian Experiments

Sixteen small-scale hybrid timber-concrete specimens, described above, were tested by the scheme of three-point bending at the action of static loading (See Fig. 6a)) in frames of the first laboratorian experiment. The specimens were tested after twenty-eight days from the specimen's formation. Intensity of the concentrated force applied in the middle of the specimen's span and maximum vertical displacements are two parameters measured during the experiment. Two mechanical indicators placed in the middle of the span (see Fig. 6a)) measured the maximum vertical displacements.

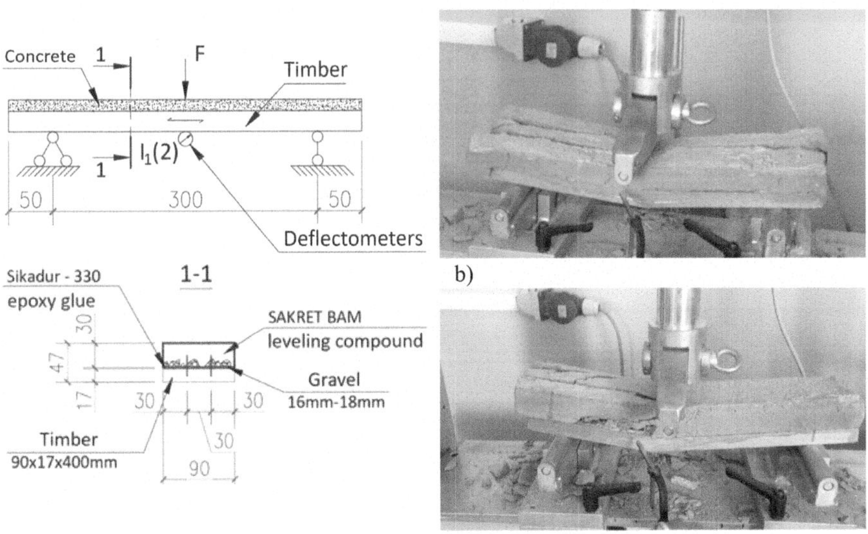

a) c)

Fig. 6. Testing of hybrid timber-concrete specimens: a) scheme of the specimens testing; b) failure mode of the specimen with the carbon fibre reinforced plastic tape applied as a shear connector in the rigid timber to concrete joint c) failure mode of the specimen with the steel tape with area of perforation equal to 10.6% [16].

The maximum vertical displacements were evaluated with the precision equal to 0.01.mm. Load-carrying capacities of all the specimens were determined by the loading until the failure. The applying of vertical load was carried out by the steps equal to 2 kN with the speed equal to 2 mm/min [16]. All the specimens were loaded until the failure.

Four hybrid timber-concrete composite panels with steel tapes applied as a shear connector in the rigid timber to concrete joint were tested in frames of the second laboratorian experiment. The areas of perforation for the steel tapes were equal to 0 and 10.6% [16].

The design scheme and apparatus placement for the timber-concrete composite panels is shown on the Fig. 7 a). Two couples of strain measuring devices and one couple of deflectometers were used during the static loading of the timber-concrete composite panels. Strain measuring devices T-2 and T-4 were placed on the surface of the finishing mass Sacret BAM in compressed zone.

Strain measuring devices T-1 and T-3 were placed in tensioned zone on the surface of the cross-laminated timber. A couple of the deflectometers I-1 and I-2 were placed in the middle of the span. Precision of the maximum vertical displacements and absolute deformations measurements was equal to 0.01 mm and 0.001 mm, correspondingly. All four specimens were loaded until the failure to determine their load-carrying capacities.

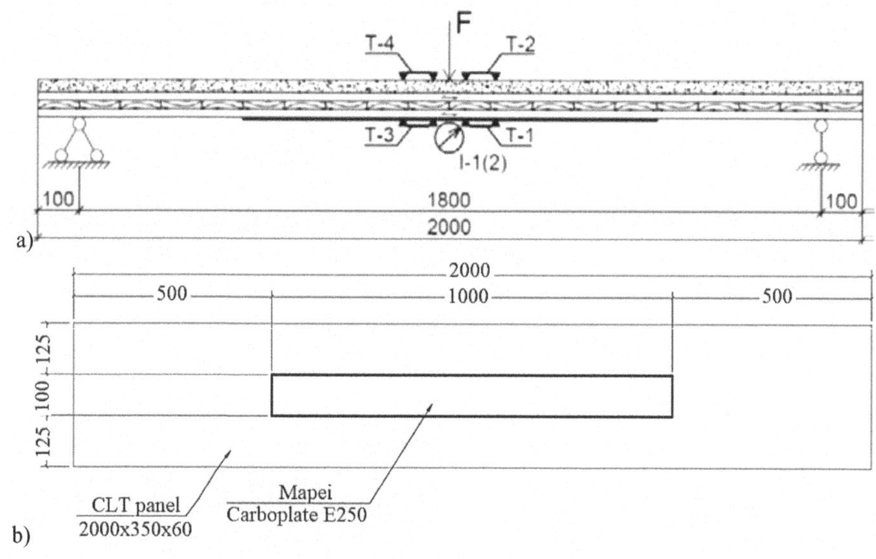

Fig. 7. The design scheme, placement of a) measuring devices for the timber-concrete composite panels; b) the carbon fibre reinforced tape on its surface in the tensioned zone [16].

3 Results

The dependences of the maximum vertical displacements of the small-scale hybrid timber-concrete specimens subjected to flexure on the intensity of the vertical applied load were obtained together with the load-carrying capacities for all sixteen specimens. The maximum vertical displacements as a function from the vertical load for the four sub-groups of tested specimens are shown on Fig. 8.

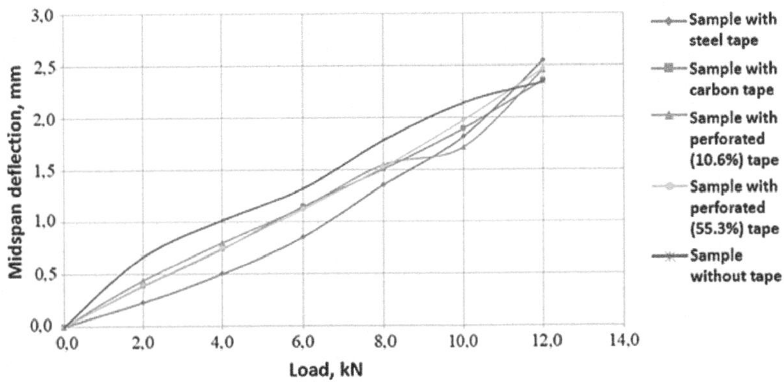

Fig. 8. The maximum vertical displacements as a function from the vertical load for the all sub-groups of tested specimens [16].

The final point for each graph corresponds to the mean load-carrying capacity of each sub-group of specimens. Each point on the graphical dependences was obtained as the mean value obtained for all specimens of the sub-group [16].

The maximum load-carrying capacities obtained for the four sub-groups of small-scale hybrid timber-concrete specimens of specimens are shown on the Fig. 9. Each from the four load-carrying capacities was determined as the mean value obtained for the four specimens of each sub-group. The values of load-carrying capacities for the sub-groups of specimens with the steel tapes with the perforation's areas equal to 10.6 and 55.3%, changes from 12.00 to 14.80 kN and from 11.80 to 16.40 kN, correspondingly. The values of load-carrying capacities for the sub-groups of specimens with the steel and carbon fibre reinforced plastic tapes without perforations, changes from 14.00 to 14.40 kN and from 12.00 to 16.40 kN, correspondingly.

The maximum load-carrying capacities in 16.4 kN were observed for the specimens with the carbon fibre tape and steel tape with the area of perforation equal to 55.3% [16].

The stresses, acting in the outer top and bottom fibres of the four timber-concrete composite panels, explained before, its maximum vertical displacements and load-carrying capacities were evaluated by the second experiment and analytically by FEM. Maximum vertical displacements of the timber-concrete composite panels were evaluated by the transformed section method also. Specimens with steel tapes without perforation will be mentioned next as first and second once. Specimens with steel tapes

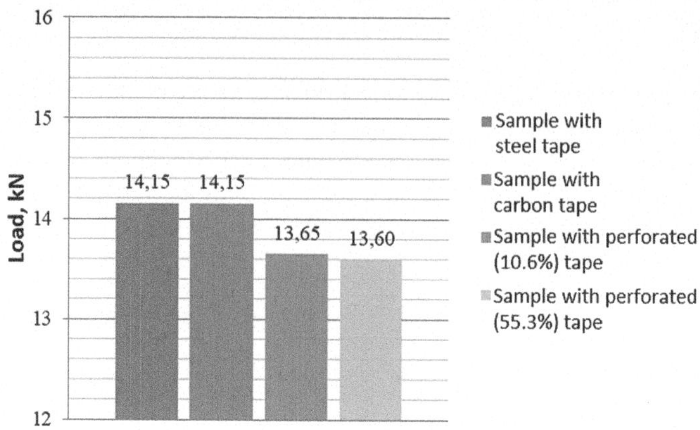

Fig. 9. Maximum load-carrying capacities obtained for the all sub-groups of specimens [16].

with perforation area equal to 55.3% will be mentioned next as third and fourth once [16]. Load-carrying capacities of four timber-concrete composite panels are shown on the Fig. 10.

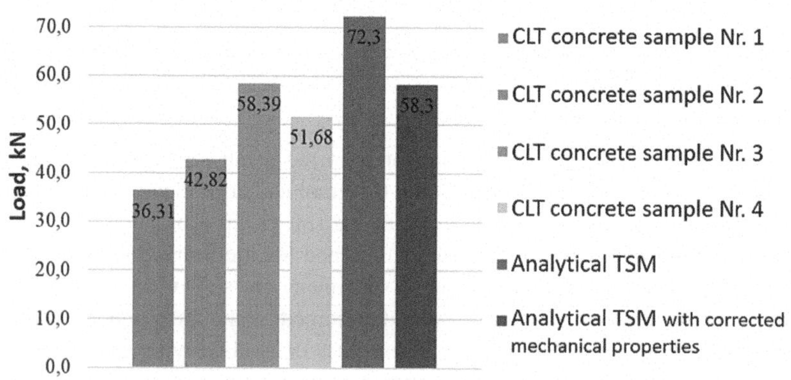

Fig. 10. Maximum load-carrying capacities obtained for the four specimens of timber-concrete composite panels [16].

The load carrying capacity in 58.30 kN, which is shown on the Fig. 10 by the green colour, was obtained by the using of the corrected mechanical properties of timber, which provide the correspondence between the load-carrying capacities, obtained by experiment and analytically.

The dependences of maximum vertical displacements on the intensity of applied vertical loads for the four specimens of timber-concrete composite panels are shown on the Fig. 11. The dependences obtained by the experiment and analytically by the FEM and transformed section method.

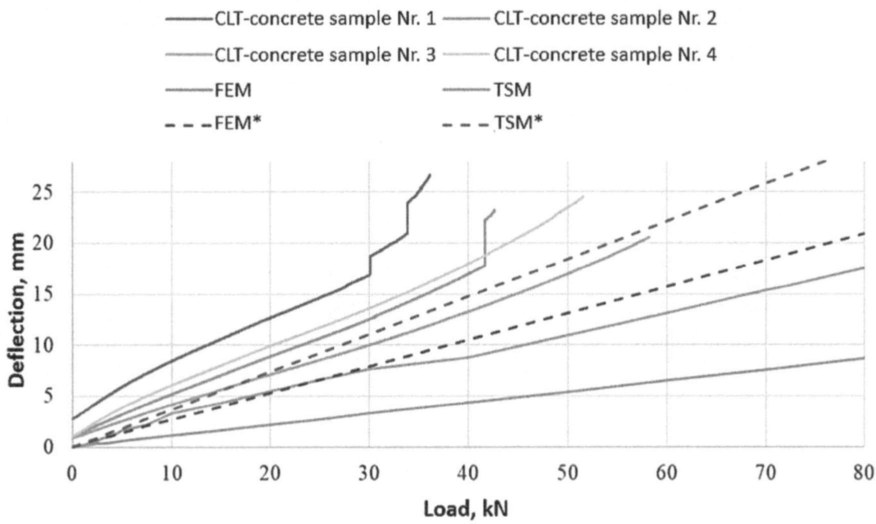

Fig. 11. The dependences of the maximum vertical displacements of the specimens on the intensity of applied vertical load: FEM – dependence obtained by the finite element method; TSM – dependence obtained by the transformed section method [16].

It was shown that the minimum values of normal stresses, acting in the outer top and bottom fibres when the vertical load was equal to 25 kN, were equal to 27.75 and 11.40 Mpa, correspondingly. The values obtained for the third specimen, which characterizes by the maximum load-carrying capacity, equal to 58.39 kN. Brittle failure of timber-concrete composite panels was not observed.

4 Discussions

Results obtained in course of two experiments, were compared with the analogous ones, and obtained for the specimens without the tape to evaluate the effect from the steel or composite tape adding as a shear connector in the rigid timber to concrete joint on the behaviour of hybrid timber-concrete composite structural members subjected to flexure. The FEM realised by the software RFEM Dlubal 5.17.02 and transformed section method were used for evaluation of load-carrying capacity of laboratorian specimens.

The maximum load-carrying capacity obtained for the sixteen small-scale specimens, was equal to 16.40 kN. The value of maximum load-carrying capacity was obtained for the specimens with the carbon fibre reinforced plastic tape and for specimens with the steel tape with perforation area in 55.3%. The maximum mean load-carrying capacity in 14.50 kN was obtained for the sub-groups of small-scale specimens with carbon fibre reinforced plastic tape and steel tape without perforation. That is the reason, why for the second experiment with timber-concrete composite panels two specimens (first and second ones) with the steel tapes without perforation were chosen as one of the tested sub-group. However, comparison of the maximum load-carrying capacities of the small-scale specimens with and without adding of the tape as a shear connector in the rigid timber to concrete joint enables to conclude, that influence of the dimensions of the crushed granite pieces is more significant. So, increase of the dimensions of the crushed granite pieces from 16–18 mm till 16–25 mm enables to increase load-carrying capacity of the small-scale specimens till the 17.80 kN. The load-carrying capacity of small-scale specimen's sub-group grows by 14.49% at the same time [9, 10, 16].

But for the timber-concrete composite panels, where placement of the tapes did not limit the dimensions of crushed granite pieces, increase of the load-carrying capacity by 1.19 times in comparison with the specimens without the tape, was stated. Decrease of the maximum vertical displacements both for the small-scale specimens and for timber-concrete composite panels from 8 to 25% was stated [9, 10, 16].

It can be stated that the main objective of the work was obtained. The influence of the steel or carbon fibre reinforced composite tape applied as a shear connector in the rigid timber to concrete joint on the behaviour of hybrid timber-concrete composite structural members subjected to flexure was investigated. Nevertheless, the question regarding the rational parameters of the rigid timber to concrete joint just is opened. It can be solved using the design of experiment methods for planning the experiment with the increased in comparison with the current study amount of the laboratorian specimens.

A special interest cause investigation the mode of failure of the hybrid timber-concrete composite structural members subjected to flexure. The hypotheses, that adding of the steel or composite tape as a shear connector to the rigid timber to concrete joint probably enables to obtain hybrid timber-concrete composite structural member with increased reliability due to the brittle failure prevention was not declined in course of conducted experiment so as a brittle failure of composite timber-concrete composite panels was not observed.

5 Conclusions

Influence of the steel or carbon fibre reinforced plastic tape applied as a shear connector in the rigid timber to concrete joint of the behaviour of hybrid timber-concrete composite structural members subjected to flexure, was investigated by the laboratorian experiment, FEM and transformed section method.

Increase of the load-carrying capacity by 1.19 times of timber-concrete composite panels with the steel tape applied as a shear connector in the rigid timber to concrete joint in comparison with the specimens without the tape, was stated. Decrease of the maximum vertical displacements both for the small-scale specimens and for timber-concrete composite panels from 8 to 25% was observed for the specimens with steel and carbon fibre reinforced plastic tape applied as a shear connector in the rigid timber to concrete joint.

Brittle failure of timber-concrete composite panels was not stated in course of conducted laboratorian experiment.

Acknowledgements. This work has been supported by Latvian Council of Science funded project "Method of correlation of coaxial accelerations in 6-D space for quality assessment of structural joints (COACCEL)" (Nr. lzp-2020/1-0240).

References

1. Dias, A., Schänzlin, J., Dietsch, P. (eds.): Design of timber-concrete composite structures. Shaker Verlag GmbH, Aachen (2018)
2. Ceccotti, A.: Composite concrete-timber structures. Prog. Struct. Mat. Eng. **4**(3), 264–275 (2002)
3. Kanocz, J., Bajzecerova, V., Steller, S.: Timber – concrete composite elements with various composite connections. Part 1: screwed connection. Wood Res. **58**, 555–570 (2013)
4. Kanocz, J., Bajzecerova, V.: Timber – concrete composite elements with various composite connections. Part 3: adhesive connection. Wood Res. **60**, 939–952 (2015)
5. Brunner, M., Romer, M., Schnüriger, M.: Timber-concrete-composite with an adhesive connector (wet on wet process). Mater. Struct. **40**, 119–126 (2007)
6. Negrao, J.H., Oliveira, F.M., Oliveira, C.L.: Investigation on timber-concrete glued composites. In: 9th World Conference on Timber Engineering (WCTE 2006). Portland, Oregon, USA (2006)
7. Al Ali, M., Bajzecerova, V., Kvocak, V.: Design methods of timber-concrete composite ceiling structure. Mag. Civil Eng. **73**(5), 88–95 (2017)
8. Yeoh, D., Fragiacomo, M., De Franceschi, M., Boon, K.H.: State of the art on timber-concrete composite structures: literature review. J. Struct. Eng. **137**(10), 1085–1095 (2011)
9. Vasiljevs, R., Serdjuks, D., Buka-Vaivade, K., Podkoritovs, A., Vatin, N.: Load-carrying capacity of timber-concrete composite panels. Mag. Civil Eng. **93**(1), 60–70 (2020)
10. Vasiljevs, R., Serdjuks, D., Gerasimova, J., Buka-Vaivade, K., Orhun Erüz, A.: Behaviour of timber-concrete joints in hybrid members subjected to flexure. In: IOP Confernce Series: Materials Science and Engineering, vol. 660, pp 1–9, 012050. IOP Publishing (2019)
11. HBV: Load Bearing Constructions using Wood-Concrete-Composite Technique with glued-in *HBV* - Shear Connectors. TiComTec GmbH, Haibach (2014)
12. Mapei Carboplate. Technical data sheet. https://www.mapei.com/in/en/products-and-solutions/products/detail/carboplate. Accessed 1 July 2020
13. Crews, K., Gerber, C.: Development of design procedures for timber concrete composite floors in Australia and New Zealand. CIB-W18/43–7-5. Karlsruhe Institute of Technology, Germany (2010)
14. Buka-Vaivade, K., Serdjuks, D., Goremikins, V., Vilguts, A., Pakrastins, L.: Experimental verification of design procedure for elements from cross-laminated timber. Procedia Eng. **172**, 1212–1219 (2016)
15. Auclair, S., Sorelli, L., Salenikovich, A.: Simplified nonlinear model for timber-concrete composite beams. Int. J. Mech. Sci. **117**, 30–42 (2016)
16. Skincs, A.: Load-carrying capacity analyse for hybrid timber-concrete composite element with steel tape reinforcement. M.S. thesis. Riga Technical university, Riga (2020)
17. European Committee for Standardization: Eurocode 5: Design of Timber Structures – Part 1-1: General-Common Rules and Rules for Buildings. CEN, Brussels (2004)
18. RFEM 5 user manual. Dlubal Software GmbH, Tiefenbach (2018)

Ensuring Effective Heat Dissipation in the Design of Light Emitting Diode (LED) Lighting Equipment

Vadim Alekseevich Chernyshov[1(✉)], Evgeny Aleksandrovich Pechagin[2], Oleg Borisovich Gladkov[3], Roman Pavlovich Belikov[3], and Nailya Kamilevna Miftakhova[4]

[1] Department of "Electrical Equipment and Energy Saving", FSBE HE «OSU name by I. S. Turgenev», Oryol, Russia
blackseam78@mail.ru
[2] Department of "Power Industry", FSBE HE «TSTU», Tambov, Russia
[3] Department of "Power Supply", FSBE HE «OSAU named after N. V. Parakhin», Oryol, Russia
[4] Department of "Power Supply of Industrial Enterprises", Kazan State Power Engineering University, Kazan, Russia

Abstract. The article touches upon a specificity of the functioning of LED lighting equipment, its design features and characteristics, as well as the situation of its production. The most important problem of thermal management in the LED industry which is connected with the service life reduction of LED light sources due to the overheating through the low efficiency of the heat dissipation and the imperfection of the housing design is specified. The literature review on this subject showed that the most advanced cooling technology of the electronic components which are used by modern IT corporations to cool the servers is immersion cooling. The idea of implementing two-phase immersion cooling of the LED light source where "dry water" – fluoroketone serves as a heat carrier inside the lamp is proposed. When the temperature of heating the LED is over 49.2 °C, fluoroketone starts boiling and absorbs the extra heat from it. After evaporation, fluoroketone condenses on the top metal part of the lamp housing and flows back into the lamp diffuser. The experiments with a prototype of a LED lamp with immersion cooling proved the high efficiency of the proposed heat dissipation. The possibility of the significant service life increase of LED light sources with immersion cooling, as well as the possibility of maintaining the optimal light flux throughout their service life are justified. The vitality of carrying out complex research connected with immersion cooling application in the industrial lighting engineering is explained.

Keywords: LED · Thermal management · The industrial lighting engineering

1 Introduction

At present, the advantageous operational and lighting characteristics of LEDs have made their production one of the leading branches of the electrotechnical industry [1].

A. Sinitsyn (Ed.): SES 2020, LNCE 141, pp. 80–88, 2021.
https://doi.org/10.1007/978-3-030-67654-4_10

On the basis of LEDs a wide variety of light sources are currently produced: spotlights, lamps, linear lamps, LED linears, signal equipment [2] and other lighting products.

The no-failure operating time of a LED light source depends not only on the quality of its assembly, but also on the conditions of its operation determined by the design features of the whole lighting device. The use of modern electrotechnical materials, as well as properly designed heat dissipation system and driver allow to achieve the no-failure operating time of a LED to the maximum value recommended by the manufacturer [3].

Maintaining the optimum temperature of p-n junction for a LED light source within the recommended limits through the efficient heat dissipation in the lighting industry is called thermal management, which covers the issues of improving not only the design of a heat sink, but the design of the housing of the LED lighting device [4].

The dust-proof and moisture-proof housing of the industrial LED lamps allows to use them in industrial spaces with severe operating conditions. However, performing its protective functions, a hermetic housing limits the free convection of the internal air, which does not provide LED light sources with an optimal temperature operation mode. At high indoor air temperatures (+50 °C and higher), typical for energy-intensive industrial production, the temperature on the cooling radiator of a LED can reach a critical point (more than + 90 °C), which will have a very negative impact on the service life of the LED itself.

The reduction in the service life of LEDs is connected with the increase in power consumption, especially at fixed sizes of a housing. There are cases when irresponsible manufacturers of lighting equipment install LEDs in outmoded housings, originally designed for the installation of fluorescent lamps. At the same time, in advertising leaflets, they indicate a longer service life of their lamps, mentioning the high reliability of the LED light sources used, whereas in the housings of their lamps the LEDs operate much less than their declared service life what is caused by the overheating.

Thus, one of the main problems of the LED industry is the reduction of the operational life of industrial LED light sources due to overheating of LEDs connected with the imperfection of the heat dissipation design and the design of their housing. Although manufacturers of LED lighting widely use aluminum radiators of different profiles for cooling LED modules of different sizes and designs, they are not able to guarantee the protection of LED light source from overheating and premature wear.

2 Aims and Tasks

The problem of thermal management is relevant for modern lighting engineering and requires additional in-depth research aimed at the development and practical implementation of more effective ways of absorbing heat from LED light sources.

According to the literature review, immersion cooling is the most revolutionary cooling technology for electronic devices used by advanced IT companies to cool their servers [5]. Liquids are the best cooling agent, they have a higher heat conductivity in comparison with air, because they have higher densities, and dense heat carriers provide a more efficient transfer of thermal energy [6]. It can be assumed that the rapid growth of scientific and technological progress will make this technology available for mass use, especially in lighting engineering.

The authors of the article propose to apply a well-known technology of immersion cooling to ensure effective heat dissipation in the designs of LED lighting equipment [7].

This measure is expected to prevent overheating of LEDs and their premature wear.

3 Conditions, Materials and Methods

In the immersion cooling systems, the heat carrier is fluoroketone or so-called "dry water". Its chemical formula is not H2O, but more complex one and it does not contain hydrogen atoms – CF3CF2C(O)CF(CF3)2 - perfluoro (ethyl-isopropyl ketone), a hexagonal substance, a class of fluorinated ketone [8].

Fluoroketone is a synthetic organic substance. In its molecule all hydrogen atoms are replaced by fluorine atoms tightly bonded to the carbon skeleton, which makes this substance absolutely inert in terms of interaction with other molecules. Fluoroketone is a colorless clear fluid with a low odor. It is 1.6 times heavier than water and most notably it does not conduct electricity.

The dielectric capacitance of fluoroketone is 2.3 (the dried nitrogen is taken as a standard per unit). One of the important properties of fluoroketone is its extremely low solubility in water, which does not allow it to pass through the cell membranes into a body, making its vapors absolutely safe for humans. Also fluoroketone easily and quickly disintegrates in the upper atmosphere by the ultraviolet light [9].

Ensuring an effective heatdissipation in the designs of LED lighting equipment when applying the immersion cooling is based on the fact that in a metal hermetic housing of the LED lamp filled with fluoroketone, in the process of heating the LEDs to a temperature of 49.2 °C, the fluoroketone starts boiling, which, when changing its aggregate state, absorbs the excessive heat from the LED. Transforming into a vapor, the fluoroketone condenses on the top of the lamp housing and flows back into the diffuser, along with it a two-phase immersion cooling technology for the LED light source takes place (see Fig. 1).

When conducting experimental studies, a prototype for the future LED lamp with immersion cooling has become an industrial fluorescent lamp designed for severe operating conditions with Ingress Protection Rating IP67, the housing of which is made of galvanized steel (see Fig. 2).

A metal housing, in contrast to a plastic one has a higher heat conductivity, which is a necessary condition for the implementation of a two-phase immersion cooling technology of LED light sources.

In this regard, any lamps with metal hermetic dust-proof and moisture-proof housings (see Fig. 3) and designed to operate with high-pressure lamps can be adapted to the installation of super bright LEDs with immersion cooling.

For the experimental studies of the efficiency of immersion cooling, some minor changes were made in the design of the industrial fluoriscent lamp chosen. These changes consisted in dismantling fluorescent lamps and ballast, placing LED linears with a driver on the bottom of the diffuser and filling the diffuser with fluoroketone up

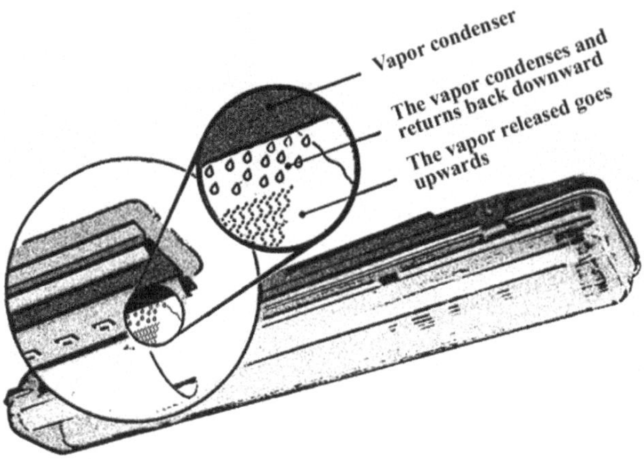

Fig. 1. Principle of realization of two-phase immersion cooling for LED light source.

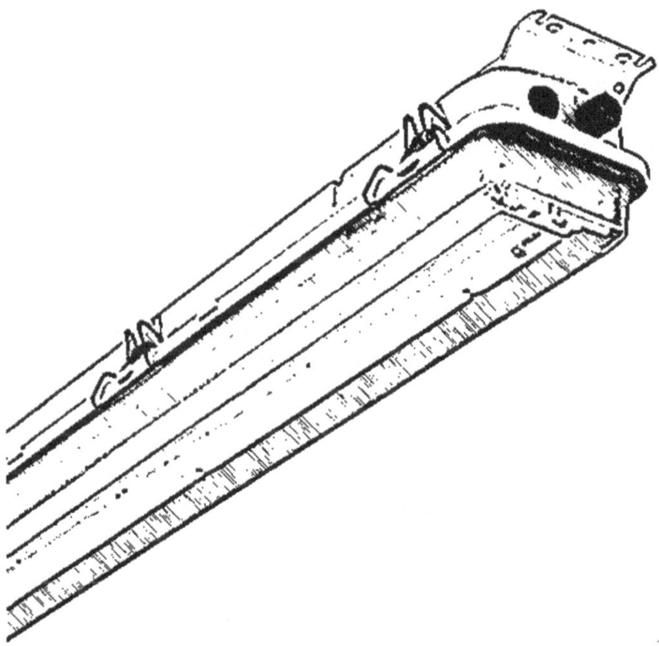

Fig. 2. Prototype of a LED lamp with immersion cooling, made on the basis of a dust-proof and water-proof fluorescent lamp in a metal housing.

to the complete immersion of the LED linears as well as additional sealing of the entry point of a power cable and exit point of control cables from the temperature and pressure sensors mounted within the housing.

When studying the efficiency of the heat dissipation during the immersion cooling, the following parameters were controlled: LED crystal temperature; LED heat

Fig. 3. Lamps suitable for installation of LED light sources with immersion cooling.

absorbing substrate temperature; LED radiator temperature, the air-vapor temperature inside the housing; the temperature of the outer surface of the upper part of the LED lamp, the air temperature under the ceiling, in the place of the LED lamp suspension [10]. Also, the air-vapor pressure inside the housing of the LED lamp and the light flux formed by it were monitored. To simulate thermal emissions from the process equipment, the air in the room was heated by an electric blow heater [11].

4 Results and Discussion

During the experimental studies on the use of immersion cooling in the design of a LED lamp, the following results were obtained:

1. When operating, the LED lamp with immersion cooling provides a more uniform light transmission from the lamp diffuser caused by the optical effect of lighting of the fluid coolant and its vapors.
2. When the LED lamp operates for a long time and the room temperature changes from +25 °C to +60 °C, the temperature of the LED module does not rise above + 49.2 °C.
3. At the maximum room temperature of +60 °C, there was no noticeable noise effect and pulsations of the light flux when the fluid coolant boils and its vapors condense.
4. At the maximum room temperature of +60 °C, the pressure of the air-vapor mixture inside the LED lamp housing did not exceed 100 kPa and did not cause a break of its airtightness.

Figure 4 shows the dependence of the service life of a LED on the temperature of its substrate (in our case, an aluminum plate) [12, 13].

When using immersion cooling, the maximum temperature of the LED crystal substrate is determined by the boiling point of the fluid coolant, i.e. +49.2 °C, and it is safe to say that the time of trouble-free operation of LEDs with immersion cooling will be at least 90 thousand hours. It should be noted that the actual service life of LEDs with natural air cooling, enclosed in a dust-proof and moisture-proof housing ranges from 10 to 30 thousand hours. Thus, the use of technology of two-phase immersion cooling of LEDs can significantly increase their service life.

When implementing air cooling of LEDs, the parameters of the light flux are largely determined by the ambient temperature (see Fig. 5) [12, 13]. For example, the

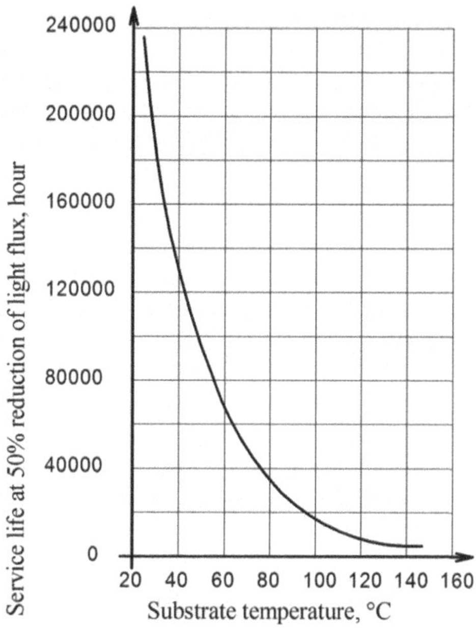

Fig. 4. Dependence of LED service life on the substrate temperature.

optimal light flux of the LED is optimal at an ambient temperature of +20–25 °C, in this case, the LED crystal temperature does not exceed 80 °C, and the substrate temperature does not exceed +50–60 °C.

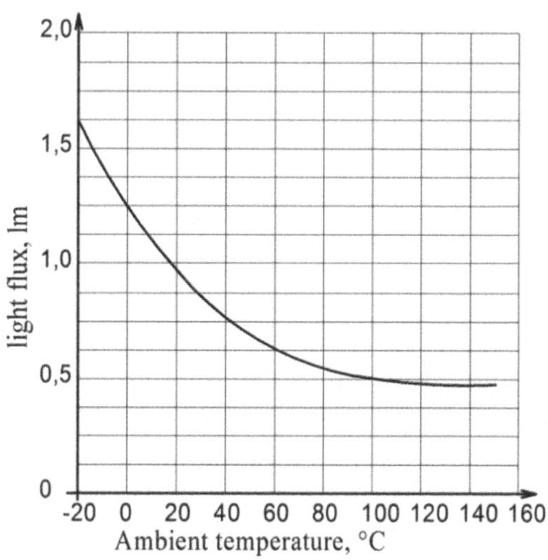

Fig. 5. Dependence of light flux on the ambient temperature.

When the ambient temperature rises from +25 °C to +60 °C, the crystal substrate temperature increases from +60 °C to +90 °C, while the temperature of the crystal itself reaches a critical value, reducing its light flux more than twice. The use of two-phase immersion cooling technology will not allow the crystal substrate temperature to rise higher than 49.2 °C, which regardless of the ambient temperature will ensure the stability of the optimal light flux of the LED throughout its service life.

Figure 6 shows the dependence of the vapor pressure of fluoroketone on the temperature inside the housing of the LED lamp. The dependence proves that at the boiling point of fluoroketone of 49,2 °C, the pressure inside the lamp will not exceed one atmosphere, and that will not require additional strengthening of the standard Ingress Protection Rating IP67. According to the chart presented in Fig. 6, at the ambient temperature of 49.2 °C (120,56 °F) the liquid phase of fluoroketone turns to vapor, creating a concentration of more than 90% by volume, which is significantly higher than typical concentrations used for firefighting [14, 15]. Thus, high concentrations of fluoroketone vapors in the housing of the LED lamp will allow to prevent a potential fire caused by a short circuit on the current-carrying parts of the lamp.

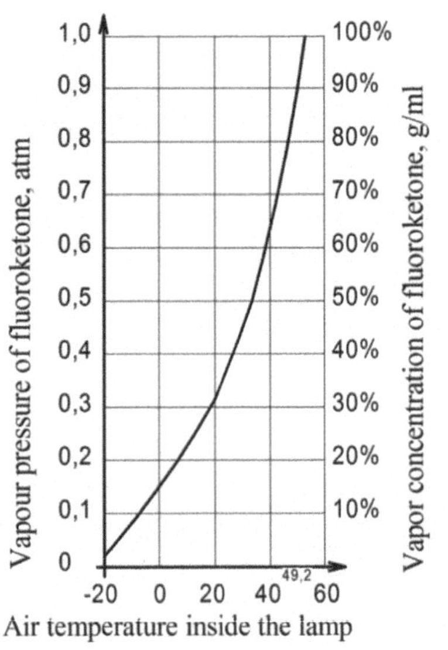

Fig. 6. Dependence of the pressure, concentration of fluoroketone vapor on the air temperature inside the lamp.

The experiment with a 80w LED lamp showed that evaporating of 1 g of fluoroketone will require about 120 J, and that means that a LED module with a power of 80w will vaporize about 0.5 g of fluid per a second, and 30 g per a minute. Taking into consideration the air temperature of +25–55 °C indoors, the operating time of lamps (more than 1 h) and the area of heat absorbing lamp surface of 0,12 sq. m., the optimum amount of fluid coolant equals 350 g.

This volume of fluid coolant provides proper immersion cooling process for a long time, where the air-vapor mixture cools below the dew point and condenses into the liquid. The similar experiments were carried out with LED lamps of lower power. They proved the dependence of the required amount of fluid coolant on the power of the lighting device (see Fig. 7).

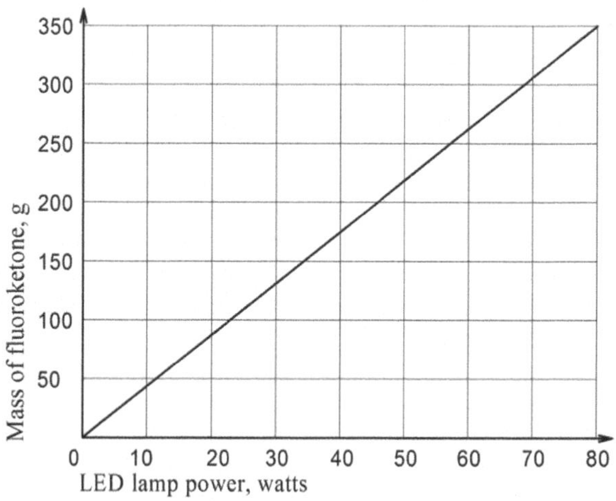

Fig. 7. Dependence of the fluid coolant mass on the LED lamp power.

5 Conclusion

Despite the theoretical justification for the engineering solution presented by the authors and confirmed by the positive results of the experimental studies, it is worth recognizing that it is too early to talk about the integrated feasibility of using immersion cooling in the industrial LED lamps. In the future, a more detailed study of a number of social, technical and economic issues related to the operation of LED lamps with immersion cooling under the conditions of modern energy-saturated industrial production will be required [16].

Nevertheless, now it is safe to say that the article touches upon a fundamentally new event aimed at effective solution of the problem of overheating of the LED light sources. Such a solution can be of great interest to potential developers and manufacturers of LED lighting and will allow to get a new generation of lamps with long service life and constant light output.

References

1. Yunovich, A.E.: LEDs as a basis for future lighting. Light Eng. **3**, 2–7 (2003)
2. Chernyshov, V.A.: The justification of the overall feasibility of using emergency lights and strobe light alarm on potentially dangerous sections of power lines of 10 kV. Bull. Agrarian Sci. **2**(71), 62–68 (2018)
3. Vasiliev, A.V.: LEDs-superagers: truth or hoax?, News of electronics+Light 0, 1, pp. 10–12 (2010)
4. Sharakshane, A.: Practical thermal management. Semi-conductor Lighting Eng. **5**, 66–74 (2012)
5. Immersion cooling of servers is gaining popularity [Electronic resource]. http://telecombloger.ru/16082
6. Dulnev, G.N.: Heat and mass transfer in electronic equipment: textbook for universities in design and manufacture of electronic equipment, High school, Moscow, vol. 247 (1984)
7. Chernyshov, V.A., Novikov, M.V.: Justification for the use of immersion cooling system in industrial LED lamps. Bull. Voronezh State Agrarian Univ. **3**(54), 88–93 (2017)
8. Dry water Novec, 1230 for protection of servers and not only: [Electronic resource]. https://habrahabr.ru/company/3mrussia/blog/200840. Accessed 9 Sept 2019
9. Material safety data sheet 3 M™ Novec™ 1230 Fire Protection Fluid [FK-5-1-12], 28 August 2013
10. Chernyshov, V.A., Novikov, M.V.: The use of microprocessor control of thermal parameters of LED lighting equipment at the stage of its design. Agric. Technol. Power Supply **3**(16), 27–33 (2017)
11. Chernyshov, V.A., Omelyanovich, A.S.: Conditions for creation, features and prospects of application of a new motor heater MP-20. Bull. Agrarian Sci. **3**(71), 101–106 (2018)
12. Narendran, N., Gu, Y., Hosseinzdeh, R.: Estimating junction temperature of high-flux white LEDs. In: Proceedings of SPIE, pp. 158–160 (2004)
13. Tringh, Q.V., Bruckner, S., Tran, Q.K.: Accurate measurement of the p-n-Junction temperature of HP-LEDs, methods and results. In: Proceedings of the International Symposium for Automotive Lighting (ISAL), Darmstadt, pp. 809–827, 28–29 September 2011
14. Chernyshov, V.A., Gladkov, O.B.: Ensuring fire safety of lighting equipment of residential and public premises, vital issues of energy in agriculture. In: Proceedings of All-Russia's scientific conference with International participation, Blagoveshchensk, Publishing house of the Far Eastern State Agrarian University, pp. 9–11, 27 February 2019
15. Gladkov, O.B., Chernyshov, V.A.: Improving fire safety of lighting equipment for industrial and civil buildings, Physics and modern technologies in agriculture. In: Proceedings of the X International Youth Conference of Young Scientists, Students and Schoolchildren, pp. 125–128 (2019)
16. Meshkov, B.N., Chernyshov, V.A.: Energy efficiency law dictated by nature. Chief Power Eng. **8**, 32–35 (2013)

Green and Environment Sustainable Supply Chain Practices in Leading Indian Manufacturing Companies

Deepak Sharma(✉), Jai Kishore Sharma, and B. K. Srivastava

Amity University, Noida, Uttar Pradesh, India
onlydeepaksharma@gmail.com

Abstract. This paper describes the development and need of Supply Chain Management (SCM) in manufacturing companies and how it gradually is trending towards Green Supply Chain Management (GSCM) in leading Indian manufacturing companies. Through elaborate literature review, the various work in the area of sustainability and GSCM have been observed and reflections on the origin of GSCM at global level have been collated. India is seemingly catching up on this global trend and so the paper gives a look out on the Indian government policies and commitment towards sustainability. The paper then synthesises on sustainability reports (2018–19) of top 10 manufacturing companies in India to check on the seriousness and extent of work being carried out by these companies who are supposedly the flag bearers of manufacturing industry in India. Finally, a discussion elaborates on the benefits of sustainability and GSCM as well as the challenges in implementing them in manufacturing environment with reference to India. In the last a conclusion is derived along with a summary to share the related observations and suggestions.

Keywords: Chain management · Manufacturing companies

1 Introduction

The supply chain can be explained as network between a company and its suppliers to produce and distribute a specific product for the final buyer. An effective and efficient supply chain ensures that the right product is made available to the end user. Innovation, Technology and Collaboration are the three widely accepted pillars of modern businesses and Supply Chain management is one of the most important technological innovations which thrives on collaboration.

As per Gunasekaran (2015) [1] supply chain collaboration is a relationship between supply chain partners developed over a period of time to achieve lower cost, higher quality, and greater product innovation, reduce risks and enhance market value.

Any manufacturing company performance in today's world is highly dependent on its fundamental supply chain design and related efficiencies. Over a period of time, it has emerged and it has become evident that the supply chains not only need to perform but need to be robust, sustainable and also caring for our ecosystem.

A. Sinitsyn (Ed.): SES 2020, LNCE 141, pp. 89–102, 2021.
https://doi.org/10.1007/978-3-030-67654-4_11

1.1 Evolution of Supply Chain

With change in dynamics, the market transformed from seller's market into buyers' market. This has been by virtue of opening up of economies and globalisation. The major changes then came in the way the companies operate, behave and perform as there emerged a pressing need for getting aligned to consumer behaviour. The products started reaching the customers very quickly. With the markets having transformed into highly flexible, competitive and had started providing wide alternates, consumers now were in a position to dictate terms. The transformation was also supported by many other drivers that shape the world even today viz. technology, globalisation, communication and government policies. The challenge thus was not to just deliver products but delivering in right time, right quality, right location, right quantity and most important at the right price.

Interestingly, industrial technology has also been upgraded and supported through evaluation of computer aided manufacturing and flexible manufacturing. This helped the companies in meeting consistency thus assuring high quality of products and also keeping the system flexible. There has been evaluation of systems like just in time which helped in meeting the flexibility and high level of quality at the same time. These types of techniques brought buyer and suppliers closer and made them more collaborative (Fig. 1).

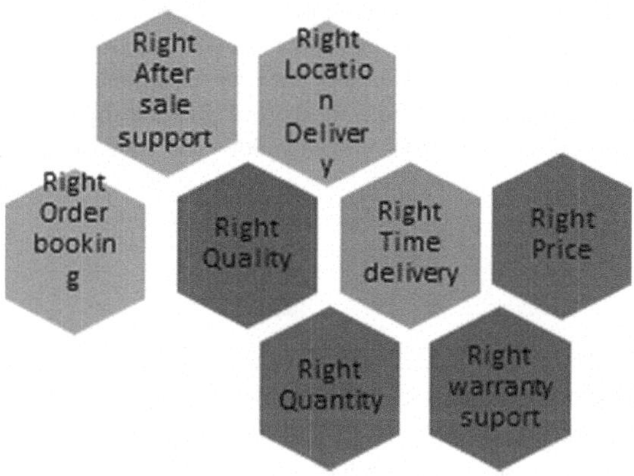

Fig. 1. Expectations of modern customer.

1.2 Greening of Supply Chain?

Since, the performance of supply chains has an impact on environment as well as wide-ranging perishable resources, thus, it is prudent that they are designed to support long term vision and have least impact on the environment.

This has developed the model of Green Supply Chain Management which is forcing the system to look at the impact of all activities on the environment, thus forcing them to design them in eco-friendly manner. As per Srivastava (2017) [2], the concept of 'Green Supply Chain' has come up which is integrating environment consideration into supply chain management, including product design material sourcing and selection manufacturing processes, delivery of the concluding product to the consumers as well as end of life management of the product after its useful life.

Further by virtue of competitiveness and globalisation, there has been word wide trend in looking at low cost countries and new markets to attain cost competitiveness as well as their markets. India being a low cost country (one of the BRIC nations) has got hugely benefitted. This new trend has helped in bringing the India companies on the global arena and to associate with the Global and multinational companies supply chains. This way the Indian companies have also got exposed to 'Green Supply Chain' expectations. India being an emerging economy and with its enhanced role in global arena has enhanced the countries commitments towards supporting issues like global warming, environmental sustainability etc..... (Fig. 2)

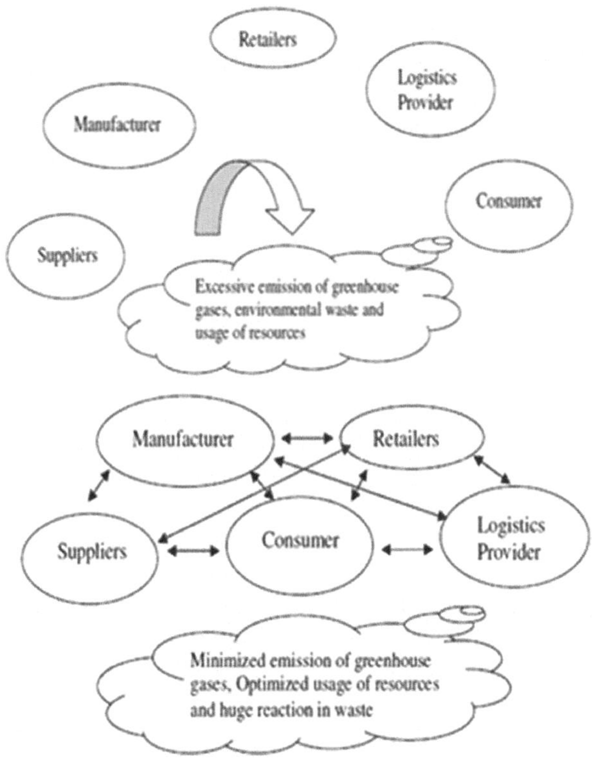

Fig. 2. Conventional SCM vs Green SCM.

2 Literature Review

An empirical study has been conducted by Laxmi et al. 2014 [3], on selected 155 manufacturing industries in Tamil Nadu in India to investigate the pressures on the companies in implementing GSCM practices. 6 external pressures were considered which would have relevance on Indian business environment. Significant positive relationship was identified which implies that Green supply chain practices are influenced by Green supply chain practices.

Govindan K et al. 2014 [4], focused on the India industries struggle in identifying the green supply chain management implementation barriers. The research paper presents 47 barriers through literature review and interviews. Analytical hierarchy process is used to identify the priorities and a sensitivity analysis check on stability of priority ranking.

Ravi V et al. 2014 [5], emphasised on reverse logistics important in GSCM practices. It has focussed on certain Indian manufacturing industries like auto, paper, food etc. to check on extent of reverse logistics implementation as volume of data entering in reverse logistics is a key driver of reverse logistics activities. The paper highlights the fact that gathering information from companies is difficult as many of the companies were reluctant in sharing the sensitive data. The paper though is able to provide a bird's eye view of the practice in India though.

Deshmukh SP et al. 2013 [6], focuses on the critical green manufacturing factors which are to be looked at while doing supplier selection. The study was done by focussing on manufacturing companies in Maharashtra India. The 5 top criteria's in sequence of relevance have been concluded. Quality came out as the top most criteria followed by environment performance assessment which was followed by green manufacturing and then green customer co-operation. Green cost emerged as fifth most important criteria which supports that cost is being looked at secondary than other criteria's

Ghobakhloo M, 2013 [7], presented an integrated GSCM implementation framework. This is done through literature review of GSCM and integrating all the aspects of key themes and sub themes. Green Product Design, Green material management, Green Marketing & Distribution and Reverse logistics. All the conventional supply chain scopes are looked along with their respective impacts on the environment. These has helped the researcher in crafting the green integrated framework.

Dheeraj N et al. 2012 [8], presents an overview of GSCM in India. It highlights the fact that how reduction, recycling, reuse and substitution of material which are key elements of Lean practices contribute into GSCM. The paper also presents a good illustrative comparison between lean and green manufacturing on 11 various factors. The conclusion is also derived that tough cost and complexity may be considered as deterrents in implementation of GSCM but it is not so. Though the pertinent question is highlighted as if it so then why many companies have not implement GSCM.

Luthra S et al. 2015 [9], looked at identifying the Critical factors in implementing the GSCM. 26 such factors have been identified through literature review and a contextual relationship has been identified among them. A hierarchy structural model has been then suggested for CFSs for implementation of GSCM. Using ISM model and MICMAC analysis, the researcher has been able to find that Scarcity of Natural resources is coming at bottom of the hierarchy wherein Enhanced brand image, firm's competitiveness and Economic benefits are coming as top three. It is ironical to see that natural resources scarcity is coming so low in the hierarchy but equally the fact that Economic benefits, competitiveness and economic benefits are at top, emphasise that

GSCM implementation somewhere has a relationship with economic performance of the company.

Rao P, 2019 [10], evaluates the extent of awareness & usage of GSCM in MSME sector in Chennai area. Second aspect kept in this paper is to check if GSCM improves the performance both economic and financial of the identified strata. Importance of MSMEs in growth of the country and the sector's contribution in GDP has been signified. The pressures on MSMEs in Chennai are on implementing GSCM comes from their customers who are majorly global MNCs. Pressures are being acceptable, having image of environmentally sustainable and being competitive. The results signify that if SMEs are able to implement green sourcing, it will have immediate positive impact on their economic as well as environmental performance.

Manimaran, 2018 [11], examine the scenario of adopting GSCM practices by considering institutional pressures. It also brings forth an overview of inter relationship between Institutional pressure, GSCM practices, and tolerance evaluation index. Institutional pressures being normative by external shareholders, coercive generated by those in power e.g. governmental agencies and mimetic e.g. need of following successful competitors. 21 GSCM practices have been listed in paper. Elements covered under tolerance evaluation matrix being Innovation & development, Information sharing Environmental performance & customer service. In conclusions, it has been suggested that the framework can be used for evaluating tolerance evaluation index for different sectors and this can be used for comparing two sectors.

Upstream engagement in GSCM has been analysed by Sharma G et al. 2019 [12], by looking at interlinkages between lean and green supply chain practices with focus on automotive sector in India. The influence of these exercise on sustainable business development has been understood and an integrated model between Green and Lean has thus been suggested. Sustainable business development is based on Economic factors viz. Operational cost, Environmental cost, inventory cost and Environmental viz. business wastage, green image, Co2 emissions etc. The research conclusion shows that lean and green upstream supply chain practices can have a substantial and positive influence on environmental & economic performance of the enterprises.

3 Objectives of Research

The world has been drastically changing in its trade practices which is evident from opening of global market, cooperation among countries, liberalization of cross border movements, growth in emerging markets and transfer of technologies and knowledge sharing.

Though this brings opportunities to Indian industry but also throws challenges as Indian companies will need catching up on global expectations and challenges. As a lot of impetus is now globally given on environmental sustainability and greening of industries through the supply chain management hence it becomes prudent for India to understand and start walking on those footsteps.

It is thus important to understand GSCM, see its evolution and to begin with, check on leading Indian manufacturing companies on their understanding and practices in this area. The wider objective thus is to understand as what defines GSCM, environmental sustainability & see it with respect to India. This study tries to gather the fundamental point being seen and raised by the countries industrial flagbearers which can help other medium and small enterprises.

4 Research Questions

- How is greening of supply chain and environmental sustainability aspect evolving in India?
- What are the top manufacturing companies in India doing in area of sustainability and green supply chain?
- What are the advantages and challenges associated with the implementation of green supply chain management and sustainability practices?

5 Research Methodology

The study has been based on literature review of related research papers, articles, journals and related government websites. The sustainability reports were looked at for company's reporting on their green supply chain and sustainability practices, their initiatives on conservation of natural resources, water preservation, waste disposal practices and their focus of renewable energy transitions. This has helped in identifying relevant questions, benefits and suggestions.

6 Origin of Sustainability at Global Level

The subject of environmental sustainable practices was formerly inducted in June 1992 at Rio de Janerio through United Nations conference on environment and development. In this summit, International organisations as well as governments decided to take useful measures to protect environment for long term economic development. Everyone had a fair understanding that this subjects of sustainability are too big and needs collaboration as cannot be dealt individually. The most important gradual achievement of this summit has been arriving at an agreement on climate change convention which has led to Kyoto protocol and Paris agreement.

Eventually, various Environmental Organisations are coming into existence and supporting the cause. Major ones being United Nation Environment Programme (UNEP), Intergovernmental Panel on climate Change (IPCC), World Resource Institute (WRI), Exnora International, Stop Climate Chaos, Global footprint Network, International Network for sustainable energy (INFORCE), World conservation union (WCU)

Various drivers have evolved which are forcing the companies to follow GSCM and sustainability practices. External pressures are from suppliers, various governments, environmental organisations, consumers, firm's rivalries, media These mounting pressures make it obligatory to comply with environmental laws, policies and requirements, hence the need of Green supply chain and its management.

6.1 India's Role and Emerging Commitment Towards Sustainability

There has been a consistent change in Indian government policies in support over last decade. Central pollution Control Board under Ministry of Environment & Forests Department of Science and Technology under Ministry of Science and Technology are actively working on green issues. Every State in India have their State pollution control boards which formulate policies for motivating green practices.

National voluntary Guidelines on Social, Environmental and economic responsibilities of business was brought by Ministry of corporate affairs to help the corporate sector in their effort towards inclusive development. This was notified by MCA in 2011 as a guidance document for businesses of all sizes to achieve the triple bottom line, an accounting framework with three parts; social, environmental and financial.

Further to this, many other initiatives have taken place like, National action plan on climate change announced in India in 2010, SEBI mandating Business Responsibility Reporting (BRR) in 2012. SEBI mandated Business responsibility reporting initially on top 100 companies, then expanding to top 500 and now it is being amended to top 1000 companies.

GRI (Global Reporting Initiative) which is an independent international organisation that has pioneered sustainability reporting since 199, and BSE (formerly known as Bombay Stock Exchange) have entered a formal MOU in 2016 to work collaboratively and support the Top 500 Indian listed companies in establishing sustainability report processes and preparing sustainability reports. This has helped organisations in coming up with transparent and detailed reporting.

7 Results- Worldwide Implementation and Indian Market

The global companies like BMW, IKEA and Walmart have been the flagbearers on adapting Green Supply Chain and sustainability practices. BMW, as an auto maker, have demonstrated that the 100% components can be recycled which demonstrates the focus of the organisation on sustainability. Similarly, IKEA has put special impetus on greening by practicing green packaging and mandating their suppliers to be environmentally friendly and aligning their environmental strategies with that of IKEA's. Walmart on the other hand uses a very stringent and comprehensive environmental score card to evaluate suppliers to ensure adherence and compliance to strict environmental standards.

The 10 Indian manufacturing companies have been picked as per BSE index based on market capitalisation as on June 2020. The websites have been explored between April 2020 to June 2020 and latest reports available on respective company websites have been referred to check on their progress in terms of environmental sustainability, environmental issues which core to green supply chain management (Table 1).

Table 1. Top 10 Indian manufacturing companies.

Name of the company	Market Capitalisation INR Cr.	Revenue FY19-20 INR Cr.	Net Profit INR Cr.	Net Profit %
Reliance industries Ltd.	973,796	350,880	30,903	8.8%
Hindustan Unilever Ltd.	495,121	38,785	6,738	17.4%
ITC ltd.	242,403	45,784	12,464	27.2%
Maruti Suzuki India Ltd.	171,822	75,610	5,650	7.5%
Nestle India Ltd.	164,108	12,368	1,969	15.9%
Asian Paints Ltd.	164,008	16,391	2,134	13.0%
Larsen & Toubro Ltd.	132,724	82,383	6,679	8.1%
Wipro Ltd.	123,353	50,407	8,680	17.2%
Sun pharmaceutical industries Ltd.	113,534	12,531	3,211	25.6%
Ultratech cement Ltd.	110,639	40,649	5,455	13.4%

7.1 Reliance (RIL)

RIL is maximising use of clean energy by using best in class technologies, reducing fresh water consumption per unit of production by maximising recycling of waste water, handling waste management through efficient use of solid catalysts including bio-catalysts with gradual replacement of solid catalysts, reducing GHG emissions intensity at all manufacturing locations and demanding a specific level of HSE compliance from all stakeholders. An effective governance mechanism has been put across all locations and impetus is given on diligent use of scarce resources.

RIL has exemplified their endeavour by making their biggest manufacturing site work without drawing a single drop of fresh water from lakes and other sources and have developed over 6,200 Ha of green belt, enriching ecological balance and biodiversity. It recycles around 60 kT/year of polyester waste with more than 2 billion post-consumer PET bottles per year. Another big step has been adapting video conferencing extensively which has replaced travelling to great extent. This helps in reducing the travel emission and also reduce the travel cost.

7.2 Hindustan Unilever (HUL)

HUL has a robust sustainable plan in place and in 2019 sustainability plan, they have taken the target of reducing environmental impact by half by 2030. HUL have benchmarked their level on greenhouse gases, water and waste level with 2008 per capita consumptions and have declared performance on CO2 emissions by −65% per tonne, −47% on water abstraction per tonne and −96% reduction in total waste per tonne of production. They have been candid in declaring not so good areas also like +2% increase in greenhouse impact per consumer and 1% water impact per consumer in comparison to 2010 levels. Sustainable sourcing is being followed having achieved 62% of all agricultural raw material sustainably sourced and target is to be at 100% level by end of 2020.

7.3 ITC

ITC has placed sustainability in its core business model, mapped its climate change risks and possible impacts and has drawn out a mitigation strategy to address the same. Board has approved Sustainability Policies, which address supply chain sustainability including environment. ITC has earned to be world's only company which is positive on Carbon (14 years in a row), Water and Solid water recycling. This has happened through innovative strategies leading to ITC's businesses linked sustainable value chains. 40.9% of ITC energy is coming from renewable sources and 25% emission reduction has been achieved through logistics optimisation. Energy conservation measures have helped ITC in avoiding 9,747 tonnes of GHG emissions. ITC has 740,491 of total area under forest cover out of which 57,068 acres of plantation has just been added under Social and Farm Forestry Initiatives (as on 31st March, 2019).

7.4 Maruti Suzuki

The Company's Green Procurement Guidelines (GPG), based on Suzuki Engineering Standards, set requirements for all component suppliers to avoid the use of Substances of Concern (SOC) such as lead, cadmium, mercury, hexavalent chromium and asbestos in manufacturing processes and products. Suppliers demonstrate compliance by submission of declarations, test reports and part-wise composition details using a globally recognised SOC control tool called the International Material Data System (IMDS). The GPG encourages suppliers to establish an Environmental Management System (EMS) at their facilities and promote EMS adoption among Tier-II suppliers. The Company follows the 3R principle of waste management. Handling, storage and disposal are carried out after proper segregation according to the waste types. A major portion of the hazardous waste (such as process sludge and ETP sludge) is sent to the cement industry for co-processing. The remaining portion of the hazardous waste (such as used oil, contaminated barrels and cloth) is sent to authorised recyclers.

7.5 Nestle

Nestle India focus on resource use and to improve the environmental performance of their food and beverages range. As per Nestle, they have achieved a recognized leadership in sustainability. Energy utilization (GJ/Ton) has reduced by 49% from 2004 to 2019, Water use (m^3/Ton) has reduced by 54% from 2004 to 2019 and CO2 (Tons eq./Ton) has reduced by 58% from 2004 to 2019. Further Waste Water Generation (m^3/Ton) has reduced by 59% from 2004 to 2019. The key brands like Maggi Noodles, Nescafe and KitKat have become plastic neutral in 2019 which could be done through focus on plastic waste management wherein an equivalent or a higher amount of plastic waste is managed (recycled) which these brands generate through their consumption in the market.

7.6 Asian Paints

Asian Paints approach for sustainable development is to incorporate into the business strategy. An integral part of that sustainable journey is the continuous endeavour to conserve water and optimise its use. The company continue to focus on reducing the energy consumption and make investments in renewable energy footprint. This has contributed to the reduction of carbon footprint at the manufacturing facilities. Asian Paints is constantly involved in waste reduction initiatives and responsible management and disposal of waste. The company has made a start in fulfilling the extended producer responsibility obligations with regard to post-consumer plastic packaging. This is a beginning to company's journey to conserve flora and fauna.

Between 2008–19, Asian paints has developed 30 Green Assure compliant products and more than 600 raw materials assessed for Green Compliance. Nearly 200 raw materials have been approved for use in products declared as Green Assure. Some of the company's contributions are like 540 million Litres of water replenished in FY2018-19, 33 million Units of electricity consumed in FY2018-19 from renewable sources, 14 million Kilograms of carbon emissions reduced in FY2018-19 from previous year, water replenishment brought to 98% in FY19 with a target of 100% by Fy2020, 54.78% electricity from renewable sources (36% from wind and 18% from Solar), 77.5% reduction in specific industrial affluent generation and 46% reduction in specific hazardous waste disposal. (All reductions are w.r.t base year FY 2013–14)

7.7 L&T

L&T offer green technology solutions to their clients, work continuously on energy intensity reduction and shrinking the carbon footprint. A 'reduce, reuse, recycle and redeem' principle has helped to reduce consumption of natural resources. Most of the campuses are water positive and many harness green energy. The environment protection initiatives focus on reducing the use of natural resources, building efficient infrastructure, cutting emissions and more importantly bringing about a behavioural change in stakeholders to sustain our efforts. Some interesting facts shared in integrated report covering sustainability are a) 90500 MT of co2 emissions at the factories avoided, b) 9.3% of total energy consumed is renewable energy, c) 33 k MT of steel and 1015 MT of Zinc recycled (cumulative), d) 17 green building in place which includes a green factory, e) Achieved Zero water discharge with high water recycling being practised. 24 of the campuses in India have turned water positive by water recycling, f) 12.4% of domestic specific water consumption reduced over the financial year.

7.8 Wipro

Wipro has been able to achieve a 40% of energy footprint from renewable energy sources. They have been able to reduce the Green House gases emissions by more than half a million tonnes over 9 years. On water consumption Wipro is using 42% water through high level recycling which indirectly has contributed in saving 5000 million litres of water over last 5 years. Wipro has been supporting urban ground water management using rigorous scientific approach and seeking citizen's support to get effect harvesting, conservation and usage in the neighbourhoods. Wipro's carbon reduction targets are aligned with the new practices and are one of the few companies to have adopted science based targets.

Wipro promotes suppliers towards ethical and responsible sourcing of green products and services. Close to 90% of total waste in India plants is recycled either in house or using the authorised vendors. There is special impetus on waste management and all campuses segregate, track and mange around 40 sub category of wastes. Further achievements have been like 29% reduction in global people based emissions intensity to 0.85 tons per person per annum, 44% increase in energy saving due to server virtualization from FY 2018 and 21% reduction in air travel footprint from FY 2018.

7.9 Sun Pharma

The company is vigilant on emerging situation and committed to take necessary step and help change the situation. The company has embraced an all-encompassing Environment, Health & Safety (EHS) policy which etches out the concerns as well as provide steps for resolving them. The company recognised that that being energy efficiency and reducing energy consumptions not only helps improves environment but also gives economic benefits.

Through greener investments the company is committed to generate more green energy and reduce the dependence on fossil fuels. This has been demonstrated by the investment in area of wind energy and testimony is in FY 19 through generation of 2.05 Million kWh of clean energy. Company having diverse product portfolio finds it challenging to calculate environmental performance per product and prefer monitoring and managing the total annual water and energy performance vis-à-vis the total annual production.

7.10 Ultratech

The company has installed first of its kind in the world, a Green Ready Mic Concrete (RMC) plant, in Mumbai which has zero discharge. Three of integrated plants have become water sufficient which is a prudent step towards becoming water positive in future. Ultratech has a water-positive score of over 2.00 across its plants. The company has developed technology to use all waste generated from the rejected concrete for recycling into RMC production. The technology is fully automated and enables the plant to be real zero discharge having no solid or water waste. This eliminates the recycling cost which would have gone in recycling waste or water.

E-procurement has resulted in more effective communication with our vendors and enabled significant reduction in paper work as well as travel hours. Company's captive renewable capacity stands at 66.23 MW. It generated 35 million units of electricity from renewable sources during the year. It is a signatory of the EP100 and committed to double its energy productivity over the next 25 years. The company has set a target to reduce its CO_2 intensity by 25% by FY21, as compared to FY06.

8 Discussion - Benefits of GSCM & Challenges in Its Implementation

The effects of GSCM expand to all areas, both tangibly and intangibly. The benefits of GSCM have been identified for different members of supply chain, whole society and the environment as shown below:

- For Environment it could be low Pollution and, safeguarding perishable resources
- For Suppliers it could be Low Cost prices, Less Rejection, Better Acceptability by bigger manufacturing organisations, lesser rejections and lower waste left out for disposals.
- For Manufacturers it could be Competitive advantage, Low Cost Prices, Easier maintenance, Better Image, Sustainability of resources, Increased efficiency, adapting to regulation and reducing risk, Brand Reputation, Return on Investment, Employee Morale.
- For Customers it could be Low Cost of Ownership, Convenience and Fun, Feel good and Quality of life
- For Society it could be Less Consumption of Resources, Better Compliance, better health

Though we have prima facia understood the benefits in deployment of Green Supply Chain Management but it is equally important to understand that its implementation is not an easy task and is full of challenges.

As per this study and with reference to Raman (2014) [13], Srivastava [14] some of the challenges are listed below

- Implementation could be deemed to be an expensive preposition. There could be negative sentiments that greening could lead to cost escalation.
- Greening needs coordination across stakeholders but complexity in supply chain and spread of stakeholders could be deterrent in its implementation.
- Limitation in getting technology or its support and fear of failure
- Certain sections of customers may not be sensitive to greening and its importance can lead to reluctance in companies in its enactment for lack of acceptance from customer.
- Top management assurance or absence of subject matter experts like architects or consultants could become another limiting factor in GSCM implementation.

The challenges are at times to collate specific sustainability reporting across sectors and across waste companies which can help analyse a sector wise, comparable sizes and type of industries.

9 Conclusion

Changing standard of living & growing population has steered to higher consumption level thus leading to ever growing industrialisation. This massive unregulated industrialisation has led to grave effect on ecology. There is a visible impact on availability of natural resources, continuous depletion of non-renewable assets, depletion of clean and fresh water resources, air quality, flora and fauna.

Greening the supply chain is a must do and is being adapted by all big companies globally and similar things need to be implemented in India. Though Indian government has taken a lot of steps but has not been able to demonstrate substantial effectiveness. The suppliers are feeling the external pressures in adapting greening principles though there has been struggle in identifying the implementation barriers and also for, researchers it has been difficult in Gathering information from industry.

Further after analysing the top Indian manufacturing companies, it is demonstrated that top companies in India are very well aware of the importance of environmental sustainability, are implementing greening in their supply chain and showing progression. The companies which have been studied for green practices were seen to be highly profitable and in their reports have consistently indicated the benefit of greening in their overall sustenance, risk mitigation and economical gains through green practices. The companies are putting greening efforts both upstream and downstream of supply chain.

Greening is not only needed across parts of supply chain but also across the various phases of life cycle of the product viz. product design, material management, sourcing, marketing, reverse logistics and distribution.

Reduction, Recycle, Reuse & substitution of materials have come out as common themes across the companies which have been studied.

10 Recommendations and Suggestions

Though implementation is seen in various areas but more visible implementation effort and results on specific upstream and downstream stakeholders will help in identifying the weak links and can highlight specific areas which may need special thrust and focus. This can help in building an effective implementation model which can be guiding document for wider industrial base in India and can help the country evolve. This could be a gift from flag bearers to rest of the industry.

References

1. Gunasekaran, A., Subramanian, N., Rahman, S.: Trans. Res. Part E: Log. Transp. Rev. **74**, 1–10 (2015)
2. Srivastava, S.: Int. J. Manage. Rev. **9**, 53–80 (2017)
3. Meera, B.L.L., Chitramani, P.: Int. J. Sci. Res. Pub. **4**(3) (2014)
4. Govindan, K., Kailyan, M., Kannan, D., Haq, A.: Int. J. Prod. Eco. **147**, 555–568 (2014)
5. Ravi, V., Shankar, R.: Benchmarking: Int. J. **22**(5), 874–899 (2014)
6. Deshmukh, S.P., Sunnapwar, V.K.: Int. J. Mod. Eng. Res. **3**(3), 1617–1622 (2013)
7. Ghobakhloo, M.H.: Int. J. Innov. Manage. Tech. **4**(1) (2013)
8. Dheeraj, N., Vishal, N.: Res. J. Recent Sci. **1**(6), 77–82 (2012)
9. Luthra, S., Garg, D., Haleem, A.: Resources Policy, 37–50 (2015)
10. Rao, P.H.: J. Supply Chain Manage. Syst. **8**(1), 15–24 (2019)

11. Manimaran, A., Muthuraman, V., Jayakumar, V.: Int. J. Mech. Eng. Tech. **9**(1), 152–159 (2018)
12. Sharma, G., Singhi, R., Mittal, A.: Int. J. Mech. Eng. Tech. **10**(03), 317–330 (2019)
13. Raman, P.: J. Supply Chain Manage. Syst. **3**(1) (2014)
14. Srivastava, S.K.: Int. J. of Manage. Rev. **9**(1), 53–80 (2007)
15. https://www.nestle.in/csv/environment
16. https://www.wipro.com/content/dam/nexus/en/sustainability/sustainability_reports/sustainability-report-fy-2018-19.pdf
17. https://www.asianpaints.com/sustainability/
18. https://www.ultratechcement.com/sustainability
19. https://www.itcportal.com/sustainability/sustainability-report-2019/sustainability-report-2019.pdf
20. Corporate websites of all top 10 Indian companies.

Environmentally Friendly Road-Building Thermal Insulating Materials Based on Zeolite-Containing Rocks

Natalia Konovalova(✉), Pavel Pankov, Elena Rush, Nadezhda Avseenko, and Dmitry Bespolitov

Irkutsk State Transport University, Irkutsk, Russian Federation
zabizht_engineering@mail.ru

Abstract. The paper shows expediency of application of natural zeolite (Transbaikal region, Russia) with ecologically safe technological additive (Na_2CO_3) as an initial raw material for foam glass production. According to the quantitative X-ray phase analysis, the clinoptilolite content is 50–55 wt%. It is established that optimal concentration of Na_2CO_3 in the composition of the charge is 15 wt%. With increasing heating temperature of dry mixture from 300 to 500 °C, it is possible to observe a systematic increase in the intensity of foaming. It was found out that the density of pellets obtained from a blend with the sodium carbonate content of 15% after firing at 900 °C is 0.25 g/cm^3. The analysis of physical and mechanical properties of the obtained foam glass shows that the samples based on natural zeolite are not inferior in properties to foam glass of known compositions. It is possible to obtain these thermal insulating materials using a low-temperature safe technology with replacement of the traditional NaOH fluxing agent by a less toxic one - Na_2CO_3. This foam glass can be used for effective thermal insulation in road construction.

Keywords: Zeolite · Clinoptilolite · Foam glass · Road-building materials · Thermal insulating materials

1 Introduction

The northern and northeastern regions of the Russian Federation are characterized by special operating conditions of roads. The permafrost factor plays a decisive role, which has a great influence on strength, stability, and durability of these structures. Under unfavorable hydrological and climatic conditions, the possibility of using thermal insulating materials for construction of structural layers of road pavements is relevant. Nowadays, for this purpose, thermal insulating compositions based on lightweight concrete, production wastes, porous aggregates (expanded clay, granulated polystyrene, shredded foamed plastic waste) are used. However, the use of many thermal insulating materials may be restricted in areas with harsh climatic conditions due to insufficient chemical resistance and drainage properties of organic highly porous materials.

One of the promising ways to solve this problem is to use foam glass as a thermal insulating layer. At present, considerable scientific and practical experience in the field

A. Sinitsyn (Ed.): SES 2020, LNCE 141, pp. 103–109, 2021.
https://doi.org/10.1007/978-3-030-67654-4_12

of technology of its production has been accumulated [1–3]. However, organization of foam glass production is slowed down due to the lack of a cheap and reliable source of raw materials. Specially welded glass granulate is a relatively expensive raw material, and technical broken glass is an unreliable source with unstable chemical composition.

In this connection, the development of alternative technologies of foam glass production with substitution of glass powder with natural mineral raw materials is relevant [4, 5]. For successful implementation of such production, the raw materials should be local, easily extracted, have a wide geographical distribution, which will allow replicating the production. In Siberia and the Far East, natural zeolite meets such conditions.

Nowadays, in the world practice, the most rational ways of using natural zeolites in industry [6–14], medicine and pharmaceutics [15–17], veterinary medicine [18, 19], and other branches of national economy are distinguished. Efficient use of natural zeolites in production of construction materials [9, 10, 20] is also a relevant task. Introduction of substandard zeolite-containing rocks into the composition of thermal insulating materials (zeolite content is less than 50%) will allow solving the problem of non-waste operation of deposits with a corresponding increase in their profitability and a decrease in the anthropogenic load on the environment due to liquidation of dumps.

The purpose of this work was to study the composition and properties of natural zeolite and thermal insulating road-building materials based on it.

2 Methods

Natural zeolite (Transbaikal region, Russia) with the specific surface of 3457 cm^2/g and a grain size of 7 mm was used as the main mineral raw material for obtaining thermal insulating materials. Na_2CO_3 and Na_2SiO_3 were used as technological additives.

The specific surface of natural zeolite was determined from nitrogen adsorption isotherms at T = 77 K on the ASAP-2400 (Micromeritics, USA). Chemical composition of zeolite was determined by X-ray fluorescent analysis using X-ray analyzer VRA-20R. The mineral composition of the original sample was determined by powder X-ray diffractometry (Thermo Scientific ARL-X'tra, CuK_α, radiation, U = 40 kV, I = 40 mA). The content of clinoptilolite in zeolite was determined with X-ray diffractometer DRON-3 by X-ray phase analysis. Microphotographs of zeolite were obtained by scanning electron microscopy using the Philips XL30-FEG microscope. Phase composition and thermal stability of samples were studied by methods of TG thermogravimetry and DTG differential thermogravimetry (Mettler TC10A). Mineral foaming was carried out in a muffle furnace KS-600. The intensity of foaming (foam formation) was estimated by the density of pellets after firing. The beginning of foaming was the temperature at which the density of pellets after firing became less than 1 g/sm^3.

3 Results and Discussion

The studied natural zeolite has an aluminosilicate composition: ω, wt%: 69.24 SiO_2; 12.72 Al_2O_3; 0.13 TiO_2; 1.17 Fe_2O_3; 0.06 MnO; 0.26 MgO; 2.51 CaO; 1.53 Na_2O; 4.58 K_2O; 0.02 P_2O_5; loses during firing 7.74.

According to X-ray powder diffractometry, the mineral composition of zeolite is presented in a series (in a decreasing order): clinoptilolite >> Ca-Na feldspar (plagioclase) > smectite > α-quartz (Fig. 1a). According to the quantitative X-ray phase analysis, the clinoptilolite content is 50–55 wt%. The extremum of low intensity observed on the DTG-curve of the studied zeolite at 500 °C indicates admixtures of clay minerals (Fig. 1b).

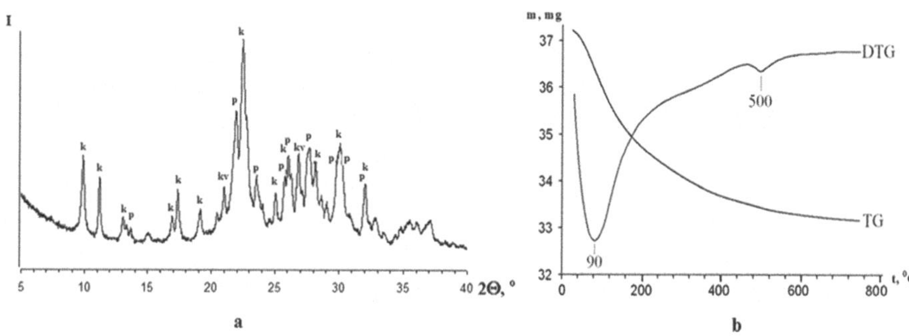

Fig. 1. Powder diffraction profile (a) and TG- and DTG-curves (b) of the zeolite: **k** – clinoptilolite; **p** – Ca-Na feldspar (plagioclase); **kv** – α- quartz

The smooth course of the TG-curve (Fig. 1b), typical for clinoptilolite, and at least at 90 °C on the DTG-curve is caused by removal of physically bound water from zeolite-containing rock. The method of scanning electron microscopy has established (Fig. 2) that crystalline clinoptilolite mineralization is characteristic for the zeolite.

Fig. 2. Microphotographs of natural zeolite obtained by scanning electron microscopy

In order to study the intensity of foaming of zeolite with Na_2CO_3 as a fluxing agent, the mixture was heated at 300 °C for 30 min and Na_2SiO_3 was injected into it. The ready mass was pelletized and dried at 90 °C. The data on granules density and their firing temperatures are presented in Table 1.

Table 1. Density of pellets after firing

Feed composition		$\rho_{pellets}$, g/sm^3 at t_{firing}, °C			
ω (zeolite), wt%	ω (Na_2CO_3), wt%	800	850	900	950
85	15	0.68	0.55	0.50	0.41
80	20	0.67	0.50	0.46	0.38

The analysis of the data (Table 1) showed that foaming intensity of compositions with Na_2CO_3 content of 15 and 20 wt% differs slightly. Therefore, further studies were conducted with mechanoactivated zeolite. Mixture of zeolite with Na_2CO_3 was shredded in a vibrating grinding mill for 10 min. Dry mixtures were heated for 30 min at 300, 400, and 500 °C. The mixture was cooled and moistened with Na_2SiO_3 aqueous solution, granules were formed and dried to air-dry state at 90 °C. The firing of granules was carried out in the range of 750–900 °C. Values of density of the fired granules ρ, temperatures of heating of a dry mix t_{ht} of investigated compositions are presented in Table 2.

Table 2. Density of pellets after firing (mechanical activation within 10 min)

Feed composition		t_{ht}, °C	$\rho_{pellets}$, g/sm^3 at t_{firing}, °C			
ω (zeolite), wt%	ω (Na_2CO_3), wt%		750	800	850	900
85	15	300	0.83	0.51	0.41	0.39
		400	0.65	0.41	0.34	0.30
		500	0.57	0.29	0.28	0.25
80	20	300	1.00	0.63	0.51	0.46
		400	0.90	0.55	0.46	0.43
		500	0.85	0.43	0.40	0.39

It is established that optimal concentration of Na_2CO_3 in the composition of the charge is 15 wt%. An increase of sodium carbonate content in the composition can cause loosening of the mass and an increase in porosity of dried granulate. It leads to a higher sintering temperature and releases more foaming gas - water vapor. With increasing heating temperature of dry mixture from 300 to 500 °C, it is possible to observe a systematic increase in the intensity of foaming. It was found out that the density of pellets obtained from a blend with the sodium carbonate content of 15% after firing at 900 °C is 0.25 g/cm^3.

The X-ray phase analysis data showed (Fig. 3) that heating of the mixture at 300 °C is not accompanied by noticeable amorphization of clinoptilolite. With increasing the temperature up to 400 and 500 °C, a decrease in the intensity of clinoptilolite reflexes is observed, which indicates the amorphization of its structure.

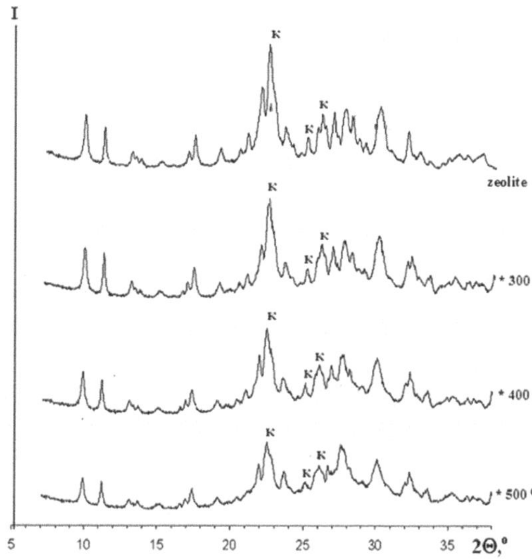

Fig. 3. Diffractograms of natural zeolite and foam glass samples: * - pellets (ω Na_2CO_3 = 15 wt %) after heating at 300, 400, and 500 °C; **k** – clinoptilolite

Thus, the production of foam glass with preset properties on the basis of the zeolite is based on the development of science-based feed compositions, conditions for pre-processing of compositions, heating and foaming modes, and application of effective methods of feed component activation.

Physical and technical properties of foam glass are presented in Table 3.

Table 3. Physical and technical properties of foam glass

Indicators	Compositions of zeolite and Na_2CO_3 based foam glass			
Average density, kg/m^3	200–350	400–590	600–750	800–850
Compressive strength limit, MPa	2.8–3.5	3.9–6.3	6.5–8.0	10.0–10.8
Water absorption, wt%	6.3–7.5	3.5–5.8	2.2–3.2	1.6–2.0
Thermal conductivity coefficient, W/m•K	0.07–0.08	0.09–0.10	0.10–0.15	0.19–0.20
Frost resistance, cycle	>30	>30	>30	>30

The analysis of Table 3 data allows us to conclude that foam glass based on natural zeolite is not inferior in its properties to the known used compositions. It is possible to obtain such thermal insulating materials using a low-temperature environmentally safe technology.

4 Conclusions

A complex of modern methods was used to study chemical and phase composition, physical and chemical properties of natural zeolite and foam glass samples based on it.

When studying intensity of foaming of natural zeolite, it was found out that the optimal composition can be considered as one containing 15 wt% of Na_2CO_3, with preliminary heating the dry mixture at 500 °C and its moisturizing with Na_2SiO_3 solution. Firing in the range of 800–900 °C. Increasing the degree of amorphization of clinoptilolite is accompanied by increasing the intensity of pellet foaming.

The analysis of physical and mechanical properties of the obtained foam glass shows that the samples based on natural zeolite (Transbaikal region) are not inferior in properties to foam glass of known compositions. It is possible to obtain these thermal insulating materials using a low-temperature safe technology with replacement of the traditional NaOH fluxing agent by a less toxic one - Na_2CO_3. This foam glass can be used for effective thermal insulation in road construction.

References

1. Silva, R.V., de Brito, J., Lye, C.Q., Dhir, R.K.: The role of glass waste in the production of ceramic-based products and other applications: a review. J. Cleaner Prod. **167**, 346–364 (2017)
2. Ji, R., Zheng, Y., Zou, Z., Chen, Z., Wei, S., Jin, X., Zhang, M.: Utilization of mineral wool waste and waste glass for synthesis of foam glass at low temperature. Constr. Build. Mater. **215**, 623–632 (2019)
3. da Silva, R.C., Kubaski, E.T., Tenório-Neto, E.T., Lima-Tenório, M.K., Tebcherani, S.M.: Foam glass using sodium hydroxide as foaming agent: study on the reaction mechanism in soda-lime glass matrix. J. Non-Cryst. Solids **511**, 177–182 (2019)
4. Ivanov, K.S.: Preparation and properties of foam glass-ceramic from diatomite. J. Wuhan Univ. Technol. **33**(2), 273–277 (2018)
5. Vereshchagin, V.I., Sokolova, S.N.: Effect of the technological parameters on the properties of granular porous crystal glass material based on zeolite-bearing rock. Glass Ceram. **66**(1–2), 46–49 (2009)
6. Naseroleslami, R., Nemati, C.M.: The effects of calcium stearate on mechanical and durability aspects of self-consolidating concretes incorporating silica fume/natural zeolite. Constr. Build. Mater. **225**, 384–400 (2019)
7. Öncü, Ş, Bilsel, H.: Effect of zeolite utilization on volume change and strength properties of expansive soil as landfill barrier. Can. Geotech. J. **54**(9), 1320–1330 (2017)
8. Sabet, F.A., Libre, N.A., Shekarchi, M.: Mechanical and durability properties of self-consolidating high performance concrete incorporating natural zeolite, silica fume and fly ash. Constr. Build. Mater. **44**, 175–184 (2013)

9. Ranjbar, M.M., Madandoust, R., Mousavi, S.Y., Yosefi, S.: Effects of natural zeolite on the fresh and hardened properties of self-compacted concrete. Constr. Build. Mater. **47**, 806–813 (2013)
10. Giosué, C., Mobili, A., Yu, Q.L., Brouwers, H.J.H., Ruello, M.L., Tittarelli, F.: Properties of multifunctional lightweight mortars containing zeolite and natural fibers. J. Sustain. Cement-Based Mater. **8**(4), 214–227 (2019)
11. Asere, T.G., Stevens, C.V., Du Laing, G.: Use of (modified) natural adsorbents for arsenic remediation: a review. Sci. Total Environ. **676**, 706–720 (2019)
12. Meshkova, I.N., Ushakova, T.M., Grinev, V.G., Nikashina, V.A., Novokshonova, L.A.: Use of synthetic and natural zeolites for fabricating immobilized olefin polymerization catalysts and polyolefin-based composite materials. Pet. Chem. **59**(8), 831–837 (2019)
13. Gezer, F., Turhan, Ş, Kurnaz, A., Ufuktepe, Y.: Radiometric characterization of zeolite minerals used in many industries and assessment of radiological risks. Appl. Radiat. Isot. **152**, 57–63 (2019)
14. Heredia, A., Gómez Avila, J., Vinuesa, A., Saux, C., Mendoza, S.M., Garay, F., Crivello, M.: Compared arsenic removal from aqueous solutions by synthetic mixed oxides and modified natural zeolites. Adsorption **25**(7), 1425–1436 (2019)
15. Golokhvast, K.S., Panichev, A.M.: Zeolites: review of biomedical data. Achiev. Life Sci. **1**, 118–152 (2009)
16. Golokhvast, K.S., Reva, I.V., Panichev, A.M., Gulkov, A.N., Rybalka, V.P., Rachkov, A.V., Borisov, S., Anisimova, A.A., Pamirskiy, I.E.: The Peculiarities of thermal burn histological skin structure after application of far eastern natural zeolite nanoparticles. Pac. Sci. Rev. **11** (2), 172–174 (2009)
17. Cerri, G., Farina, M., Brundu, A., Daković, A., Giunchedi, P., Gavini, E., Rassu, G.: Natural zeolites for pharmaceutical formulations: preparation and evaluation of a clinoptilolite-based material. Microporous Mesoporous Mater. **223**, 58–67 (2016)
18. Schneider, A.F., Almeida, D.S.D., Yuri, F.M., Zimmermann, O.F., Gerber, M.W., Gewehr, C.E.: Natural zeolites in diet or litter of broilers. Br. Poult. Sci. **57**(2), 257–263 (2016)
19. Superchi, P., Saleri, R., Ossiprandi, M.C., Riccardi, E., Passaglia, E., Cavalli, V., Beretti, V., Sabbioni, A.: Natural zeolites (chabazite/phillipsite) dietary supplementation influences faecal microbiota and oxidant status of working dogs. Ital. J. Anim. Sci. **16**(1), 115–121 (2017)
20. Ramezanianpour, A.A., Mousavi, R., Kalhori, M., Sobhani, J., Najimi, M.: Micro and macro level properties of natural zeolite contained concretes. Constr. Build. Mater. **101**, 347–358 (2015)

Experience in Reactive Power Compensation in the Power Supply System of Industrial Facilities

Elena Ivanovna Gracheva[1(✉)], Muhayo Islomovna Toshkhodzhaeva[2], Okhunbobo Saifidinovich Rakhimov[2], Ayubjon Jumaevich Vohidov[2], and Ismoiljon Ilkhomovich Ismoilov[2]

[1] Department Power Supply of Industrial Enterprises, Kazan State Power Engineering University, Kazan, Russia
elyor.saitov@mail.uz
[2] Department of Power Supply and Automation, Khujand Polytechnic Institute of the Tajik Technical University Named after Academician M.S. Osimi, Khujand 735700, Tajikistan

Abstract. It is known that the main consumers of reactive power are electrical machines and electric drives, which are widely used in various mechanisms of industrial enterprises. Reactive power is necessary to create magnetic fields in electric machines, but not accounting or excessive consumption of reactive power leads to a decrease in the power factor of the power supply system, which will negatively affect the power system as a whole. Therefore, the study of methods and methods of reactive power compensation is an urgent issue, for the solution of which an integrated approach is required. The article presents the results of the experience of reactive power compensation at industrial enterprises, and also proposes a methodology for carrying out measures to compensate for reactive power. The article analyzes in detail the main problems in industrial enterprises due to consumers of reactive power. A methodology for calculating and choosing reactive power compensators for industrial enterprises is proposed, which differs from other methods with simplicity and availability in practical conditions for use.

Keywords: Industrial facilities · Power supply system · Reactive power compensation

1 Introduction

Compensation of reactive power ka we know is especially relevant for industrial enterprises, the main electrical consumers of which are asynchronous motors, as a result of which the power factor without taking compensation measures is 0.7–0.85. Compensation of reactive power, at present, is an important factor in solving the issue of energy saving and reducing the load on the power grid [1–4].

This inductive load during operation is a source of reactive electricity (reactive power), which oscillates between the load and the source (generator), is not associated with the performance of useful work, but is spent on creating electromagnetic fields and

A. Sinitsyn (Ed.): SES 2020, LNCE 141, pp. 110–118, 2021.
https://doi.org/10.1007/978-3-030-67654-4_13

creates an additional load on the power supply lines. Therefore, the reactive power compensator is very important [5–9].

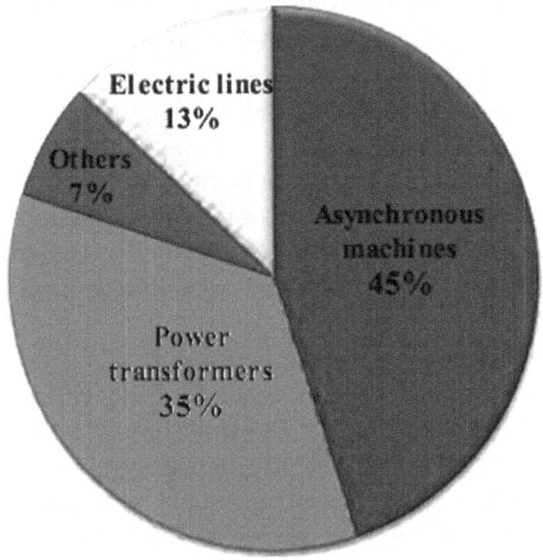

Fig. 1. Structure of reactive power consumers in power systems (by installed active power).

As we can see from Fig. 1, the main load in industrial power grids is induction motors and distribution transformers. The most effective and efficient way to reduce the reactive power consumed from the network is to use reactive power compensation units (capacitor units) [10–16].

Measures for reactive power compensation, the use of capacitor units as a reactive power compensator for an enterprise allows:

- reduce the load on transformers, increase their service life;
- use wires, cables of a smaller cross-section by reducing the load on them;
- improve the quality of electricity at electrical receivers (by reducing the distortion of the voltage waveform);
- to reduce the load on the switching equipment by reducing the currents in the circuits;
- to reduce energy costs;
- unload power supply lines, transformers and switchgear;
- when using a certain type of installations, reduce the level of higher harmonics;
- suppress mains interference, reduce phase imbalance;
- to make distribution networks more reliable and economical.

2 Methodology

The research object was the North-East industrial zone of Khujand City, Sughd region, Republic of Tajikistan. Power supply to the North-East industrial zone of Khujand City is carried out through two independent power sources. The first is the «Zarechnaya-110/10» substation, and the second is the «Radiy-110/10» substation. Both substations are located not far from the North-East industrial zone and were operated in the last century. Substation «Zarechnaya-110/10» has two transformers of the TDN-16000/110/10 type. The power of each transformer is 16,000 kVA, and the Radiy-110/10 substation also has two transformers of the TDN-10000/110/10 type, the power of each transformer is 10,000 kVA. From the substation to consumers, the distribution of electricity is carried out through overhead high-voltage power lines, wires of the AABLU-240–3 type. The characteristics of the transformers of the Zarechnaya-110/10 and Radium-110/10 substations are shown in Table 1.

As we have noted, the main power sources of our enterprises are substations «Zarechnaya - 110/10» and «Radiy-110/10». Both substations are powered from the «Khujand - 220/110» substation.OJSC JV «Javoni» has two power transformers with a capacity of 1600 kVA each. Power is supplied from the Zarechnaya - 110/10 substation through 10 kV lines with AABLU-3x240 wire with a distance of 3.35 km and through the Radiy-110/10 substation of a 10 kV line with AABLU-3x240 wire with a distance of 1.43 km. OJSC«MMK» has one 630 kVA transformer in its balance sheet. The power source for this enterprise is only the substation «Zarechnaya - 110/10» with a wire of the type AABLU-3x240. The distance from the source to our consumer is 1.87 km.A detailed description of the types of transformers in industrial enterprises is given in Table 2.

OJSC «Niku Khujand» has two power transformers with a capacity of 630 kVA each. And power is supplied from two substations, ie, «Zarechnaya - 110/10» and «Radiy-110/10» through 10 kV lines with a wire of the type AABLU-3x240. The distance from the «Radiy-110/10» substation to OJSC «Niku Khujand» is 1.45 km, and from the «Zarechnaya - 110/10» substation is about 2.13 km.To reduce the mains voltage from 10 kV to 0.4 kV, industrial enterprises have their own step-down transformers.

The scheme of power supply to consumers without taking into account distributed generation shown in Fig. 2.

Taking into account the prospects for the development of industrial enterprises and justifying the fact that our enterprises will increase production in the future, we take the actual load equal to 20% as a potential load [17–22].

Table 1. Characteristics of the transformer substation «Zarechnaya - 110/10» and «Radiy - 110/10».

№	Substation	Transformer type	Rated power S, kVA	Number of transformers	Frequency f, Hz	Rated voltage (H.V.), kV	Rated voltage (L.V), kV
1	«Zarechnaya-110/10»	TDN-16000/110/10	16000	2	50	110	10
2	«Radiy-110/10»	TDN-10000/110/10	10000	2	50	110	10

Table 2. Power supply of the industrial enterprise and the type of transformers.

№	Industrial enterprise	Number of transformers	Transformer type	Line	Power substation
1	OJSC JV «JAVONI»	2	TM -1600/10/0,4	AABLU -3x240	«Zarechnaya 110/10» and «Radiy 110/10»
2	OJSC «MMK»	1	TM -630/10/0,4	AABLU -3x240	«Zarechnaya 110/10»
3	OJSC «Niku Khujand»	2	TM -630/10/0,4	AABLU -3x240	«Zarechnaya 110/10» and «Radiy 110/10»

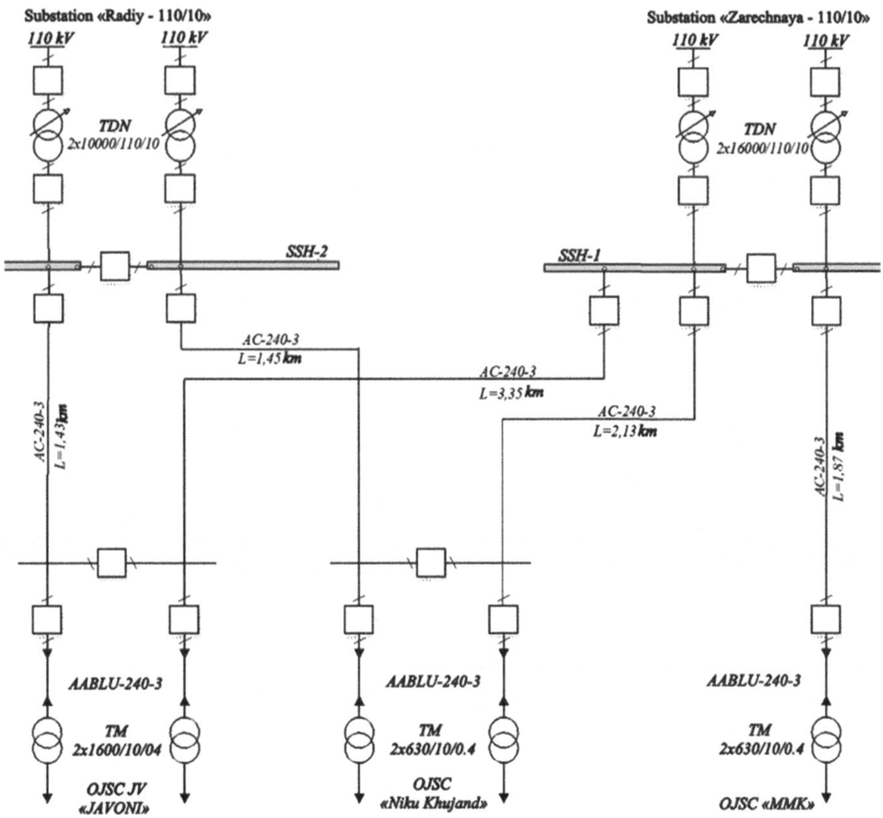

Fig. 2. Power supply system of industrial enterprises from the substation «Zarechnaya - 110/10» and «Radiy-110/10».

Using this formula, we will calculate the current of an industrial enterprise:

$$I = \frac{S}{\sqrt{3} \cdot U_n \cdot cos\varphi} \tag{1}$$

where I - current of industrial enterprises, A; S - full power of industrial enterprises, kVA; U_n - rated voltage, kV; $cos\varphi$ - power factor.

The following is used to calculate the reactive power of industrial plants, kvar

$$Q_r = P_r \cdot tg\varphi \tag{2}$$

where P_r - rated active power, kW; Q_r - rated reactive power, kvar; $tg\varphi$ - active power factor.

The active power factor is calculated using the following formula:

$$tg\varphi = \sqrt{\frac{1 - cos^2\varphi}{cos^2\varphi}} \tag{3}$$

Using the following formula, we will calculate the estimated capacity of the enterprise, kW:

$$P_{\text{p}} = I \cdot U \tag{4}$$

With the formula below, we will now calculate the total capacity of our industrial plants, kVA:

$$S_{\text{p}} = \sqrt{P_{\text{p}}^2 + Q_{\text{p}}^2} \tag{5}$$

3 Calculation Results

Due to the fact that 10 kV is taken on the low voltage side of the supply transformers, the voltage by distributed generation will be taken as 10 kV.

The possibility of using distributed generation sources provides good conditions for the integration of both centralized and decentralized power generation sectors, both for a common network with centralized sources and for individual consumers. In the decentralized sector, distributed generation sources have the advantage, although a local power supply system can also be installed here, combining several distributed sources to work together on the same network.

When introducing distributed generation systems, a potential load is assumed at each power source.

Determination of the compensated reactive power is made by the following formula

$$Q' = P_{\text{p}} \cdot (tg\varphi\pi - tg\varphi e) \tag{6}$$

where Q' - compensated reactive power, kvar; P_p - calculated active power, kW; $tg\varphi\pi$- consumer reactive power factor; $tg\varphi e$ - reactive power factor of the network.

The determination of the compensated reactive power can be calculated using the following formula:

$$Q'' = Q_{\text{p}} - Q' \tag{7}$$

where Q'' - compensated reactive power, kvar; Q_p - rated reactive power, kvar (Table 3).

Table 3. Calculation of the forecast for active and reactive loads, taking into account the length of the power lines of these enterprises.

№	Industrial enterprises	L, km	$cos\varphi$	$tg\varphi$	I, A	P_{p}, kW	Q_{p}, kvar	S_{p}, kVA
1	OJSC JV «JAVONI»	3,35	0,89	0,51	207,83	2078,33	1064,76	2335,2
		1,43						
2	OJSC “Niku Khujand”	1,45	0,88	0,54	82,76	827,64	446,71	940,5
		2,13						
3	OJSC “MMK”	1,87	0,86	0,59	42,34	423,44	251,26	492,38

We calculate the total power according to the following formula, kVA

$$S'' = \sqrt{P_{\text{p}}^2 + Q''^2} \tag{8}$$

where S'' - apparent power, kVA; P_p - calculated active power, kW; Q'' - compensated reactive power, kvar.

The power factor is determined by the following formula

$$cos\varphi = \frac{P_{\text{p}}}{S''} \tag{9}$$

Now let's calculate the current of industrial enterprises after reactive power compensation according to the following formula

$$I'' = \frac{S''}{\sqrt{3} \cdot U_H \cdot cos\varphi} \tag{10}$$

The calculation results and the proposed choice of compensating devices are shown in Table 5.

Table 4. Data after reactive power compensation.

№	Industrial enterprises	*cosφ»*	*tgφ*	*I»*, A	P_p, kW	*Q'*, kvar	$Q»_p$, kvar	S_p», kVA
1	OJSC JV «JAVONI»	0,9	0,51	147,81	2078,33	67,16	997,60	2305,35
2	OJSC "Niku Khujand"	0,9	0,54	58,86	827,64	49,45	397,27	918,05
3	OJSC "MMK"	0,9	0,59	30,12	423,44	48,00	203,25	469,7

Table 5. Selection of power and type of compensating devices (CD).

№	Industrial enterprises	P_p, kW	*Q''*, kvar	*S''*, kVA	*Q'*	Type CD
1	OJSC JV «JAVONI»	2078,33	997,60	2305,35	67,16	AUKRM -0,4–75-10
2	OJSC "Niku Khujand"	827,64	397,27	918,05	49,45	AUKRM -0,4–50-10
3	OJSC "MMK"	423,44	203,25	469,70	48,00	AUKRM -0,4–50-10

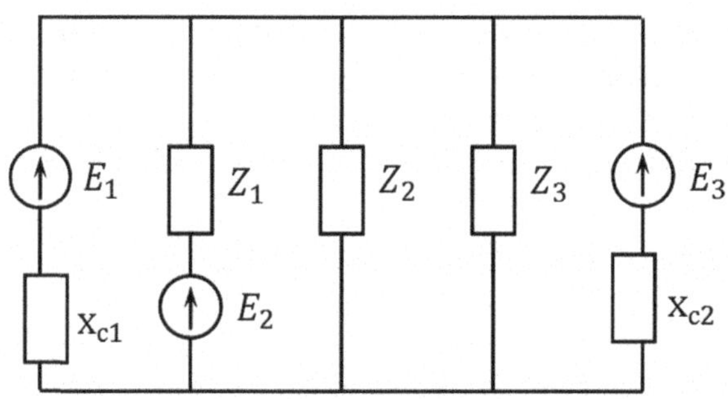

Fig. 3. Scheme taking into account distributed generation sources: Z_1 - total resistance of OJSC JV «JAVONI» Ohm; Z_2 - total resistance of OJSC "NikuKhujand" Ohm; Z_3 - total resistance of OJSC "MMK" Ohm; X_{c1}, X_{c2} - system resistance.

4 Conclusion

For our industrial enterprises, we choose capacitor units of the AUKRM type - a reactive power compensation unit with automatic regulation, for a nominal voltage on the side of 0.4 kV. Based on the location of our factories and sources of electrical energy, the distribution scheme is changed i.e. a common nodal load with three sources of electrical energy is formed between industrial enterprises (Fig. 3 and Table 4).

We will use the two-node method to calculate electrical loads taking into account distributed generation. This method of calculating electrical circuits is based on the adoption of voltage between two nodes of the circuit. When calculating currents by the method of two nodes, the voltage between the nodes is first determined, and then, according to Ohm's law, currents are found for a section of the circuit.

References

1. Hazra, J., Sinha, A.K.: Congestion management using multiobjective particle swarm optimization. IEEE Trans. Power Syst. **22**(4), 1726–1734 (2007)
2. Huang, W., et al.: A review on microgrid technology containing distributed generation system. Power Syst. Technol. **9**, 006 (2009)
3. Kundur, P., Balu, N.J., Lauby, M.G.: Power System Stability and Control, p. 7. McGraw-hill, New York (1994)
4. Khasanov, S.R., Gracheva, E.I., Toshkhodzhaeva, M.I., Dadabaev, S.T., Mirkhalikova, D.S.: Reliability modeling of high-voltage power lines in a sharply continental climate. In: E3S Web Conferences, vol. 178, p. 01051 (2020). https://doi.org/10.1051/e3sconf/202017801051
5. Varganova, A.V.: On methods of optimizing the operating modes of electric power systems and networks. Bull. South Ural State Univ. Ser. Energy **17**, 3 (2017)
6. Filippova, T.A., Rusina, A.G.: Modern concepts for optimizing the regimes of electric power systems. Mater. Russ. Conf. Energy Russ. **21**, 1–4 (2010)
7. Gracheva, E.I., Naumov, O.V.: Vliyanie rezhima raboty naveroyatnostnye kharakteristiki elektricheskikh ustroystv (Operating mode influence on probability characteristics of electric devices). J. Eng. Appl. Sci. **11**(1), 2934–2938 (2016)
8. Gracheva, E.I., Naumov, O.V.: Estimation of power losses in electric devices of the electrotechnical complex. In: International Conference on Industrial Engineering. Applications and Manufacturing (ICIEAM), 6 (2019)
9. Tolibjonovich, D.S., Islomovna, T.M., Saidulloevna, M.D.: Modeling of starting transition processes of asynchronous motors with reduced voltage of the supply network. Eur. J. Electr. Eng. **22**(1), 23–28 (2020). https://doi.org/10.18280/ejee.220103
10. Toshkhodzhaeva, M.I.: Sravnitel'nyy analiz mekhanicheskikh svoystv traditsionnykh i vysokotemperaturnykh provodov vlep – 110 kV (Comparative analysis of the mechanical properties of traditional and high-temperature vlep wires – 110 kV). Izvestiya Tul'skogogo sudarstvennogo universiteta, Tekhnicheskienauki **3**, 169–175 (2017)
11. Toshkhodzhaeva, M.I., Khodzhiev, A.A.: Osobennosti diagnostirovaniya vlep-110 kv v usloviyakh rezkokontinental'nogo klimata (Features of diagnosing high-voltage power lines of 110 kV in a sharply continental climate). Izvestiya Tul'skogogo sudarstvennogo universiteta, Tekhnicheskienauki **2**, 364–369 (2019)
12. Rakhimov, O.S., Khodzhiev, A.A., Toshkhodzhaeva, M.I.: Povyshenie nadezhnosti vlep-110 kv na stadii proektirovaniya i ekspluatatsii (Improving the reliability of 110 kV high-voltage power lines at the design and operation stages). Elektrooborudovanie: ekspluatatsiya i remont **3**, 55–57 (2017)
13. Amin, M., Stringer, J.: The electric power grid: today and tomorrow. MRS Bull. **33**(04), 399–407 (2008)
14. Athas, W.C., et al.: Low–power digital systems based on adiabatic–switching principles. IEEE Trans. Very Large Scale Integr. (VLSI) Syst. **2**(4), 398–407 (1994)
15. Barker, P.P., De Mello, R.W.: Determining the impact of distributed generation on power systems. I. Radial distribution systems. In: Power Engineering Society Summer Meeting, IEEE, vol. 3, pp. 1645–1656 (2000)
16. Benini, L., Bogliolo, A., De Micheli, G.: A survey of design techniques for system-level dynamic power management. IEEE Trans. Very Large Scale Integr. (VLSI) Syst. **8**(3), 299–316 (2000)

17. Gregorio, P., Ahmadi, M., Buehler, M.: Design, control, and energetics of an electrically actuated legged robot. IEEE Trans. Syst. Man Cybern. Part B (Cybern.) **27**(4), 626–634 (1997)
18. Henkel, J.: A low power hardware/software partitioning approach for core-based embedded systems. In: Proceedings of the 36th Annual ACM/IEEE Design Automation Conference, Berlin, pp. 122–127. ACM (1999)
19. Kiessling, F., et al.: Overhead Power Lines: Planning, Design, Construction. Springer, Heidelberg, p. 321 (2014)
20. Lakervi, E., Holmes, E.J.: Electricity distribution network design. IET **21**, 130–142 (1995)
21. Reddy, T.A.: Applied data analysis and modeling for energy engineers and scientists. Springer, Boston, 320 (2011)
22. Schlichting, J.: Molybdenum disilicide as a component of modern high-temperature composites. High Temp. High Pressures **10**(3), 241–269 (1978)

Composite Rubber-Polymer Binder in the Design and Construction of Amur Bridge

S. I. Dubina[1], A. V. Sorokin[1], V. A. Lobachev[1], R. M. Dzhafarov[2], V. A. Kudryavtsev[3], L. V. Bezshtanko[3], V. G. Nikolskii[4], I. A. Krasotkina[4], and T. V. Dudareva[4(✉)]

[1] "Energotex" JSC, Kurchatov, Russian Federation
[2] "Institute Giprostroymost" JSC, Moscow, Russian Federation
[3] "Asfalt" JSC, Blagoveshchensk, Russian Federation
[4] N.N. Semenov Federal Research Center for Chemical Physics, Russian Academy of Sciences, Moscow, Russian Federation
yanadva@mail.ru

Abstract. In 2019 composite rubber-polymer bitumen binder (CRPBB) was used in the design and construction of bridge road pavement over the Amur River (Heilongjiang) near the cities of Blagoveshchensk (Russia) and Heihe (China). CRPBB was prepared by low-shear mixing of bitumen BND 100/130 and 11% powder rubber-polymer modifier (PRPM) at a temperature of 135 ÷ 140 °C for 1 ÷ 1,5 h. PRPM was obtained by high-temperature shear co-grinding of Crumb Tire Rubber and thermoplastic elastomer – the styrene-butadiene-styrene block copolymer. CRPBB had storage stability and was used to prepare polymer-rubberized stone mastic asphalt SMA-15 without any stabilizing additive. The temperature-time regimes for mixture preparation, construction and compaction of the road pavement with CRPBB were the same as for mixture without PRPM. Comparative rheological studies of CRPBB and bitumen BND 100/130 were carried out using the Superpave methodology and standards. Based on the results of laboratory tests and monitoring data, it is possible to predict the high resistance of the pavement to plastic deformation (rut), fatigue and low-temperature cracking.

Keywords: Design and construction of bridge road pavement · Composite rubber-polymer bitumen binder · Powder rubber-polymer modifier · Stone mastic asphalt mix

1 Introduction

Modern bridge structures on highways and roads to them, which are part of federal highways, operate under significant loads caused by heavy traffic with a predominance of heavy vehicles. When developing design documentation for the construction of such structures, it is necessary to search and substantiate optimal technical and economic solutions, including the choice of modern innovative road building materials.

In 2018–2019, the Russian Federation carried out the design and construction of a road bridge over the Amur River (Heilongjiang) near the cities of Blagoveshchensk (Russia) - Heihe (China).

A. Sinitsyn (Ed.): SES 2020, LNCE 141, pp. 119–128, 2021.
https://doi.org/10.1007/978-3-030-67654-4_14

The Amur Region, in which the object is located, is predominantly characterized by a continental climate, with significant daily temperature drops up to 25 °C, long and cold winters (cold five-day −38 °C) and short but hot summers with temperatures on the surface of the pavement up to 63 °C. Road clothes for the Amursky bridge were designed taking into account the high traffic intensity with a predominance of heavy trucks, while it was necessary to take into account the presence of sections characterized by low speed, with frequent braking, stopping and accelerating. Therefore, it was proposed the design solution proposed to use a modified binder with increased shear stability.

In world practice, improvement of the rheological properties of bitumen is achieved through the use of various modifiers [1, 2], among which the most common are polymer-modified binders based on block copolymers of the styrene-butadiene-styrene (SBS) type and rubber-bitumen binders. Various solutions are known for the application of the so-called hybrid binders representing various combinations of SBS, tire rubber and other modifiers [3–8]. The production of modified binders, as a rule, is carried out for a long time (1 ÷ 3 h) at high temperatures (170 ÷ 190 °C) using special equipment [9–12]

Before the start of work on the bridge over the Amur River, w carried out preliminary studies of stone mastic asphalt SMA-15 samples containing various binders, including: BND 100/130 bitumen, PMB90 and composite binders based on powder rubber modifiers (PRM) and powder rubber-polymer modifier (PRPM).

At the same time were studied rheologycal properties of bitumen BND 100/130 and modified binder. Based on the obtained results were accepted the decision to use a composite rubber-polymer binder based on the PRPM.

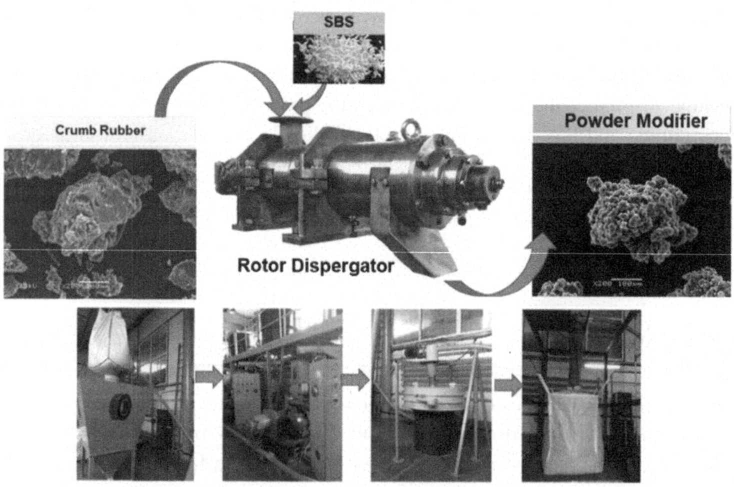

Fig. 1. Production line for obtaining the powder rubber-polymer modifier (PRPM).

The basis of the PRM modifier is an active powder of discretely devulcanized tire rubber (APDDR), obtained by the method of high-temperature shear grinding (HTSG) of crump rubber of waste tire in a rotor dispergator [13]. PRPM additionally contains block copolymers of the styrene-butadiene-styrene (SBS) type and is a homogeneous product of co-grinding of crumb rubber and SBS by the HTSG method (see Fig. 1). The presence of SBS in the composition of PRPM provides increased, in comparison with PRM, the stability of the binder to plastic deformation (rutting). The technology of high-temperature shear grinding and co-grinding is characterized by energy efficiency in comparison with other methods of rubber processing of used car tires.

The typical images of the surface of powder modifiers particles obtained using the method of scanning electron microscopy are shown in Fig. 2. Modifiers are characterized by a structure with a large number of macro and micro pores, which allows them to quickly decomposition in hot bitumen into micro and nanofragments, thereby providing improving the resistance to rutting, fatigue and low-temperature cracking resistance of bitumen [14–16].

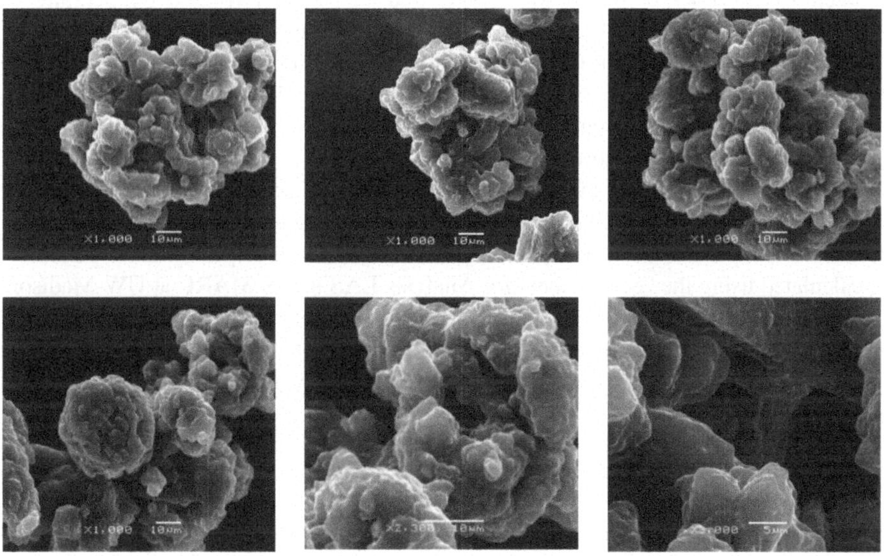

Fig. 2. CEM images of powder modifier particles at different magnifications.

The special structure of PRM and PRPM, obtained by the HTSG, allows them to be used for modifying bitumen directly at the time of mixing aggregates and bitumen, or used for preparing a modified binder.

This paper presents the results of comparative rheological tests of bitumen and a composite rubber-polymer binder based on PRPM using the Superpave methodology and standards. The technology for preparing CRPBB and construction the pavement on the bridge over the Amur River are described.

2 Materials and Methods

It was studied air-blown bitumen BND 100/130, the properties of which are presented in Table 1.

Table 1. Physical properties of bitumen BND 100/130.

Penetration@ 25 °C, 0.1 mm	Soft. Point, R&B, °C	Fraass Breaking Point,°C
120	45	−25

It was used powder rubber-polymer modifier POLYEPOR-RP (PRPM) obtained by High-Temperature Shear co-Grinding Crumb Rubber of waste tire and thermoplastic elastomer – the styrene-butadiene-styrene block copolymer. An average particle size of modifier particles is 0.25 mm, max size 0.63 mm.

Model composite rubber-polymer bitumen binder (MB) were prepared by blending bitumen BND 100/130 heated to 160 °C and 12,5% PRPM at a temperature of 160 ± 3° C for 3 min at a speed of rotation of a blade mixer 600 ± 50 rpm.

Rheological tests were performed for original, RTFO-aged [17] and PAV-aged [18] binders using “Anton Paar” “Smart “Pave 101” dynamic shear rheometer (DSR) with parallel plate geometry: 25 mm for rutting tests [19, 20], 8-mm for fatigue tests [19, 22] and 4-mm for low temperature tests [23].

The Linear Amplitude Sweep test (LAS-test) was performed according to the Johnson [21] starting from 19 °C and then with decreasing temperature with a step of 3 °C. A new sample was used at each temperature. Binder fatigue performance parameter N_f calculated using the “Spreadsheet for Analysis LAS test of MARC at UW-Madison. T_{CrLAS} was determined by the deviation from the straight line of the Deflection angle (α) versus the amplitude of strain (γ) in the range γ from 0.1 to 5% and by the appearance of a plateau on the curve of shear stress (τ) versus γ.

3 Experimental Results

Figure 3 shows the results of the rutting resistance test (MSCR test) for bitumen and Model composite rubber-polymer bitumen binder (MB) at 58 °C. It can be seen that even at this temperature bitumen does not provide the required un-recoverable creep compliance (Jnr3.2 < 0.5) for roads with extreme traffic. The introduction of the PRPM reduces Jnr3.2 and Jnr10 to values less than 0.5 kPa^{-1} at temperatures up to 70 °C.

Figure 4 shows the expansion of the temperature range of plasticity according to the results of comparative tests of bitumen and MB. The rutting parameter temperature ($G^*/\sin\delta_{RTFO}$ = 2,2 kPa) increased from 63,4 to 93,6 °C.

Determination of Intermedia temperature (T_{Int}) (RTFO + PAV)-aged binders corresponding to the value of the loss modulus $G^*\cdot\sin\delta_{PAV}$ = 5 MPa showed that the introduction of the modifier led to its decrease from 7,1 °C to 0, 5 °C. Decrease in sample cracking temperature during the LAS test (T_{CrLAS}) by 6 °C is also seen.

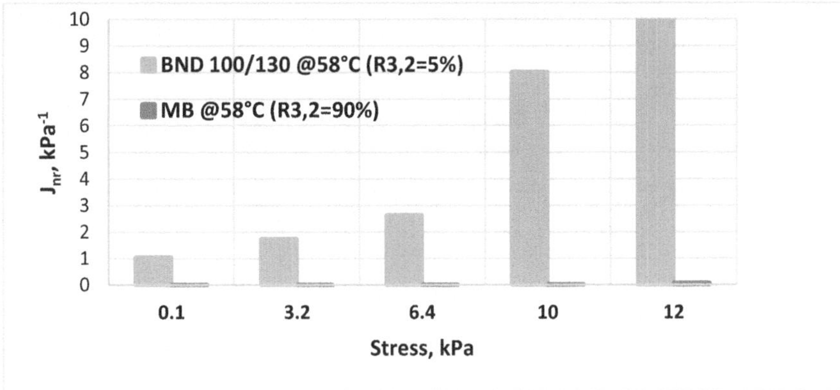

Fig. 3. Un-recoverable creep compliance (Jnr) versus load.

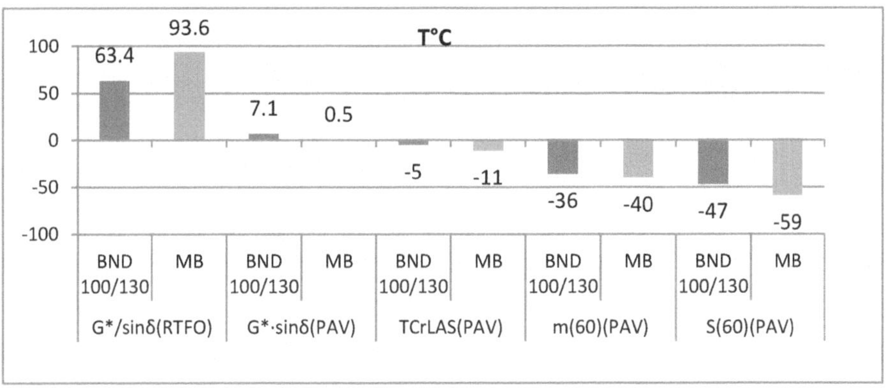

Fig. 4. Change in plasticity interval as a result of modification.

Also, as a result of the modification, there was a decrease low PG temperature determined by the m-value and the creep stiffness S(60) by 6 and 12 °C, respectively.

Figure 5 shows the data the relaxation modulus G(60) and the mr-value obtained during low-temperature tests (DCR 4 mm test) of (RTFO + PAV)-aged binders. The values of the relaxation modulus G(60), which determines the rigidity for MB, are lower than for bitumen in the entire temperature range of tests. Note that the m_r-value for MB at a temperature of −28 °C was higher (better) than that of bitumen. In addition, it can be seen that G(60) and m_r MB are less sensitive to temperature changes compared to bitumen.

The binder fatigue performance parameter N_f for the RTFO-aged and PAV-aged binders (LAS-test) was estimated in the temperature range from −2 to 16 °C. The increase in N_f as a result of the modification was from 3 to 12 times.

Figure 6 shows the curves of the complex modulus (G*) (a) and shear stress (τ) (b) for (RTFO + PAV)-aged bitumen and MB obtained in the LAS test at temperatures of 4 and 16 °C. It can be seen that a sharp decrease in G* at $\gamma > 10 \div 15\%$ which is

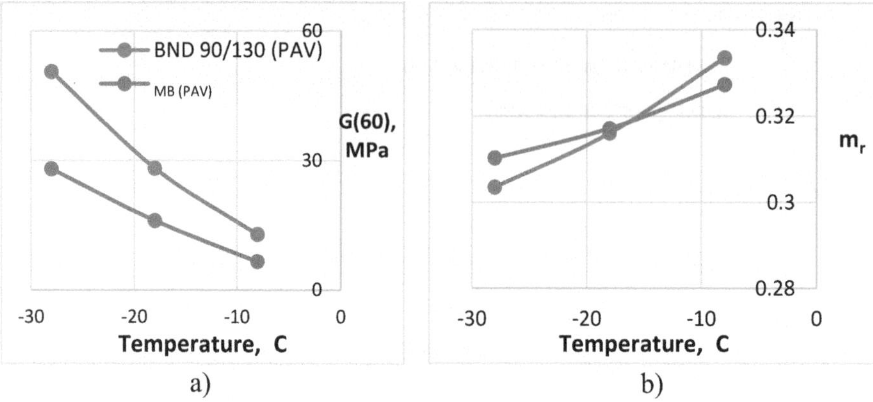

Fig. 5. Relaxation modulus G(60) (a) and m_r-value (b) versus temperature of (RTFO + PAV)-aged binders.

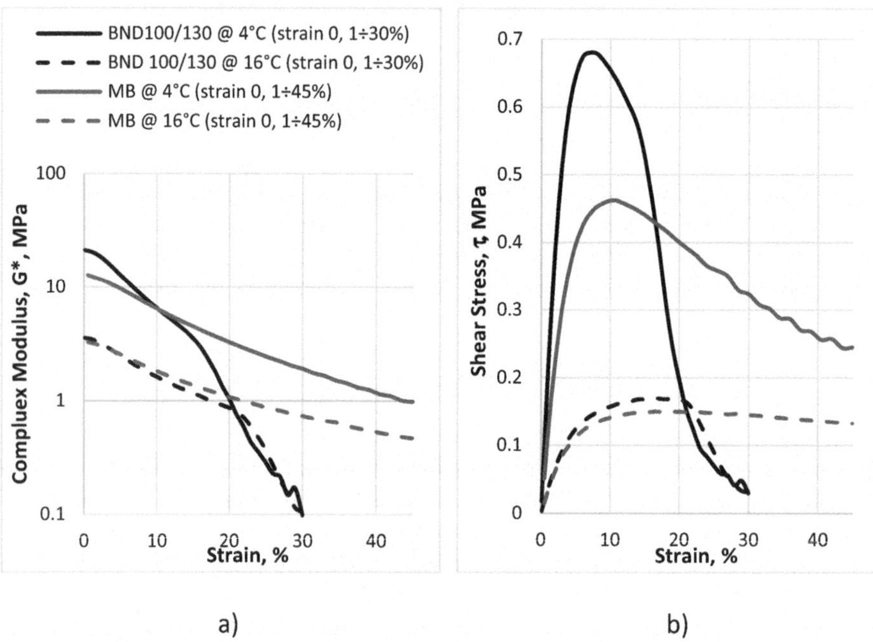

Fig. 6. Complex modulus (G*) (a) and shear stress (τ) (b) for (RTFO + PAV)-aged bitumen and MB at temperatures of 4 and 16 °C (LAS test).

observed for bitumen, is absent in MB even at 45% strain. Therefore, it can be suggested about the strength of the elastic mesh created by the modifier in the binder.

Table 2 shows the PG grade for bitumen and MB taking into account the traffic intensity at a load of 3.2 kPa and 10 kPa.

Table 2. PG Grade.

Binder	Low-temperature by m-value	Low-temperature by S(60)
BND 100/130	PG 58H(-)-34	PG 58H(-)-46
MB	PG 64E(E)-40	PG 64E(E)-58

The using of PRPM in the composition of stone-mastic asphalt concrete was preceded by preliminary work on the organization of the bitumen base of the asphalt-concrete plant. The original "wet method" for CRPBB preparation at 135 ÷ 140 °C for 60 ÷ 90 min was developed. A tank with a volume of 30 m^3 was used, equipped with 3 planetary mixing units. The modifier was dosed by a dispenser. Thus, a high-quality and continuous daily production of modified stone-mastic asphalt concrete mixture was ensured (see Fig. 7).

Fig. 7. Dosing of POLYEPOR-RP (a); asphalt-concrete plant with tank (b) (JST "Asphalt").

The asphalt concrete mixture with CRPB was prepared without changes in temperature and time conditions. Compaction of the mixture was carried out according to a typical technological scheme with an asphalt paver with compaction with 8 t and 13.5 t smooth drum rollers. Figure 8 shows the pavement SMA-15 with CRPBB on the Amur Bridge immediately after compaction.

The results of laboratory tests of stone mastic asphalt concrete SMA-15 with CRPBB are presented in the Table 3.

Fig. 8. Pavement with CRPBB on Amur bridge.

Table 3. The results of laboratory tests of stone mastic asphalt concrete SMA-15 with CRPBB.

No	Characteristic	Unit	Requirement, GOST 31015-2002	Actual value
1	Porosity of the mineral part	%	15÷19	17,3
2	VFB value (in place air voids)	%	1,5-4,5	4,5
3	Dry density (on recipe)	g/cm³	not standardized	2,33
4	Water Absorption Test	%	1,0-4,0	2,2
5	Compressive strength at a temperature of 20°C 50°C	MPa	 no less than 2,2 no less than 0,65	 2,9 1,1
6	Water resistance grade	-	no less than 0,85	0,93
7	Shear resistance according to the coefficient of internal friction	-	no less than 0,93	0,95
8	Shear grip at a temperature of 50°C	MPa	no less than 0,18	0,28
9	Crack resistance: breaking tensile strength at a temperature of 0°C	MPa	2,5-6,0	3,1
10	Grip of bitumen on minerals	-	must withstand	withstand
11	The quality of adhesion of the bituminous binder to the surface of crushed stone	mark	no less than 4 (good)	4
12	Resistance of the mixture to delamination in terms of bitumen runoff	%	0,2	0,07

Table 4 shows the test results of stone-mastic asphalt concrete cores with CRPBB.

Table 4. The test results of stone-mastic asphalt concrete cores with CRPBB.

No	Characteristic	Unit	Requirements, GOST 31015-2002	Actual value
1	Average density of cores	g/cm^3	Not standardized	2,43
2	Core water saturation	%	No more than 3,5	1,6
3	Layer thickness	cm	5	5,4
4	Adhesion of layers	–	+	+

The minimum temperature in the Amur Bridge region during the winter 2019/2020 was −33 °C, and the temperature change during the day was observed from 2 ÷ 5 °C to −20 °C. Pavement monitoring did not reveal any defects.

4 Conclusion

A distinctive feature in the construction of the pavement of the international road bridge between China and Russia was the using of the powder rubber-polymer modifier obtained by the method of high-temperature shear co-grinding of rubber crumb of worn-out tires and a styrene-butadiene-styrene block copolymer.

For the first time in Russia, the technology of obtaining a composite rubber-polymer-bitumen binder (CRPBB) at temperatures not exceeding 135 ÷ 140 °C was applied. CRPBB has an extended plasticity range, which allows it to be used in the climatic conditions of this region.

SMA-15 with CRPBB was prepared without changes in temperature and time regimes. The pavement was compacted according to a typical technological scheme.

Based on the results of laboratory tests and monitoring data, it is possible to predict the high resistance of the pavement to plastic deformation (rutting), fatigue and low-temperature cracking.

5 Acknowledgements

This work was supported into frameworks of the state task for ICP RAS 0082-2019-0009. State registration number. AAAA-A20-120021390044-2.

References

1. Porto, M., Caputo, P., Loise, V., Eskandarsefat S., Teltayev, B., Rossi, C.O.: Bitumen and bitumen modification: a review on latest advances. Appl. Sci. **9**(4), 742 (2019). https://doi.org/10.3390/app9040742
2. Yildirim, Y.: Polymer modified asphalt binders. Constr. Build. Mater. **1**(21), 66–72 (2007). https://doi.org/10.1016/j.conbuildmat.2005.07.007
3. Senise, S., Carrera, V., et al.: Hybrid rubberised bitumen from reactive and non-reactive ethylene copolymers. Polymers (Basel). **11**(12), (2019). https://doi.org/10.3390/polym11121974

4. Ezzat, E.N., Abed, A.H.: Enhancement rheological properties of asphalt binder modified with hybrid polymers according to superpave system. Mater. Today Proc. **20**, 572–578 (2020). https://doi.org/10.1016/j.matpr.2019.09.191
5. Rath, P., Love, E.J., Buttlar, W.G., Reis, H.: Performance analysis of asphalt mixtures modified with ground tire rubber modifiers and recycled materials. Sustainability **11**, 1792 (2019). https://doi.org/10.3390/su11061792
6. Sazbocs, B., et al.: Asphalt rubber versus other modified bitumens. In: Proceeding of the Asphalt Rubber 2012 Conference, Germany (2012)
7. Roque, R., Lopp, G., et al.: Evaluation of Hybrid Binder for Use in Surface Mixtures in Florida. University of Florida, 181 p. (2009)
8. Chen, Z., Pei, J., Amirkhanian, S.: Low-temperature characteristic evaluations of base and hybrid asphalt binders. J. Test. Eval. **48**(3), 1962–1974 (2020). https://doi.org/10.1520/JTE20190154
9. Han, L., Zheng, M., Wang, C.: Current status and development of terminal blend Tire rubber modified asphalt. Constr. Build. Mater. **128**, 399–409 (2016)
10. Presti, D.L.: Recycled Tire rubber modified bitumens for road asphalt mixtures: a literature review. Constr. Build. Mater. **49**, 863–881 (2013)
11. Lu, X., Isacsson, U.: Modification of road bitumens with thermoplastic polymers. Polym. Test. **20**, 77–78 (2001)
12. Fawcett, A.H., McNally, T.: Blends of bitumen with polymer having a styrene component. Polym. Eng. Sci. **41**, 1251–1264 (2001)
13. Sorokin, A.V., i dr.: Patent RF 2612637 Metod polucheniya visokodispersnogo materiala i ustroistvo dlya ego osuschestvleniya. [RF Patent 2612637 A method for producing a highly dispersed polymer material and a device for its implementation.] (2017). (in Russian)
14. Berlin, A.A., et.al.: Recycling of waste rubber and the active powder of discrete devulcanized rubber. Polym. Sci. Ser. D. Glues Sealing Mater. **3**(11), 323–329 (2018). https://doi.org/10.3303/cet1865062
15. Gorbatova, V.N., et al.: Composite bitumen binders based on an active powder of discretely devulcanized tire rubber. In: Key Engineering Materials. Materials Science Forum, XVI International Scientific and Practical Conference "New Polymer Composite Materials", vol. 869, pp. 146–151. Trans Tech Publications Ltd., Switzerland (2020). ISSN 1662–9795. https://doi.org/10.4028/www.scientific.net/KEM.869.146
16. Chekunaev, N.I., Kaplan, A.M.: On increase of strength and fatigue life of asphalt composites by introducing ultra-small particles of elastomers. Procedia Mater. Sci. **3**, 698–705 (2014)
17. AASHTO T 240-13 Standard Method of Test for Effect of Heat and Air on a Moving Film of Asphalt (Rolling Thin-Film Oven Test) (2017)
18. AASHTO R 28-12 Accelerated Aging of Asphalt Binder Using a Pressure Aging Vessel (PAV) (2016)
19. AASHTO T315-12 Standard Method of Test for Determining the Rheological Properties of Asphalt Binder Using a Dynamic Shear Rheometer (DSR) (2016)
20. AASHTO TP 70-13 Standard Method of Test for Multiple Stress Creep Recovery (MSCR) Test of Asphalt Binder Using a Dynamic Shear Rheometer (DSR) (2013)
21. Jonson, C.M.: Estimating asphalt binder fatigue resistance using an accelerated test method. Ph.D. thesis, University of Wisconsin-Madison, Madison (2011)
22. AASHTO TP 101-12 Standard Method of Test for Estimating Fatigue Resistance of Asphalt Binders Using the Linear Amplitude Sweep (2018)
23. Farrar, MJ., Sui, C., Salmans, S., Quin, Q.: Determining the low-temperature rheological properties of asphalt binder using a dynamic shear rheometer (DSR), Technical White Paper, Laramie, WY (2015)

Study of the Influence of Various Factors on the Air Temperature in the Greenhouse with Radiant Heating

Mikhail Pavlov[1(✉)], Denis Karpov[1], Anton Mikhin[2], and Timur Akhmetov[3]

[1] Vologda State University, Lenina Street 15, Vologda 160000, Russia
pavlov_kaftgv@mail.ru
[2] Peter the Great St. Petersburg Polytechnic University, Saint Petersburg, Russia
[3] Kazan State Power Engineering University, Str. Krasnoselskaya, 51, 420066 Kazan, Russia

Abstract. The main role in obtaining a rich harvest of vegetables and fruits in the winter period of the year is played by the parameters of the greenhouse microclimate: the internal air temperature, its relative humidity and speed. The favourable conditions for growing plants in greenhouses can be achieved only through the use of modern, high-tech heating systems that will compensate for the loss of thermal energy and provide a given thermal regime in the building. Radiant heating can be successfully used for these purposes. This type of heating, using ceiling infrared emitters as heat sources, directs the required heat flux directly to plants and soil, thus heating them. At the same time, direct or indirect emissions of harmful substances into the environment are minimal, and in view of the absence of heat losses during the production of heat and transportation of the coolant, this type of heating is efficient from both energy-saving and economic points of view. This paper presents a developed mathematical model of the greenhouse radiant heating, and considers the influence of meteorological, aerodynamic, thermal engineering and other factors on the air temperature in the cultivation structure is investigated.

Keywords: Radiant heating · Green-house microclimate

1 Introduction

Industrial greenhouses are widely used for year-round cultivation of plants in protected ground [1–4]. It is obvious that vegetable growing can be efficiently carried out only if favourable temperature and humidity conditions are created in cultivation facilities during the entire cold season [5–7]. For this, various types of heating systems are used in greenhouses: water (or steam), air, combined water or steam-air and gas-air. Greenhouses can also be heated using electrical energy. As for every building and construction facility, the issues of energy savings for greenhouse powering are highly relevant now [8–11]. So, in addition to the listed traditional heating systems, alternative options for heat supply to greenhouses are also organized, for example, using heat pumps [12] or solar collectors (solar greenhouses) [13].

A. Sinitsyn (Ed.): SES 2020, LNCE 141, pp. 129–135, 2021.
https://doi.org/10.1007/978-3-030-67654-4_15

Currently, radiant heating systems are of great interest in agriculture for a number of reasons [14, 15]:

- It is an artificial source of infrared radiation;
- There is no need for a boiler plant, heating networks and related equipment;
- The technological process is automatically controlled;
- The heat energy is uniformly distributed on the soil surface;
- The carbon dioxide (CO2) from combustion products using gas infrared emitters can be potentially used for plant nutrition.

To predict, maintain and control the thermal regime of the greenhouse one should calculate all incoming and outgoing flows of heat and mass from the construction. The existing mathematical models, for example [16], describe the cases of heat and mass transfer without taking into account a number of important factors, such as the ability of the soil surface to absorb heat energy or the irrigated area.

The aim of this work is to study the influence of various kinds of factors (external, internal) on the temperature of the internal air in the greenhouse. The achieved results will enable preventing the decrease and increase in the air temperature relative to the required value and, as a consequence, will protect the grown plants from possible diseases and death.

2 Materials and Methods

The algorithm for calculation the internal air temperature in the greenhouse, realized using the Mathcad software, includes the following main stages:

1. Setting the initial calculation parameters (greenhouse size, materials, type of ventilation, meteorological conditions, etc.).
2. Calculation of design indicators according to the initial data (for example, the area of the greenhouse enclosure or the flow rate of external air entering the greenhouse).
3. Determination of heat and mass flows (for example, heat loss through the greenhouse enclosure or moisture evaporation from the soil surface).
4. Solution of the system of interrelated equations of heat and material balances of the greenhouse premises, its enclosure and soil.
5. Finding the required parameters, the key of which is the temperature of the internal air in the greenhouse.

The developed mathematical model was used to study the influence of various factors on the temperature of the internal air of the greenhouse. The consideration was carried out using the example of the industrial greenhouse "Farmer 7.5" of the company "Volya" (Krasnogorsk district, Moscow region).

3 Results and Discussion

The results of the software calculation indicate a linear relationship between the temperature of the internal air in the greenhouse T_{int_air}, °C, and the temperature of the external air T_{ext_air}, °C at different air exchange rates in the room n_{air}, m^3/h per 1 m^2, Fig. 1.

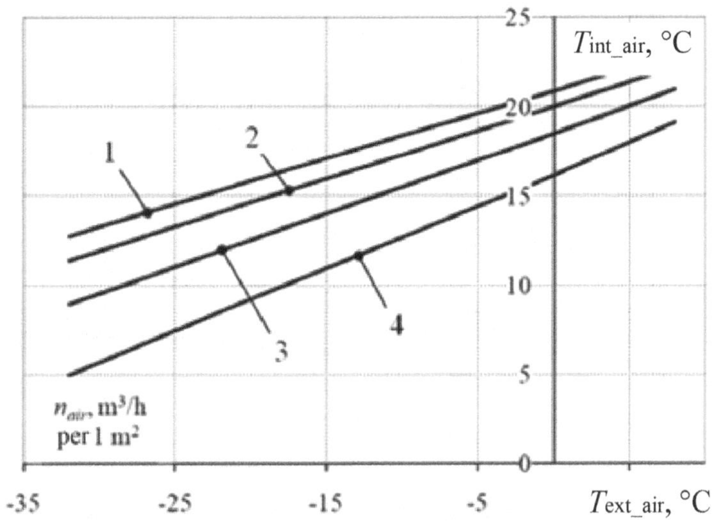

Fig. 1. Internal air temperature as a function of external air temperature for various air exchange rates: 1 – n_{air} = 2.5 m^3/h per 1 m^2; 2 – n_{air} = 5.0 m^3/h per 1 m^2; 3 – n_{air} = 10.0 m^3/h per 1 m^2; 4 – n_{air} = 20.0 m^3/h per 1 m^2.

Figure 1 clearly shows that with a decrease of external air temperature T_{ext_air}, the internal air temperature in the greenhouse T_{int_air} decreases too, and vice versa. For example for the air exchange rate n_{air} = 2.5 m^3/h per 1 m^2 and the external air temperature T_{ext_air} = 8 °C, the internal air temperature in the greenhouse is T_{int_air} = 23.0 °C. For the external air temperature of $T_{ext\_air} = -32$ °C it will be T_{int_air} = 12.8 °C, i.e. will decrease by 10.2 °C. With an increase in the air exchange rate the change in internal air temperature will occur more intensively, and vice versa. For the external air temperature $T_{ext\_air} \in [-32; 8]$ °C the following results were obtained: 0.26 °C per 1 °C of change in external air temperature for n_{air} = 2.5 m^3/h per 1 m^2; 0.27 °C per 1 °C of change in external air temperature for n_{air} = 5.0 m^3/h per 1 m^2; 0.30 °C per 1 °C of change in external air temperature for n_{air} = 10.0 m^3/h per 1 m; and 0.35 °C for n_{air} = 20.0 m^3/h per 1 m^2.

In traditional heating systems the maintenance of the internal air temperature at the level of optimal or permissible values can be performed in two ways. The first implies changing the surface temperature of the heating devices by changing the temperature of the coolant (water heating). The second way is by regulating the temperature of the preheated supply air (air heating). Unlike traditional heating systems, during radiant heating without any additional sources of heat, it is quite difficult to achieve a constant thermal regime of the room. Here, provided that the specified temperature of the irradiated surface (in this case, the soil surface) is maintained, the temperature of the internal air is a dependent value. And here, as in any power engineering system, the issue of heat losses should be precisely studied [17, 18]. With a decrease in the external air temperature, heat losses through the room enclosures increase in the first instance, which leads to a decrease in the internal air temperature. With an increase in the greenhouse air exchange rate, there is a significant increase in heat losses with the ventilation air leaving the room, which also affects a decrease in the temperature of the internal air.

The thermal resistance of the enclosure R_{enc}, $m^2{\cdot}k/W$, plays an important role in creating the required microclimate in the greenhouse (Fig. 2).

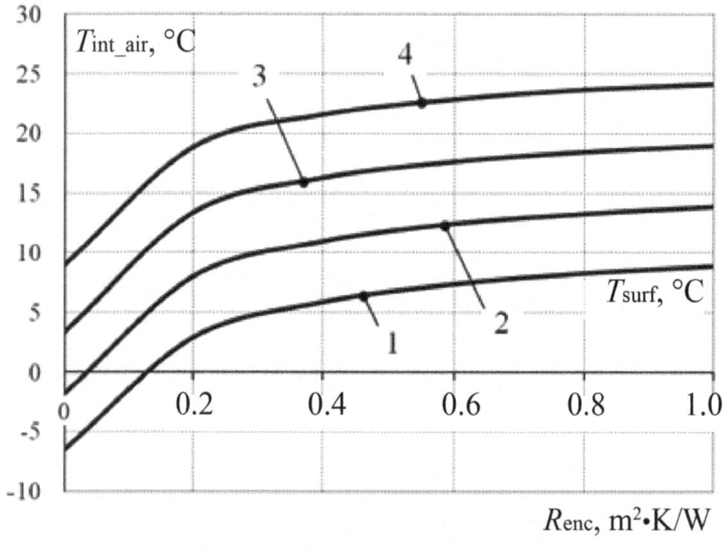

Fig. 2. The internal air temperature as a function of the thermal resistance of the enclosure R_{enc} for: 1 – T_{surf} = 15 °C; 2 – T_{surf} = 20 °C; 3 – T_{surf} = 25 °C; 4 – T_{surf} = 30 °C.

Figure 2 shows that the use of ultra-thin translucent covering materials in the cold season is impractical. For example, polyethylene film, regardless of its thickness (usually from 100 to 200 m for greenhouses) has $R_{enc} \approx 0$. For the calculated external air temperature $T_{ext\_air} = -32$ °C, the following calculation results were obtained: for the soil surface temperature T_{surf} = 15 °C, the internal air temperature will be T_{surf} = −6.5 °C; for T_{surf} = 20 °C, $T_{int\_air} = -1.8$ °C; for T_{surf} = 25, T_{int_air} = 3.4; for T_{surf} = 30 °C, T_{int_air} = 9.0 °C. It is obvious that in all cases such a thermal regime in a greenhouse is unacceptable for growing plants.

If we consider cellular polycarbonate with a thickness of 10 mm as a covering material, then the results of calculations for the cold season are quite acceptable for some cases (at $R^{r}_{enc} = 0.2$ $m^2{\cdot}k/W$): for T_{surf} = 15 °C the internal air temperature will be $T_{int\_air} = 2.9$ °C; for $T_{surf} = 20$ °C, $T_{int\_air} = 8.0$ °C; for $T_{surf} = 25$ °C, $T_{int\_air} = 13.4$°C; for $T_{surf} = 30$ °C, $T_{int\_air} = 18.9$ °C. In the range of values $R_{enc} \in [0; 0,2]$ m^2·k/W, there is a linear increase in the internal air temperature T_{int_air}. In the rest part of the graph, a gradual increase in air temperature in the greenhouse with an increase in the heat-shielding qualities of its enclosing structures can be noted. This situation can be explained as follows: the temperature of the internal air depends, first of all, on the heat losses by heat transfer. For example, in the range of $R_{enc} \in [0; 0.2]$ m^2·k/W, in all cases, they decrease by more than two times.

Further, the influence of the enclosure thermal resistance on the heat loss from the room is much less. During the design of traditional heating, at the stage of heat engineering calculation the internal air temperature is preset, after which the required thermal resistance of the enclosure is determined under design conditions (usually it is its thickness with a known thermal conductivity of the material). In case of radiant heating, as mentioned earlier, the room temperature is a dependent variable. Therefore, the construction of a graph (Fig. 2) at the design temperature of the external air for a given construction area can become the final part of the thermal engineering calculation of the enclosing structures.

Consider this variant of the heat engineering calculation using the example of the Farmer 7.5 industrial greenhouse. Assume that the following temperature conditions in the greenhouse are required: $T_{surf} = 25$ °C and $T_{int\_air} = 15°$ C. According to Fig. 2, the thermal resistance of the greenhouse enclosure should be $R_{enc} \approx 0.27$ m^2·k/W. If cellular polycarbonate is supposed to be used as a covering material, then its thickness should be at least 16 mm.

With an increase in the irrigation coefficient k_{irr}, a noticeable increase in the air temperature in the greenhouse is observed (Fig. 3). For example, at a fixed value $A_1 = 65\%$, the following results are obtained: for $k_{irr} = 0$, the internal air temperature will be equal to $T_{int\_air} = 4.4$°C; for $k_{irr} = 0.4$, $T_{int\_air} = 13.4$ °C; for $k_{irr} = 0.8$, $T_{int\_air} = 17.5$ °C; for $k_{irr} = 1.0$, $T_{int\_air} = 18.9$ °C. Thus, the temperature of the internal air in the range of values of the soil irrigation coefficient $k_{irr} \in [0; 1]$ for the absorption coefficient of its surface $A_1 = 65\%$ will increase by 14.5 °C.

With an increase in the area of soil irrigation, moisture on its surface will begin to actively evaporate into the greenhouse, since the temperature of the soil surface with radiant heating is higher than the temperature of the internal air. As a result, the air will be saturated with water vapour and its enthalpy will increase (therefore, along with this, heat losses with the ventilation air leaving the greenhouse will also increase). Since the air temperature in the greenhouse is lower than the temperature of the soil surface, some of water vapour will condense. This will lead to saturation of the air with water vapour and an increase in its temperature. The absorption coefficient of the soil surface A does not have such a strong influence on the change in the temperature of the indoor air, as the irrigation coefficient k_{irr}. In any case, the more heat the soil surface absorbs, the lower the air temperature in the greenhouse will be at a constant value of k_{irr}. For example at $k_{irr} = 0.4$ and for $A = 65\%$, the air temperature will be equal to $T_{int\_air} = 13.4$ °C, for $A_1 = 95\% - T_{int\_air} = 10.9$ °C.

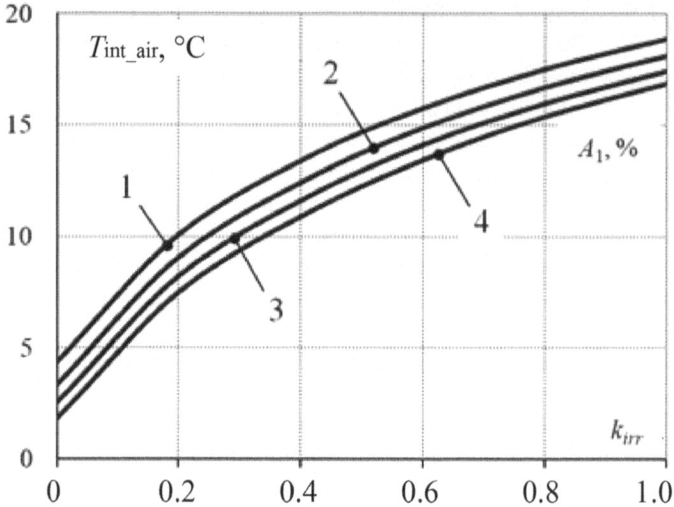

Fig. 3. The internal air temperature as a function of irrigation coefficient k_{irr} for: 1 – A_1 = 65%; 2 – A_1 = 75%; 3 – A_1 = 85%; 4 – A_1 = 95%.

4 Conclusion

The performed work allows one to assess and, possibly, predict the nature of the behaviour of the internal air temperature in a winter greenhouse under the influence of constantly changing environmental factors. It is obvious that an uncontrolled drop in air temperature in a greenhouse can lead to a decrease in yield or even to a plant death. At the stage of the greenhouse design, the obtained dependencies make it possible to determine the optimal space-planning and design solutions, taking into account the meteorological conditions of the region under consideration. They also allow one to select a soil option that meets the necessary requirements for the absorption of infrared radiation, to develop the correct soil irrigation technology, and, possibly, to apply, in addition to infrared emitters, additional sources of thermal energy (for example, convective heaters for heating air).

References

1. van Straten, G., van Willigenburg, G., van Henten, E., van Ooteghem, R.: Optimal Control of Greenhouse Cultivation (2010)
2. Resh, H.: Soilless Culture (2013)
3. Ghoulem, M., El Moueddeb, K., Nehdi, E., Boukhanouf, R., Calautit, J.K.: Biosyst. Eng. **183**, 121 (2019)
4. Ghani, S., Bakochristou, F., ElBialy, E.M.A.A., Gamaledin, S.M.A., Rashwan, M.M., Abdelhalim, A.M., Ismail, S.M.: Eng. Agric. Environ. Food **12**, 48 (2019)
5. Srbinovska, M., Gavrovski, C., Dimcev, V., Krkoleva, A., Borozan, V.: J. Clean. Prod. **88**, 297 (2015)

6. Ganguly, A., Ghosh, S.: Energy Build. **41**, 521 (2009)
7. Kolokotsa, D., Saridakis, G., Dalamagkidis, K., Dolianitis, S., Kaliakatsos, I.: Energy Convers. Manag. **51**, 155 (2010)
8. Kulachinskaya, A., Akhmetova, I.G., Kulkova, V.Y., Ilyashenko, S.B.: J. Open Innov. Technol. Mark. Complex. **6** (2020)
9. Markov, V. et al.: IOP Conference Series: Earth and Environmental Science, vol. 337 (2019)
10. Sultanguzin, I.A., et al.: Development strategy of the energy supply system for MPEI campus blocks based on green building. In: IOP Conference Series: Earth and Environmental Science, vol. 288, no. 1, p. 012071 (2019)
11. Akhmetova, I.G., Kalyutik, A.A., Fedukhin, A.V., Derevianko, O.V., Mukhametova, L.R.: IOP Conference Series Earth and Environmental Science, vol. 288, p. 012076 (2019)
12. Aye, L., Fuller, R.J., Canal, A.: Int. J. Therm. Sci. **49**, 202 (2010)
13. Saravia, L., Echazú, R., Cadena, C., Condorí, M., Cabanillas, C., Iriarte, A., Bistoni, S.: Renew. Energy **11**, 119 (1997)
14. Teitel, M., Shklyar, A., Dikhtyar, V., Jerby, E., Elad, Y.: Acta Hortic. **534**, 189 (2000)
15. Pavlov, M., Lukin, S., Derevianko, O.: MATEC Web Conferences, vol. 193, p. 03006 (2018)
16. Kavga, A., Panidis, T., Bontozoglou, V., Pantelakis, S.: Trans. ASABE **52**, 2055 (2009)
17. Sitas, V.I., Fedyukhin, A.V., Akhmetova, I.G., Mitrofanov, A., Makoev, S.O., Asadpoori, A., Sinitsyn, A.A., Kikot, E.A.: Int. J. Civ. Eng. Technol. **10**, 2106 (2019)
18. Pavlov, M., Karpov, D., Akhmetova, I., Monarkin, N.: E3S Web Conferences, vol. 178 (2020)

Innovative Use of Sustainable Solar Power in Environment Protection and Emergency Response

Shailja Sinha[1(✉)], Sumeet Singh Jasial[1], and Manik Sinha[2]

[1] Amity Business School, Amity University, Noida, India
shailjasinha@yahoo.com
[2] Avadh University, Faizabad, UP, India

Abstract. Growing trend of environmental damage due to fossil fuel reduction and greenhouse gas emission, and advancement in solar power generation technology, calls for using solar power as an emergency response and disaster recovery. In any disaster circumstances, restoring electricity becomes one of the most crucial aspects of emergency response. Rapid deployment of solar energy could potentially bring quick normalcy in the disrupted lives of the people. Time, and not cost or capacity, is the most critical factor why solar power bears a better alternative to gensets. When recovery needs extended time, the generator will need large amount of fuel. Need of continuous re-fuelling generators gives solar its major advantage: However, severe health and environment impediments will pop-up in Solar too, which needs to be addressed and further research to be continued to place it as a fool-proof and sustainable solution. To address the current hurdles and leverage its full potential we are recommending the development of Smart solar distribution system. This would empower communication controls and new technologies working together to make the grid more secure, reliable, efficient and cleaner. It is expected that these innovative solutions will strengthen the "environment protection" and "disaster management" evolutionary processes in a focused manner using solar technology.

Keywords: Solar power · Smart solar · Distribution system

1 Introduction

Solar PV and related technologies have advanced considerably, but even today diesel gensets continue to dominate disaster relief efforts. This is the time for considering the solar energy when planning for disaster management for vast cities, primarily the areas prone to natural disasters like flood, storms, earthquakes causing severe power outages and damages to vast areas and emergency response needed to save lives [1].

Solar power is best suited solution for its sustainable and stand-alone capabilities. It is not a question of whether solar power can be used for disaster relief, moreover it is a recommendation to switch to this alternative looking into the other disastrous consequences of depending on generator power and fossil fuel damaging the environmental stability and causing more threats for the disasters. This article intends to explore and identify possible solar applications for disaster relief and preventing power outages

A. Sinitsyn (Ed.): SES 2020, LNCE 141, pp. 136–147, 2021.
https://doi.org/10.1007/978-3-030-67654-4_16

during disaster. The community is also interested in having solar powered schools and hospitals which are shelters and life savers during disasters [2].

The study is done by identifying and evaluating widespread solar equipment's which can be used in different situations of disaster mainly to recover the power outage caused by it. The effectiveness of various equipment's is also evaluated from disaster data and statistics available on the scale of damage caused and the amount spent on disaster rescue and relief. In addition, a pilot implementation of DC equipment based solar setup is done in a school which also acts as a shelter during disaster situations [3].

The data and statistics have been collected form the regional and national bodies from the energy and emergency management agencies [4, 5].

2 Disaster Planning

Preparedness for disaster is a key challenge for the densely populated country like India with 1.3 billion residents. India has been traditionally vulnerable to natural disasters because of its unique geo-climate conditions. Floods, droughts, cyclones, earthquakes and landslides have been a recurrent phenomenon. About 60% of the landmass is prone to earthquakes of various intensities; over 40 million hectares is prone to floods; about 8% of the total area is prone to cyclones and 68% of the area is susceptible to drought [5]. Also, at the global level, there has been a considerable concern over natural disasters including floods, dust storms, cyclone, hurricane and earthquakes causing power outages in vast areas for a prolonged period [6]. Blackouts due to electricity problems itself becomes a disaster situation and needs a serious attention on improving the reliability and sustainability of the electricity sources thus calling the need for renewable energy sources.

In a disaster situation, the key response lies on rescue, shelter, medical and communication capabilities as the lifesaving resources by the disaster response teams. All these disaster relief services need electric power. In most of the disasters, the normal electricity power is disrupted thus dependent on alternative power sources primarily diesel generators to provide these services and resources.

A key development is the National Disaster Management Plan, recently launched by the Government of India [7]. This is the first national-level plan in the world which is aligned to the Sendai Framework for Disaster Risk Reduction. It represents an important step in making India prepared to manage disasters and extreme climate-event risks, by balancing response and preparedness activities and strategies. Worldwide too, the disaster management programs are devised which provides training and knowledge to the disaster response teams on various ways of rescue and relief operations depending upon the conditions and damage. Communities are more concerned on environment too, and hence these programs need to take into consideration the conservation of natural resources and prevent further damage to the environment by restricting usage of fossil fuels.

The Pandit Jawaharlal Nehru National Solar Mission [8], also known as the National Solar Mission, is an initiative of the Government of India and State Governments to promote solar power. The mission is one of the several initiatives that are part of the National Action Plan on Climate Change. The objective of the National

Solar Mission is to establish India as a global leader in solar energy, by creating the policy conditions for its diffusion across the country as quickly as possible. Under the original plan, the Government aimed to achieve a total installed solar capacity of 100 GW by 2022 [9]. Across the globe, the communities are working for building environmentally friendly infrastructures and implementing renewable energy sources for sustainable development. A "sustainable city" is a city designed, constructed, and operated to efficiently use its natural resources, minimizing waste and to manage and conserve them for the use of present and future generations.

3 Research on Evolution of Electricity and Energy Sources

Before entering the discussion of solar energy, let us first go through the long-time researches and discoveries done by great scientists and engineers regarding evolution of electricity, which we are using in our modern era. Prior to the discovery of electric light, we were using lamps like lantern illuminated by kerosene oil, candles, torch stick with burning materials at one end known in Hindi as mashal, a wick in oil or fire. In the history of electricity, the names of three great scientists and engineers namely Thomas Alva Edison, (1847–1931) of America, Serbian-American inventor Nikola Tesla, (1856–1943) and George Westinghouse (1846–1914), an American entrepreneur and engineer based in Pennsylvania have become scientists immemorial… Edison was the first to invent electrical energy apart from mass communication, sound recording and motion picture. Prior to him, the power was generated by gas light. After about 50 years of continued research by the scientists, finally Edison was able to trump over the era of gas-light system and created electric power. After having become successful in harnessing electric power, he established his own company namely Edison Electric Light Company and made a surprise to the World by manufacturing electric bulb with the use of platinum filament and subsequently by using carbonized bamboo as a source. However, George Westinghouse propounded theory of (Alternating Current) AC current and subsequently Edison merged his company Edison General Electric Company with George Westinghouse's company and became General Electric Company in 1892. Nikola Tesla, the renowned electric and mechanical engineer introduced the idea of DC (Direct Current) who discovered and patented the rotating magnetic field, the basis of most alternating-current machinery.

Traditionally, we have seen the source of power generation was fossils, coal, lignite, petroleum, gas etc., but since these sources are finite in nature and shall exhaust in near future, the world turned to discover alternative sources of energy. And, thus the scientists discovered hydroelectric, wind, biogas, geo-thermal and hot spring water as the power production sources. Of late, the great scientist "Homi Jahangir Bhabha" of India, discovered for the first time atomic energy generated from two elements known as 'uranium' and 'thorium' which are found in the sands of coastal areas of South India, which Bhabha invented and then the era of Nuclear energy came into being and presently the World is generating Nuclear power, but that cannot be accepted at the cost of lives, So, Solar Power generation came into production, however the technology is still evolving.

The first solar power plant in India was initially installed in village Bhadla in Thar Desert situated at about 200 km away from Jodhpur (Rajasthan) fitted with solar panels on an arid stretch spread over 45 sq km to produce fuel for Rajasthan's ambitious plan to generate 2,255 MW, making it "the largest solar park in the world". At present it is generating 746 MW and under phase 3 stage it will further generate 1,000 MW power. Similarly, in Pokharan situated in district Jaisalmer in Rajasthan 100-MW solar power plant is in operation. Apart from that several other projects have also been commissioned like Kurnool Ultra Solar Park in Andhra Pradesh and Pavagada Solar Power Park in Karnataka targeting to generate 1000 MW and 2700 MW respectively. The tariff is also very cheap i.e. about Rs. 2.45 per unit. Apart from the cost aspect of the Solar energy, this technology is much more safer than Nuclear technology, whose gigantic catastrophe we have seen in Fukushima tragedy in Japan in 2011. In this way, the Solar power generation technology is by all means safer and cheaper than any other technology.

So far generation of the solar power is concerned; India is at the top in the World. Initially it was producing total 7,568 MW, but it was increased within a period of two years target. At present Rajasthan, Tamil Nadu, Gujarat and Andhra Pradesh are generating 1285.932 MW, 1267 MW, 1120 MW and 864 MW respectively [10].

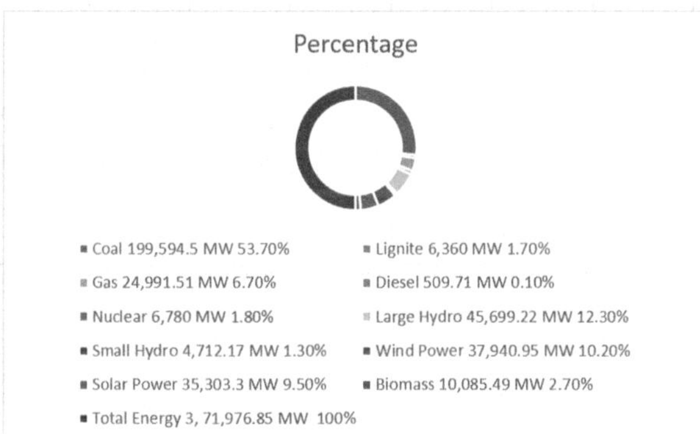

Fig. 1. Electricity production in India from different sources.

The above picture shows that out of all sources of power production, the highest production comes from Coal i.e. 53.7%, but since Coal production is at the verge of extinction for the purpose of Electricity generation, hence this system is likely to retire from the field sooner or later, so we have to abandon the power production from Coal and the consequent loss of 53.7% of the energy produced from it has to be compensated from other sources and that it is the main concern. Then comes Hydro-electric technology - both large and small totalling to 13.6%, thereafter comes Wind power technology which stands at the fourth place i.e. producing 10.2% and the last but not the least is the Solar energy production which stands at 95% of the total production. The question is that after the exit of Coal production system, we are left with nine systems mentioned in the above table and in that condition which system is to be adopted and in harnessing any other technology, two factor are the main consideration- firstly the safety and secondly the cost of production. Productions from Lignite (1.7%), Gas (6.7%), Diesel (0.1%) and Biomass (2.7%) are very limited and we must abandon these power sectors soon. Thus, after excluding Coal, Lignite, Gas, Diesel and Biomass; power sectors, we are left with only four sectors- namely, Hydro-electric, Wind, Nuclear and Solar power systems. The Country's largest nuclear power plant was installed at Gorakhpur in (Haryana) by the Nuclear Power Corporation of India Ltd., whose foundation stone was laid by Dr, Manmohan Singh, the then Prime Minister of India on 13th January 2014 with its installed capacity of 2,800 MW. So far as production from nuclear technology is concerned, the production from this sector is 1.8%, which is very meagre. In view of the Fukushima Daiichi massacre and all round agitation against new set ups of nuclear power plants in India not only by the farmers but also the political parties, Parliamentarians, intellectuals like social activists, retired bureaucrats and judges, academicians, media and State Governments like West Bengal, Maharashtra etc., the future of nuclear technology is bleak despite the fact that the Government of India had assured that all necessary precautions and safeguards would be ensured so that there may not occur any disaster or adverse effect on the human health, nature or change, but the fact remains that the effect of such project is violative of the Fundamental Rights of the People of India and the same is in contradiction with the principle of Sustainable Development. Thus, the nuclear power is presently ruled out. After eliminating nuclear technology, we see that Hydro-electric and Wind technology taken together produce only 23.8%, which is enough to cater the need of our Country. Thus, after eliminating all the power sectors discussed above, we are left with the solar sector only, which at present gives us only 9.5%, thus, the emphasis is on the solar sector to depend upon.

The unique aspect of this sector is that it can be produced by the individuals also for their domestic consumption that too free of cost, which is not possible in other sectors. It is also generated centrally and transmitted through its Grids to power houses, then to the Sub-Stations and ultimately to the consumers like the individuals as well as for the agricultural, residential, industrial, commercial, educational institutions, roads, parks, train services purposes etc.

4 Solar for Emergency Services

In the developed countries like USA, PV frameworks have of late provided emergency power in the consequence of numerous past disasters:

- Northridge Earthquake, 1994: The quake caused boundless power blackouts all through the Los Angeles region and past. The whole Los Angeles Department of Water and Power frameworks went down. This prompted 1.3 million clients to be without power. Half of the clients recuperated power within six hours, and within 24 h, 95 percent of the clients were back to normalcy. Notwithstanding the utility's convenient rebuilding of intensity, near 100,000 homes and business were without power for over 24 h [12]. Water supply was additionally disturbed because of breaks in the supply and conveyance lines. The utilization of PV kept a few correspondences joins working. It likewise provided capacity to Southern California inhabitants that had installed systems in their houses (U.S. DOE 1995) [13].

- Hurricane Andrew, 1992: The tempest left around 3 million homes and organizations without power, including 1.4 million Florida Power [11] and Light clients. Water and wastewater utilities were additionally hit hard, as trees fell over water mains and dissemination lines. The power blackouts likewise crippled the water siphons. A few sections of Dade County (Miami territory) had power back inside about fourteen days however different territories, for example, the southern end, did not have control for no less than multi month. Solar PV frameworks gave capacity to havens and therapeutic facilities. They additionally powered road lights and communication systems (U.S. DOE 1995) [13].
- Hurricane Hugo, 1989: On September 21, 1989, the classification 4 sea tempest hit South Carolina. Charleston, Folly Beach, Sullivan's Island, Isle of Palms and McClellan Ville were hit hard. Sea tempest Hugo left 26 individuals dead in South Carolina alone. The harm gauge would go as high as $7.2 billion (in 1990 dollars) [12]. Sixty structures in downtown Charleston were levelled. 5,100 homes were obliterated. 11,928 additional homes were left appalling. Harm was accounted for in twenty-nine regions, the clear majority of which were assigned as government disaster zones. Ninety-eight percent of the city's occupants lost power, and for a few, fixes were not made for over about fourteen days. Power blackouts Building Industry Trends - 10.293 made a lot of crude sewage sidestep treatment plants and stream into streams all through Mecklenburg County. (Sea tempest Hugo 1997) A versatile solar electric generator—controlled a network community for about a month and a half after the tempest (U.S. DOE 1995) [13].

5 Growing Natural Disasters and Increased Power Outages

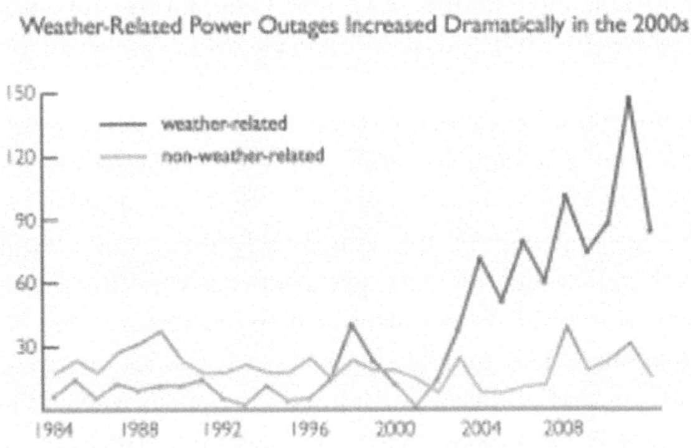

Fig. 2. Weather-related power outages graph [14].

- India Blackout, 2012: Two serious power outages influenced a large portion of northern and eastern India on 30th and 31st July 2012. The 30th July 2012 power outage influenced more than 300 million individuals and was quickly the biggest power blackout in history by number of individuals influenced, beating the January 2001 power outage in Northern India (230 million influenced). An expected 32 GW of creating limit was taken disconnected. An article in The Wall Street Journal expressed that of the influenced populace, 320 million at first had power, while the rest needed direct access. Electric administration was re-established in the influenced areas between 31st July and 1st August 2012 [15].

- NORMAN Storms, 2019—Nearly 4,000 Norman habitations endured power blackouts Thursday 3rd Jan 2019 as the consequence of ice and snow. OG&E representative Karen Kurtz said roughly 3,000 OG&E clients were influenced, and, starting at 4 p.m. Thursday, 230 Norman inhabitants stayed without power. Over the state, OG&E detailed approximately 7,800 blackouts [16] (Fig. 2).

6 Solar is the Solution

Now, if we look at the history of 200 years back of our civilization, we will find that there was no mechanical system to produce energy. The only source of energy was burning of wood or plants etc. As we are aware the fossil fuel is reducing at a very high rate and if we do not switch to other form of energy at this stage, we will leave nothing for our coming generations. Fossil fuel on the earth is finite. People won't be able to burn and derive energy any more beyond the lasting periods which is as close as the following for some of the fuels [17]:

(a) Coal… 120 years
(b) Oil… 250 years
(c) Nuclear fuel… 200 years

Moreover, how far is it justifiable for the energy packs, which have taken millions of years to form to consume in the next few generations. Just because the future generation is not present to bid today, we are claiming our stake on the entire coal blocks and the oilfields.

Solar energy being abundant in nature is a suitable alternative at this stage, since we now have the technology to channelize this solar energy for power generation. Half of the earth is always receiving incessant energy in the form of solar radiation.

Everybody realizes that regions hit with cataclysmic events require some investment, exertion, cash and frequently rely upon repetitive political negotiations before starting the recovery. Re-establishing power in such territories is one of the more urgent parts of disaster response. The pioneer innovation for solar energy called the Rapid Roll [18] framework instead of diesel generators, guarantees to be a Game-Changer for Disaster Relief and for environment protection from future disasters.

Rapid Roll framework provides a system to unfurl the solar panels like a carpet from behind a truck. Making it a mobile solar technology.

These mobile high-proficiency solar panels could help produce instant energy in disaster territories. The Rapid Roll solar panels produce an average 11 kW of power, associated with batteries that can store 24 kW/h. That is about a day's worth of energy for a family. The best part is that the Rapid Roll solar panels can last up to 10 years [18].

The Rapid Roll panels are pressed in 4 × 4 trailers, which convey enough solar panels to power a mobile clinic with 120 beds required for disaster help.

The Rapid Roll framework was structured most importantly for quick deployment. That is especially valuable in territories that don't have prepared access to a standard energy source, or for zones effected by catastrophic events. Without a doubt it can possibly spare lives by – for instance, lessening or dispensing with the requirement for military fuel escorts or by powering medical facilities in disaster zones.

Re-establishing power is pivotal in disaster-stricken territories. Solar energy powers communication networks and gives a bit of normalcy in the disrupted lives of the general population influenced by such disasters. A stunning 55% of Puerto Rico's populace were accounted for without drinkable water following Hurricane Maria [19]. Although it might be past the point of no return this time, rapid-roll solar panels could possibly mitigate a lot of this sort of affliction, sooner rather than later.

With an innovation like Rapid Roll solar panels, re-establishing power to tropical storm and hurricanes hit territories won't be as troublesome. It's not important to have solar designers demonstrate the explicit site and compute specific solar field positions and setups before sending – the Rapid Roll will work instantly and will produce the best output possible in that situation.

Other Applications of Solar in Disaster Relief besides Rapid Roll system are:

1. Hybrid Solar Project
2. Solar Charged Battery Backups
3. Mobile Solar Clinics
4. Solar Panelled Fire Brigade, etc.

7 Developing Smart Solar Distribution System

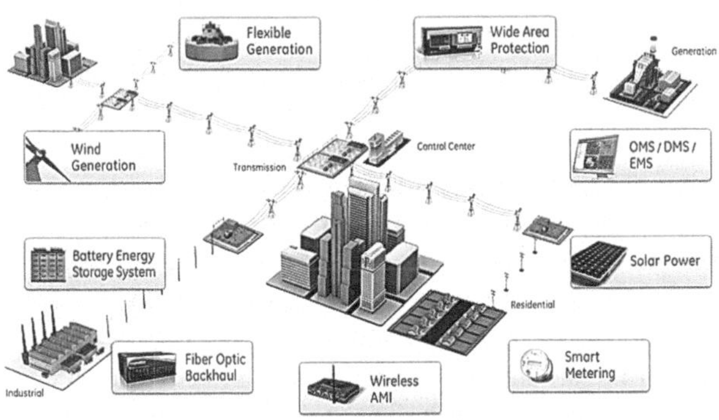

Fig. 3. Smart solar distribution system.

In comparison to the earlier modes of energy production, the solar power technology is the cheapest, because we don't have to purchase Sunlight, a great gift of nature to the entire civilization of the world including human, atmosphere and agriculture. The unique feature of this technology is that it can be produced by an individual also not only for his domestic purpose rather free also, because the raw basic element i.e. Sunrays are free for all and only installation cost is required which is also not very costly comparing with the recurring cost of electricity purchased by the electricity providers. Apart from generating the power individually and privately and without any intervention from the State authorities or agencies, it can also be generated centrally by the State instrumentalities or agencies on the large scale like the traditional power generating plants to serve a bulk of the society. We are aware that the power generated by the Sunrays, the nature of current is of DC, but it is then converted into AC for its transmission to its Grids wherefrom it is further transmitted to its various Sub-Stations and ultimately to the individual consumers, shops, commercial establishments, industries, schools, colleges, Universities etc. like generation. Since the electricity cannot be transmitted on DC current, hence it is initially converted into AC current. However, the individuals generating their own electricity, the nature of current they receive is DC. The technology used, in short, is that Solar panels are installed on the house roofs, which absorbs the Sunrays which is thereafter converted into DC current, which is connected with a Battery, which is then charged by it and from the charged battery, we use the electricity on DC lines. The rooftop solar can directly electrify the electronic equipments since they run on DC. However, for electrical equipments too, if we switch from AC equipments to DC equipments, then the need of transformer and invertor is also eliminated, and solar solution becomes cost effective too. But the real advantage would come when several buildings combine such DC networks. Pooling supply, demand and storage gives you the makings of a "smart grid"—a low-voltage DC power networks which is more resilient and energy-efficient for people in rural areas who have inadequate possibility of connecting to the AC mains grid (Fig. 3).

Developing Smart Solar distribution system can empower you to intelligently think for the consumer. It allows for a two-way interaction between the supplier and the consumer an enabling real-time exchange of electricity and information 24/7. The smart grid is a digitally controlled network of advanced metering communication controls and new technologies working together to make the grid more secure, reliable, efficient and cleaner. IT enabled customer care centres help in tracking consumer complaints and queries and provide solutions in a time-bound manner. The utilities are also using data acquisition and control systems along with GIS mapping, this provides a visual monitoring of the entire power system. Associated with it are the smart micro grid which can connect and disconnect from the grid to enable it to operate in an economic efficient and sustainable manner. in villages smart micro grids will help ensure power supply when required, making energy available and affordable and thereby improving life of the rural population. Smart grids also help in minimizing power cuts, a state-of-the-art automated surveillance system alerts the utility as soon as there is a fault or loss in the distribution system. Workmen are then immediately dispatched to fix the fault. The government of India is building a strong institutional structure to support the foundation of a dynamic smart grid network across the nation intervention through smart trades will enable efficient use and management of one of the most valuable resources on earth energy. A national transition to a smarter and inclusive grid is taking place smart grids are laying the foundation for transforming our cities to small cities for a harmonious world that is connected smarter and intelligent. Smart grid will be energizing smart cities for a smarter future.

The inevitable transformation of the centralized electrical grid to a distributed generation configuration requires solar system capabilities much beyond simple net metering and grid-connected approaches [20]. Time of use and peak demand rate structures needs much more sophisticated system design which integrates solar radiation management and power storage into its design system and architecture. Controlling the power flow to and from the utility grid needs to ensure the grid is reliable and stands strong in power quality. Alternative security and protection strategies are needed to accommodate large number of distributed solar panels as energy sources. It further needs to be Integrated with advanced communication technology.

8 Intelligent Utility Network (IUN) Communication Services

IBM has developed Intelligent Utility Network Communications (IUN) Services which helps you plan, design, and implement a smart grid communication solution to support your utility operations [21]. IBM IUN Communications Services includes end-to-end strategy, design, assessment, integration, and deployment services of our utility communications networks to enable a smart solar grid. The IUN communication service focuses on network scalability and security, enables sending and receiving more information throughout the solar electricity grid, and it helps in reducing outages and faults, thus improving the responsiveness. By providing a communications network infrastructure, this service helps in maintaining service levels, managing costs, increasing the operational efficiency and responding to immediate and forecasted demands.

9 Further Research to Be Continued

This section provides an overview of the research and development needs and describes few pathways to a promising solution. The solutions, in most of the cases, will require research and development of new components, solar radiation management system, innovative inverters and controllers, innovative power storage and a suite of advanced technical methodologies control algorithms, protocols and the associated communications. As of now, we leave this problem to be addressed by our scientists and experts and it is hoped one day or other, our scientist shall find out the solution. It is expected that these innovative solutions will help to push forward the "environment protection" and "disaster management" evolutionary processes in a fast and focused manner using Solar Technology.

Acknowledgements. I would like to thank my guide Dr Sumeet Singh Jasial and co-author Prof. Manik Sinha for guidance during my studies and in personal development. Additionally, I would like to show my appreciation to the fellow colleagues of my university Amity Noida and my organization IBM India for help and cooperation in research activities. My sincere appreciation to my family for support in all the activities and challenges that I have taken in my life.

References

1. Hausfather, Z.: Analysis: Fossil-fuel émissions in 2018 increasing at fastest rate for seven years. CarbonBrief (2018)
2. Gore, A.: When will fossil fuels run out? Ecotricity (2019). https://www.ecotricity.co.uk/our-green-energy/energy-independence/the-end-of-fossil-fuels
3. Bohra, S.S.: DC-current sensor-less MPPT based grid-fed single-phase photovoltaic (PV) micro-inverter. Appl. Solar Energy **56**(2), 85–93 (2020)
4. State Agency, Federal Emergency Management Agency (FEMA). https://www.fema.gov
5. Ministry of Home Affairs, India: Disaster Management in India - A Status Report, UNDP GOI (2012). https://www.in.undp.org/content/india/en/home/library/environment_energy/disaster_managementinindiaastatusreport.html
6. Dillinger, J.: Countries Most Prone to Natural Disasters. WorldAtlas (2018). https://www.worldatlas.com/articles/countries-with-the-deadliest-natural-disasters.html
7. Meadows, D.H., Meadows, D.L., Randers, J., Behrens III, W.W.: The Limits to Growth. A Potomac Associated Book (1972). https://donellameadows.org/wp-content/userfiles/Limits-to-Growth-digital-scan-version.pdf
8. Goyal, P.: India's solar capacity to cross 20GW in next 15 months. Money Control (2018). https://en.wikipedia.org/wiki/National_Solar_Mission
9. Menghani, V.: Plant wise details of Renewable Energy Installed Capacity. GOI (2020). https://en.wikipedia.org/wiki/Renewable_energy_in_India
10. Avezova, N.R., Avezov, R.R., Vokhidov, A.U., et al.: Influence of ambient temperature, wind speed, emissivity, and average working temperature of light-absorbing heat-exchange panels of flat-plate solar water-heating collectors on their thermal losses through translucent coatings. Appl. Solar Energy **55**(1), 30–35 (2019)
11. Young, W.R., Jr.: History of Applying Photovoltaics to Disaster Relief. Florida Solar Energy Center (1996). https://www.fsec.ucf.edu/en/publications/pdf/fsec-cr-934-96.pdf
12. Sinha, S., Jasial, S.S., Sinha, G.P.: integration of disaster management with solar technology for disaster relief. Int. J. Sci. Technol. Res. **9**, 4037–4045 (2020). https://www.ijstr.org/research-paper-publishing.php?month=feb2020
13. DOE, U.S.: Budget FY 1995. Office of Science (1995). https://science.energy.gov/budget/fy1995/
14. Mehta, J.: India has 5 million diesel generators, Loudcell has raised $1 million to improve their efficiency. Yourstory (2014). https://yourstory.com/2014/11/loudcell-1-million
15. Robles, F., Bidgood, J.: Three Months After Maria, Roughly Half of Puerto Ricans Still Without Power. The New York Times (2017). https://en.wikipedia.org/wiki/List_of_major_power_outages
16. Burke, M.: Storm brings power outages. The Norman Transcript (2019). https://www.normantranscript.com/news/local_news/storm-brings-power-outages/article_3dd7142d-f78b-5116-972a-0342b748a519.html
17. N.D.M.A, Government of India: National Disaster Management Plan, 2016. New Delhi (2016). https://ndma.gov.in/images/policyplan/dmplan/National%20Disaster%20Management%20Plan%20May%202016.pdf
18. Galeon, D., Bergan, B.: Roll-Up Solar Panels May Be Game-Changer for Disaster Relief. MACH (2017). https://www.nbcnews.com/mach/science/roll-solar-panels-may-be-game-changer-disaster-relief-ncna808111

19. Renewable Energy World: Getting Serious about Solar for Disaster Response and Recovery. https://www.renewableenergyworld.com/articles/2017/10/getting-serious-about-solar-for-disaster-response-and-recovery.html
20. SEGIS: Solar Energy Grid Integration System (2007)
21. IBM: IUN Communication Services (2010). https://www-01.ibm.com/common/ssi/rep_ca/0/897/ENUS610-060/ENUS610-060.PDF

Use of Forcing Algorithms in Search Methods in Systems with Direct Moment Control

Tatyana Sinyukova[1(✉)], Alexey Sinyukov[1], and Alfred Safin[2]

[1] Lipetsk State Technical University, 30 Moskovskayastreet, Lipetsk, Russia
stw0411@mail.ru

[2] Kazan State Power Engineering University, Str. Krasnoselskaya, 51, 420066 Kazan, Russia

Abstract. Ensuring energy saving in asynchronous electric drive control systems is an urgent topic for research. In order to achieve energy-saving mode by reducing power consumption, it is proposed to use search algorithms, the advantage of which is independence from the properties of the object, which is controlled, and reliability. When a signal for data correction appears, the search algorithm needs time to find the optimal point, so its use in devices with cyclic action with short-term cycles or on objects with sharply variable load patterns is impractical. The solution to this problem is possible by introducing forcing algorithms that reduce the time between cyclic surveys in search methods and reduce the transition time. In this paper, the mechanism of movement of the bridge crane trolley is used as the object under study.

Keywords: Asynchronous electric drive · Control systems · Energy saving

1 Introduction

One of the ways to ensure energy-saving mode in an asynchronous electric drive is the use of search algorithms implemented on the basis of fuzzy logic, using a stepwise measured voltage level for calculations, which allows you to find the points of minimum energy consumption. The search criteria can also be the minimum current consumption or the maximum $\cos\varphi_1$.

Using search algorithms is a reliable way to achieve energy savings that is independent of the properties of a particular control object. The search algorithm may take several tens of seconds to find the optimal point, which makes it unsuitable for use in cyclical mechanisms with small cycles, or in drives with sharply variable load. Optimal from the point of view of efficiency is the implementation of an energy-saving mode using a mathematical apparatus with minimal dependence on changes in the parameters of the substitution scheme. In this case, the search algorithms will only perform a small correction of the control effect.

A. Sinitsyn (Ed.): SES 2020, LNCE 141, pp. 148–154, 2021.
https://doi.org/10.1007/978-3-030-67654-4_17

2 Maintaining the Optimal Angle Between Moment-Forming Vectors

2.1 Theory

One of the simplest algorithms implemented in vector control systems is to maintain the optimal angle between the moment-forming vectors. The optimal operating modes of the electric drive can be achieved by single-criteria optimization: optimization to minimize current or optimization to minimize losses, or by using multi-criteria optimization. However, these search algorithms for optimal operating modes of the electric drive [1] have a serious drawback, which is expressed in the duration of the system polling cycles and their number. After making a corrective action, the search engine must wait for the end of transients in the electromagnetic circuit of the engine. In fact, the process can be considered complete if a flow coupling of the rotor is formed, the modulus of which depends only on the input values of the electromagnetic variables that affect its formation.

2.2 Mathematical Apparatus

One of the fundamental equations of the vector control system [2] describes the relationship between the rotor flow coupling vector ψ_2 and the stator current component in-phase with this flow coupling I_{1d}.

$$I_{1d} = \frac{|\psi_2|}{L_m}(1 + T_2 p) \tag{1}$$

where $T_2 = L_2/R_2$- rotor time constant; L_m - inductance of the magnetizing circuit; L_2- full inductance of the rotor; R_2 - active resistance of the rotor phase windings.

In the direct torque control system (PUM), there are no direct elements for calculating the flow coupling of the rotor and calculating the components of the stator current that affect it. Therefore, to simplify the mathematical apparatus, it is necessary to find the relationship between the stator flux coupling module calculated in the PUM $|\psi_1|$ and determining the time of the transition process by the flow coupling of the rotor $|\psi_2|$.

An equation from [1] is known, obtained by solving the original system of equations for the spatial vectors of an asynchronous motor, describing the flow coupling vector as a function of the stator current $\tilde{I}_1$ and flux of the rotor $\tilde{\psi}_2$:

$$\tilde{\psi}_1 = L_1\tilde{I}_1 + \frac{L_m}{L_2}(\tilde{\psi}_2 - L_m\tilde{I}_1) = \sigma L_1\tilde{I}_1 + k_2\tilde{\psi}_2, \tag{2}$$

where $\sigma = 1 - L_m^2/(L_1L_2)$; $k_2 = L_m/L_2$, $\tilde{\psi}_1$- the stator flux linkage vector; L_1- full inductance of the stator.

As a reference vector, we take the rotor flow coupling vector and Orient the coordinate system along it so that its real axis coincides with the direction ψ_2 ($|\psi_2| = \psi_{2d}$).

Then Eq. (2) will take the following form:

$$\frac{|\Psi_1|}{\cos\Theta} = \sigma L_1 I_{1d} + k_2|\Psi_2|, \tag{3}$$

where Θ- the angle between the stator and rotor flow coupling vectors. The value of this angle depends on the magnitude of the stator and rotor currents and the scattering inductances of the stator windings $L_{1\sigma}$ and rotor $L_{2\sigma}$. Since for General industrial engines the ratio is $L_m/L_{1\sigma} \geq 10$ and $L_m/L_{2\sigma} \geq 10$, even in starting modes, this angle does not exceed 10-15 degrees. As a result, the maximum error in calculations, when reducing $\cos\Theta \approx 1$, it will be 1–3%, which is within the permissible error of the control system.

Substituting (1) in (3), we get the desired dependency:

$$|\psi_2| = |\psi_1|\frac{k_\psi}{1+T_\psi p} \tag{4}$$

where $T_\psi = \frac{T_2}{1+\frac{L_m^2}{\sigma L_1 L_2}}$ - is the time constant of the aperiodic link; $k_\psi = \frac{L_2 L_m}{L_m^2 + \sigma L_1 L_2}$- dimensionless scaling factor.

As can be seen from Eq. (4), the rate of change in the rotor flow coupling depends on the value of the module $|\psi_1|$. And since the rate of formation of the stator flux coupling exceeds that of the rotor flux coupling, it is possible to use the magnetic flux forcing mode.

To provide an automatic controlled boost mode, an additional block for generating a task for stator flow coupling was developed (Fig. 1).

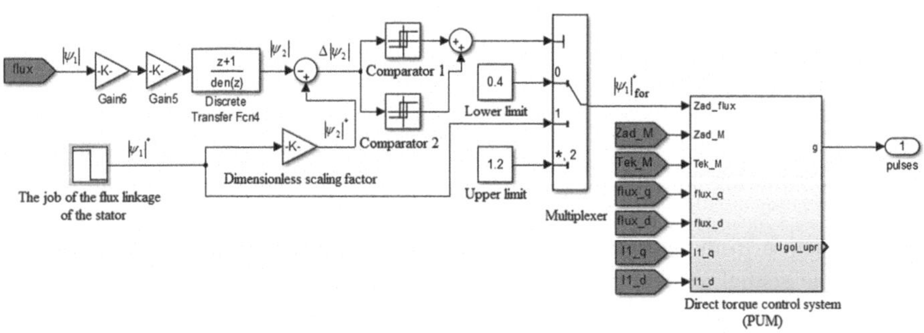

Fig. 1. Block for automatic generation of the stator flow coupling task with boost.

The job of the flux linkage of the stator $|\psi_1|^*$ first, it is converted to the signal of the specified rotor flow coupling without inertia $|\psi_2|^*$. This value should be formed at the end of the transition process in the engine. The received signal is sent to the second input of the adder. The real value of the modulus of the stator flux linkage $|\psi_1|$ through the amplifiers, it is fed to the transfer link described by Eq. (4), at the output of which we get the current value of the rotor flow coupling module $|\psi_2|$, which is received at the first input of the adder. As a result, a signal is generated at the output of the adder $\Delta|\psi_2| = |\psi_2|^* - |\psi_2|$ proportional to the difference between the set and current value of the rotor flow coupling. The output signal of the comparator can take three different States: 0, 1, 2. In our case, the threshold values of the comparator setpoint are set as 0.4 and 1.2. If $\Delta|\psi_2|$ the comparator setpoint is lower than the threshold value, the output signal is 0, which corresponds to the state in which the output should be forced along the lower permissible limit, regardless of the current task. If $\Delta|\psi_2|$ the upper threshold value of the comparator setpoint is higher, the output should be forced along the upper permissible limit, which corresponds to signal 2. In the latter case $\Delta|\psi_2|$ it lies within the acceptable limits and the output must be transmitted to the task signal $|\psi_1|^*$. In other words, for small signal deviations, the mode of unjustified signal amplification is excluded. The switch functions as an output multiplexer, which connects one of the three inputs depending on the signal from the comparator. As a result, the output of the multiplexer generates a task signal for stator flow coupling, taking into account the boost $|\psi_1|^*_{for.}$

Automatic controlled boost mode is possible by using a dynamic controller with hysteresis and saturation, the current signal of which depends on the value of the output signal at the previous time.

3 Modeling

3.1 Starting the Engine with a Task for Nominal Flow Coupling

To test the effectiveness of the proposed method, a mathematical model was compiled in the Simulink shell of the Matlab software package [3–22]. The results of comparative modeling of the engine with a power of 37 kW are shown in Fig. 2, 3, 4 and 5. The upper and lower limits for forcing the magnetic flux are selected from the condition of stable operation of the engine. The value of the upper limit of the stator flux linkage $\psi_{1\max}$ it is 1.2 of the nominal flow coupling of the stator ψ_{1nom}, value of the lower limit of stator flow coupling $\psi_{1\min}$ compose 0,4 ψ_{1nom}.

Figure 2 and 3 shows the process of starting the engine with a task for nominal flow coupling $\psi_1 = 0,96$ Wb. As can be seen from the graphs, the flow coupling of the rotor without forcing took a steady value after 0.0316 s, while in the proposed solution it is twice as fast, that is, after 0.0152 s.

3.2 Investigation of the Harmonic Components of the Network with an Optimally Selected Transformation Coefficient

A typical case for the search algorithm is shown in Fig. 4 and 5. The engine is loaded at 20% of the rated load and at a time of 0.6 s, a step signal is received to attenu-ate the flow from 0.96 VB to 0.7 VB. Without forcing, the transition process for the rotor flow coupling is 0.0172 s. Forcing when the flow is weakened accelerates the process by 38%, i.e. the steady-state process is reached in 0.0107 s after the control signal is applied.

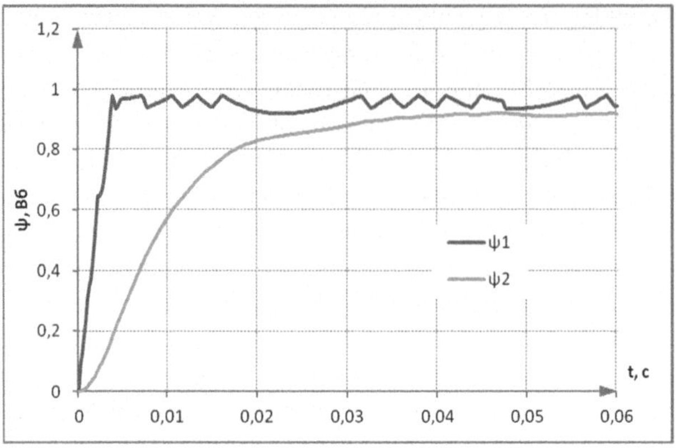

Fig. 2. Graph of changes in the stator and rotor flow coupling during start-up without forcing the magnetic flux.

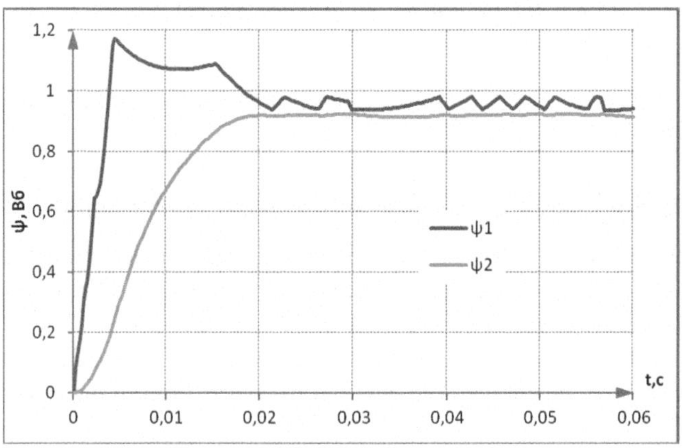

Fig. 3. A graph of the flux linkage of the stator and rotor at start-up by forcing the magnetic flux.

The proposed method makes it possible to significantly speed up the formation of the rotor flow coupling without adding sensors and significantly complicating the mathematical apparatus.

In self-tuning search systems, the minimum or maximum quality measures (minimum current, minimum power, etc.) are found using specially organized search signals. The lack of a priori information is compensated by the object's reaction to artificially entered search (trial, test) effects. The problem arises of unambiguous interpretation of the electric drive response to as minimal impacts as possible. Therefore, further improvement of the proposed system is possible through the introduction of neural networks.

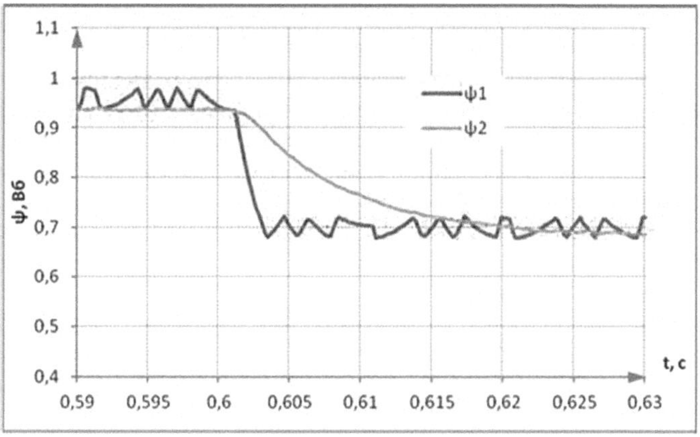

Fig. 4. Graph of changes in the stator and rotor flow coupling when the flow is weakened at 20% load without forcing the magnetic flux.

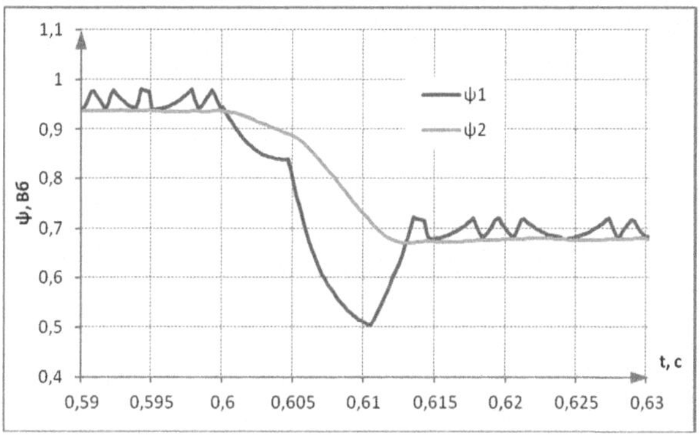

Fig. 5. Graph of changes in the stator and rotor flow coupling when the flow is weakened at 20% load with magnetic flux forcing.

References

1. Meshcheryakov, V.N., Levin, P.N.: Multi-criteria optimization of asynchronous electric drive based on fuzzy logic. Electrotechnical systems and complexes: international collection of scientific papers, Magnitogorsk, pp. 53–58 (2012)
2. Sokolovsky, G.G.: Electric Drives of Alternating Current with Frequency Regulation: A Textbook for Students of Higher Educational Institutions, Publishing Center Academy, Moscow, p. 272 (2006)
3. Gracheva, E.I., Alimova, N.A.: Calculating methods and comparative analysis of losses of active and electric energy in low voltage devices. In: International Ural Conference on Electrical Power Engineering (UralCon), pp. 361–367 (2019)
4. Gracheva, E.I., Naumov, O.V.: Estimation of power losses in electric devices of the electrotechnical complex. In: International Conference on Industrial Engineering, Applications and Manufacturing (ICIEAM), p. 6 (2019)

5. Sinykova, T.V., Sentsov, E.V., Sinyukov, A.V.: Neural network speed observers. In: Proceedings 2019 1st International Conference on Control Systems. Mathematical Modelling, Automation and Energy Efficiency (SUMMA), Lipetsk, Russia, p. 288 (2019)
6. Sinykova, T.V., Gladyshev, V.E., Sinyukov, A.V.: Methods for reducing electromechanical oscillations in conveyor control systems. In: Proceedings 2019 1st International Conference on Control Systems, Mathematical Modelling, Automation and Energy Efficiency (SUMMA), Lipetsk, Russia, p. 435 (2019)
7. Fedotov, E.A., Gracheva, E.I., Fedotov, A.I., Chernova, N.V.: Local fourier transformation application for mathematic modeling of synchronous machine valve actuator. J. Eng. Appl. Sci. **11**(13), 2939–2945 (2016)
8. Meshcheryakov, V., Sinykova, T., Sinyukov, A., Boikov, A., Mukhametzhanov, R.: Modeling and analysis of vector control systems for asynchronous motor. In: High Speed Turbomachines and Electrical Dreves Conference 2020 (HSTED-2020), Prague, Czech Republic, 09 July 2020, p. 01001 (2020)
9. Meshcheryakov, V.N., Sibirtsev, D.S.: Frequency asynchronous electric drive with correction of phase shift between moment-forming vectors. Control Syst. Inf. Technol. **2** (68), 48–57 (2017)
10. Meshcheryakov, V., Sibirtsev, D., Mikhailova, E.: Mathematical simulation of synchronized asynchronous electric drive. In: High Speed Turbomachines and Electrical Dreves Conference 2020 (HSTED-2020), Prague, Czech Republic, 09 July 2020, p. 01021 (2020)
11. Shtovba, S.D.: Design of Fuzzy Systems by Means of MATLAB, p. 288. Hotline-Telecom, Moscow (2007)
12. Pegat, A.: Fuzzy Modeling and Control, p. 798. BINOM, Laboratory of knowledge, Moscow (2012)
13. Gracheva, E.I., Petrov, V.N., Russova, N.V., Samuilov, D.V., Svintsov, G.P., Protosovitsky, I.V.: On modeling on a single-winding forced valve electromagnet in a circuit with a ballast resistor. E3S Web Conf. **124**, p. 02012 (2019). SES-2019, pp. 5–9
14. Kozyaruk, A.E.: Energy-efficient electromechanical complexes of mining and transport machines. Notes Min. Inst. **218**, 261–269 (2016)
15. Alekseev, V.V., Kozyaruk, A.E.: Comparative analysis of models of automated asynchronous electric drive systems with vector control. Electr. Eng. **12**, 47–51 (2013)
16. Verma, P., Saxena, R., Chitra, A., Sultana, R.: Implementing fuzzy PI scalar control of induction motor. In: IEEE International Conference on Power, Control, Signals and Instrumentation Engineering (ICPCSI), Chennai, India, 21-22 September 2017, pp. 1674–1678 (2017)
17. Pugachev, A.: Efficiency increasing of induction motor scalar control systems. In: International Conference on Industrial Engineering. Applications and Manufacturing (ICIEM), St. Petersburg, Russia, 16-19 May 2017, pp. 1–5 (2017)
18. Fiser, J., Sitek, P., Vyhlidal, T.: Neutral PID control loop investigated in terms of similarity theory. IFAC-Papersonline **51**(14), 212–217 (2018)
19. Fiser, J., Sitek, P., Vyhlidal, T.: Robust PID controller design for plants with delay using similarity theory. IFAC-Papersonline **51**(14), 236–241 (2018)
20. Hu, Z., Bodyanskiy, Y.V., Tyshchenko, O.K.: A multidimensional adaptive growing neuro-fuzzy system and its online learning procedure. In: Advances in Intelligent System and Computing, vol. 689, pp. 186-203 (2018)
21. Gracheva, E., Alimova, A.: Calculating probability of faultless work of shop nets with the help of coefficients of ratio. In: 2019 International Russian Automation Conference, RusAutoCon 2019, p. 8867727 (2019)
22. Gracheva, E.I., Naumov, O.V.: Evaluation criteria of contact group technical state concerning electrical appliances. Int. J. Pharm. Technol. **8**(4), 26763–26770 (2016)

Experimental Study of the Relationship Between the Technical State of a Power Transformer and the Parameters of the Higher Harmonic Components of Currents and Voltages Generated by It

A. Nemirovskiy[1(✉)], E. Gracheva[2(✉)], and M. Bashirov[3,4]

[1] Vologda State University, Lenina Street 15, Vologda 160000, Russia
[2] Kazan State Power Engineering University, str. Krasnoselskaya, 51, Kazan 420066, Russia
gracheva.i@bk.ru
[3] Salavat Branch of Ufa State Petroleum Technological University, Gubkina Street 67, The Republic of Bashkortostan Salavat 453250, Russia
[4] Sterlitamak branch of the Institute for Strategic Studies of the Academy of Sciences of the Republic of Belarus, Kirova Street, 15, Ufa 450008, Russia

Abstract. The aim of this work is to study the relationship between the technical state of a power transformer and the parameters of the higher harmonic components of currents and voltages generated by it. An experimental setup was developed to study the parameters of the higher harmonic components of currents and voltages generated by a dry three-phase power transformer. The research results are given in the form of diagrams reflecting the change in the spectrum of harmonic components (from 3 to 13 inclusively) of phase currents and voltages when simulating various transformer damages. Regularities of changes in these parameters are revealed and the possibility of their use for assessing the technical condition and predicting the resource of power transformers is shown.

Keywords: Power transformer · Higher harmonic components · Voltages

1 Introduction

Power transformers are one of the most critical elements of power supply systems of enterprises, the technical condition of which directly affects the continuity and safety of technological and production processes [1, 2]. Any power transformer contains high and low voltage windings, winding taps, magnetic core, insulation, cable entry, cooling system. Oil-immersed transformers have a more complex design compared to that of dry-type power transformers; they additionally contain a casing (tank) filled with oil, an expansion tank, an on-load tap-changer, a system of natural or forced oil circulation, including pipe connections, radiators, electric drives of pumps and air fans. The proper operation of transformer cannot be achieved without a relay protection system and

A. Sinitsyn (Ed.): SES 2020, LNCE 141, pp. 155–166, 2021.
https://doi.org/10.1007/978-3-030-67654-4_18

automation. Failure of any of these elements leads to failure of the entire transformer [3–12].

The most damaged transformer nodes are bushings, windings, taps and voltage regulators (up to 80%); a significant proportion of failures are caused by leaks and leakage of transformer oil. The most typical type of damage in the windings is the turn-to-turn fault (about 80%), which is caused by aging destruction of the insulation due to its natural wear or prolonged overloads of the transformer with insufficient cooling of windings. Failure of insulation turns can also occur due to mechanical damage during short circuits. According to statistics, all other elements that form the transformer structure are also susceptible to damage, although to a lesser extent. A fairly high percentage of damage is associated with improper actions of personnel (about 25%). The primary reasons are the unsatisfactory transformer operation and poor-quality repairs. Repair work without proper diagnostics and refinement of the technology for restoring characteristics often leads to a decrease in the transformer reliability.

To assess the technical condition of power transformers, a whole range of methods and means of diagnostic examination, testing, measurement and control is used: chromatographic analysis of gases dissolved in transformer oil; thermal imaging of the transformer temperature field; vibration diagnostics of the active part and auxiliary components of the transformer; acoustic control; measurement of winding insulation parameters; measurement of winding resistance to direct current; measurement of no-load losses; measurement of short circuit resistance; measurement of tgδ of oil; electrical and acoustic measurements of partial discharges; checking the transformation ratio; physical and chemical testing of oil; testing the insulation of windings and bushings with increased voltage; visual control. In addition, it is necessary to carry out diagnostics of protective and measuring equipment, switching devices, tank and pipe connections, cooling radiators, oil pumps and fans in conjunction with an electric drive and control and protection devices [13, 14].

All these methods of diagnostics, testing, measurement, and control of power transformers are focused on identifying the already developed defects, most of them need the withdrawal of the inspected transformer from the operating mode, its partial or complete disassembly. The obtained results cannot be used in modern intelligent control systems for the technical condition, energy efficiency and operational safety of electrical and power equipment; they do not provide detecting and identifying defects in transformers at an early stage of their development.

To ensure the necessary reliability of power supply, technical and organizational measures are currently being developed to prevent the occurrence and development of emergency scenarios caused by the failure of power transformers. First of all, it is a redundancy of transformers in combination with an effective relay protection system and a fast automatic transfer switch. This approach to ensuring the reliability of power supply has reached a high level of development by now. However it no longer fits into the modern concept of automated management of enterprises and electric grids, focused not on eliminating, but on preventing emergency situations based on the use of comprehensive equipment diagnostics systems. Therefore, at present, the task of developing new methods and diagnostic tools for electrical and electric power equipment, elements and automation systems, focused on preventing their failure, capable of integrating into automated control systems of enterprises and electrical networks, is urgent. Improving the methods and means for assessing the technical condition and resource of safe operation of electric power equipment, including power transformers and their components, also requires the transition to an economically more profitable system of maintenance and repair based on the actual technical condition.

In recent years, there has been an intensive development of diagnostic methods for transformer equipment with a pronounced tendency to choose effective methods for online assessment of their technical condition without disconnecting from the network. One of these methods is a method based on spectral analysis of transformer phase currents and voltages. Due to the nonlinearity of the electrical and magnetic characteristics of the constituent elements, the principle of operation and design features of three-phase transformers, they initially generate higher harmonic components of phase currents and voltages, which form a frequency image of the transformer's serviceable technical condition. During operation, the electrical, magnetic and geometric parameters of the transformer change, which is the reason for the change in the spectrum of the generated higher harmonic components of currents and voltages.

The purpose of this work is to study the relationship between the technical state of a power transformer and the parameters of the higher harmonic components of currents and voltages generated by it. The revealed regularities can be used to solve the problems of assessing the technical condition and predicting the resource of safe operation of power transformers.

Table 1. Technical characteristics of the TSZP-25/07 transformer.

Type	Rated power, kVA	Rated voltage of windings, V		Dimensions, mm			Mass, kg
		HV	LV	Length	Width	Height	
TSZP-25/07	25	380	160	590	320	530	140

2 Materials and Methods

For experimental studies, a dry three-phase power transformer TSZP-25/07 (Russia) was selected. According to its abbreviation, it has the following characteristics:

T - three-phase;

S - dry;

Z - protected execution;

P - converting;

25 - transformer power, kVA;

0.7 - voltage class of high voltage winding, kV.

TSZP transformers are widely used at enterprises of oil and gas production, mechanical engineering, power engineering. They are designed to convert electricity in three-phase AC power grids for use in stationary thyristor rectifiers of galvanic installations, excitation devices of synchronous machines and other converter equipment. They have a continuous operating mode from a network with 380 V voltage, and rated frequency of 50 Hz.

The TSZP transformer consists of a minimum set of basic elements typical for any three-phase power transformer - a magnetic circuit, high and low voltage windings, taps. So the research results can be generalized for a wide class of power transformers. The rated voltages and currents, the overall and mass indicators of the transformer allow conducting research in laboratory conditions. Table 1 shows the technical characteristics of the transformer.

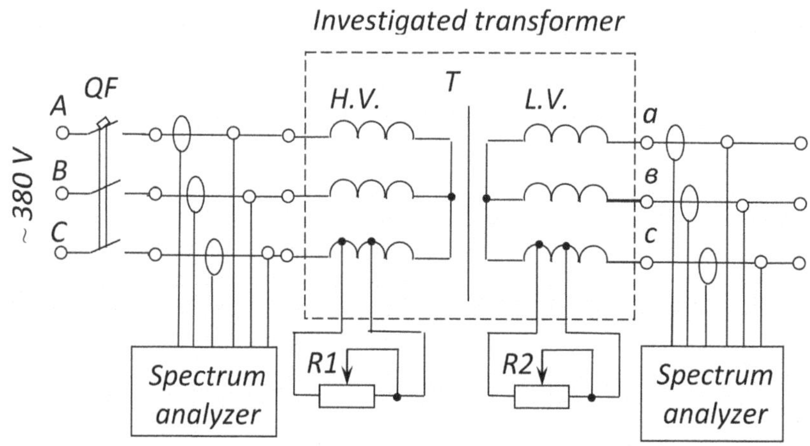

Fig. 1. Experimental setup for studying dry three-phase power transformer TSZP-25/07.

To perform the examination, an experimental setup was developed, the structural diagram of which is shown in Fig. 1. During the study, typical transformer damages were artificially created: turn-to-turn faults in the low (LV) and high voltage (HV) phase windings, breakage of one and two phases on the LV and HV side, the active resistances of the windings were changed by turning on the rheostats R1 and R2, the magnetic circuit was weakened. An asynchronous electric motor was used as the transformer load. Based on the results of measuring the phase currents and voltages of the transformer, power quality (spectrum) analyzers Resurs-UF2M and Energotester PKE determined the coefficients of the n-th harmonic components of the phase currents K_{In} and voltages K_{Un} as a percentage of the amplitude of the first (fundamental) harmonic. The results of the decomposition of the curves of phase currents and voltages in a Fourier series were presented using a specially developed computer program in the form of graphs - spectra of harmonic components of phase currents and voltages [15–17].

During the study, measured were the initial phases of the harmonic components of phase currents $\psi_{i(n)}$ and voltages $\psi_{u(n)}$ relative to the initial phase of the fundamental harmonic and the phase shift angles $\varphi_{ui(n)}$ between the corresponding harmonic components of phase currents $i_{(n)}$ and voltages $u_{(n)}$.

The spectrum of the transformer phase currents includes harmonics from the network, caused by the quality of the supply voltage, which is not perfectly sinusoidal. Filtration of harmonic components coming from the network was carried out as follows [17–19]:

- If the phase shift angle $\varphi_{ui(n)}$ between $u_{(k)}$ and $i_{(k)}$ is more than 90° or less than −90°, then the transformer on the considered averaging interval contains a source of the n-th harmonic component $i_{(k)}$ of current;
- If the phase shift angle $\varphi_{ui(n)}$ between $u_{(k)}$ and $i_{(k)}$ is less than 90° or more than −90°, then the transformer on the considered averaging interval does not contain a source of the n-th harmonic component i (k) of the current.

3 Results and Discussion

When simulating various damages of the TSZP-25/07 transformer, the most sensitive were the odd harmonic components of phase currents and voltages from 3 to 13 inclusively. Subsequent experimental research was focused on the observation and analysis of changes in these harmonic components of phase currents and voltages. The obtained results are presented in Figs. 2, 3, 4, 5, 6, 7, 8 and 9.

Figure 2 shows the harmonic components of the transformer phase currents with turn-to-turn fault in phase C of the HV winding. An increase in the content of all observed current harmonic components of phase C is seen. The 3rd harmonic showed a most significant increase (2.5 times), and the 5th, 7th, 9th and 11th harmonics increased approximately 2 times. At the same time, there is a change in the content of these harmonics in currents and voltages of the undamaged phases A and B, both upward and downward, and the 5th harmonic of the current in phase B increased by almost 3 times.

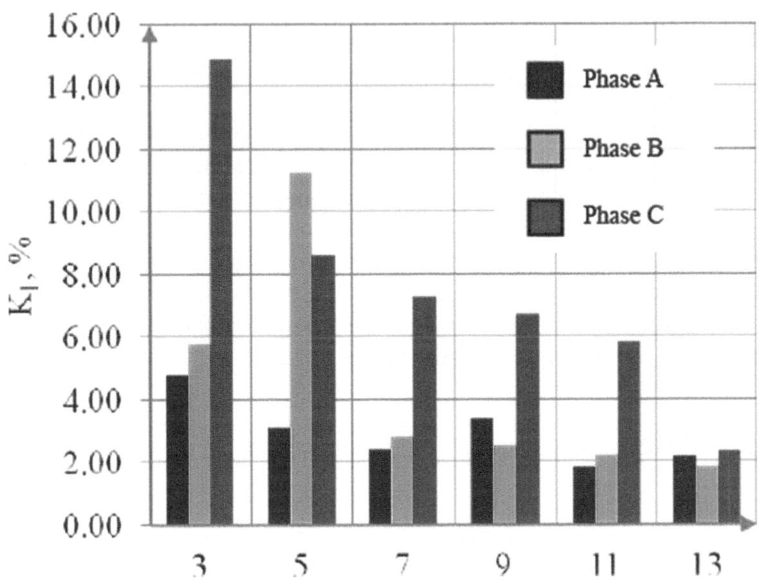

Fig. 2. Harmonic composition of phase currents with turn-to-turn fault in phase C of the HV transformer winding.

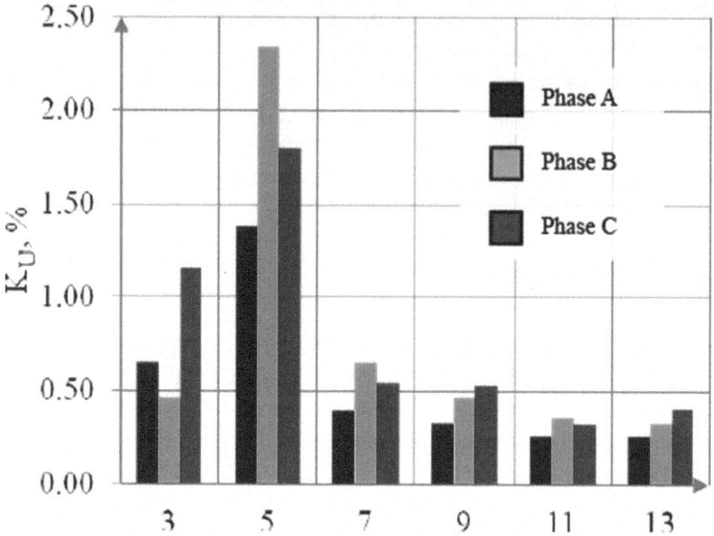

Fig. 3. Harmonic composition of phase voltages with turn-to-turn fault in phase C of the HV transformer winding.

Figure 3 shows the harmonic components of the transformer phase voltages with turn-to-turn fault in phase C of the HV winding. As one can see, there is an increase in all harmonic components of the damaged phase, the 3rd harmonic of the phase C voltage increases by about 2 times, while an increase of 2–2.5 times is observed in the 5th harmonic of voltages of all three phases of the transformer. The 7th and 13th harmonics increase in the B phase voltage. The harmonic content in the A phase voltage changes slightly.

Figure 4 shows the harmonic components of phase currents on the HV side with a turn-to-turn fault in the C phase of the LV transformer winding. It can be seen that the regularity of an increase in all harmonics of the current of phase C is preserved and an increase in the 5th, 7th, 11th and 13th harmonics of the current of phase B occurs. The 3rd harmonic of phase C current increases most noticeably. Harmonics of A phase current change insignificantly. It should be noted that the relative change in all harmonics with turn-to-turn fault in the HV winding is approximately 2 times greater than with turn-to-turn fault in the LV winding.

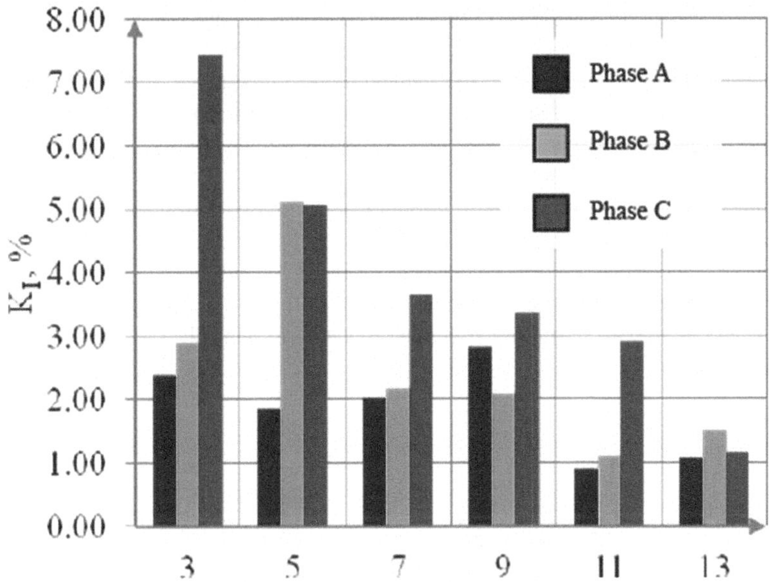

Fig. 4. Harmonic composition of phase currents on the HV side with turn-to-turn fault in phase C of the LV transformer winding.

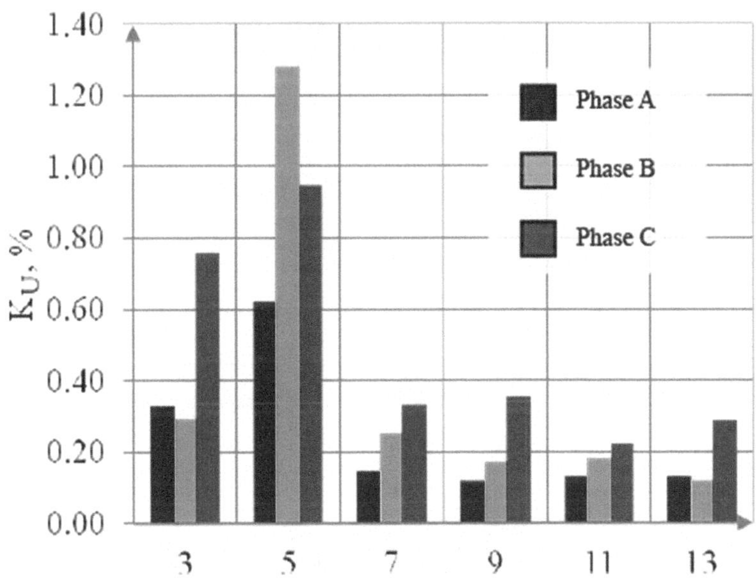

Fig. 5. Harmonic composition of phase voltages on the HV side with turn-to-turn fault in phase C of the LV transformer winding.

Figure 5 shows the harmonic components of phase voltages on the HV side with a turn-to-turn fault in the C phase of the LV transformer winding. The general regularity of change in voltage harmonics in the damaged and undamaged phases remains, as in the case of a turn-to-turn fault in the HV winding, while the relative change in all harmonics is approximately 2 times less. In the damaged phase C, the 3rd and 5th harmonics change most significantly.

Figure 6 shows the harmonic components of phase currents generated by the transformer during a short-circuit of the HV winding of phase C. Such damage to the transformer is accompanied by a sharp increase in all current harmonics in the damaged phase, the increase in the 3rd harmonic is especially noticeable - 5 times. In intact phases A and B, the 3rd harmonic also increases, and in phase B, the 5th harmonic also increases.

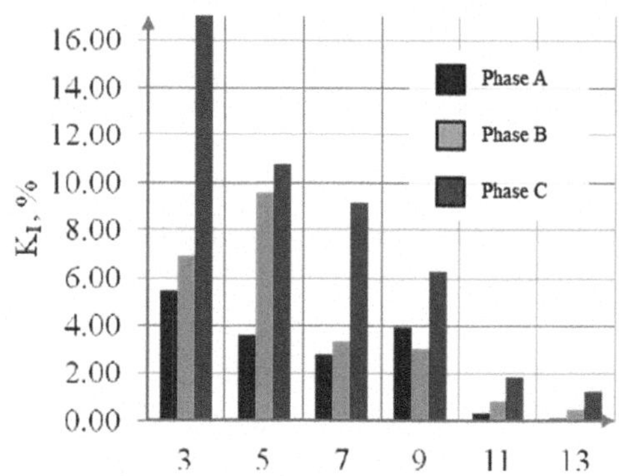

Fig. 6. Harmonic components of phase currents on the HV side in case of short circuit in phase C of the LV winding.

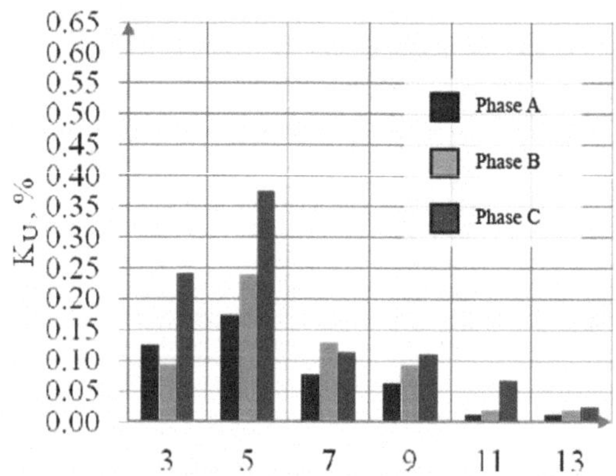

Fig. 7. Harmonic components of phase voltages on the HV side in case of short circuit in phase C of the LV winding.

Figure 7 shows the harmonic components of phase voltages generated by the transformer during a short-circuit of the phase C winding. As one can see, with such a fault, all harmonics of the phase C voltage increase sharply. The 5th, 7th and 9th harmonics of voltage of the undamaged phase B increase as well as the 3rd and 5th harmonics of voltage of the intact phase A.

It should be noted that, depending on the nature and location of damage, along with an increase in some harmonic components, a decrease in other harmonic components of currents and voltages of the undamaged phases is observed.

Figures 8 and 9 show the harmonic components of phase currents and voltages on the HV side when loosening the tie of the transformer magnetic circuit, accompanied by a change in its magnetic properties. An increase in all harmonic components of currents and voltages of all three phases is observed, but it is most pronounced in phase C.

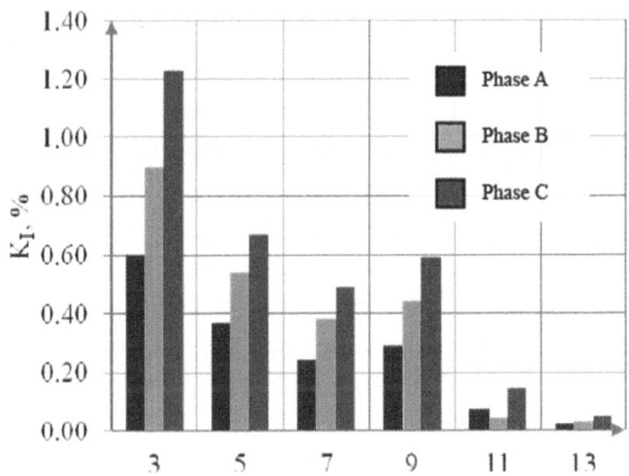

Fig. 8. Harmonic composition of phase currents on the HV side when loosening the tie of the transformer magnetic circuit.

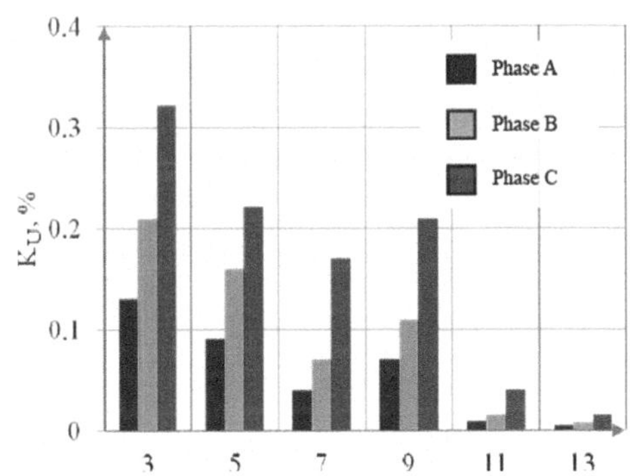

Fig. 9. Harmonic composition of phase voltages on the HV side when loosening the tie of the transformer magnetic circuit.

During the study, registered were the initial phases of the harmonic components of phase currents $\psi_{i(n)}$ and voltages $\psi_{u(n)}$ relative to the initial phase of the fundamental voltage harmonic and the phase shift angles $\varphi_{ui(n)}$ between the corresponding harmonic components of phase currents $i_{(n)}$ and voltages $u_{(n)}$, which can significantly increase the information content of the harmonic components of the phase currents and voltages of the transformer when performing spectral analysis using a special computer program [15].

The reason for the generation of higher harmonics by transformers is the nonlinearity of the magnetization curve of their magnetic circuits and the presence of a hysteresis loop. The nonlinearity of the magnetization curve and the presence of a hysteresis loop lead to a distortion of the sinusoidality of the magnetizing current of the transformers and, as a consequence, to the appearance of higher harmonics in current consumed by the transformer from the network. Due to the asymmetry of the magnetic circuit of three-phase three-rod transformers, the magnetizing currents of the outermost phases (A and C) exceed the current of the middle phase (B). For this reason, all odd harmonics, including multiples of three, are present in the magnetizing currents. The 3rd, 5th and 7th harmonics have the highest specific gravity.

Simulation of damage to the transformer phase windings in the form of turn-to-turn faults and short circuits leads to an increase in its electrical and magnetic asymmetry, and, accordingly, to the generation of higher harmonic components. Turn-to-turn faults can lead to resonant circuits and a corresponding sharp increase of the presence of certain harmonics of phase currents and voltages. Simulation of changes in the magnetic properties of the magnetic circuit by loosening its tie also leads to a change in the spectrum of harmonic components of phase currents and voltages - the presence of the 3rd, 5th, 7th and 9th harmonics has increased. These harmonics are sources of additional losses and noise in transformers. The difference in the spectra of the harmonics of currents and voltages of different phases indicates the unevenness of the weakening of the tie of the cores of the magnetic circuit, in the future this phenomenon can be used to identify and localize damage.

When simulating various transformer damages, a change in the spectra of harmonic components of currents and voltages of all three phases was observed. This indicates that a specific technical condition can be represented by a set of harmonic components of all phase currents and voltages - its frequency model (frequency image). To solve the problem of recognizing the technical state of the transformer and predicting its safe operation life based on the analysis of changes in the parameters of the harmonic components of phase currents and voltages, the most optimal is the use of an artificial neural network [15].

4 Conclusion

An experimental setup was developed consisting of a dry three-phase power transformer TSZP-25/07, two analyzers of spectrum of harmonic components of currents and voltages, two rheostats and a load in the form of an asynchronous motor. It was used to perform laboratory studies of the relationship between the technical state of a power transformer and the parameters of higher harmonic components of currents and voltages generated by it. During the study, typical transformer damage was artificially created - turn-to-turn short circuits in the phase windings, the break of one and two phases, the active resistances of windings were changed, the magnetic circuit was weakened.

During experimental studies, it was found that when simulating damage in one of the transformer phases, the spectra of harmonic components of currents and voltages of all three phases change. The parameters of the odd harmonic components of phase currents and voltages from the 3rd to the 13th inclusively were the most sensitive to changes in the technical state of the transformer. The percentage of these harmonics in the current and voltage of the damaged phase increased up to 5 times, and in undamaged phases, the content of these harmonics changed both upward and downward, depending on the nature and severity of the damage. Hence, we can conclude that a specific set of harmonic components of phase currents and voltages corresponds to a specific technical state of the transformer and this state can be represented by its frequency model (frequency image). To solve the problem of recognizing the technical state of a transformer and predicting its safe operation life based on the analysis of changes in the parameters of the harmonic components of phase currents and voltages, the most optimal solution is to use an artificial neural network. In this case, the power transformer itself can be considered as a highly sensitive electromagnetic active sensor that converts all changes in the technical state of its elements into the parameters of the generated higher harmonic components of phase currents and voltages.

References

1. Alyunov, A.N., Vyatkina, O.S., Akhmetova, I.G., Pentiuc, R.D., Sakipov, K.E.: E3S Web Conf. **124**, 02015 (2019)
2. Hariharasudan, A., Kot, S.: Soc. Sci. **7**, 227 (2018)
3. Dolin, A.P.: Energetic **7** (2001)
4. Gun, I.G., Salganik, V.M., Evdokimov, S.A., Sarlybaev, A.A.: Bull. Magnitogorsk State Tech. Univ. n.a. G.I. Nosov **1**, 102 (2012)
5. Udaratin, A., Alyunov, A., Krutikov, A., Mukhametova, L.R., Zaripov, O.O., Bochkarev, I. V.: E3S Web Conf. **124**, 02020 (2019)
6. Ryder, B.S., Communication 1 (2006)
7. Nemirovskiy, A. E.,. Kashin, A. I, Kosmach, V. F., Titovec, Y. F., Toptygin, I. N., and. Zaripova, D. A.: IOP Conf. Ser. Earth Environ. Sci. **337**, 012071 (2019)
8. Ydaratin, A., Akhmetov, T., Makoev, S.: Int. J. Civ. Eng. Technol. **10**, 2001 (2019)
9. Vanin, B. V.: Electr. Stn. **9** (2001)
10. Rashevskaya, M., Yanchenko, S., Tsyruk, S.: 2018 20th International Symposium on Electrical Apparatus and Technologies, SIELA 2018 - Proceedings (2018)
11. Zatsepina, V., Zatsepin, E., Shachnev, O., Shachnev, A., Petrov, T.: E3S Web Conf. **178**, 01019 (2020)
12. Akhmetshin, A., Marin, G., and Mendeleev, D.: E3S Web Conf. **178**, 01015 (2020)
13. Kuvshinov, A.A., Khrennikov, A.Y., Vakhnina, V.V., Chernenko, A.N.: International Conference on Actual Problems of Electron Devices Engineering's, APEDE 2018 47 (2018)
14. Shabanov, V.A.: Diagnostics of the Technical Condition of Electrical Equipment of Power Supply Systems. USPTU, Ufa (2016)

15. Bashirov, M.G., Mironova, I.S.: Sci. Work. NIPI Neft. SOCAR **1**, 46 (2015)
16. Bashirov, M.G., Prakhov, I.V., Gabbasov, R.S.: Transp. Storage Pet. Prod. Hydrocarb. Raw Mater. **2**, 25 (2013)
17. Bashirov, M. G.: Ind. Power Eng. **7** (2018)
18. Ruling Document RD 153–34.0–15.501–2000: Guidelines for the Control and Analysis of the Quality of Electrical Energy in General-Purpose Electrical Networks. Part 1, Moscow (2000)
19. Ruling Document RD 153–34.0–15.502–2002: Guidelines for the Control and Analysis of the Quality of Electrical Energy in General-Purpose Electrical Networks. Part 2, Moscow (2002)

A Survey-Based Approach to Investigate the Relation Between Open-Plan-Office Environments and Occupants' Sleep Quality in India

Dherya Mehta[1(✉)] and Gnanasambandam Subbaiyan[2]

[1] Independent Researcher, Ujjain 456010, India
dheryamehta@gmail.com
[2] National Institute of Technology, Tiruchirappalli 620014, India

Abstract. India expects a significant number of people working in the formal sector under the roof of Open-Plan-Offices (OPO's) by 2030. A wide range of Indoor Environmental Quality factors affecting occupants and their effect on human health is a reality now. A critical aspect of measuring human health is sleep quality. This study relates surveys (n = 102) of occupants from four different OPO's in India about their sleep quality with the indoor workplace environment. Pittsburgh Sleep Quality Index (PSQI) was used to measure sleep quality. Analysis of variance (ANOVA) was applied to test whether these office groups differ significantly on their average Global PSQI Score. The study uses Welch's adjusted F ratio as the assumption of homogeneity (through Levene's Test) of variance did not meet. Post hoc follow-up procedures (Games-Howell) tested the difference between all unique pair-wise comparisons. The results reveal that the value of carpet area per person, use and personal control of curtains, surrounding features and outdoor view, type of plan, and lack of glare from windows affect the sleep quality of occupants. Building orientation, Window-to-Wall ratio, interior partition height, climatic zone, and use of personal computers have also shown an association with sleep quality.

Keywords: Open-plan-offices · Occupants' sleep quality · Indoor environmental quality

1 Introduction

People in the developed nations spend their maximum time indoors [1]; likewise, people working in the formal sector spends most of the time within their office buildings. Historically daylight guided building design but the technological advancements in energy-efficient fluorescent lighting and artificial cooling during the mid 19th century led the way for low-heightened rooms (which avoided natural light from clerestories and provided better artificial cooling) eventually evolving into Open-Plan-Offices (OPO's) [2], which are still in practice.

The design of a building along with its physical features, is directly linked to occupant comfort [3] and satisfaction [4] which affects their performance and

A. Sinitsyn (Ed.): SES 2020, LNCE 141, pp. 167–181, 2021.
https://doi.org/10.1007/978-3-030-67654-4_19

productivity [5, 6]. Indoor Environmental Quality (IEQ) is a good indicator of the performance of a building [7]. "IEQ refers to the environmental conditions inside regularly occupied space that is determined by many factors, including Indoor Air Quality (IAQ), thermal comfort, visual comfort, and acoustic comfort" [8]. The latest review studies done on criteria for occupant comfort in workplaces have also mentioned biophilia [9, 10], office location, layout, amenities and feel [9], appearance and colour, privacy, flexibility, and personal control as criteria for occupant well-being [10]. Now, IEQ affects the health and well being of occupants [11] along with their productivity in the workplace [12].

Along with the factors mentioned above, better sleep quality may also ensure better work productivity [13]. A concern about sleep quality in the working population has already been shown [14]. Sleep quality affects physical and mental health [15], well-being, and productivity [16] in office workers. Circadian rhythm in the human body affects sleep, alertness, mood, and behaviour [17]. Disrupting these normal human rhythms may affect performance, safety, and health [18]. Hence, one can postulate that IEQ of offices, sleep quality of occupants and their work productivity are inter-related.

India being a developing nation, holds an immense possibility of learning from the past deeds of the already developed nations. As per the census report of 2011, 31% of the total population (1.2 Billion) lived in the urban area of the country [19]. India had a significant (20–30%) portion of working-class people, of which 30% belonged to the main-worker category [19, 20]. Around 78% of the total main worker category were from the age 25–59 [21], out of that, 87% were literate, and 28% possessed higher qualifications [22]. India with a current population of 1.36 Billion (as of 20-October-2020) [23] and to fulfil the United Nations 2030 Agenda for Sustainable Development (with a focus on health & well-being, quality education, industry, innovation & infrastructure, and sustainable cities) [24] a significant amount of young educated population will work in the formal sectors in the future which will definitely require office spaces to cater the infrastructure needs. Hence there is a need of the study which investigates the link between sleep quality of the occupants in OPO's and the latest researched IEQ parameters, the results from which may guide the future workplace environment design.

Table 1. Architectural data of offices.

Office code	Office 1	Office 2	Office 3	Office 4
Office type/use	Administrative	Research & development	Bank	Marketing
City	Tiruchirappalli	Pune	Indore	Indore
Climatic zone	Hot and Humid	Moderate	Composite	Composite
Total carpet area (square meter)	352.8	583.54	168.48	94.5
Carpet area (sqm)/ occupant	6.53	7.88	3.66	2.25
Office floor	First	Third	Ground	Ground
The facade of window location	Longer	Longer	Longer	Shorter
Orientation & WWR (%)	N = 21.08, S = 21.08	E = 26.58, W = 26.58	E = 44.61	NW = 80

(*continued*)

Table 1. *(continued)*

Office code	Office 1	Office 2	Office 3	Office 4
Nearby features within 6 m	Yes	Yes	No	No
Type of nearby features	N-Building Block, S-Trees	E-Nothing, W-Solar panels	Nil	Nil
Were curtains drawn during the survey?	No	Yes	Yes	No
Personal computers on all table	No	Yes	Yes	Yes
Partition height in the interior	No	Waist height	Waist height	Waist height

2 Materials and Methods

2.1 Research Instrument - Questionnaire Form

A self-reported survey form was prepared based on two parts. The first part was about their necessary information such as name (optional), age (open-ended), and gender (female/male/other). It also included the question about their perceived seating distance from the window (less than one feet/1feet to 10 feet/10 feet to 20 feet/ more than 20 feet). Part two included the validated Pittsburgh Sleep Quality Index (PSQI) [25], which related to the occupant's usual sleep habits during the past one month only. PSQI is a set of nineteen self-rated questions and five questions rated by the bed partner or roommate. The 19 self-rated questions have seven components, namely subjective sleep quality, sleep latency, sleep duration, habitual sleep efficiency, sleep disturbances, use of sleeping medication, and daytime dysfunction. Each component is giving a grade which ranges from 0 to 3. Towards the end, the total grading from all these seven components provides the Global Pittsburgh Sleep Quality Index Score (GPSQIS), which ranges from 0 to 21, "0" indicating no difficulty and "21" indicating severe difficulties in all areas [26]. Scoring only from the self-rated questions is considered.

2.2 Principal Information

2.2.1 About the Buildings

The study considers four different air-conditioned OPO's from India. All these offices were a part of an existing building. The assumption is that each office room is an independent unit. At least one side of these office rooms was directly open to the outdoor view and/or was directly accessible from outdoors. Table 1 compiles the primary architectural data of the offices.

The first setting (Office 1) was an administrative room located in the hot and humid climate of Tiruchirappalli. Office 2 holds research and development activity in the moderate climate of Pune. The other two offices (Office 3 and Office 4) were from Indore, which experiences composite climatic conditions. Office 3 is of bank premises whereas Office 4 is a marketing office. The maximum ratio of carpet area(square meter) per occupant was in Office 2 (7.88), followed by Office 1 (6.53), Office 3 (3.66), and Office 4 (2.25).

The interior view of office 1, 2, 3 & 4 is shown in Figs. 1, 3, 5 and 7, respectively. The furniture layout of office 1, 2, 3 & 4 (with dimensions in mm) are shown in Figs. 2, 4, 6 and 8 respectively. The Office 1 room studied is located on the first floor of the building. It is a north-south oriented room with windows on both the long sides. There are an adjacent block and trees on the north side within 6 m, and similarly, there are trees on the south side. The Office 2 room is east-west oriented, located on the third floor of the building with windows on both the long sides. Within the 6 m range, there are solar panels on the first floor level of the west side block, whereas there is no obstruction on the east side. The rooms studied for Office 3 and Office 4 are both located on the ground floor. Office 3 is east facing with no obstructions on the east within 6 m, whereas Office 4 is North-west facing with no obstruction in 6 m. Office 2, 3 and 4 had waist height partitions with computer systems on every desk.

Fig. 1. Interior view of office 1.

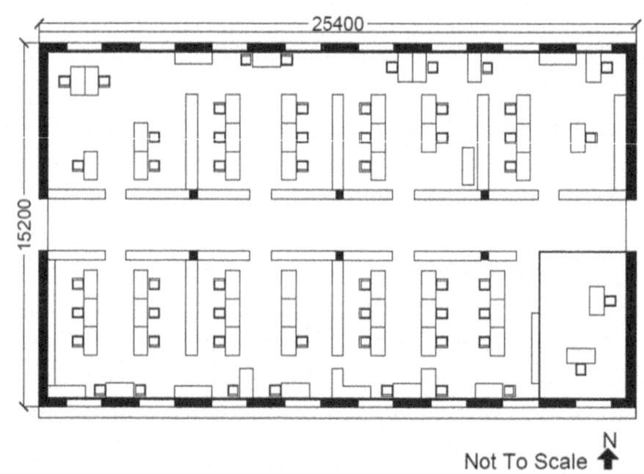

Fig. 2. Plan with furniture layout of office 1.

Fig. 3. Interior view of office 2.

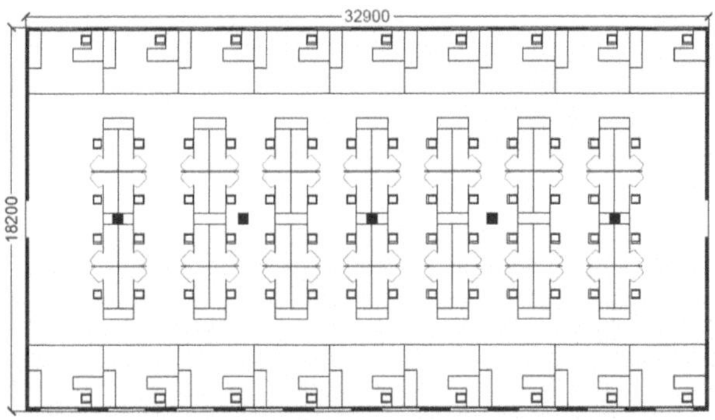

Fig. 4. Plan with furniture layout of office 2.

Fig. 5. Interior view of office 3.

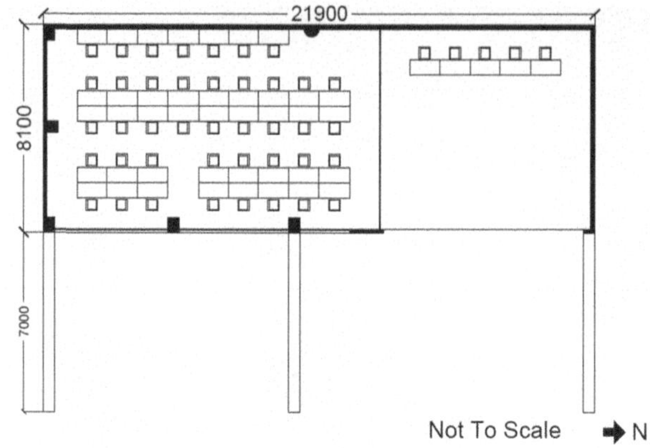

Fig. 6. Plan with furniture layout of office 3.

Fig. 7. Interior view of office 4.

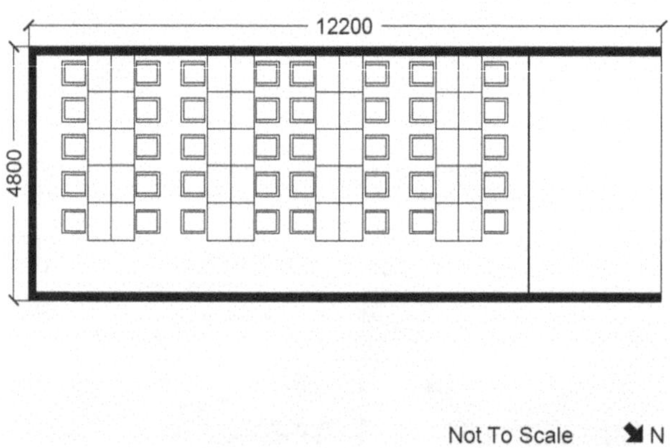

Fig. 8. Plan with furniture layout of office 4.

On the contrary, Office1 did not have personal desktops on every workstation and was devoid of any partitions, but had waist height partition in the central aisle. Occupants in Office 1 did not draw the curtains, whereas people in Office 2 and 3 pulled out the curtains to screen themselves from low angle morning and/or evening sun respectively. Occupants from Office 4 were devoid of any blinds.

The Window-to-Wall Ratio (WWR) in Office 1 was 21.08 on the north and south wall each. Office 2 had a WWR of 26.58 on both east and west wall. The value of WWR on the east wall of Office 3 was 44.61 and on the north-west wall of Office 4 was 80.

2.2.2 Respondents Data

A total of 102 responses from 4 different offices in India during March 2018 were collected. Figure 9 shows the office wise total number of occupants and the total number of participants. Figure 10 represents the office wise participants' gender data. Figure 11 shows the age data of the participants, and Fig. 12 explains the seating distance of participants from windows.

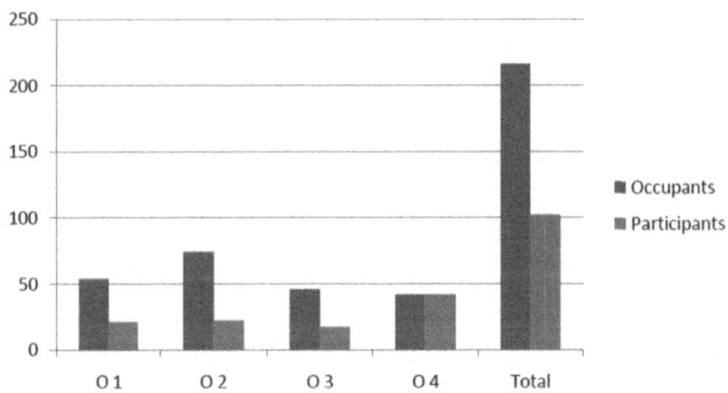

Fig. 9. Office wise occupant and participant data.

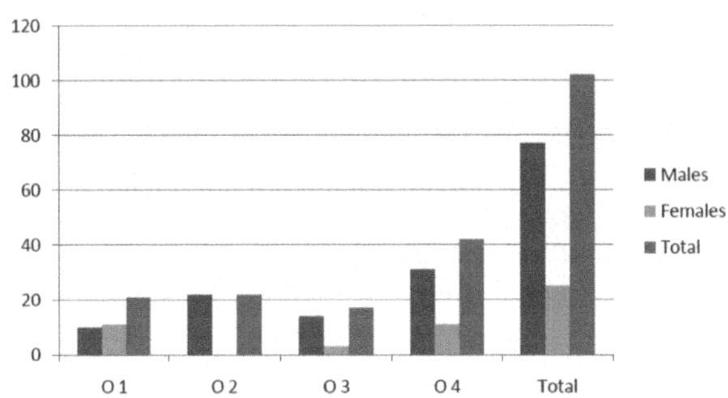

Fig. 10. Office wise participants' gender data.

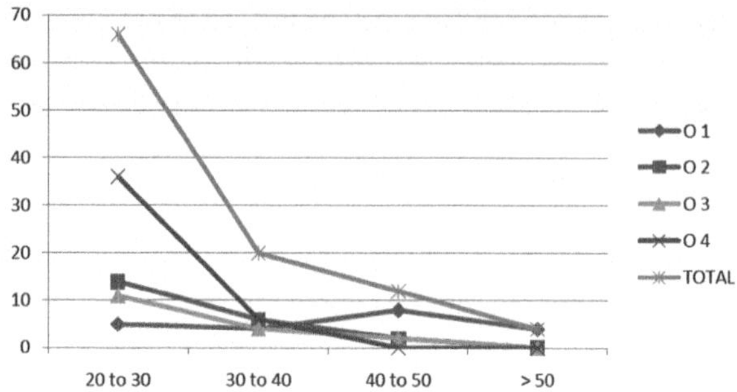

Fig. 11. Office wise participants' age data.

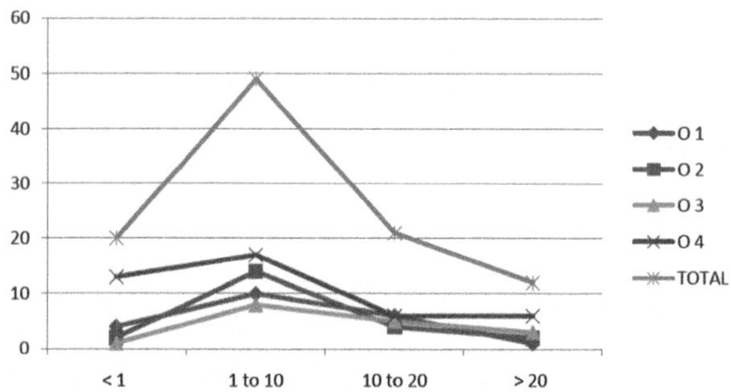

Fig. 12. Office wise participants' seating distance from window.

Overall, almost half (47.2%) of the total number of occupants participated in the survey. There were more male participants (75.5%) than female participants (24.5%). The majority of the participants (64.7%) belonged to the younger age group (20 to 30 years). While 19.6%, 11.8%, and 3.9% of the participants were in the age group of 30–40, 40–50, and above 50, respectively. A significant number of participants (67.6%) were seated near (<10 feet) to the windows.

All the occupants of Office 4 participated in the survey. Whereas only 30–40% of occupants from Office 1, 2, and 3 responded to the questionnaire. A good percentage of females (52.4%) from Office 1 respondents was part of the survey, followed by Office 4 (26.2%) and Office 3 (17.6%). All the participants of Office 2 were males. From the number of participants in each office, a maximum (85.7%) portion of young participants was from Office 4, followed by Office 2 & 3 (almost 64%) and Office 1 (23.8%). Similarly, people between 30 to 40 year age group were in the almost same number (14–27%) in all the offices, but a significant portion of middle-aged people (40 years and above) could be found in Office 1 (57.1%) only.

2.3 Procedure

The survey forms were printed and distributed to the participants during the lunch break. They voluntarily filled up the form. The survey data collected was then manually feed in MS Excel to generate GPSQIS.

2.4 Data Analysis

A total of 102 occupants from 4 different offices participated. The data analyses used SPSS version 21 (IBM Corp. Released 2013. IBM SPSS Statistics for Windows, Version 21.0. Armonk, NY: IBM Corp.). Post data analysis, this study relates to occupants' sleep quality of the past one month through GPSQIS with the architectural features of their respective offices. Architectural features considered for the analysis are building and window orientation, WWR, use of curtains & personal control, floor area per user, and indoor partition height. Other factors considered for analysis were surrounding features & outdoor view, climatic zone, and the use of personal computers.

3 Results

Box plot was plotted for GPSQIS for all the four offices, as shown in Fig. 13. It shows a significant difference in the mean. Overall it could be found that no occupant from any Office records Mean GPSQIS greater than 15. Further analysis from the box plot mentions that office 4 records the highest value of the mean, and contrary to this, the mean value in office 2 is lowest. Office 1 and 3 with almost similar mean value lies somewhere in between. Table 2 provides the Mean (M), and Standard Deviation (SD) values of individual offices.

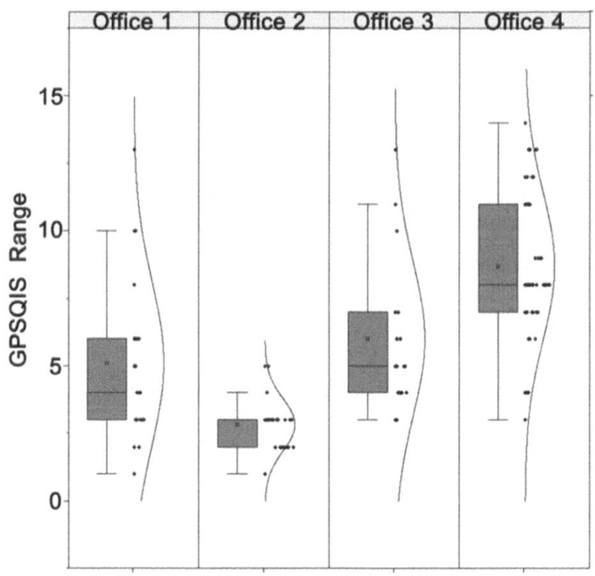

Fig. 13. Box plot for office wise GPSQIS.

Table 2. Office wise Mean and Standard deviation of GPSQIS.

Office	n	Mean	SD
1	21	5.09	3.03
2	22	2.81	0.95
3	17	6.00	2.85
4	42	8.69	2.75
Total	102	6.23	3.42

A One-way Analysis of Variance (ANOVA) examines the question of whether office 1, 2, 3 or 4 differ concerning occupants' sleep quality. The independent variable represented the offices with four groups: 1) Office 1; 2) Office 2; 3) Office 3, and 4) Office 4. The dependent variable was the average GPSQIS that occupants made on a measure of sleep quality with a range of 0 (indicating no difficulty) to 21 (indicating severe difficulties). However, the assumption of homogeneity did not meet (Table 3), because of p (.002) < α (.05) which is indicated by Levene's Test of Homogeneity of Variances, $F(3, 98) = 5.481$, $p = .002$. With an alpha level of .05, p (.002) < α (.05), which indicates significance, the null hypothesis (no variance difference) is rejected – as such, indicating that the assumption of homogeneity of variance did not meet.

Table 3. Test of homogeneity of variances.

PSQI score			
Levene statistic	df1	df2	Sig.
5.481	3	98	.002

Since the assumption of homogeneity of variance did not meet for this data, we used the obtained Welch's adjusted F ratio (53.58) (refer Table 4), which was significant at the .05 alpha level (p shown as .000, that is $p < .05$) reported as Welch's $F(3, 41.788) = 53.58$, $p < .05$, we can conclude that at least two of the four office groups differ significantly on their average GPSQIS.

Table 4. Robust tests of equality of means.

PSQI score				
	Statistic[a]	df1	df2	Sig.
Welch	53.578	3	41.788	.000

[a]Asymptotically F distributed

However, beyond that, post hoc follow-up procedures (Games-Howell) was conducted (Table 5) to test the difference between all unique pair-wise comparisons. The discussion considers the office pairs with a significant mean difference (p < α (.05)). Refer to Table 2 for M and SD values. The results indicate that occupants of office 4

(*M* = 8.69, *SD* = 2.75) had a significantly higher average score on the measure of GPSQIS than the occupants of office 3 (*M* = 6.00, *SD* = 2.85), office 2 (*M* = 2.81, *SD* = 0.95) and office 1 (*M* = 5.09, *SD* = 3.03). Further, the occupants of office 1 (*M* = 5.09, *SD* = 3.03) and office 3 (*M* = 6.00, *SD* = 2.85) had a significantly higher average score on the measure of GPSQIS than the occupants of office 2 (*M* = 2.81, *SD* = 0.95). The office (1 & 3) pairs ($p < \alpha$ (.05)) do not have a significant mean difference which implies that the average GPSQIS of these two offices was the same.

Table 5. Post hoc test.

Multiple comparisons						
Dependent variable: PSQI score						
Games-Howell						
(I) Office no.	(J) Office no.	Mean difference (I-J)	Std. error	Sig.	95% confidence interval	
					Lower bound	Upper bound
1.00	2.00	2.27706*	.69236	.015	.3659	4.1882
	3.00	−.90476	.95686	.781	−3.4848	1.6753
	4.00	−3.59524*	.78628	.000	−5.7105	−1.4799
2.00	1.00	−2.27706*	.69236	.015	−4.1882	−.3659
	3.00	−3.18182*	.72087	.002	−5.2107	−1.1529
	4.00	−5.87229*	.47149	.000	−7.1206	−4.6240
3.00	1.00	.90476	.95686	.781	−1.6753	3.4848
	2.00	3.18182*	.72087	.002	1.1529	5.2107
	4.00	−2.69048*	.81150	.013	−4.9025	−.4785
4.00	1.00	3.59524*	.78628	.000	1.4799	5.7105
	2.00	5.87229*	.47149	.000	4.6240	7.1206
	3.00	2.69048*	.81150	.013	.4785	4.9025

*The mean difference is significant at the 0.05 level

4 Discussion

As discussed earlier that when the value of GPSQIS is "0" it indicates no difficulty and when the value is "21" it indicates severe difficulties in all areas. Overall it could be found that the mean GPSQIS is not greater than 15 in any office, which means that all the occupants had no to moderate difficulties while sleeping. None of the occupants faced severe issues while sleeping. While comparing all the 4 offices, the results, based on Mean GPSQIS shows that Office 2 (M = 2.81, SD = 0.95) has the least GPSQIS followed by office 1 (M = 5.09, SD = 3.03) and office 3 (M = 6.00, SD = 2.85). Office 4 (M = 8.69, SD = 2.75) has the highest Mean GPSQIS among all.

The best sleep quality of the occupants from office 2 may exist because of the moderate (comfortable) climatic condition the city and its people experience. The staff of office 2 also enjoys the maximum carpet area per occupant (7.88 sq m per person). The glare from low angle sun on the east and west of the third floor utterly unobstructed by any surrounding features may have created issues, but the use of blinds may have ensured visual comfort. Further, the controlled presence of daylight penetrating through these translucent curtains may be a reason for the right circadian stimulus among the participants. At the same time maximum of the occupants (72.7%) who responded to the survey were seated close (<= 10 feet) to the windows. Hence, the personal control on the blinds, which the occupants may have used for outdoor view in selective times, must have affected the sleep quality. However, only a small (29.7%) portion of the entire staff participated in the study out of which all were males with a significant (63.6%) section of them in the younger age (20–30 years) group.

The staff from office 1 has also shown better sleep quality. The most significant factor contributing to this may be the north-south-oriented building with trees on the south and mutual shading on the north enabling occupants to avoid the use of curtain proving a glare less good view outside. The people in office 1 also have a higher (6.53 sq m per person) carpet area per occupant. These results are trustworthy because a good number (66.6%) of people were seated close to the (<= 10 feet) windows, further, an equal distribution of gender, male (47.6%) and females (52.4%) from all the age groups participated. At the same time, a small but significant amount (29.7%) of the total occupants participated in the study.

The participants from office 3 also have good sleep quality. The people have comparatively less carpet area per person (3.66) and use the blind to curtail the low angle east sun in the morning hours. However, the manually operated blinds would have provided an open view on the west post noon but might have caused glare on the monitor screens. Only a considerable portion (37%) of participants from the total occupancy of the office filled the forms. The better sleep quality evaluation may be because a maximum (82.4%) of them were males and most (64.7%) of them being comparatively younger.

The occupants of office 4 have shown difficulties while sleeping. All the occupants participated in the study. A deep plan office, with a window only on one shorter side, causing glare and limited daylight may be a reason for slightly higher GPSQIS. People from office 4 have the least (2.25) carpet area on individual and are devoid of any curtain on the window. A good (71.5%) percentage of people sat near (<= 10 feet) the window and most (85.7%) of them being young (20–30 years) have still shown low sleep quality. A higher (73.8%) number of occupants were male.

Sleep quality is better in males compared to females; further, it reduces as humans grow older [27]. The Mean GPSQIS of all four offices suggests that occupants from office 2 have the best sleep quality, followed by office 1 & 3. Office 4 have affected sleep quality. These results are impressive because if the better sleep quality in office 2 is because of the younger population, the same age group from office4 does not identify the same. Since furniture placement can also ensure better Circadian Stimulus (CS) [28], it implies that the occupants from office 4 are disturbed by the glare from the curtain devoid window and reflections in monitors screens which is not the case in office 2 because of monitor orientation and personal control on curtain even if the room is east-west oriented. The higher WWR value, deep plan office layout, and less area per occupant in office 4 may affect the overall sleep quality.

If the occupants of office 1 who belong to all age groups have shown similar sleep quality than those of office 3 who are mostly young, then it implies that the more carpet area per occupant, outdoor view, lack of glare, and diffused daylight penetration within interiors support better circadian rhythm among the inhabitants. The absence of a personal desktop on all the seats and waist height partition may also be a reason for better sleep quality in office 1. Office 2, 3 & 4 have a maximum of young males sitting near the window participated, but the respective mean GPSQIS and carpet area per person shows that the more the carpet area better the sleep quality.

5 Conclusion

The study investigates a link between the sleep quality of office workers and their respective working environments. The discussion from the results reveals that the value of carpet area per person, use and personal control of curtains, surrounding features and outdoor view, type of plan, and lack of glare from windows may affect the Circadian Stimulus of the inhabitants eventually affecting their sleep quality. Building orientation, WWR, interior partition height, climatic zone, and use of personal computers have also shown an association with sleep quality.

The limitation of the study is the variety in the type of work performed in all four offices. Since workers' perception of space and their IEQ demands in an OPO depends upon the type of activity performed [12], the generalizability of results is critical. Further personal characteristics like gender and age also affect occupants perception of a space which may affect their satisfaction [29]. Evaluation of sleep quality in workers should include office environments. However, there may be many factors beyond the workplace environment which may affect the sleep quality of a person like consumption of alcohol and caffeine, smoking habits, meal timings and patterns, exercise, and exposure to light before bed like the use of cell phone before sleep [16].

Hence there is a scope for future studies that link individual applicable parameters of the office environment in open-plan offices with the sleep quality of the workers to find the extent of their effect.

Acknowledgements. The author conducted the study during his post-graduation in between 2016–2018 from National Institute of Technology, Tiruchirappalli, 620014, India.

References

1. Frontczak, M., Wargocki, P.: Literature survey on how different factors influence human comfort in indoor environments. Build. Environ. **46**(4), 922–937 (2011). https://doi.org/10.1016/j.buildenv.2010.10.021
2. Çakir, A.E.: Daylight for Health and Efficiency, p. 9 (2009)
3. Andargie, M.S., Touchie, M., O'Brien, W.: A review of factors affecting occupant comfort in multi-unit residential buildings. Build. Environ. **160**, 106182 (2019). https://doi.org/10.1016/j.buildenv.2019.106182
4. Hill, M.C., Epps, K.K.: The impact of physical classroom environment on student satisfaction and student evaluation of teaching in the university environment. Acad. Educ. Leadersh.ss J. **14**(4), 16 (2010)

5. Samani, S.A., Samani, S.A.: The impact of indoor lighting on students' learning performance in learning environments: a knowledge internalization perspective. Int. J. Bus. Soc. Sci **3**(24), 10 (2012)
6. Shafaghat, A., Keyvanfar, A., Ferwati, M.S., Alizadeh, T.: Enhancing staff's satisfaction with comfort toward productivity by sustainable open plan office design. Sustain. Cities Soc. **19**, 151–164 (2015). https://doi.org/10.1016/j.scs.2015.08.001
7. Mihai, T., Iordache , V.: Determining the indoor environment quality for an educational building. Energy Procedia **85**, 566–574 (2016). https://doi.org/10.1016/j.egypro.2015.12.246
8. ISHRAE_IEQ_Feb_26_2019_public_draft. https://ishrae.in/Content/Download/ISHRAE_IEQ_Feb_26_2019_public_draft.pdf.
9. Al Horr, Y., Arif, M., Kaushik, A., Mazroei, A., Katafygiotou, M., Elsarrag, E.: Occupant productivity and office indoor environment quality: a review of the literature. Build. Environ. **105**, 369–389 (2016). https://doi.org/10.1016/j.buildenv.2016.06.001.
10. Mansor, R., Sheau-Ting, L.: Criteria for occupant well-being: a qualitative study of Malaysian office buildings. Build. Environ. **186**, 107364 (2020). https://doi.org/10.1016/j.buildenv.2020.107364
11. Boyce, P.R.: Review: the impact of light in buildings on human health. Indoor Built Environ. **19**(1), 8–20 (2010). https://doi.org/10.1177/1420326X09358028
12. Lou, H., Ou, D.: A comparative field study of indoor environmental quality in two types of open-plan offices: open-plan administrative offices and open-plan research offices. Build. Environ. **148**, 394–404 (2019). https://doi.org/10.1016/j.buildenv.2018.11.022
13. Lawson, K.M., Lee, S.: Better previous night sleep is associated with less next day work-to-family conflict mediated by higher work performance among female nursing home workers. Sleep Health **4**(5), 485–491 (2018). https://doi.org/10.1016/j.sleh.2018.07.005
14. Visvalingam, N., et al.: Prevalence of and factors associated with poor sleep quality and short sleep in a working population in Singapore. Sleep Health **6**(3), 277–287 (2020). https://doi.org/10.1016/j.sleh.2019.10.008
15. Soprovich, A.L., et al.: A systematic review of workplace behavioural interventions to promote sleep health in men. Sleep Health **6**(3), 418–430 (2020). https://doi.org/10.1016/j.sleh.2020.04.008
16. Shimura, A., et al.: Which sleep hygiene factors are important? Comprehensive assessment of lifestyle habits and job environment on sleep among office workers. Sleep Health **6**(3), 288–298 (2020). https://doi.org/10.1016/j.sleh.2020.02.001
17. Webb, A.R.: Considerations for lighting in the built environment: non-visual effects of light. Energy Build. **38**(7), 721–727 (2006). https://doi.org/10.1016/j.enbuild.2006.03.004
18. Bellia, L., Bisegna, F., Spada, G.: Lighting in indoor environments: visual and non-visual effects of light sources with different spectral power distributions. Build. Environ. **46**(10), 1984–1992 (2011). https://doi.org/10.1016/j.buildenv.2011.04.007
19. Census Info India 2011 Final Population Totals. https://www.dataforall.org/dashboard/censusinfoindia_pca/files/profiles/profiles/PDF/IND_India.pdf.
20. Main workers, Marginal workers, Non-workers and those marginal workers, non-workers seeking/available for work classified by age and sex. https://censusindia.gov.in/2011census/B-series/DDWCT-0000B-01.xls.
21. Main workers classified by age, industrial category and sex - 2011. https://censusindia.gov.in/2011census/B-series/DDW-B04CT-0000.xls.
22. Main workers classified by industrial category, educational level and sex - 2011. https://censusindia.gov.in/2011census/B-series/DDW-B07CT-0000.xls.
23. The World bank. https://data.worldbank.org/indicator/SP.POP.TOTL?locations=IN.

24. Sustainable Development Goals National Indicator Framework Progress Report 2020 Version 2.1. https://www.mospi.gov.in/sites/default/files/publication_reports/Sustainable_Development_Goals_National_Indicator_Framework_Progress_Report_2020_Version2.1.pdf.
25. Del Rio João, K.A., Becker, N.B., de Neves Jesus, S., Isabel Santos Martins, R.: Validation of the Portuguese version of the Pittsburgh sleep quality index (PSQI-PT). Psychiat. Res. **247**, 225–229 (2017). https://doi.org/10.1016/j.psychres.2016.11.042.
26. Pittsburgh Sleep Quality Index. https://www.opapc.com/uploads/documents/PSQI.pdf.
27. Meers, J., Stout-Aguilar, J., Nowakowski, S.: Sex differences in sleep health. In: Sleep and Health, pp. 21–29. Elsevier (2019)
28. Figueiro, M.G., et al.: The impact of daytime light exposures on sleep and mood in office workers. Sleep Health **3**(3), 204–215 (2017). https://doi.org/10.1016/j.sleh.2017.03.005
29. Andargie, M.S., Azar, E.: An applied framework to evaluate the impact of indoor office environmental factors on occupants' comfort and working conditions. Sustain. Cities Soc. **46**, 101447 (2019). https://doi.org/10.1016/j.scs.2019.101447

Materials for Facade Glazing

Egor Chakin and Olga Gamayunova(✉)

Peter the Great St. Petersburg Polytechnic University, Polytechnicheskaya, 29, St. Petersburg 195251, Russia
gamayunova@inbox.ru

Abstract. More and more buildings are beginning to be designed using stained glass. The purpose of the study is to select the optimal design of facade glazing to ensure energy efficiency and durability of the building. The work was an analysis of the properties and characteristics of facade glazing, as well as calculations taking into account the climatic conditions of construction, structural, space-planning and architectural solutions of the building. As a result of the analysis of the building envelope market, the advantages and disadvantages of various types of structures were identified, which were analyzed in terms of their impact on the energy efficiency of the building. According to the results of the analysis, rack-and-beam glazing (modular) with a single-chamber double-glazed window using i-glass and filling the inter-glass space with argon is the most energy-efficient.

Keywords: Construction · Glass · Glazing · Building materials · Energy efficiency · k-glass · i-glass

1 Introduction

Facades made of stained glass glazing can have various shapes, colors, which allows you to make unique and unusual structural, architectural and space-planning decisions. Stained-glass glazing is a translucent structure that acts as a building envelope; it can fully replace the characteristics of external walls.

Buildings using facade glazing have the following main advantages:

- Light transmission (natural illumination of the building).
- Thermal insulation (creating a comfortable microclimate in the building, by introducing excess heat from the outside, and reducing heat loss inside the building).
- Soundproofing (creating the optimal level of noise inside the room coming from outside the building).
- Modern look.

One of the conditions for modern buildings is indoor comfort, that is, the creation of an optimal microclimatic regime, which is achieved by the integrated use of energy-saving technologies. One of the effective energy-saving technologies can be attributed to the choice of the correct glazing, which determines heat loss in winter and maintaining the optimal indoor temperature in summer.

A. Sinitsyn (Ed.): SES 2020, LNCE 141, pp. 182–189, 2021.
https://doi.org/10.1007/978-3-030-67654-4_20

2 Materials and Methods

2.1 Front Glazing Profiles

Profiles of stained glass systems are made in most cases from aluminum and plastic. Advantages and disadvantages of aluminum and plastic profiles are shown in Table 1.

Table 1. Advantages and disadvantages of aluminum and plastic profiles.

Type of profile	Advantages	Disadvantages
Plastic	• Lightness is the main advantage of the material • Convenience of transportation and installation. Light weight plastic profiles greatly simplifies the installation of facade systems • Low cost (compared to aluminum systems, plastic is cheaper by ~20%)	• Loss of color, delamination (plastic constructions lose their original appearance after a few years under the influence of the external environment) • Deformation (blowing, leaking and freezing of structures is possible) • The scope of application of plastic systems is limited (due to the low resistance of plastic to mechanical stress, it is impossible to install any anti-burglary devices)
Aluminum	• Durability (aluminum is not susceptible to corrosion, which positively affects the strength and appearance of the facade) • Reliability (aluminum is well resistant to temperature extremes, it is environmentally friendly and safe to use) • Versatility and practicality (aluminum profiles can be given a different shape, while the material does not lose its physical properties and ensures safe fastening of the stained-glass window structure)	• Higher cost compared to PVC profile frames

Thus, translucent structures made of aluminum profile are the most popular type of frame for stained glass facades.

2.2 Classification of Facade Glazing by the Type of Fastening of Double-Glazed Windows

- ***Column-beam facade system.*** This system consists of an internal frame made primarily of aluminum (crossbars and racks) and external clamps. A design feature is that on its outside there are special aluminum plates (covers) and they are needed in order to overlap the main clamping strips, which are fasteners for a double-glazed window.

- ***Column-beam facade half-closed system.*** Here there is only one type of decorative covers (pads) - horizontal or vertical. Instead of clamping strips in the absent direction, special sealants are used that reliably hold the glazing. The space between the mask strips is sealed with sealant. This type of seal is resistant to any weather conditions, and also provide good sound and thermal insulation.
- ***Structural glazing system.*** The structural facade is similar to a column-beam system, the main difference is the absence of a visible clamping profile from the outside. The frame consists of vertically mounted racks and horizontal jumpers (crossbars). A double-glazed window is held in a given plane by a special sealant. The structural facade system is different from all other types of glazing, since the walls of the building from the outside do not protrude beyond the plane of the facade.
- **Spider glazing system.** Glass in this type of glazing is mounted on spiders. Spiders are called special fasteners that connect double-glazed windows to each other and fix them to the supporting structure of the building. Each glass has through holes through which the spider is attached using hinges. Spiders are attached to the structural elements of the building using anchors or pivot bolts. After the glazing is mounted, all the cracks between the glasses are filled with sealant, which fully ensures the tightness of the glasses.
- ***Modular glazing system.*** This technology is based on the installation of prefabricated glass blocks. When performing glazing, individual modules are fixed on special brackets, and then they are insulated and docked with adjacent elements from inside the building. A modular system costs more than a column-beam system, but the difference in price is fully compensated by the almost complete absence of waste and savings on the cost of installation services. Ultimately, this glazing system is the fastest during installation.

2.3 Energy Efficiency Measures for Facade Glazing

Energy-efficient glazing is glazing that allows you to reduce heat loss from the room and (or) reduce the influx of excess solar radiation into the room through the use of various glasses and double-glazed windows with the appropriate characteristics.

Energy-saving glass is used to reduce heat loss. Windows with glasses of this type protect the room from overheating with increased solar radiation, and do not let heated air out. A metallized layer is applied to the surface of the window, which transmits visible rays of sunlight and delays thermal rays.

According to their technology, energy-efficient glasses are divided into k-glass and i-glass:

- K-glass is the ordinary glass but with a reduced coefficient of thermal emission (radiation). The glass is coated with a film resistant to mechanical stress, reflecting infrared (thermal) rays. K-glass has a large, compared to the usual, reflective effect. During the heating period, low-emission k-glass returns to the apartment up to 70% of the heat waves emitted by heating devices, and in the summer reflects the thermal, infrared part of solar radiation. As a result, it gets warmer indoors in winter, cooler in summer.

- I-glass has an even greater reflectivity than k-glass. I-glass by its heat-saving properties is 1.5 times higher than k-glass. The difference between k-glass and i-glass is the emissivity, as well as the technology for its preparation. During the heating period, low-emission i-glass returns up to 90% of the heat waves emitted by heating devices to the apartment, and in the summer it reflects the thermal, infrared part of solar radiation.

Another way to increase the energy efficiency of glazing is to increase the resistance to heat transfer, this can be achieved by filling the glass between the glass panes with inert gases instead of air. These gases are safe for humans and are not subject to burning. Typically, argon and krypton are used. Inert gases have a higher density compared to air. Argon is among them the lightest, which reduces its effectiveness, but it was he who received the greatest distribution due to the low cost relative to other inert gases. Also, the use of inert gases in double-glazed windows increase the noise insulating characteristics.

One of the main requirements that the building envelope must meet is that the reduced heat transfer resistance of the individual enclosure structures must be no less than the normalized value $(R^{req} \geq R^{n})$. In our case $R^{N} = 0.51\ \left(\frac{m \cdot {}^{\circ}C}{W}\right)$.

Compare of the main types of standard and low-emission ones with filling the inter-glass space with inert gases of double-glazed windows with different glass thicknesses is shown in the Table 2.

Table 2. Comparative characteristics of standard and low-emission types of double-glazed windows.

Type of double-glazed window	Variant of double-glazed window	Double-glazed window formula	Directional transmittance (not less),%	Heat transfer resistance, R^{req}, $\frac{m \cdot {}^{\circ}C}{W}$
Standard	Single chamber	4M1-16-4M1	83	0.32
	Single chamber	4M1-24-4M1	83	0.34
	Double chamber	4M1-10-4M1-10-4M1	72	0.47
	Double chamber	4M1-16-4M1-16-4M1	72	0.52

(*continued*)

Table 2. (*continued*)

Type of double-glazed window	Variant of double-glazed window	Double-glazed window formula	Directional transmittance (not less),%	Heat transfer resistance, R^{req}, $\frac{m \cdot {}^{\circ}C}{W}$
Low emission with filling the inter-glass space with air and inert gas	Single chamber with i-glass	4M1-16-I4	73	0.59
	Single chamber with k-glass	4M1-16-K4	75	0.53
	Single chamber with i-glass, Ar	4M1-16Ar-I4	73	0.66
	Double chamber with i-glass, Ar	4M1-12Ar-4M1-12Ar-I4	66	0.75
	Double chamber with i-glass, Ar	4M1-10Ar-4M1-10Ar-I4	66	0.71
	Double chamber with i-glass, Ar	4M1-16Ar-4M1-14Ar-I4	66	0.80

Standard single-chamber and two-chamber double-glazed windows are not further considered in connection with a low value of heat transfer resistance, with the exception of a two-chamber double-glazed window with the formula 4M1-16-4M1-16-4M.

The calculation of transmission losses of thermal energy for heating is calculated by the formula:

$$Q = 0.0024 \cdot D_h \cdot \sum_{i=1}^{m} \frac{A_i}{R_i} \cdot n,$$

where:
Q - transmission heat loss;
D_h - degree days of the heating period;
A_i - area of the i-type of the external building enclosing structures, taken for the building in question according to the specification of the elements for filling stained-glass windows;
R_i - reduced resistance to heat transfer of the i-th type of the external enclosing structure, taken equal to the required;
n - a coefficient taking into account the dependence of the position of the outer surface of the enclosing structures in relation to the outside air and reducing the

temperature difference for a separate fence that does not come in contact with the outside air; for external walls is taken equal to 1.

The values of transmission heat loss and the cost of thermal energy are given in Table 3.

Table 3. The values of transmission heat loss and the cost of thermal energy.

Double-glazed window	Transmission heat loss, Gcal/year	Cost of heat energy transmission losses, USD
4M1-10-4M1-10-4M1	276.20	6,970.5
4M1-16-I4	243.43	6,143.5
4M1-16Ar-I4	217.62	5,491.9
4M1-12Ar-4M1-12Ar-I4	202.29	5,105.2
4M1-10Ar -4M1-10Ar-I4	191.50	4,832.9
4M1-16Ar -4M1-16Ar-I4	179.53	4,530.8

The total costs of heat energy of transmission losses and the cost of stained-glass windows for the entire building depending on the time (1 year, 10 years, 30 years, 50 years) for 25-floor building are given in the Table 4.

Table 4. The total cost of facade glazing, depending on the type of double-glazed windows.

Double-glazed window	Energy costs for 1 year + cost of glazing, USD	Energy costs for 10 years + glazing cost, USD	Energy costs for 30 years + glazing cost, USD	Energy costs for 50 years + glazing cost, USD
4M1–10-4M1–10-4M1	404,845.1	467,591.5	607,000.0	746,408.5
4M1–16-I4	296,732.4	352,028.2	474,901.4	597,774.6
4M1-16Ar-I4	305,281.7	354,704.2	464,549.3	574,380.3
4M1-12Ar-4M1-12Ar-I4	452,028.2	497,971.8	600,084.5	702,183.1
4M1-10Ar -4M1-10Ar -I4	433,366.2	476,859.2	573,521.1	670,183.1
4M1-16Ar -4M1-16Ar -I4	463,718.3	504,493.0	595,112.7	685,732.4

3 Results and Discussion

Double-glazed windows with the formulas: 4M1-16-4M1-16-4M1, 4M1-12Ar-4M1-12Ar-I4, 4M1-10Ar-4M1-10Ar-I4, 4M1-16Ar-4M1-16Ar-I4 are will not pay back even after 50 years of use, so the choice of these options for double-glazed windows is not economically feasible.

Choosing between double-glazed windows 4M1-16-I4 and 4M1-16Ar-I4), we can conclude that a double-glazed window 4M1-16-I4 has a lower cost, but due to saving money on the cost of heat energy of transmission losses, a double-glazed window 4M1-16Ar-I4 pays off after 24 years. Also, the directional light transmittance of these double-glazed windows is almost the same, so we can conclude that a single-chamber energy-saving double-glazed window with argon 4M1-16Ar-I4 is more profitable from a technical point of view, since it has a greater resistance to heat transfer of the structure. The higher this coefficient, the higher its thermal insulation properties.

As a result of the analysis, the advantages and disadvantages of various glazing options were identified.

The technological process does not stand still and every year more and more new technologies appear, offering new options for facade glazing systems. However, architects and designers still often opt for a column-beam glazing system. The successful combination of low cost with high functionality, attractive appearance, durability and other advantages of the system help make the right choice in favor of the column-beam glazing system. The fastening of double-glazed windows is done in a modular way, which can significantly simplify the work and increase the speed of installation.

High-rise buildings using facade glazing are characterized by a large area of translucent walling, so it is important to evaluate the heat loss of various glazing options, light transmission, as well as the cost per square meter of a double-glazed window.

According to the results of the analysis, the most energy-efficient is column-beam glazing system (modular) with a single-chamber double-glazed window using I-glass and filling the inter-glass space with argon (4M1-16Ar-I4).

References

1. Vatin, N.I., Nemova, D.V., Rymkevich, P.P., Gorshkov, A.S.: Influence of building envelope thermal protection on heat loss value in the building. Mag. Civil Eng. **34**(8), 4–14 (2012). https://doi.org/10.5862/MCE.34.1
2. Lazarev, Yu., Akhatuly, A., Alzhanova, R., Panfilova, A.: Experimental studies of steel corrugated constructions. In: MATEC Web of Conferences, p. 01065 (2016). https://doi.org/10.1051/matecconf/20165301065
3. Gravit, M., Dmitriev, I., Lazarev, Y.: Validation of the temperature gradient simulation in steel structures in Sofistik. Adv. Intell. Syst. Comput. **983**, 929–938 (2019). https://doi.org/10.1007/978-3-030-19868-8_92
4. Rybakov, V.A., Kozinetc, K.G., Vatin, N.I., Velichkin, V.Z., Korsun, V.I.: Lightweight steel concrete structures technology with foam fiber-cement sheets. Mag. Civil Eng. **82**(6), 103–111 (2018). https://doi.org/10.18720/MCE.82.10
5. Zubarev, K.P., Gagarin, V.G.: Experimental comparison of construction material vapor permeability in case of horizontal or vertical sample position. IOP Conf. Ser. Mater. Sci. Eng. **463**, 032082 (2018)
6. Musorina, T.A., Gamayunova, O.S., Petrichenko, M.R., Soloveva, E.: Boundary layer of the wall temperature field. Adv. Intell. Syst. Comput. **1116**, 429–437 (2020)
7. Gamayunova, O., Spitsov, D.: Technical features of the construction of highrise buildings. In: E3S Web of Conferences, p. 08008 (2020). https://doi.org/10.1051/e3sconf/2020164 08008
8. Zemitis, J., Terekh, M.: Optimization of the level of thermal insulation of enclosing structures of civil buildings. In: International Scientific Conference on Energy, Environmental and Construction Engineering (EECE-2018) Electronic Edition, MATEC Web of Conferences, p. 06002 (2018)

9. Pavlenko, A., Mishakova, A., Pertseva, O., Ivanova, V., Olekhnovich, Y.: Feasibility of using of accelerated test methods for determination of frost-resistance for concrete. In: E3S Web Conference Key Trends in Transportation Innovation (KTTI-2019), vol. 157, p. 06035 (2020). https://doi.org/10.1051/e3sconf/202015706035
10. Talipova, L., Shonina, E., Strelets, K., Lapteva, S.: Reconstruction of the gray belt objects based on energy efficiency clusters. In: E3S Web of Conferences, p. 01021 (2019). https://doi.org/10.1051/e3sconf/201911001021
11. Kutsenko, D.I., Naboka, A.A., Petrochenko, M.V.: Methods improvement of technological and price audit for construction investment projects. Constr. Unique Build. Struct. **68**(5), 62–74 (2018). https://doi.org/10.18720/CUBS.68.6
12. Rybakov, V., Smirnov, A., Volkova, A., Seliverstov, A., Petrov, D.: Strength characteristics of foam concrete samples with various additives. In: International Scientific Conference on Energy, Environmental and Construction Engineering (EECE-2018) Electronic Edition, MATEC Web of Conferences, p. 03015 (2018). https://doi.org/10.1051/matecconf/201824503015
13. Vatin, N., Sultanov, S., Krupina, A.: Comparison of thermal insulation characteristics of pir, mineral wool, carbon fiber, and aerogel. Adv. Intell. Syst. Comput. **983**, 877–883 (2019). https://doi.org/10.1007/978-3-030-19868-8_86
14. Nefedova, A., Bykova, J., Kukolev, M., Kosov, S., Zajacs, A., Borodinecs, A.: Possibility of thermal storage system use with different accumulating material in SPBSTU. In: MATEC Web of Conferences, p. 02010 (2016)
15. Korniyenko, S.V.: Renovation of residential buildings of the first mass series. In: IOP Conference Series: Materials Science and Engineering, p. 022060 (2018)
16. Soldatenko, V.S., Smagin, V.A., Gusenitsa, Y.N., Gera, V.I., Soldatenko, T.N.: The method of calculation for the period of checking utility systems. Mag. Civil Eng. **2**(70), 72–83 (2017). https://doi.org/10.18720/MCE.70.7
17. Gagarin, V.G., Akhmetov, V.K., Zubarev, K.P.: Mathematical model using discrete-continuous approach for moisture transfer in enclosing construction. In: IOP Conference Series: Materials Science and Engineering, vol. 463, p. 022023 (2018)
18. Gamaunova, O., Musorina, T.: Return on investment in the implementation of energy-efficient solutions for facade insulation. In: MATEC Web of Conferences, p. 05041 (2018). https://doi.org/10.1051/matecconf/201825105041
19. Ptuhina, I., Spiridonova, T., Musorina, T., Kanyukova, S., Rezvaia, A.: Efficiency of urban development under high-rise construction of districts. In: Murgul, V. (ed.) MATEC Web of Conferences, p. 01049 (2016). https://doi.org/10.4028/www.scientific.net/AMM.725-726.153
20. Sergeev, V.V., Petrichenko, M.R., Nemova, D.V., Kotov, E.V., Tarasova, D.S., Nefedova, A.V., Borodinecs, A.B.: The building extension with energy efficiency light-weight building walls. Mag. Civil Eng. **84**(8), 67–74 (2018). https://doi.org/10.18720/MCE.84.7

Development of Customized Simulation Models for High Voltage Generators

Ivan Kirillov[1(✉)], Nikolay Kuznetsov[2], Ivan Morozov[1,2], Oleg Federov[3], and Leonid Dolomanyuk[4]

[1] Federal State Budgetary Educational Institution of Higher Education, "Murmansk Arctic State University" (MASU), 183038 Murmansk, Russia
fov52@mail.ru

[2] Northern Energetics Research Centre – Branch of the Federal Research Centre, "Kola Science Centre of the Russian Academy of Sciences" (NERC KSC RAS), 184209 Apatity, Russia

[3] Federal State Budgetary Educational Institution of Higher Education, «Nizhny Novgorod State Technical University Named After R.E. Alekseeva», 603950 Nizhny Novgorod, Russia

[4] Kazan State Power Engineering University, Street Krasnoselskaya, 51, 420066 Kazan, Russia

Abstract. The article is devoted to the topical issue of using simulation modeling in the educational process of training specialists in the field of electric power industry and for the design of electric power networks. The issue of creating and using models of high-voltage test equipment is considered on the example of single-stage and multi-stage pulse voltage generators. The simulation models presented in the article can be used not only for educational purposes, but also for conducting research in energy systems.

Keywords: High voltage generators · Simulation models · Power networks

1 Introduction

The modern educational process associated with the training of specialists in technical areas in higher educational institutions cannot be imagined without the use of digital technologies.

Such technologies include multimedia tools, software environments for programming in languages of various levels, and electronic library systems. One of the most popular types of these technologies are software systems that allow simulation modeling.

The most widely used software systems for modeling electrical circuits are Matlab, ElectronicsWorkbench, Micro-Cap V, etc.

When studying high-voltage installations in universities in such areas of training as 13.03.02 "Electric power and electrical engineering" with training profiles "Electric drive and automation" or "High-voltage electrical power and electrical engineering", teachers are faced with the problem of lack of necessary laboratory equipment or the

A. Sinitsyn (Ed.): SES 2020, LNCE 141, pp. 190–196, 2021.
https://doi.org/10.1007/978-3-030-67654-4_21

inability to use it in due to the lack of a qualification group for students, which allows work with a certain tension.

Simulation tools can be used to solve this problem. When using simulation models for educational purposes, it is possible to conduct experiments, the physical implementation of which is difficult or impossible even with the presence of special equipment.

This article presents an assessment of the possibility of using simulation modeling in order to implement high voltage generator circuits used in high voltage test installations.

It should be noted that the use of simulation and the Matlab program in particular, has proven itself well for solving various kinds of technical problems [1, 2].

An additional advantage of the use of simulation models is the ability to quickly calculate transients in power supply systems [3–6] and electric drive [7, 8], as well as during numerical experiments that provide a certain resource saving at industrial enterprises [9].

2 Simulation of a Schematic Diagram of a Single-Stage Pulse Voltage Generator

As the first example of the developed simulation model of a high-voltage installation, the circuit of a single-stage pulse voltage generator with separated front and snubber resistors is shown in Fig. 1.

The mathematical description of the operation of the presented circuit is as follows:

$$\begin{aligned} U_{C1} &= U_m(1 - \exp(-t/R_1C_2)) \\ U_{C2} &= U_m(1 - \exp(-t/R_2C_1)) \end{aligned} \tag{1}$$

Resistor R3 plays the role of a vibration damper and eliminates the influence of the parasitic parameters of the discharge circuit.

This circuit is suitable for obtaining standard high voltage pulse shapes and for simulating the effect of these pulses on power plants.

Directly with the help of the presented model, at the Department of Physics, Biology and Engineering Technologies of the Murmansk Arctic State University, transient processes are simulated in electrical networks in the presence of lightning overvoltages.

The installation scheme is simple to implement, but it has a number of features. For this scheme, a graphical interface has been created that allows the user to change the settings of the links in the electrical circuit, as well as a block that simulates the change in resistance parameters in the circuit during prolonged operation of the installation has been implemented.

The MATLAB Function block changes the resistances R1–R4 after a while. The fact that such installations operate in short-term modes, in which there is no fundamental change in the resistance values from overheating, is not taken into account. This feature of the presented model is implemented for educational purposes, so that students have an idea of the influence of changes in parameters in power plants on the transient processes occurring in them.

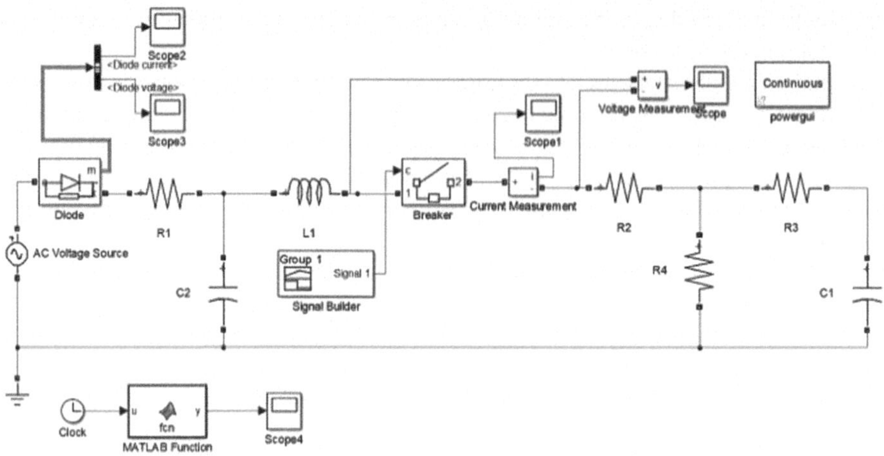

Fig. 1. Simulation model of a single-stage pulse voltage generator circuit with separated front and damper resistors.

With this model, it is possible to successfully obtain impulses of various shapes, for example: lightning impulses, switching impulses with oscillations, aperiodic impulses with single surges or superimposed oscillations.

Figure 2, for an example of the simulation result, shows a lightning impulse obtained using this model (upper graph) and also an aperiodic impulse with superimposed oscillations (lower graph). In this case, all the pulses that can be obtained using the model can have the necessary parameters, since the model has the ability to adjust.

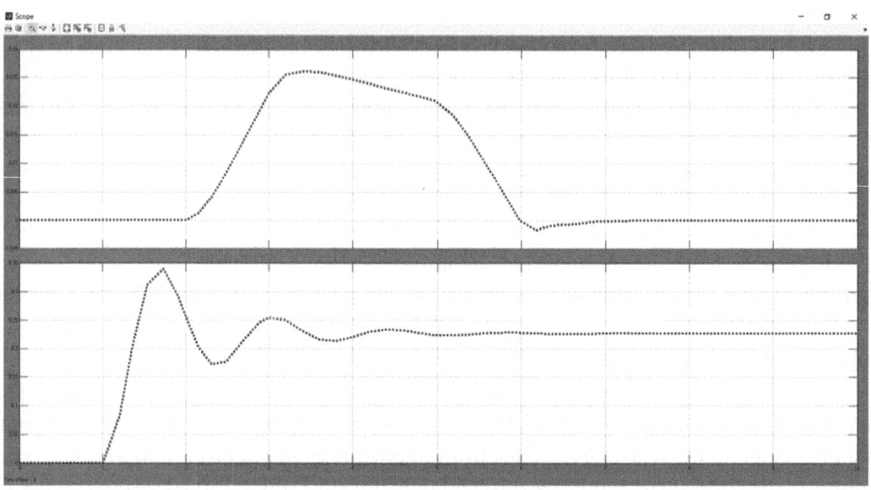

Fig. 2. Impulse voltage simulation results.

An important fact that prompts the use of such circuits is that the result of their work will be not just physical quantities, but numerical values of currents and voltages. This contributes to the possibility of using previously designed circuits in models of large energy systems of high complexity, for example, as in [3].

3 Modeling the Schematic Diagram of a Multistage Pulse Voltage Generator

The second example presented for consideration in this article is the model of a multistage pulse voltage generator (MPVG). Such generators are used to obtain pulse voltages with an amplitude of up to several million volts, which completely excludes the possibility of their use in a training laboratory. This fact once again confirms the need to use simulation models.

The direct purpose of this device is to test the insulation of electrical equipment at industrial enterprises or in scientific laboratories. The training course for specialists in the field of electric power engineering includes training in disciplines related to the methods of selection and testing of insulation, and therefore the relevance of the developed model for educational purposes is high.

The scheme of this generator was proposed by the professor of Moscow University Vladimir KonstantinovichArkadiev in 1914.

The model shown in Fig. 3 shows a two-step scheme. In real MPVGs, schemes with several stages are used, but for modeling transient processes for educational purposes, two stages are sufficient. As in the previous example, users have the ability to customize the parameters of the model elements.

The circuit of the considered MPVG contains a structure of high voltage capacitors that provide high pulse voltages during discharge. Capacitors are switched using spark gaps (usually ball gaps). The Simulink application of the MatLab software environment lacks elements for modeling the spark gaps. The imitation of arrester operation was carried out using disconnector blocks (Bracker).

The MPVG model includes an ammeter and a voltmeter as measuring devices, which make it possible to evaluate transient processes at the extreme spark gap.

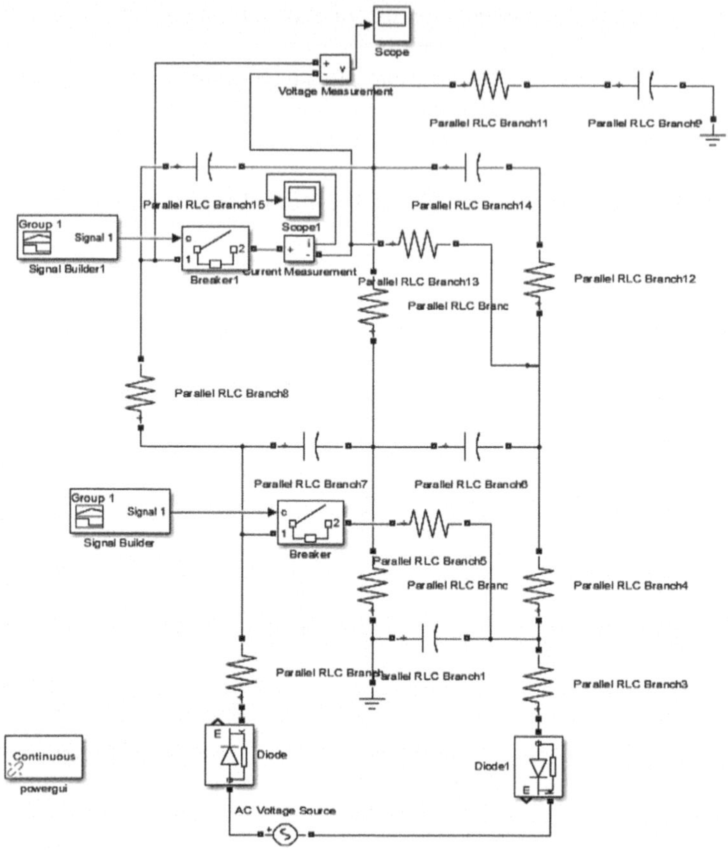

Fig. 3. Simulation model of a multistage pulse voltage generator circuit.

By changing the settings of this model, it is possible to obtain chains of pulses of any shape, which allows them to be used for the implementation of simulation models of other systems. The use of several models with modified block parameters is a clear example of demonstrating the influence of the parameters of the circuit elements on the form of the transient process. Figure 4 shows two transients obtained using the circuit in Fig. 3, but with changed capacitor parameters.

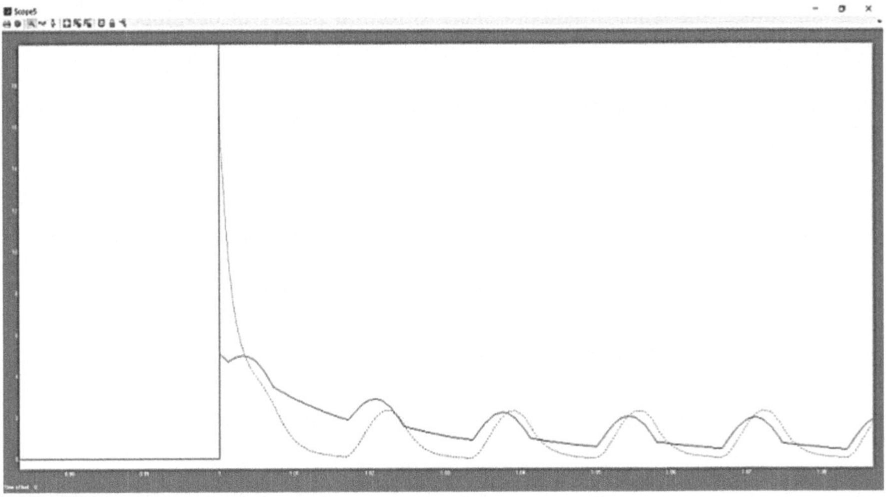

Fig. 4. Transient processes obtained using the multistage MPVG model.

The team of authors of this publication has developed a whole complex of models of test installations, which is successfully used in the educational and research process of the Department of Physics, Biology and Engineering Technologies of the Murmansk Arctic State University.

The list of developed models, in addition to various voltage generator circuits presented in this publication, includes models:

- single-stagerectificationcircuits;
- voltagemultipliers;
- electronicvoltageregulators;
- generatorsofimpulsecurrents.

4 Conclusion

In conclusion, it should be noted that the use of simulationhigh-voltage electrophysical installations allows:

- avoid the use of equipment that for a number of reasons cannot be present in educational laboratories;
- carry out a wider range of experiments due to the ability to adjust the parameters of the model and simulate conditions that, when carrying out a real physical experiment, lead to equipment damage;
- improve the quality of teaching students, due to the ability to observe changes in the operation of models with a slight change in the settings of the blocks;
- to simulate transient processes in complex energy systems, not only within the educational process, but also in research and development and production.

In the future, the team of authors plans to form a library of developed and developed models.

References

1. Kirillov, I.E., Kolesnikov, E.B., Morozov, I.N., Shpreher, D.M.: Proactive system for controlling the speed of movement of the coal cleaning combine in the conditions of the Arctic region. In: 4th International Scientific Conference Arctic: History and Modernity, IOP Conference Series: Earth and Environmental Science, vol. 302, p. 012042 (2019). https://doi.org/10.1088/1755-1315/302/1/012042
2. Morozov, I.N., Kirillov, I.E.: Promyshlennaya energetika **9**, 10–14 (2019)
3. Morozov, I.N., Kuznetsov, N.M., Belova, L.A., Borozdina, Ye.D.: Vestnik Chuvashskogo universiteta **1**, 113–122 (2020)
4. Semenov, A.S., Semenova, M.N., Bebikhov, Yu.V.: Proceedings – 2019 International Russian Automation Conference, RusAutoCon, p. 8867704 (2019)
5. Kopylov, K.N., Reshetnyak, S.N., Kubrin, S.S.: Gornyy informatsionno-analiticheskiy byulleten' (nauchno-tekhnicheskiy zhurnal) **12**, 40–50 (2016)
6. Morozov, I.N., Kirillov, I.E., Naznachilov, A.V.: Analysis of arc surges in electric networks of the Arctic region. In: 5th International Scientific Conference Arctic: History and Modernity, IOP Conference Series: Earth and Environmental Science, vol. 539, p. 012148 (2020). https://doi.org/10.1088/1755-1315/539/1/012148
7. Semenov, A.S., Egorov, A.N., Khubieva, V.M.: International Conference on Industrial Engineering, Applications and Manufacturing (ICIEAM), p. 8743025 (2019)
8. Fashchilenko, V.N., Reshetnyak, S.N.: Gornyyzhurnal **7**, 80–83 (2017)
9. Sedelnikova, I.M., Fedorov, O.V., Yakusheva, A.M.: Int. J. Econ. Finan. Manag. Sci. **7**(6), 210–214 (2019)

RETRACTED CHAPTER: Thermal Conductivity of Geopolymer Concrete with Different Types of Aggregate

Oleg Krotov[1(✉)], Pavel Gromyko[1], Marina Gravit[1], Svetlana Belyaeva[1], Shuhrat Sultanov[1], Yuriy Lazarev[1], Marina Petrochenko[1], Viktor Velichkin[1], and Marcelo de Andrade Romero[2]

[1] Peter the Great St. Petersburg Polytechnic University, 195251 St. Petersburg, Russia
Krotov.om@edu.spbstu.ru

[2] Universidade de Sao Paulo - USP, Sao Paulo, Brazil

Abstract. In this article the influence of aggregate of geopolymer concrete used in concrete printing of buildings and structures, architectural elements on its thermal conductivity is studied. All tests were conducted on device for measuring thermal conductivity called PIT-2 designed to measure the thermal conductivity of building and heat-insulating materials under stationary thermal conditions in accordance with ISO 7345:1987* and ISO 9251:1987. As the fine aggregate most commonly used river sand, quartz sand, ceramic foam, as well as one new coniferous shavings were taken. The results of 4 tests to determine the thermal conductivity of geopolymer concrete with various aggregates are listed in the summary table in the study. According to the test results, the lowest coefficient of thermal conductivity has geopolymer concrete on ceramic foam. In this study geopolymer concrete is considered as the main material for concrete printing of the houses and architectural forms. This material reduces CO2 emissions in the atmosphere.

Keywords: Thermal conductivity · Geopolymer concrete · Thermal conductivity coefficient · Aggregates · Concrete printing

1 Introduction

The necessity of reliable building materials is greatly increasing nowadays. Currently, one of the durable and most widely used material is Portland cement concrete (PCC). Its manufacturing technology was invented in 19 century and PCC is applied in construction for decades. However, there are sharp debates about the influence of PCC production on the environment [1]. The total energy consumption is critically increasing all over the world. It is highly connected with building energy consumption and, evidently, significant part of this energy is used for heating and cooling purposes [2]. A great number of recent research studies are dedicated to reduction of energy consumption, pollution in all industrial enterprises, in global manufacturing processes and to application of alternative cementitious eco-friendly materials with less energy consumption [3]. Manufacture of PCC has a disastrous effect on the environment.

The original version of this chapter was retracted: The retraction note to this chapter is available at https://doi.org/10.1007/978-3-030-67654-4_49

A. Sinitsyn (Ed.): SES 2020, LNCE 141, pp. 197–205, 2021.
https://doi.org/10.1007/978-3-030-67654-4_22

Production of Portland cement lead to the CO2 emission, which contributes to about 5–8% of the total CO2 emissions, and is the third man-made CO2 source after transport and energy generation [4]. Obviously, PCC has beneficial properties, such as low thermal conductivity, high specific heat capacity, high density, and high mechanical strength. Nevertheless, PCC exhibits a negative effect on the environment due to the emission of carbon dioxide (CO2) and a great amount of energy consumption during the production of cement, that is why it is vital to think about innovative and environmentally friendly material [5–13]. Geopolymer concrete (GPC) is an innovative building material that is applied in construction and 3D printing. The composition of GPC includes components based on natural raw materials, for instance, kaolinite and clay, as well as substances based on recyclable materials that are considered to be industrial waste. Geopolymer is synthesized by alkali activation of materials rich in silica and alumina (from industrial waste materials such as fly ash, coal ash, rice-husk ash, red, mud and ground granulated blast furnace slag [14–24]. Geopolymer can be used as an alternative binder for concrete to decrease carbon dioxide emissions. Moreover, high pressures or temperatures are not required for production of GPC during which there are no dust wastes, while there are dust wastes during manufacturing of PCC from raw material factories, kilns, clinker refrigerators, cement mills. GPC has a great number of advantages over PCC such as high initial strength, small drying shrinkage, high fire resistance, superior acid resistance, shorter setting time, better thermal performances. Therefore, phase change materials can be integrated in GPC to reduce indoor temperature fluctuation, maintain thermal comfort, and minimize the peak of the cooling and heating loads [5–9, 18–24].

Low thermal conductivity means good thermal insulation, and, hence, energy efficiency in operation. The thermal conductivity of concrete largely depends on the type of aggregate that is used. Apparently, lightweight concrete has less thermal conductivity by reason of density reduction and porosity increase, pores are filled with air that is considered to be a good heat insulator, that's why lightweight concrete is suitable for construction of exterior walls. Thus, lightweight concrete blocks have been widely used around the world due to its low unit weight, high thermal insulation, and good freeze/thaw stability [24–27]. River sand, quartz sand, ceramic foam, coniferous shavings were chosen as aggregates due to their diverse weight, density and different coefficient of thermal conductivity. This study is focused on the lab scale experiment by mixing geopolymer cement, geosilicate with different fine aggregates. In this research the thermal conductivity of 3 GPC specimens were determined and compared.

The purpose of this study is to compare the thermal conductivity of geopolymer concrete with commonly used aggregates (river sand, quartz sand, ceramic foam) and new one (coniferous shavings) and define suitable aggregates for geopolymer matrix.

To achieve this goal it is necessary to solve the following tasks:

1. Prepare samples of geopolymer concrete with various aggregates;
2. To test samples on the installation of PIT-2.1;
3. Compare the values of the thermal conductivity of the samples.

2 Methods

2.1 Materials

To conduct the experiment with GPC 4 samples were prepared. The dimension of each concrete sample was 30 × 250 × 250mm. Mix proportions for GPC are shown in the Table 1.

Table 1. Mix proportions for geopolymer concrete.

Component	Sample 1	Sample 2	Sample 3	Sample 4
Fine aggregate, kg	Quartz sand, 2.871	Ker-wood 0.375	Quartz sand, 1.435	Coni-ferous shavings, 0.700
		Cera-mic foam, 0.375	River sand, 1.435	
Geosilicate, kg	0.754	0.700	0.754	0.700
Geopolymer cement, kg	2.175	1.100	2.175	1.000
Density, kg/m^3	1981	952	1[illegible]0	1180

After mixing 4 samples of GPC, they were filled in the molds designed for samples with dimension 30 × 250 × 250 mm. [illegible]h later samples were extracted from the molds and put into the drying chamber. After full drying specimens were ready for experiments on PIT-2.1 – thermal conductivity meter.

2.2 Principle of Operation of the Measuring Device

PIT consists of two refrigerators (upper and lower), lower heater (cold plate), protective heater, protective ring, measuring heater, heat insulator (Fig. 1).

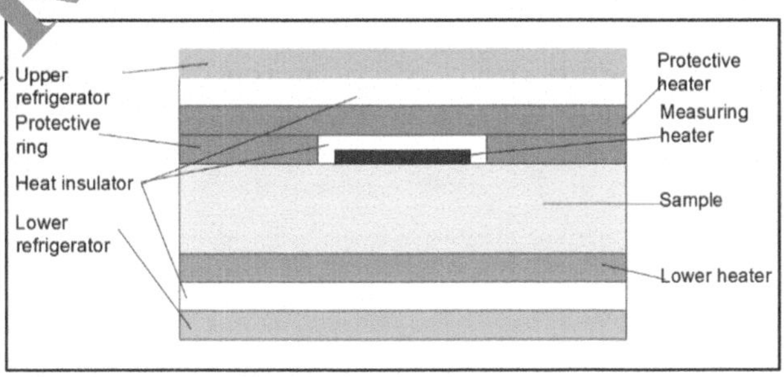

Fig. 1. Measuring device.

Refrigerators are designed to provide PIT operation at average sample temperatures below ambient temperature. In the device refrigerators are made on thermoelectric elements Peltier with the possibility of additional cooling due to the flow of liquid. The safety heater and the safety ring are designed to eliminate losses from the measuring device.

The temperature of the security heater and the security ring is maintained at the same temperature measuring heater. The measuring heater is designed to determine the power P required to maintain set temperature difference ΔT between the measuring and lower heaters.

The essence of the method is to create a stationary heat flow passing through a flat sample of a certain thickness and directed perpendicular to the face (largest) faces of the sample, measuring the power required to create this heat flow, the temperature of the opposite face faces and the thickness of the sample. The coefficient of thermal conductivity is determined by the formula (1), where S is the area of the measuring heater and h is the height of the sample.

$$\lambda = P * h/(S^{*}\Delta T) \tag{1}$$

2.3 Experiment Conduction

Before experiment specimens were measured with campers in four corners at a distance of (50.0 ± 5.0) mm from the top of the corner and in the middle of each side.

PIT-2.1 is designed for one sample. Firstly, specimen was installed in PIT-2.1 between top and bottom plates, after that the meter was turned on. The device display shows: the current thermal conductivity of the sample (in digital and graphical forms), the power "P" allocated in the measuring heater, the temperature difference "ΔT" between the measuring and lower heaters, the average temperature of the sample "t" and the thickness of the sample «h». It was necessary also to specify average temperature and thickness of each sample in the meter. The discreteness of changes in the average sample temperature is 0.1 °C, and the sample thickness is 0.1 mm. The discreteness of changes in the average sample temperature is 0.1 °C, and the sample thickness is 0.1 mm (Fig 2).

Experiment lasts 2.5–3 h for each sample. The device signals the output to stationary mode by changing the sign " ≠" to the sign " =" before the number of measured thermal conductivity. For accurate determination of the device's output to the stationary mode a graph of the current thermal conductivity of the sample is displayed on the PIT display.

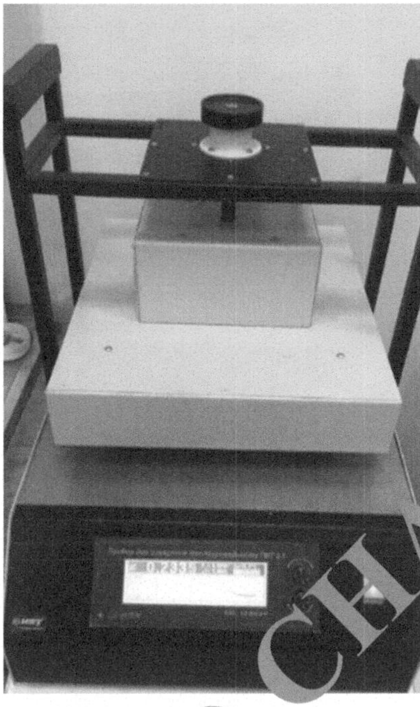

Fig. 2. Laboratory-scale plant PIT-2.1.

3 Results and Discussion

Environmentally-friendly geopolymer concrete specimens with same dimensions (30 × 250 × 250 mm) and different proportions of geopolymer cement and geosilicate were achieved. Moreover, diverse fine aggregates were used for manufacturing of samples. The humidity of sample has an important impact on the measured thermal conductivity. Thus, all samples were cured at 80 °C in the drying chamber for one month. Experiments were conducted on 4 specimens with various weight and density. The results of the experiments are shown in the Table 2. Evidently, coefficient of thermal conductivity directly depends on the type of fine aggregate. Thus, the heavier were the aggregate, the higher thermal conductivity were achieved. The lowest coefficient of thermal conductivity had specimens number 2 and number 4 ($\lambda_2 = 0.233$; $\lambda_4 = 0.243$), in which ceramic foam and coniferous shavings respectively were used as a fine aggregate. Such geopolymer concrete are very light and cannot be used as a load-bearing structure without reinforcing. Specimens number 1 and number 3, in which quartz sand and river sand were used, had higher coefficient of thermal conductivity ($\lambda_1 = 0.777$; $\lambda_3 = 0.823$). Samples 1 and 3 were heavier and had higher density. Geopolymer concrete with such fine aggregates is suitable for 3D printing and can be applied in construction. Portland cement concrete has coefficient of thermal conductivity within the limits of 0.2 to 1.0 W/m*K (and higher), reinforced concrete has even more coefficient value, results are shown in Table 2 (Fig. 3).

Fig. 3. Achieved samples.

Table 2. The results of the experiment.

	Sample1	Sample2	Sample3	Sample4
Density, kg/m^3	1981	952	1900	1180
Weight, kg	3.7[illegible]4	1.784	3.860	2.212
Coefficient of thermal conductivity, W/m*K	[illegible]77	0.233	0.823	0.243

Thermal conductivity coefficient of porous GPC, which was prepared using fly-ash, water glass and additive amount of sodium water glass and H2O2, is 0.0744 W/m*K, its density is 240–335 kg/m^3 [5]. Comparing such GPC with sample 2 and 4, which are the most porous in current study, it has a lower coefficient of thermal conductivity, however, it has a very low density and compressive strength, which is 0.82 MPa [6], and can be applied only as insulation material. The near dried geopolymer foam concrete thermal conductivity coefficients are in range of 0.15 to 0.48 W/m*K and their densities change from 720 to 1600 kg/m^3 [6]. Such GPC is similar to sample 2, determined parameters of which (density, thermal conductivity coefficient) are within specified limits. The thermal conductivity coefficient of oil palm shell foamed geopolymer concrete, where low-calcium fly ash and palm oil fuel ash as cementitious materials, and oil palm shell as lightweight coarse aggregate were used, is 0.[illegible]7 W/m*K, which is lower than coefficients of samples 1 and 3 due to lower density, that was in range of 1300–1700 kg/m^3 and higher porosity, that was varied between 25 and 40% [7]. Lightweight geopolymer concrete based on recycled expanded polystyrene as aggregate has coefficient of thermal conductivity in range of 0.121 to 0.207 W/m*K and density in range from 500 to 800 kg/m^3 [8], which is lower than prepared specimens. Thermal conductivity coefficients of the materials for the multi-layer walls were determined in [10]. These coefficients are higher than the obtained values. Thus, thermal conductivity coefficient of GPC, which density was 2199 kg/m^3, was 1.35 W/m*K. Thermal conductivity coefficient of GPC with microencapsulated phase change materials, which density was 1875 kg/m^3, was 0.74 W/m*K. Thermal conductivity coefficient of GPC with pure phase change materials, which density was 825 kg/m^3, was 0.2 W/m*K.

4 Conclusion

An experimental investigation on the thermal conductivity of GPC with various types of fine aggregate was presented in this study. From the experimental results, the following conclusions can be drawn:

1) Conducted experiments have shown that thermal insulation properties of GPC are greatly connected with the type of fine aggregate. Thus, GPC with kerwood, ceramic foam and with coniferous shavings has low coefficients of thermal conductivity 0.233 W/m*K and 0.243 W/m*K relatively, whereas GPC with quartz sand and river sand has higher coefficient value 0.777 W/m*K and 0.823 W/m*K respectively.

2) The density of prepared samples varied from 952 to 1981 kg/m^3 and the low the density is the higher the porosity of the specimen is, the lower its compressive strength is. As a result, samples with high density have higher coefficient of thermal conductivity.

3) Thermal conductivity of GPC meets the building requirements and, apparently, can be improved, that is why the replacement of PCC by GPC in terms of thermal insulation properties is feasible. Eventually, it is clear that there is a necessity to focus on further research using mixtures characterized by different geopolymer cement and geosilicate ratio; by diverse fine aggregates and their proportions in composition and, then, apply achieved and analyzed results in calculation software complexes.

Acknowledgement. This research work was supported by the Academic Excellence Project 5–100 proposed by Peter the Great St. Petersburg Polytechnic University.

References

1. Cao, V., Bui, T., Kjøniksen, A.: Thermal analysis of multi-layer walls containing geopolymer concrete and phase change materials for building applications. Energy **186** (115792), 1–2 (2019)
2. Cao, V., Pilehvar, S., Salas-Bringas, C., Szczotok, A., Bui, T., Carmona, M., Rodriguez, J., Kjøniksen, A.: Thermal performance and numerical simulation of geopolymer concrete containing different types of thermoregulating materials for passive building applications. Energy Build. **173**, 678–688 (2018)
3. Cao, V., Pilehvar, S., Salas-Bringas, C., Szczotok, A., Bui, T., Carmona, M., Rodriguez, J., Kjøniksen, A.: Thermal analysis of geopolymer concrete walls containing microencapsulated phase change materials for building applications. Sol. Energy **178**, 295–307 (2018)
4. Cao, V., Pilehvar, S., Salas-Bringas, C., Szczotok, A., Rodriguez, J., Carmona, M., Al-Manasir, N., Kjøniksen, A.: Microencapsulated phase change materials for enhancing the thermal performance of Portland cement concrete and geopolymer concrete for passive building applications. Energy Convers. Manag. **133**, 56–66 (2016)

5. Cao, V., Pilehvar, S., Salas-Bringas, C., Szczotok, A., Valentini, L., Carmona, M., Rodriguez, J., Kjøniksen, A.: Influence of microcapsule size and shell polarity on thermal and mechanical properties of thermoregulating geopolymer concrete for passive building applications. Energy Conver. Manag. **164**, 198–209 (2018)
6. Feng, J., Zhang, R., Gong, L., Li, Y., Cao, W., Cheng, X.: Development of porous fly ash-based geopolymer with low thermal conductivity. Mater. Des. **65**, 529–533 (2014)
7. Zhang, Y., Zhang, Y., Liu, G., Yang, Y., Wu, M., Pang, B.: Fresh properties of a novel 3D printing concrete ink. Constr. Build. Mater. **174**, 263–271 (2018)
8. Liu, M., Alengaram, U., Jumaat, M., Mo, K.: Evaluation of thermal conductivity, mechanical and transport properties of lightweight aggregate foamed geopolymer concrete. Energy Build. **72**, 238–245 (2013)
9. Colangelo, F., Roviello, G., Ricciotti, L., Ferrándiz-Mas, V., Messina, F., Ferone, C., Tarallo, O., Cioffi, R., Cheeseman, C.: Mechanical and thermal properties of lightweight geopolymer composites. Cement Concr. Compos. **86**, 266–272 (2017)
10. Zhang, Y., Zhang, Y., She, W., Yang, L., Liu, G., Yang, Y.: Rheological and harden properties of the high-thixotropy 3D printing concrete. Constr. Build. Mater. **201**, 278–285 (2018)
11. Patel, Y., Shah, N.: Development of self-compacting geopolymer concrete as a sustainable construction material. Sustain. Environ. Res. **28**(6), 412–421 (2018)
12. Ramos, F., Reis, R., Grafova, I., Grafov, A., Monteiro, S.: Eco-friendly recycled polypropylene matrix composites incorporated with geopolymer concrete waste particles. Mater. Res. Technol. **9**(3), 3084–3090 (2020)
13. Petrillo, A., Cioffi, R., Ferone, C., Colangelo, F., Borrelli, C.: Eco-sustainable geopolymer concrete blocks production process. Agric. Agric. Sci. Procedia **8**, 408–418 (2016)
14. Kruger, J., Zeranka, S., van Zijl, G.: 3D concrete printing: a lower bound analytical model for buildability performance quantification. Autom. Constr. **106**(102904), 516–522 (2019)
15. Naghizadeh, A., Ekolu, S., Musonda, I.: High temperature heat - treatment (HTHT) for partial mitigation of alkali attack in hardened fly ash geopolymer binders. Case Stud. Constr. Mater. **12**(e00341), 1–1 (2020)
16. Pilehvar, S., Cao, V., Szczotok, A., Valentini, L., Salvioni, D., Magistri, M., Pamies, R., Kjøniksen, A.: Mechanical properties and microscale changes of geopolymer concrete and Portland cement concrete containing micro-encapsulated phase change materials. Cem. Concr. Res. **100**, 341–349 (2017)
17. Nuaklong, P., Wongsa, A., Sata, V., Boonserm, K., Sanjayan, J., Chindaprasirt, P.: Properties of high-calcium and low-calcium fly ash combination geopolymer mortar containing recycled aggregate. Heliyon **5**(e02513), 1–9 (2019)
18. Chitrala, S., Jadaprolu, G., Chundupalli, S.: Study and predicting the stress-strain characteristics of geopolymer concrete under compression. Case Stud. Constr. Mater. **8**, 172–192 (2018)
19. Rybakov, V., Ananeva, I., Pichugin, E., Garifullin, M.: Heat protective properties of enclosure structure from thin-wall profiles with foamed concrete. Mag. Civil Eng. **94**(2), 11–20 (2020)
20. Ivanov, E., Semenov, K., Manovitskij, S., Barabanshchikov, Y., Vavilova, A., Mushchanov, V.: Crack resistance criteria of massive concrete and reinforced concrete structures during the construction period. In: Borodinecs, A., Vatin, N., Sergeev, V. (eds.) CONFERENCE 2019, EECE, vol. 70, pp. 575–584. Springer, Heidelberg (2019)
21. Suksiripattanapong, C., Krosoongnern, K., Thumrongvut, J., Sukontasukkul, P., Horpibulsuk, S., Chindaprasirt, P.: Properties of cellular lightweight high calcium bottom ash-portland cement geopolymer mortar. Case Stud. Constr. Mater. **12**(e00337), 1–3 (2020)

22. Svatovskaya, L., Shershneva, M., Baydarashvily, M., Sychova, A., Sychov, M., Gravit, M.: Geoecoprotective properties of cement and concrete against heavy metal ions. Procedia Eng. **117**(1), 345–349 (2015)
23. Svatovskaya, L., Sychov, M., Sychova, A., Gravit, M.: New geoecoprotective properties of the construction materials for underground infrastructure development. Procedia Eng. **165**, 1771–1775 (2016)
24. Cui, Y., Wang, D., Zhao, J., Li, D., Ng, S., Rui, Y.: Effect of calcium stearate based foam stabilizer on pore characteristics and thermal conductivity of geopolymer foam material. Build. Eng. **20**, 21–29 (2018)
25. Błyszko, J.: Comparative analysis of creep in standard and fibre reinforced concretes under different load conditions. Procedia Eng. **193**, 478–485 (2017)
26. Buswell, R., de Silva, W.L., Jones, S., Dirrenberger, J.: 3D printing using concrete extrusion: a roadmap for research. Cem. Concr. Res. **112**, 37–49 (2018)
27. Assaedi, H., Shaikh, F., Low, I.: Effect of nano-clay on mechanical and thermal properties of geopolymer. Asian Ceram. Soc. **4**(1), 19–28 (2015)

Study of Starting Transition Processes of Asynchronous Motor at a Lowered Mains Voltage Frequency

Shakhboz Tolibjonovich Dadabaev[1](✉), Elena Ivanovna Gracheva[2], and Zainura Anvarkhonovna Dadabaeva[1]

[1] Department of Power Supply and Automation, Khujand Polytechnic Institute of the Tajik Technical University named after Academician M.S. Osimi, Khujand 735700, Tajikistan
shahbozdadoboev@mail.ru

[2] Department Power Supply of Industrial Enterprises, Kazan State Energy University, Kazan 420066, Russian Federation

Abstract. The aim of the work is to study the starting transient processes of asynchronous motor with a squirrel-cage rotor at different values of the mains voltage frequency. The main research method in the article is computer modeling, with the help of which a model is created to study the starting transient processes of asynchronous motor. The results of the study are presented in the form of graphs of transient processes of asynchronous motor, which shows the behaviour of this type of machine when starting from the network with a reduced frequency. The novelty of the research lies in the computer simulation of asynchronous motor with a squirrel-cage rotor in absolute units and in a detailed graphical display of the results. Prior to that, to study this issue, modeling in relative units was mainly used, and at the same time, the output of the results was generalized, not detailed.

Keywords: Asynchronous motor · Lowered mains voltage frequency · Starting transition processes

1 Introduction

World technical progress requires constant improvement of work and expansion of functions of mechanisms and machines, which is impossible without controlled drives. Therefore, the mechanisms provided are continuously improved and the control becomes more complicated. It has long been known that the ability to easily adjust the speed of rotation of asynchronous motors (AM) can be achieved only by controlling the frequency of the voltage supplying them. But the theoretical possibility of wide regulation of the speed of rotation of AM was difficult and expensive to implement technically [1]. Because of this, the use of a variable frequency drive in technology has long been in the nature of expensive control systems. Relatively recently, this theoretical possibility was achieved and implemented in wide industrial practice thanks to the advances in electronics, namely the successes of power electronics. The asynchronous squirrel-cage motor is the simplest, most technologically advanced and the

A. Sinitsyn (Ed.): SES 2020, LNCE 141, pp. 206–213, 2021.
https://doi.org/10.1007/978-3-030-67654-4_23

cheapest one to manufacture and operate. Therefore, this type of motor is considered the most attractive for developers of electric drives for any purpose [2].

According to general estimates, electric AC drives in the world are now 60% adjustable. As already noted, an electric drive with frequency control is considered to be an adjustable asynchronous electric drive, and therefore the dependence of the network frequency with other parameters of an asynchronous electric drive is a very attractive topic for scientists and researchers in this field.

2 Methodology

The generalized AM contains a three-phase winding on the stator and rotor, which are connected to symmetrical three-phase voltage sources, example shown in Fig. 1 [1, 3].

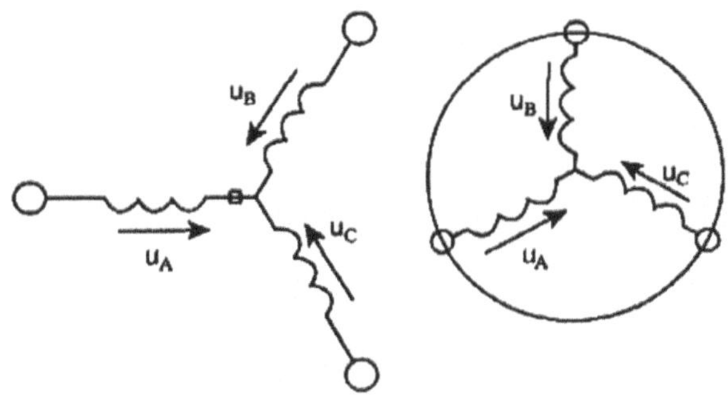

Fig. 1. Generalized asynchronous machine.

The equations of equilibrium of electromotive force on the stator and rotor windings of AM have the next form [3]:

For stator

$$u_A = R_A i_A + \frac{d\Psi_A}{dt},$$

$$u_B = R_B i_B + \frac{d\Psi_B}{dt},$$

$$u_C = R_C i_C + \frac{d\Psi_C}{dt}. \quad (1)$$

For rotor

$$u_a = R_a i_a + \frac{d\Psi_a}{dt},$$

$$u_b = R_b i_b + \frac{d\Psi_b}{dt},$$

$$u_c = R_c i_c + \frac{d\Psi_c}{dt}. \tag{2}$$

Equations 1 and 2 feature instantaneous voltages, currents and flux linkages of the stator and rotor, as well as the active resistances of the windings [4–8].

In AM, the induction flux of the stator and rotor windings with their currents are described by the following equations:

For stator

$$\Psi_A = L_{AA}i_A + L_{AB}i_B + L_{AC}i_C + L_{Aa}i_a + L_{Ab}i_b + L_{Ac}i_c,$$

$$\Psi_B = L_{BA}i_A + L_{BB}i_B + L_{BC}i_C + L_{Ba}i_a + L_{Bb}i_b + L_{Bc}i_c,$$

$$\Psi_C = L_{CA}i_A + L_{CB}i_B + L_{CC}i_C + L_{Ca}i_a + L_{Cb}i_b + L_{Cc}i_c. \tag{3}$$

For rotor

$$\Psi_a = L_{aA}i_A + L_{aB}i_B + L_{aC}i_C + L_{aa}i_a + L_{ab}i_b + L_{ac}i_c,$$

$$\Psi_b = L_{bA}i_A + L_{bB}i_B + L_{bC}i_C + L_{ba}i_a + L_{bb}i_b + L_{bc}i_c,$$

$$\Psi_c = L_{cA}i_A + L_{cB}i_B + L_{cC}i_C + L_{ca}i_a + L_{cb}i_b + L_{cc}i_c. \tag{4}$$

From Eq. 3–4 it is seen that the induction flux of AM depends from the currents of the windings.

For completeness, it is necessary to use a law that serves to bind the vector values of the moment, flux linkage and current. This is Lenz's law that has the following form:

$$\vec{M} = k(\vec{\Psi} \times \vec{i}) \tag{5}$$

Note that, despite the full and rigorous mathematical description, the use of Eqs. 1–5 for the study of the machine has serious difficulties. To solve such complex problems today, various computer programs are widely used that can simplify the process of mathematical calculations or solving equations [9].

The supply voltage of the AM excites a rotating magnetic field in the air gap between the stator and the rotor [10]. This magnetic field rotates with the angular velocity described with expression 1.

$$\omega_1 = \frac{2\pi f_1}{p} \tag{6}$$

Where f_1 is the mains voltage frequency, Hz; p is the number of pole pairs of the electric motor.

Formula 6 shows that the speed of AM is proportional to the frequency f_1 of the voltage applied to the windings and inversely proportional to the number of pole pairs p of the stator winding [11–13].

Taking the EMF to be approximately equal to the supply voltage U, the magnetic flux of the motor can be expressed by the following ratio:

$$\Phi \cong \frac{U}{4,44 f_1 w_1 k_{ob}} \quad (7)$$

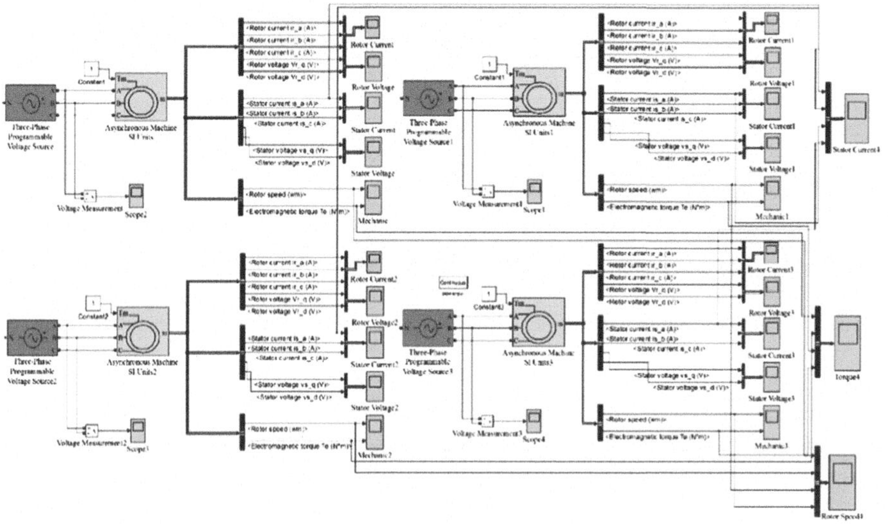

Fig. 2. Computer model of an asynchronous electric drive.

The current in the rotor conductors appears when the magnetic field moves relative to the rotor, after which its own magnetic field is created in the rotor. The magnetic flux interacting with the rotor current creates an electromagnetic torque M on the rotor shaft.

$$M \equiv \Phi I_2 \quad (8)$$

A frequency-controlled AM is usually understood as an electric machine with two control parameters, one of these parameters is frequency, and the other is the effective value of the voltage supplying the motor. Because of this, the control is considered two-parameter, which means that when controlling any parameter of the IM load, it is necessary to control the frequency and voltage too.

We are investigating the behavior of AM at a low quality of the mains voltage, namely at a reduced mains frequency. For a detailed analysis of such processes in AM,

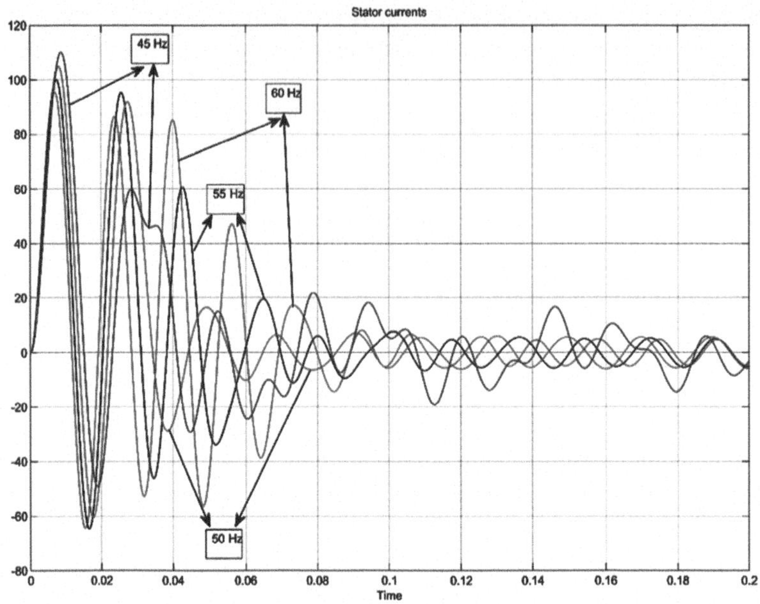

Fig. 3. Graphs of the transient processes of the AM stator currents at different network frequencies.

it is necessary to carry out computer simulation. However, the exact mathematical description of an AM, like any electrical machine, is very difficult. The main difficulty lies in the fact that all the main parameters of the operation of AM (voltage, currents and flux linkages) are variable vector quantities, characterized by frequency, amplitude and phase. That is why computer systems can help us in this problem, which can simulate transient processes of AM when the frequency of the mains voltage changes [14].

3 Modeling and Simulation Result

For computer simulation, we use the MATLAB system, which is designed to simulate engineering and technical systems [1–4, 13–16]. In the course of modeling and the operation of an AM, we make some assumptions, this is necessary to simplify the modeling process. The created model and simulation results are shown in Figs. 2, 3, 4 and 5.

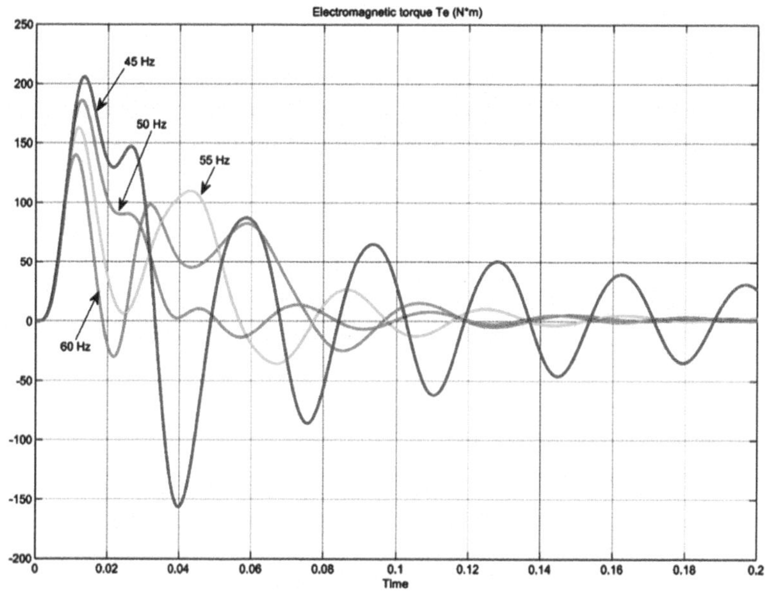

Fig. 4. Graphs of transient processes of the electromagnetic torque of AM at different network frequencies.

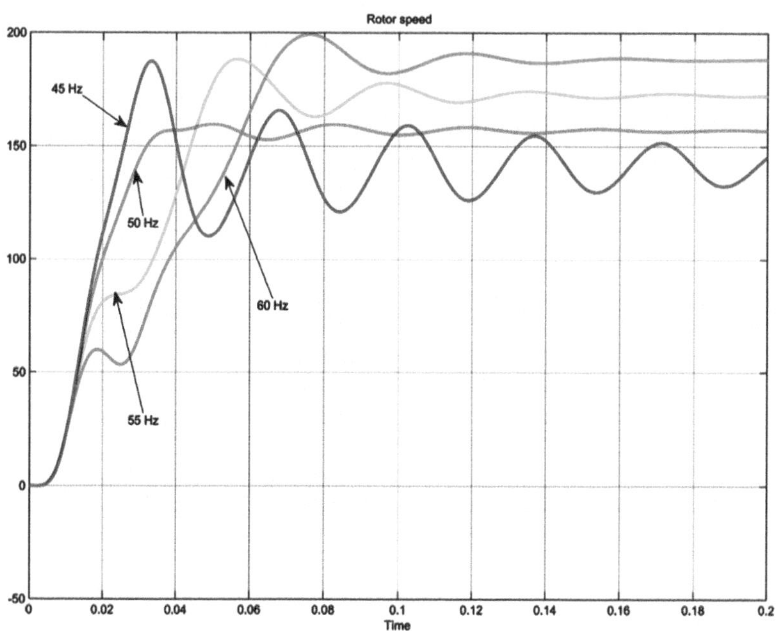

Fig. 5. Graphs of transient processes of the speed of AM at different network frequencies.

4 Conclusion

As the simulation results show, when the frequency of the supply network changes, AM have big problems, and in all the parameters of the machine, like the torque, speed and current of the stator. Such a decrease in the network frequency is naturally impossible in general cases, but with loaded power systems or in rural areas, this phenomenon is not uncommon. Therefore, to eliminate this type of problem, it is necessary to act in an integrated manner, i.e. it is necessary to modernize both in the energy and electrical engineering. The graphs shown in Figs. 4 and 5 show well that when the mains frequency decreases by 20%, the speed and torque of the AM switch to an oscillatory nature and the control of such a machine or electric drive will be a difficult task.

References

1. Tolibjonovich, D.S., Islomovna, T.M., Saidulloevna, M.D.: Modeling of starting transition processes of asynchronous motors with reduced voltage of the supply network. Eur. J. Electr. Eng. **22**(1), 23–28 (2020) https://doi.org/10.18280/ejee.220103
2. Dadabaev, S.T.: Simulation of soft start of synchronous electric drives, CAD/EDA. In: Modeling and Simulation in Modern Electronics: Collection of Scientific Papers of the III International Scientific and Practical Conference, pp. 140–145. https://doi.org/10.30987/conferencearticle_5e0282116891e4.18084453
3. Terekhin, V.B.: Modelirovanie system elektroprivoda v Simulink (Matlab 7.0.1) [Simulation of electric drive systems in Simulink (Matlab 7.0.1)]: uchebnoe posobie (Tomsk, Izd-vo Tomskogo politekhnicheskogo universiteta, 2008), p. 320. (2008). (in Russian)
4. Dadabaev, S.T.: Optimization of starting modes of high-voltage electric drives of irrigation pumping station in hot climate. Russ. Electromech. **2**, 86–91 (2018). https://doi.org/10.17213/0136-3360-2018-2-86-91. (in Russian)
5. Dadabaev, S.T.: Modelirovanie rezhimov puska vysokovol"tnogo sinkhronnogo elektroprivoda s ustroystvom plavnogo puska [Modeling starting modes of a high voltage synchronous electric drive with a smooth start device]. In: Sbornike, V.: SAPR i modelirovanie v sovremennoy elektronike, Sbornik nauchnykh trudov I Mezhdunarodnoy nauchno-prakticheskoy konferentsii, Pod red. L.A. Potapova, A.Yu. Drakina, Bryansk, pp. 91–94 (2017). (in Russian)
6. Dadabaev, S.T., Dadabaeva, Z.A.: Komp"yuternoe modelirovanie sposobov puska elektroprivodov s ventilyatornoy nagruzkoy [Computer simulation of ways to start electric drives with fan load]. In: Sbornike, V.: Elektroprivod na transporte i v promyshlennosti trudy II Vserossiyskoy nauchno-prakticheskoy konferentsii, pp. 323–327 (2018). (in Russian)
7. Bruce, F.M., Graefe, R.J., Lutz, A., Panlener, M.D.: Reduced-voltage starting of squirrel-cage induction motors. IEEE Trans. Ind. Appl. IA-20 **1**, 46–55 (1984)
8. Nevelsteen, J., Aragon, H.: Starting of large motors – methods and economics. IEEE Trans. Ind. Appl. **25**(6), 1012–1018 (1989)
9. Chapman, S.: Electric Machinery Fundamentals, 5th edn. Mcgraw Hill Higher Education (2011)
10. Krause, P.C., et al.: Analysis of Electric Machinery and Drive Systems, 3rd edn. Wiley-IEEE Press, Hoboken (2013)

11. Gracheva, E.I., Naumov, O.V.: Operating mode influence on probability characteristics of electric devices. J. Eng. Appl. Sci. **11**(1), 2934–2938 (2016)
12. Gracheva, E.I., Naumov, O.V.: Estimation of power losses in electric devices of the electrotechnical complex. In: International Conference on Industrial Engineering. Applications and Manufacturing (ICIEAM) p. 6 (2019)
13. German-Galkin, S.G., Kardonov, G.A.: Elektricheskie mashiny: laboratornye raboty na PK [Electric machines: Lab work on a PC], Saint-Petersburg, KORONA print, p. 256 (2003). (in Russian)
14. Chernikh, I.V., Modelirovanie elektrotekhnicheskikh ustroystv v MATLAB, SimPowerSystema i Simulink [Modeling of electrical devices in MATLAB, SimPowerSystema and Simulink], p. 288, DMK Press, Saint-Petersburg, Piter, Moscow (2008). (in Russian)
15. Khasanov, S.R., Gracheva, E.I., Toshkhodzhaeva, M.I., Dadabaev, S.T., Mirkhalikova, D.S.: Reliability modeling of high-voltage power lines in a sharply continental climate. In: E3S Web Conference, vol. 178, p. 01051 (2020). https://doi.org/10.1051/e3sconf/202017801051
16. Mathworks, SimPowerSystems, User's Guide, 2010–2017

Prevention, Detection, and Proof of Offenses in Road and Bridge Construction Under the Management of Geoinformation System

Aleksey Y. Mironov[1](✉), Anna Y. Mironova[2], and Vyacheslav G. Burlov[3]

[1] Russian State Hydrometeorological University, St. Petersburg, Russia
wakepolarbear@gmail.com
[2] National Research University of Technologies, Mechanics and Optics, St. Petersburg, Russia
[3] Peter the Great St. Petersburg Polytechnic University, St. Petersburg, Russia

Abstract. In the conditions of narrow departmental management, the administrative practice integrity in road and bridge construction is being destroyed by a lack of completeness and authenticity of production on the affairs about administrative offenses within a reasonable time. The timely application of the geoinformation system in order to prompt prevention, reliable revealing and proper proving of offenses ensures a preventively guaranteed efficiency and security of road and bridge construction. This paper proposed a synthesis of the administrative practice management model in construction, based on the natural-scientific approach to the adoption of managerial decisions. The condition for the existence of guaranteed management, which relates the characteristics of processes and performance indicators, is concretized by the Kolmogorov-Chapman equations system. Network models of interacting processes of the revealing and proving of offenses, the origin, and identification of violations, them neutralization through the managerial impact on the limited resources of administrative practice in construction and of geoinformation system are built. Based on the simulation results, the requirements for the geoinformation system that ensure guaranteed management and maximum security to the road and bridge construction within a reasonable time are determined.

Keywords: Guaranteed management · Administrative practice · Road and bridge construction · Geoinformation system · Natural-scientific approach

1 Introduction

The huge latency, estimated up to 3/4 of the really committed, administratively punishable misconducts, indicates an unsatisfactory ensure of engineering and ecological security in road and bridge construction in terms of incomplete revealing and inauthentic proving of administrative offenses [1]. The problems are associated with unauthorized development of protected and dangerous zones, illegitimate subsoil use, non-compliance with environmental standards, deviations from the design and regulatory documents, environmental and landscaping damage. The destruction of the

A. Sinitsyn (Ed.): SES 2020, LNCE 141, pp. 214–223, 2021.
https://doi.org/10.1007/978-3-030-67654-4_24

integrity of administrative practice is caused by the random nature of the reasons used to initiate affairs in accordance with article 28.1 of the Code Russian Federation about Administrative Offenses (Code) [2]. The sporadic and post-factum response of law enforcement officials to the flow of administrative offenses gives a vicious punitive-fiscal meaning by production on the affairs about them. The administrative law enforcement agencies are focused on collecting fines for committed acts and are indifferent to the damage caused [3]. Overcoming the incompleteness and unreliability of administrative practice is especially relevant in relation to administrative offenses, the traces of which are hidden behind the features of the terrain or covered by offenders. A mechanism for early fixing, proper investigation and minimization of consequences is required for 1/7 of the Special Part compositions of Code. Administrative prevention, revealing and proving in road and bridge construction within a reasonable time regularly requires the geographical coordinates with the signs of preparation or commission of offenses to administrative law enforcement agencies in the territory under their jurisdiction, for carryout targeted inspections immediately and for prevent or suppress of harmful consequences [4].

To implement the geocoordinate support of preventive and operational-search activities, it is necessary to systematically analyze cartographic information on tracking changes in the position, shape, and structure of anthropogenic or natural objects with an eye to the ascertainment of event and of elements on the administrative offense [5]. Addresses for targeted inspections can be obtained using the geoinformation system during the monitoring of territories under the jurisdiction of the administrative law enforcement agency [6]. Dozens of geoportal solutions at the federal, regional and local levels operate in government or municipal departments, commercial structures of Russia [7]. But due to the failure to timely prevent and suppress of administrative offenses, the catastrophic consequences of destructions, fires, floods, pollutions, misuses of land and forests annually arise. Therefore, in road and bridge construction, at the junction of administrative practice and geoinformatics, an adequate model for managing administrative production using a geoinformation system is gaining scientific and practical interest [8].

According to the theory of functional systems, a decision-maker implements the management process based on the model, which is manifested in the chain of basic elements of its formation: "excitement" - "recognition" - "reaction to the situation" [9]. The relevance of this paper to the automation of management is determined by the lack of an adequate mathematical model for managerial decision making, based on ensuring a balanced unity of the functioning of social and geoinformation systems, the basic laws of the world order within the framework of a single approach to the universal formalized criterion.

In contrast to analysis, synthesis forms processes with predetermined properties [10]. The synthesis of production management will allow guaranteed to ensure the revealing and proving of administrative offenses in road and bridge construction within a reasonable time [11].

Thus, in this paper, the following tasks are solved:

- Synthesis of a Mathematical Model for Managerial Decision Making;
- Substantiation of the criterion for guaranteed management of administrative production using a geoinformation system;
- Construction of structural-functional technology for automated management of administrative practice in road and bridge construction.

2 Materials and Methods

2.1 Synthesis of a Model for Managerial Decision Making

In order to exclude arbitrariness in reasoning and conflicting conclusions, the axiomatic-deductive method is used [12]. To ensure the adequacy of the model for managerial decision making, its synthesis is based on the law of object integrity conservation, as a stable repetitive relationship between the properties of the object and the properties of its actions for a fixed mission [13].

In accordance with the natural-scientific approach, integrating the properties of the world around us, consciousness and cognition, the process of managerial decision making is considered in the light of its three properties at each of the three levels of world cognition (see Fig. 1) [14].

The formation of an adequate model for managerial decision making is to establish formal analytical relations between the three technological components that are characterized by temporary resources, irreversible for the decision-maker (see Fig. 2):

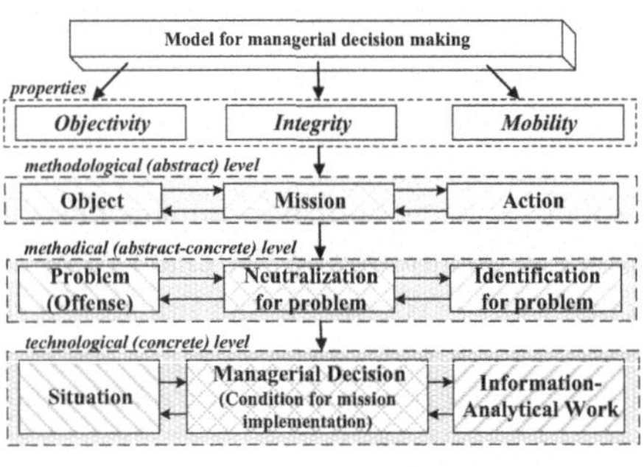

Fig. 1. Deployment of the natural-science approach.

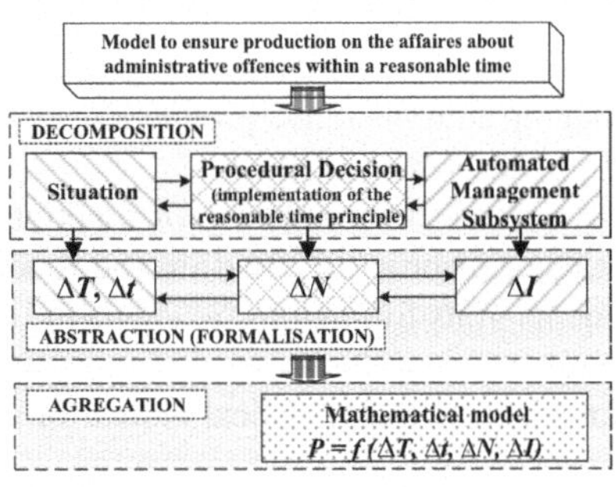

Fig. 2. Synthesis of a mathematical model.

1. Situation (Object) is a set of current state characteristics of the administrative production, factors, and conditions of decision-maker activity, which is identified with the period of the target process ΔT and the periodicity of offense occurrence (average time to emergence problem) Δt;
2. Procedural Decision (Mission) is the provision by the decision-maker of the condition for administrative production implementation within a reasonable time in the

current Situation to achieve the management goal, which is identified with an adequate periodicity of response to offenses (average time to neutralize problem) ΔN;
3. Information-Analytical Work (Action) is the continuous extraction, accumulation, generalization, analysis of geodata about the current Situation, which is generalized into periodicity of offenses detection (average time to identify the problem) ΔI.

Using the methods of decomposition, abstraction, and aggregation, the process of managerial decisions making is formalized into the mathematical aggregate of the model [15], where:

$$P = f(\Delta T, \Delta t, \Delta N, \Delta I) \tag{1}$$

- P is the probability of finding administrative proceedings during the management in each of its basic states: Initial, Target, Identification or Neutralization;
- ΔT is the generalized characteristic (average period) of the regular implementation of the target process, functionally consolidating acts (works) on the transition through its states to achieve the target task;
- Δt is the generalized characteristic (average time) of problem occurrence, functionally uniting the works for moving through its states to problem maturation;
- ΔI is the average time of problem identification, functionally linking information-analytical works on the passage of process states to identify the problem;
- ΔN is the average time of problem neutralization, functionally linking the acts of decision-makers on advance through process states to eliminate the problem.

2.2 Criterion for Guaranteed Management of Administrative Production

In a real environment of road and bridge construction, the decision-maker is focused on the implementation of targeted activities for the prevention, detection, and proof of offenses within a reasonable time due to the procedural methods defined by law and departmental methods. The implementation of proven administrative practice schemes based on normative time and other resources provided is complicated by a stream of objective and subjective circumstances that make it difficult to quickly establish the geographical coordinates of the administrative offense place where there are signs of the event and corpus delicti [16]. Stochastically emerging problems can and should be proactively identified using a geoinformation system and guaranteed to be eliminated with the involvement of additional resources in the conditions of restrictions on their availability [17]. If the goals of reliable fixing and proving the administrative offense are not achieved within a reasonable time, the decision-maker is forced to delay, procedurally extend or terminate the affair investigation, which is tantamount to the target failure and to the breakdown of production management [18].

Considering the three-component nature of the basic model of managerial decision making to protect from the interference of problems (offenses) in the target activity, the management of administrative production includes four interacting processes. The target process during the execution of administrative practice in the normal mode is objectively accompanied by a Poisson stream of problems leading to failures of a reasonable time. When a reasonable period of implementation is delayed, the target process breaks back to the initial state by the extension or termination of administrative proceedings. In order to prevent the non-fulfillment of the target task and the failure of management, the identification and neutralization processes due to the works of the decision-maker in the automated management subsystem based on the geoinformation system to diagnose the problems occurrence and to take the remedial measures for elimination them with the comparable time and the resources deficiency [19]. In cases of insoluble problems, part of the identification and neutralization processes can also be disrupted to the initial state (see Fig. 3).

The administrative production management model in the form of a continuous Markov chain is characterized by the probabilities of being in one of four basic states associated with the intensities of target process $\zeta = 1/\Delta T$, of problem occurrence $\lambda = 1/\Delta t$, of its identification $\nu = 1/\Delta I$ and neutralization $\omega = 1/\Delta N$, as well as the failure rates of target process $\xi = f(\Delta T)$, of identification $\mu = f(\Delta I)$ and neutralization $\tau = f(\Delta N)$ (see Fig. 4):

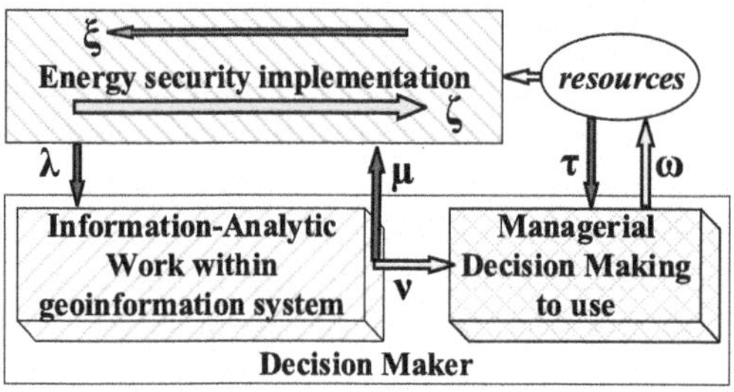

Fig. 3. Management scheme of administrative production.

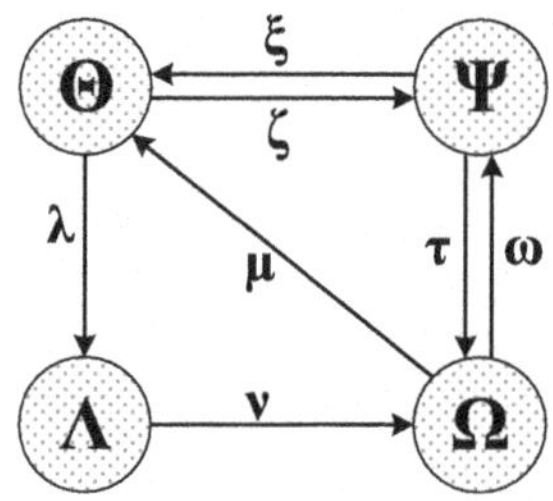

Fig. 4. States graph of management.

- Θ is the probability of initial state when the administrative practice of the decision-maker is in the initial state of administrative proceedings, which does not require the identification or neutralization of problems;
- Ψ is the probability of target completing when the decision-maker fulfilled the target task of administrative proceedings with the guarantee of neutralizing the associated problems (offenses) within a reasonable time;
- Λ is the probability of identification when the decision-maker identifies problems leading to a breakdown of a reasonable time for their subsequent neutralization;
- Ω is the probability of neutralization when the decision-maker, based on the results of identification, neutralizes problems that cause disruptions in ensuring a reasonable time.

The relationship (1) of the probability $P = (\Theta, \Psi, \Lambda, \Omega)$ of finding administrative proceedings in each of its basic states, respectively, with the intensities ζ, ξ, λ, ν, μ, ω, τ of continuous Markov transitions between states of the graph is specified Kolmogorov-Chapman differential equations system:

$$\begin{aligned}
d\Theta(t)/dt &= -\Theta(t)[\zeta+\lambda]+\Psi(t)\xi+\Omega(t)\mu \\
d\Psi(t)/dt &= \Theta(t)\zeta-\Psi(t)[\xi+\tau]+\Omega(t)\omega \\
d\Lambda(t)/dt &= \Theta(t)\lambda-\Lambda(t)\nu \\
d\Omega(t)/dt &= \Psi(t)\tau+\Lambda(t)\nu-\Omega(t)[\mu+\omega] \\
\Theta(t)&+\Psi(t)+\Lambda(t)+\Omega(t)=1
\end{aligned} \tag{2}$$

When, over time, the Poisson streams of interacting processes tend to the limiting stationary mode, the Kolmogorov-Chapman differential Eqs. (2) are transformed into a system of linear homogeneous algebraic equations, the solution of which using the Cramer method is system-forming factors of management:

$$\begin{aligned}
\Theta &= \frac{\xi\nu\omega+\nu\mu(\xi+\tau)}{\xi\omega(\lambda+\nu)+\mu(\lambda+\nu)(\xi+\tau)+\nu(\zeta+\lambda)(\omega+\tau)+\nu(\zeta\mu+\xi\lambda)} \\
\Psi &= \frac{\nu\omega(\zeta+\lambda)+\zeta\nu\mu}{\xi\omega(\lambda+\nu)+\mu(\lambda+\nu)(\xi+\tau)+\nu(\zeta+\lambda)(\omega+\tau)+\nu(\zeta\mu+\xi\lambda)} \\
\Lambda &= \frac{\xi\lambda\omega+\lambda\mu(\xi+\tau)}{\xi\omega(\lambda+\nu)+\mu(\lambda+\nu)(\xi+\tau)+\nu(\zeta+\lambda)(\omega+\tau)+\nu(\zeta\mu+\xi\lambda)} \\
\Omega &= \frac{\nu\tau(\zeta+\lambda)+\xi\lambda\nu}{\xi\omega(\lambda+\nu)+\mu(\lambda+\nu)(\xi+\tau)+\nu(\zeta+\lambda)(\omega+\tau)+\nu(\zeta\mu+\xi\lambda)}
\end{aligned} \tag{3}$$

3 Results

In the current Situation, characterized by the intensities of target process ζ and of problem occurrence λ, at the normatively established levels of the maximum allowable disruption frequencies ξ, μ, τ and of the minimum sufficient efficiency Ψ, guaranteed criterion (3) of production management allows the decision-maker to control the sufficiency and to optimize the intensities of identification ν due to the Information-Analytical Work within geoinformation system and of neutralization ω due to the Managerial Decision by rationalizing the durations of transitions by their events [20].

The dependence of management efficiency Ψ on the failure rate of the target process ξ is close to a decreasing exponent (see Fig. 5). At the same time, a step-by-step twofold increase in the intensities of identification ν and neutralization ω causes an exponentially decaying increase of efficiency Ψ within identical slice ξ. Therefore, the failure rate of the target process ξ sets the status of compliance (suitability) of administrative practice by the signs of a reasonable time. It is within the framework of the suitability indicator ξ that the necessary management efficiency Ψ is adjusted by varying the intensities of identification ν and neutralization ω.

The management efficiency Ψ decreases more significantly from the increase in the share of neutralization failure τ than from a similar identification failure μ (see Fig. 6). Technical errors of the geographic information system can be partially corrected by other means of information and analytical work. A limited set of procedural means of neutralizing offenses does not leave by the decision-maker the opportunity to compensate for irreparable losses of a reasonable time.

When changes the situation, characterized by the intensity of the target process ζ and the problem flow λ, the intensities ν, ω, of activity of the geographic information system and of the decision-maker must be strengthened by the available reserves to identify and neutralize offenses within a reasonable time taking into account the failure frequencies μ, τ.

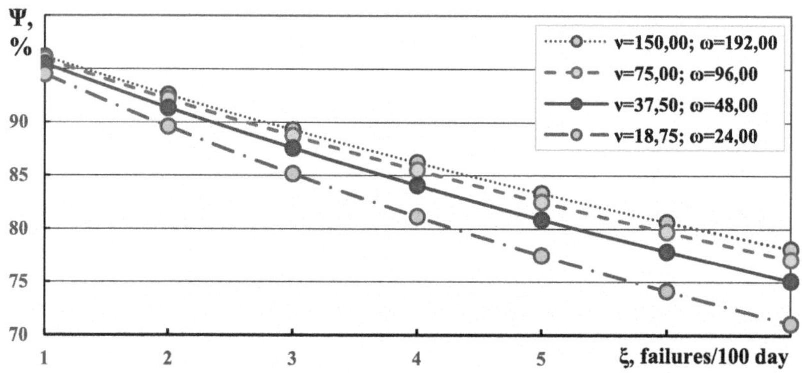

Fig. 5. Dependence of efficiency Ψ on the failure rate ξ during double extension ν, ω.

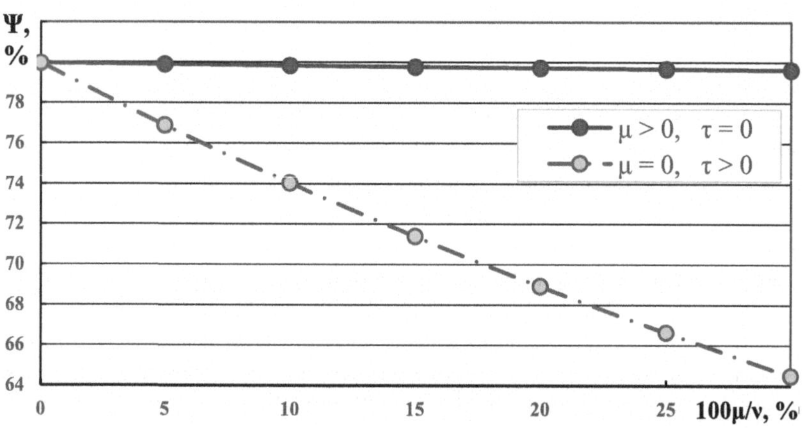

Fig. 6. Dependence of efficiency Ψ on the failure rates μ = *f* (ν), τ = *f*(ω).

4 Discussion

The intensity of the stream depends on the structure and duration of transitions by states (events) within the process. Therefore, it is advisable to obtain the intensity of target process $\zeta = 1/\Delta T$, the intensity of problem occurrence $\lambda = 1/\Delta t$, the intensities of its identification $\nu = 1/\Delta I$ and neutralization $\omega = 1/\Delta N$ by the structural-functional method [21]. When using it, by clearly linking the works on transitions and the times spent on them in the network model of each process, network analysis makes it possible to evaluate periods ΔT, Δt, ΔI, ΔN through the critical path of each process [22]. According to the observations of administrative statistics regarding the timeliness of the issuance and execution of ordinances on the affairs about administrative offenses, $\xi \approx 0{,}25\ \zeta$ and $\tau \approx 0{,}05\ \omega$. The probability of technical errors in the functioning of the geographic information system does not exceed $\mu \approx 0{,}10\ \nu$.

Based on the average statistical durations of transition between process events, the following indicators are calculated in each network model [23]:

- the earliest possible time of event j

$$T_{\mathrm{p}}(j) = \max_{i \subset j}[T(i) + t_{ij}] \quad (4)$$

where $i \subset j$ occurs when event i precedes the next event j;
t_{ij} is the duration of process transfer from event i to j;

- the latest allowed time of event i

$$T_{\Pi}(i) = \min_{i \subset j}[T(j) - t_{ij}] \quad (5)$$

- the time reserve of event i

$$R_i = T_{\Pi}(i) - T_{\mathrm{p}}(i) \quad (6)$$

- the full reserve of time for transfer from event i to j

$$r(i,j) = T_{\Pi}(j) - T_{\mathrm{p}}(i) - t_{ij} \quad (7)$$

- the duration of the critical path of the process, considered as the maximum total duration to transfer the process from the initial to the final event of the network model or as the durations sum of work for which the full reserves of time are zero.

5 Conclusion

A numerical assessment of system-forming factors (3) on network models of the processes of administrative practice in road and bridge construction confirmed that, based on the average statistical duration of the process procedures, the proposed management model is capable of monitoring critical changes in the situation according to the conditions of efficiency and security. It also helps to reconfigure the structure and functionality of prevention, detection, and proof of administrative offenses within a reasonable time under the current Situation.

The leadership of the enforcement agencies and road and bridge construction was proposed a geoinformation mechanism that, within the regulatory acceptable range of suitability for administrative production, allows to maintain sufficient efficiency in the prevention, detection, and proof of offenses by timely attracting the necessary resources.

References

1. Derjuga, A.N., Motrovich, I.D.: The reasons of the latency administrative delictology. Adm. law Process. **7**, 57–62 (2013)
2. Mironov, A.Y.: Problems to implementing a reasonable time of criminal production in the stages of initiating a criminal affair and its preliminary investigation. In: Boer, V.M. (ed.) The 70th International Scientific Conference of State University of Aerospace Instrumentation, vol. 3, pp. 126–131. State University of Aerospace Instrumentation, St. Petersburg (2017)
3. Mironov, A.Y.: Automated management system of production on affairs about administrative offense at the regional level. In: Boer, V.M. (ed.) The 69th International Scientific Conference of State University of Aerospace Instrumentation, vol. 1, pp. 314–318. State University of Aerospace Instrumentation, St. Petersburg (2016)
4. Mironov, A., Mironova, A., Sipovich, D.: Information security of revealing and proving administrative offenses when using geoinformation system. Reg. Informatics Inf. Secur. **7**, 402–406 (2019)
5. Bondur, V.G., Gordo, K.A.: Satellite monitoring of burnt-out areas and emissions of harmful contaminants due to forest and other wildfires in russia. Izv. Atmos. Ocean. Phys. **54**, 955–965 (2018). https://doi.org/10.1134/S0001433818090104
6. Savinykh, V.P., Oznamets, V.V., Selmanova, N.N., Tsvetkov, V.Y.: System-categorical analysis used while monitoring of the land according to the remote sensing data. Geod. Aerophotosurveying. **62**, 106–113 (2018). https://doi.org/10.30533/0536-101X-2018-62-1-106-113
7. Kuznetsov, A.M., Trusov, S.V., Baraboshkin, O.I., Bobrovskij, S.A., Romanov, A.A., Romanov, A.A.: Analysis of the results obtained over three years of operation of AIS vessel monitoring equipment based on the resurs-P No. 2 Spacecraft. Rocket. Device Eng. Inf. Syst. **5**, 80–87 (2018). https://doi.org/10.30894/issn2409-0239.2018.5.4.80.87
8. Kashnitskii, A.V., Lupyan, E.A., Balashov, I.V., Konstantinova, A.M.: Technology for designing tools for the process and analysis of data from very large scale distributed satellite archives. Atmos. Ocean. Opt. **30**, 84–88 (2017). https://doi.org/10.1134/S1024856017010080

9. Sudakov, K.V.: Theory of functional systems: a keystone of integrative biology. In: Cognitive Systems Monographs, pp. 153–173 (2015). https://doi.org/10.1007/978-3-319-19446-2_9
10. Andreev, A.V., Burlov, V.G., Grachev, M.I.: Information technologies and synthesis of the management process model in the enterprise. In: 2019 International Science and Technology Conference "EastConf.", pp. 1–5. IEEE (2019). https://doi.org/10.1109/EastConf.2019.8725428
11. Burlov, V.G., Mironov, A.Y.: Synthesis of automated management. In: Lepesha, G.V. (ed.) The All-Russian Scientific and Practical Conference on Innovative technologies and security issues of the real economy, pp. 27–37. St. Petersburg State University of Economics, St. Petersburg (2019)
12. Mironov, A.Y.: Model of optimal management by administrative production. In: Boer, V.M. (ed.) The 70th International Scientific Conference of State University of Aerospace Instrumentation, vol. 2, pp. 139–144. State University of Aerospace Instrumentation, St. Petersburg (2017)
13. Burlov, V., Andreev, A., Gomazov, F.: Mathematical model of human decision - a methodological basis for the realization of the human factor in safety management. Procedia Comput. Sci. **145**, 112–117 (2018). https://doi.org/10.1016/j.procs.2018.11.018
14. Burlov, V., Grachev, M.: Development of a mathematical model of traffic safety management with account for opportunities of web technologies. Transp. Res. Procedia. **20**, 100–106 (2017). https://doi.org/10.1016/j.trpro.2017.01.023
15. Burlov, V.G., Mironov, A.Y., Baranov, Y.V.: Synthesis of model of management of affairs about administrative offenses. Inf. Measur. Manag. Syst. **3**, 22–34 (2019). https://doi.org/10.18127/j20700814-201903-03
16. Burlow, V.G., Grachev, M.I., Shlygina, N.S.: Adoption of management decisions in the context of the uncertainty of the emergence of threats. In: 2017 XX IEEE International Conference on Soft Computing and Measurements (SCM), pp. 107–108. IEEE (2017). https://doi.org/10.1109/SCM.2017.7970510
17. Istomin, E.P., Abramov, V.M., Burlov, V.G., Sokolov, A.G., Fokicheva, A.A.: Risk management metod in parametric geosystems. In: 18th International Multidisciplinary Scientific Conference on EARTH & GEOSCIENCES, pp. 377–384. SGEM, Albena (2018)
18. Burlov, V.G., Gomazov, F.A., Uvarova, A.O.: Technology for managing the processes of ensuring labor safety at municipal waste processing enterprises. In: 2018 IEEE International Conference "Management of Municipal Waste as an Important Factor of Sustainable Urban Developmen" (WASTE), pp. 72–74. IEEE (2018). https://doi.org/10.1109/WASTE.2018.8554168
19. Burlov, V., Mironov, A., Kostareva, K.: Synthesis of automated management by production on affairs about administrative offenses. Inf. Technol. Syst. Manag. Econ. Transp. law. **3**(35), 110–114 (2019)
20. Burlov, V.G., Mironov, A.Y., Mironova, A.Y.: Guaranteed management of production on affairs about administrative offences when using geoinformation system. Glob. Reg. Res. **1**, 200–210 (2019)
21. Lu, M., Li, H.: Resource-activity critical-path method for construction planning. J. Constr. Eng. Manag. **129**, 412–420 (2003). https://doi.org/10.1061/(ASCE)0733-9364(2003)129:4(412)
22. Burlov, V.G., Grobitski, A.M., Grobitskaya, A.M.: The management of construction production taking into account the indicator of successful fulfillment of a production task. Mag. Civ. Eng. **63**, 77–91 (2016). https://doi.org/10.5862/MCE.63.5
23. Burlov, V.G., Mironov, A.Y., Mironova, A.Y.: Application of geoinformation system in prevention, identification and evidence about administrative offenses. Proc. Russ. State hydrometeorol. Univ. **57**, 126–146 (2019). https://doi.org/10.33933/2074-2762-2019-57-126-146

Technology of 3-D Printing of Fiber Reinforced Mixtures

S. V. Klyuev[1(✉)], A. V. Klyuev[1], N. I. Vatin[2], and E. S. Shorstova[1]

[1] Belgorod State Technological University named after V.G. Shukhov, Belgorod, Russia
Klyuyev@yandex.ru
[2] Peter the Great St. Petersburg Polytechnic University, Saint Petersburg, Russia

Abstract. Due to the introduction of reinforcing fibers into concrete, it is possible to overcome one of the main disadvantages of concrete-low tensile and bending strength. This is due to the fact that the reinforcing fibers take on a tensile stress, resulting in increased tensile resistance (in percentage terms, this is about 250%). In addition, reinforcing fibers contribute to the uniform distribution of moisture in the concrete due to which internal loads are reduced, crack resistance is approximately doubled and the impact strength of the concrete is 12 times increased. In addition, the deformability, frost resistance, heat resistance, water resistance and corrosion resistance of concrete structures are significantly increased. What makes fibroblast ideal material for use in 3D printing.

Keywords: Silica fume · Additive · 3-D technology · Concrete · Fiber

1 Introduction

One of the important motives for the use of additive technologies in the construction industry is also the safety of industry workers, especially in extreme conditions. When construction is carried out in harsh conditions, the degree of complexity and risks for the worker inevitably increases, which negatively affects both the quality of construction and human safety [1–7]. For example, when conducting construction works in permafrost areas, the main problem is the excavation and pouring of concrete. In areas with high ambient temperatures, the main problems in construction are dehydration of construction workers. There are frequent cases of construction on sites subject to chemical or nuclear contamination, which can also pose a serious threat to human health. According to the author of this study, the only solution, for such problems, would be the manufacture of structural parts for the collection of the construction structure outside the construction site. It is obvious that the parts and components delivered from the place of manufacture and assembled directly on the construction site reduce the amount of labor on site, in addition, this approach to construction will also improve the quality and stability of building structures [8–13].

Another opportunity for additive technologies in the construction industry is to shorten the supply chain, especially for parts that require expedited delivery. Additive technologies allow you to print ordered parts from a 3D model without spending a lot of time on the order, thus reducing the number of stages in the supply chain. For a

A. Sinitsyn (Ed.): SES 2020, LNCE 141, pp. 224–230, 2021.
https://doi.org/10.1007/978-3-030-67654-4_25

construction company, this in turn means that instead of collaborating with several companies producing different structural or non-structural components, such components can be produced by the construction company itself directly using 3D printing techniques immediately after their development. This eliminates performance issues caused by delays in the delivery of building structures to the construction site.

Another motivation for the use of additive technologies is the reduced availability of skilled labor. Contractors somehow face difficulties in hiring labor with the necessary skills (for example, experienced carpenters, heavy equipment operators, welders and manufacturers), using additive technologies will need only workers who are qualified to work with the appropriate equipment [14–24].

2 Methods and Materials

When designing the composition of fine-grained concrete for 3D printing, it is necessary to take into account the consistency of the concrete mixture to obtain the required workability. The technological scheme of production of a concrete mix begins with reception of composite binders and includes a set of operations as a result of which receive the required product.

The technology of layering the mixture is realized through extrusion printing technique, which produces three-dimensional objects from a computer model, as in a typical rapid prototype process. Detailed models can be obtained from computed tomography, magnetic resonance imaging, or model data created from three-dimensional object digitization systems and used in this technique. In the surfacing simulation method, fiber concrete is extruded and deposited as a semi-molten polymer onto a platform in a layer-by-layer cycle. Figure 1 shows how each layer is created. The base of the platform is lowered, and the next layer is laid on top of the other, as prescribed by the computer model.

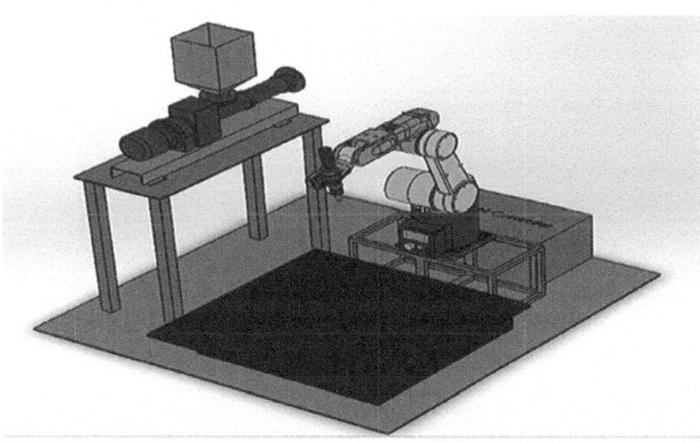

Fig. 1. Graphic explanation of extrusion printing

A comparative analysis of the technical and physico-mechanical characteristics of the fiber types considered is given in Table 1.

Table 1. Technical and physico-mechanical characteristics of various types of fibers

	Characteristics and values of fiber characteristics indicators							
Types of fiber	Strength on Stretching (MPa)	Diameter fibers	Length fibers (mm)	Module of elasticity GPa	Coefficient elongations (%)	Temperature melting (°C)	Density, (g/cm^3)	Resistance to corrosion and alkalis
Metal wire made of carbon steel	600–1500	0,5–1,2 mm	30–50	190	3–4	1550	7,8	Low
Polypropylene	150–600	10–25 mkm	6–18	35	20–150	160	0,91	High
Fiberglass	1500–3500	13–15 mkm	4,5–18	75	4,5	860	2,6	Average

The analysis of technical and physico-mechanical characteristics of different types of fiber, given in Table 1 allows us to conclude that each type of fiber has certain advantages.

3 Results and Discussions

The reason for the reduced travel speed and fill speed was mainly to avoid problems such as skipping steps during printing, which occurs in most cases due to high speeds during printing. The fill type was chosen square grid and the density was set to 70%, filling so that the printed parts were solid.

In the last few years, various methods of 3D concrete printing have been used based on two technologies, namely extrusion-based and powder-based. Consider extrusion technology. The extrusion-based method is similar to the slurry deposition simulation method, which extrudes cement material from a nozzle mounted on a portal, crane, or 6-axis robotic arm to print the structure layer by layer. This method was aimed at application in on-site construction and is used for large-scale construction components with complex geometries, and has great potential for use in the construction industry. The scheme of extrusion technology is illustrated in Fig. 2.

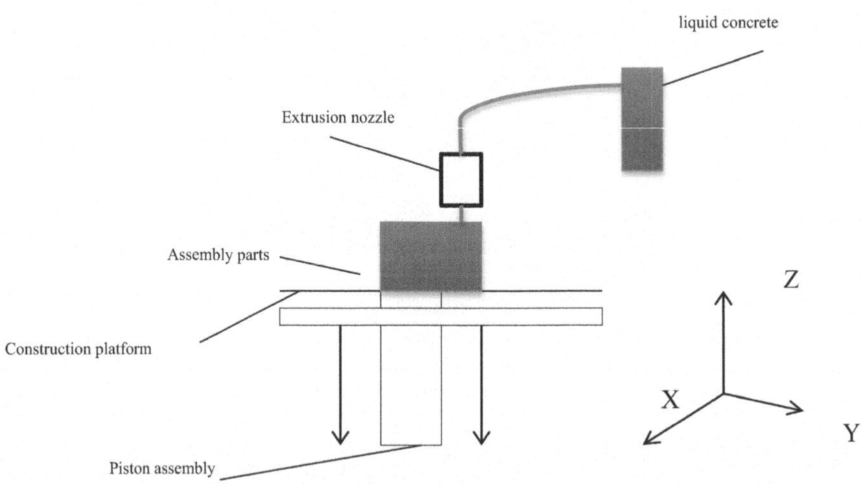

Fig. 2. Diagram of the extrusion technology 3D concrete printing.

The second technique is contouring. Contour Crafting technology was developed at the University of southern California, USA. At the heart of this technology, the extrusion method is used in order to press 2 layers of cementing mixture to build a vertical concrete form-formwork. Custom reinforcement ties are manually inserted between the layers (at every 30 cm horizontally and 13 cm vertically) while the machine continually extrudes layers of spatula-like blades, attached to the printhead create smooth extruded surfaces. Once the extruded formwork is complete, the concrete is then poured by hand to a height of 13 cm and an hour later a second portion is poured on top of the first portion. Each batch of concrete is laid with a delay of one hour, in order to control the lateral pressure of the concrete, allowing it to partially harden. And harden. The shape of the concrete wall made by the machine is shown in Fig. 3.

Fig. 3. Concrete wall shape, manufactured with contour machining machine, with reinforced screeds, custom made, manually inserted between the layers

The main advantages of the technology are excellent surface quality and significantly increased production speed. Another key advantage of this technology is the ability to integrate with other robotic techniques to install internal components, such as pipes, electrical conductors and amplification modules, to improve mechanical properties. Contouring technology currently allows vertical elements to be produced mainly in compression. When a doorway or window is required, a jumper is installed to bridge the gap.

In particular sources, it is possible to meet arguments about the shortcomings of this technology, among which are:

- limiting vertical extrusion, which gives a 2.5 D topology (vertical expansion of the flat shape);
- initial formwork and spatula can be quite difficult to implement depending on the size and shape of the printed object;
- intermittent sequential pouring of concrete inside the formwork due to hydrostatic pressure and weak mechanical properties of extruded concrete can lead to weakening of the interfacial zones between the layers.

The use of fiber concrete requires additional support to create protrusions and other functions of arbitrary shape: the use of another material besides the concrete itself, for example, a composite binder. The disadvantage of this process is that an additional deposition device is required for the second material.

The following disadvantages of the technology of the analyzed printing technology are distinguished:

- time to maintain dimensional accuracy, makes the printing process quite slow, so the prospects for the intended industrial application are uncertain;
- using a second material to support overhangs reduces the efficiency and flexibility of the process while increasing the cost of the material;
- dimensions and possibilities in terms of shaping are limited by the size of the printed frame.

4 Conclusion

It should be noted that 3D printing is mostly done for fast operations or prototyping, which means low fill volume. A typical fill percentage is about 20% for 3D printing depending on the actual use of the 3D printed part. If the part is to be used for construction, the fill percentage increases, if the part is printed as a prototype to test the design, sometimes a lower fill percentage is used, such as 10%. Printing a part with 100% filling adversely affects the purpose of 3D printing using material extrusion and significantly increases the printing time.

References

1. Fediuk, R.S., Lesovik, V.S., Liseitsev, Yu.L., Timokhin, R.A., Bituyev, A.V., Zaiakhanov, M.Ye., Mochalov, A.V.: Composite binders for concretes with improved shock resistance. Mag. Civ. Eng. **85**(1), 28–38 (2019)
2. Erofeev, V.T., Rodin, A.I., Kravchuk, A.S., Kaznacheev, S.V., Zaharova, E.A.: Biostable silicic rock-based glass ceramic foams. Mag. Civ. Eng. **84**(8), 48–56 (2018)
3. Dmitrieva, T.V., Markova, I., Strokova, V.V., Bezrodnykh, A.A., Kutsyna, N.P.: Efficiency of stabilizers of various compositions for strengthening soils with mineral binders. Constr. Mater. Prod. **3**(1), 30–38 (2020). https://doi.org/10.34031/2618-7183-2020-3-1-30-38

4. Fediuk, R.S., Lesovik, V.S., Svintsov, A.P., Mochalov, A.V., Kulichkov, S.V., Stoyushko, N.Y., Gladkova, N.A., Timokhin, R.A.: Self-compacting concrete using pretreatmented rice husk ash. Mag. Civ. Eng. **79**(3), 66–76 (2018)
5. Build. Struct. **73**, 7–15 (2018). https://doi.org/10.18720/CUBS.73.1
6. Korniyenko, S., Glukhoverya, D.: Influence of placement of the window block on wall thickness on heat losses. Constr. Unique Build. Struct. **67**, 62–71 (2018). https://doi.org/10.18720/CUBS.67.5
7. Denisova, Yu.V.: Additive technology in construction. Constr. Mater. Prod. **1**(3), 33–42 (2018)
8. Kozhukhova, N.I., Strokova, V.V., Kozhukhova, M.I., Zhernovsky, I.V.: Structure formation in alkali activated aluminosilicate binding systems using natural raw materials with different crystallinity degree. Constr. Mater. Prod. **1**(4), 38–43 (2018)
9. Klyuev, S.V., Klyuev, A.V., Khezhev, T.A., Pucharenko, Y.: Technogenic sands as effective filler for fine-grained fibre concrete. J. Phys. Conf. Ser. **1118**, 012020 (2018)
10. Klyuev, S.V., Klyuev, A.V., Shorstova, E.S.: The micro silicon additive effects on the fine-grassed concrete properties for 3-d additive technologies. In: Materials Science Forum, vol. 974, pp. 131–135 (2019)
11. Klyuev, S.V., Klyuev, A.V., Khezhev, T.A., Pucharenko, Yu.V.: High-strength fine-grained fiber concrete with combined reinforcement by fiber. J. Eng. Appl. Sci. **13**(8 SI), 6407–6412 (2018)
12. Klyuev, S.V., Khezhev, T.A., Pukharenko, Yu.V., Klyuev, A.V.: The fiber-reinforced concrete constructions experimental research. In: Materials Science Forum, vol. 931, pp. 598–602 (2018)
13. Klyuev, S.V., Khezhev, T.A., Pukharenko, Yu.V., Klyuev, A.V.: Fiber concrete on the basis of composite binder and technogenic raw materials. In: Materials Science Forum, vol. 931, pp. 603–607 (2018)
14. Klyuev, S.V., Klyuev, A.V., Shorstova, E.S.: Fiber concrete for 3-d additive technologies. In: Materials Science Forum, vol. 974, pp. 367–372 (2019)
15. Poluektova, V.A.: Theoretical concepts of the modification effect on the interparticle energy of the polymer-mineral dispersions for 3D printing. In: Materials Science Forum, vol. 974, pp. 107–112 (2020)
16. Usanova, K.: Properties of cold-bonded fly ash lightweight aggregate concretes. Lecture Notes in Civil Engineering, vol. 70, pp. 507–516 (2020)
17. Tolstoy, A.D.: Fine-grained high-strength concrete. Constr. Mater. Prod. **3**(1), 39–43 (2020). https://doi.org/10.34031/2618-7183-2020-3-1-39-43
18. Travush, V.I., Karpenko, N.I., Erofeev, V.T., Rodin, A.I., Rodina, N.G., Smirnov, V.F.: Development of biocidal cements for buildings and structures with biologically active environments. Power Technol. Eng. **51**(4), 377–384 (2017)
19. Fedyuk, R.S., Baranov, A.V., Mugahed Amran, Y.H.: Effect of porous structure on sound absorption of cellular concrete. Constr. Mater. Prod. **3**(2), 5–18 (2020). https://doi.org/10.34031/2618-7183-2020-3-2-5-18
20. Poluektova, V.A.: Electrokinetic properties and aggregative stability of polymer-mineral dispersions for 3D printing in building. Russ. J. Phys. Chem. A **93**(9), 1783–1788 (2019)
21. Erofeev, V., Emelyanov, D., Tretiakov, I., Kalashnikov, V., Balathanova, E., Erofeeva, I., Smirnova, O., Matvievskiy, A.: Biological resistance of cement composites filled with limestone powders. In: Materials Science Forum, vol. 871, pp. 22–27 (2016)
22. Erofeev, V., Dergunova, A., Piksaikina, A., Bogatov, A., Kablov, E., Startsev, O., Matvievskiy, A.: The effectiveness of materials different with regard to increasing the durability. In: MATEC Web of Conferences 04021 (2016)

23. Drozdyuk, T.A., Ayzenshtadt, A.M., Frolova, M.A., Rama, S.V.: Mineral wool composite with the use of saponite-containing mining industry waste. Constr. Mater. Prod. **3**(3), 21–27 (2020). https://doi.org/10.34031/2618-7183-2020-3-3-21-27
24. Ahmed, A.A., Fedyuk, R.S., Liseytsev, Y., Timokhin, R.A., Murali, G.: Use of Iraq concrete scrap as filler and aggregate of heavyweight and lightweight concrete. Constr. Mater. Prod. **3** (3), 28–39 (2020). https://doi.org/10.34031/2618-7183-2020-3-3-28-39

Analysis of the Efficiency of Development Scenarios for the Energy Complex of the North-East of Russia

Alexander Semenov[1(✉)], Natalya Volotkovskaya[1], Yuriy Bebikhov[1], Ilya Yakushev[1], Oleg Fedorov[2], and Elena Gracheva[3]

[1] Ammosov North-Eastern Federal University, Yakutsk 677000, Russia
as.semenov@s-vfu.ru
[2] Alekseev Nizhny Novgorod State Technical University, Nizhny Novgorod 603155, Russia
[3] Kazan State Power Engineering University, Kazan 420066, Russia

Abstract. Scenarios for the development of the energy complex of one of the regions of the North-East of Russia for the period up to 2030 with a target vision up to 2050 are discussed. The current state of the energy complex is analysed. The latest changes in the power supply system of the Western Energy Region of the Republic of Sakha (Yakutia) are shown, associated with the appearance of a ring circuit with four nodal substations. The advantages for the organization of the ring power supply circuit are indicated. Prospective directions for the transfer of electricity from the Republic and back are proposed. The strategic and moderate scenarios for the development of the energy sector of the Republic of Sakha (Yakutia) are assessed in modern conditions, taking into account the increase in the energy efficiency of the region's economy. The indicators of electricity consumption throughout the Republic, the Western Energy Region and the decentralized power supply zone were evaluated. The obtained long-term forecasts do not confirm the development of events according to a more stringent strategic scenario. At the same time, the results for the moderate scenario are fully correlated with both the current state of the energy sector and the prospects for its development.

Keywords: Energy complex · Electric power industry · Energy efficiency · Electricity consumption · Power supply system · Substation · Development scenario

1 Introduction

The Russian energy policy is aimed at the most efficient use of natural energy resources and the potential of the energy sector for sustainable economic growth, improving the quality of life of the country's population and helping in strengthening of its foreign economic positions [1–3]. The implementation of the energy policy at the subfederal level affects the objective processes of the socio-economic development of the Republic, including taking into account the sharp changes in external and internal

A. Sinitsyn (Ed.): SES 2020, LNCE 141, pp. 231–240, 2021.
https://doi.org/10.1007/978-3-030-67654-4_26

factors that determine the main parameters of functioning of the country's fuel and energy complex [4, 5].

The main directions of the energy policy are used in the development and adjustment of the socio-economic development strategy, program documents for the industrial development of the territory of the Republic. They are also used in the preparation and adjustment of the parameters of investment programs and large projects of the energy sector companies [6, 7].

The implementation of projects related to the development of natural resources, with the development of manufacturing industries, projects in the service sector and agriculture requires advanced planning for the development of the energy infrastructure, the fuel and energy complex [8, 9].

The aim of this work is to analyze and assess the implementation of measures of the approved “Energy Strategy of the Republic of Sakha (Yakutia) for the period up to 2030”, to determine the strategic directions for further development of the energy sector of the Republic of Sakha (Yakutia) in modern conditions, taking into account the increase in energy efficiency of the regional economy and improving the life quality of population.

The following tasks are solved: to analyse the current state of the energy complex of the Republic; to assess the resource base for the electric power industry; to analyse the efficiency of the developed measures (scenarios) for the development of the energy complex of the Republic for the period up to 2030 with a target vision until 2050.

2 Materials and Methods

The object of research is electric power indicators in the energy sector of the Republic of Sakha (Yakutia), including the Western Energy Region (WER) and taking into account the decentralized power supply.

The Republic of Sakha (Yakutia), as a subject of the Russian Federation, is part of the Far Eastern Federal District. It occupies almost the entire north-eastern part of the Asian continent, overlooking the coast of the two seas of the Arctic Ocean. Over 40% of the Republic territory is located beyond the Arctic Circle. In the east, the Republic borders the Chukotka Autonomous District, Magadan Region and Khabarovsk Territory, in the south - with the Trans-Baikal Territory, Amur and Irkutsk Regions, in the west - with the Krasnoyarsk Territory, in the north it is washed by the Laptev and East Siberian Seas. The total length of the sea coastline exceeds 4.5 thousand km.

3 Results and Discussion

3.1 Analysis of the Current State of the Electric Power Industry

The energy system of the Republic of Sakha (Yakutia) consists of three energy regions - Western, Central and South-Yakutsk [10]. The South-Yakutsk and Western Energy Regions are electrically connected via two single-circuit 220 kV overhead lines “Olekminsk - OPS-15 - Nizhniy Kuranakh” with spikes to oil pumping stations

(OPS) OPS-14 and OPS-16. The South-Yakutsk and Central Energy Regions are electrically connected via the 220 kV "Nizhny Kuranakh – Maya" overhead line. The South Yakutsk Energy Region provides energy to the South Yakutsk territorial-industrial complex, the Neryungrinsky and Aldan industrial and agricultural centers, and is connected by two 220 kV power lines to the United Energy System (UES) of the East. The Central Energy Region provides energy to the central industrial center and a group of central regions. The Western Energy Region unites the Aikhalo-Udachninsky, Mirninsky, Lensky industrial centers and a group of Vilyui agricultural regions, and also has a connection with the Olekminsky region (Fig. 1).

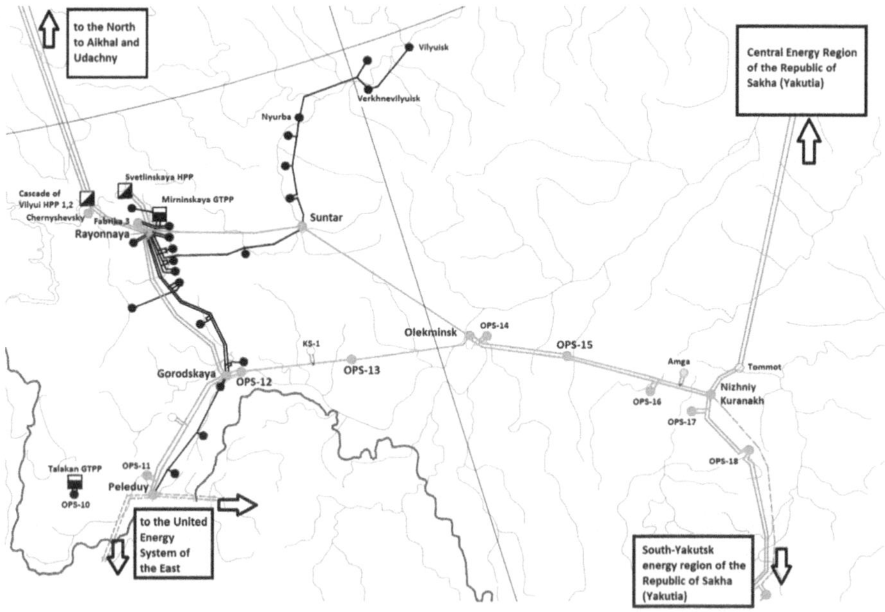

Fig. 1. Power system map of the Western Energy Region: green line – 220 kV, black line – 110 kV.

On January 2, 2019, the Western and Central Energy Regions became part of the United Energy System of Russia included for parallel operation with the United Energy System of the East. The Republic of Sakha (Yakutia) has external electrical connections with the Amur, Irkutsk and Magadan power systems, as well as with the Chaun-Bilibino power center of the Chukotka Autonomous Okrug.

The decentralized power supply zone includes the vast territory of the Republic with a large number of autonomous power plants that supply individual villages and mining enterprises [11]. The zone of operation of autonomous energy covers an area of 2.2 mln km^2 (64%) with 15% of the population living in the Republic. The main part of the autonomous power plant capacity (about 200 MW) is located on the territory of the so-called Northern Energy Region and accounts for about 29% of the Republic electricity consumption [12].

In the conditions prevailing in the Western Energy Region, given the large power generation capacities and very long distances for the transmission of electrical energy from sources to consumers, as well as adverse climatic conditions, new 220 kV transmission lines have been put into operation (in 2019) in addition to the 110 kV networks [13].

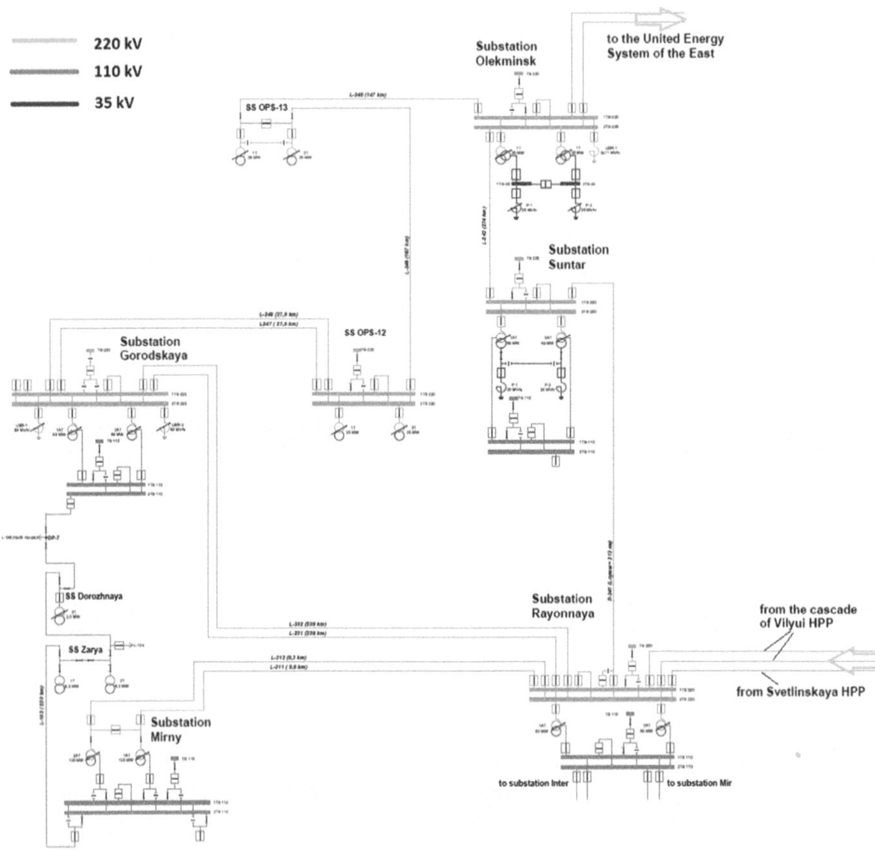

Fig. 2. Single-line ring diagram of power supply of the Western Energy Region.

The 110 and 220 kV transmission lines were earlier made in single-circuit, two-circuit and three-circuit (from the cascade of Vilyui HPPs at the Aikhal settlement) versions. However the newly launched 220 kV electrical network is made according to the ring scheme (Fig. 1) and covers an area of more than 50,000 km^2. It significantly increases one of the main requirements for the power supply of consumer facilities, namely, reliability or continuity of operation [14]. The single line diagram presented in Fig. 2 clearly shows the execution of electrical connections of nodal substations.

The advantage of the ring circuit, in addition to the reliability of power supply, is the independence of the flows in the substation sections from the flows in the high voltage network itself, which is especially important in post-emergency modes and makes it possible to use simpler relay protection and automation devices, and, as a rule, lower levels of short-circuit currents. Such system is quite flexible, adapting to different modes of power distribution arising as a result of changes in loads at consumer nodes, as well as in case of planned or emergency outages in the network. The configuration and parameters of the network enable its further development without fundamental changes, namely, the inclusion of newly introduced walk-through substations in the cut of the existing ring power transmission line, for example, oil pumping and compressor stations (OPS-11,14,16,17; CS-1,2). All of them are 220/6-10 kV deep input substations with a large concentration of loads.

Currently, the ring line is made in reinforced concrete supports of 220 kV overhead lines and is connected to the cascade of Vilyui HPPs 1 and 2, as well as to the Svetlinskaya HPP (total installed capacity of 957.5 MW) through the Rayonnaya nodal substation. It is a single-loop ring circuit with several nodal points (a circuit point with more than two lines connected). Three more hub substations of the ring are located in Suntar settlement (Suntar Substation), Olekminsk (Olekminsk Substation) and Lensk (Gorodskaya Substation). The previously existing high-voltage lines in wooden supports with a voltage of 110 kV are also in operation from the Mirny Substation to the Gorodskaya Substation, which goes further to the Peleduy substation, and from the Mirny Substation to the Suntar Substation, which goes further to villages of the Vilyui group of uluses.

From the Suntar Substation, the 110 kV transmission line goes further to the Nyurba Substation, the Verkhnevilyuysk Substation and the Vilyuysk Substation, thereby enabling connecting the Western Energy Region with the Central Energy Region system of the Republic in the future. The 220 kV transmission line and the previously existing 110 kV transmission line from the Gorodskaya Substation to the Peleduy Substation transfer electrical energy to the gold mining enterprises of the Bodaibo District of the Irkutsk Region, as well as to the ESPO and Power of Siberia oil pipelines. Power plants of OJSC "Surgutneftegas" with a total installed capacity of 194.4 MW also operate autonomously on the territory of WER.

At the end of 2019, the construction of the 220 kV Peleduy - Sukhoi Log overhead line was completed with the 220 kV Sukhoi Log substation and further to the Mamakan substations 1 and 2, which made it possible to complete the formation of the 220 kV Peleduy - Sukhoi Log - Taksimo - Mamakan - Ust-Kut - OPS-9 - Peleduy" (indicated by double dashed lines in Fig. 1).

3.2 Analysis of the Proposed Scenarios for the Development of the Energy Complex

Over the past 10 years, energy consumption in the Republic of Sakha (Yakutia) has grown by 56%, and the annual energy supply potential has increased by 12% due to an increase in electricity consumption, electricity losses and the development of power converter technology. According to forecasts, electricity production in 2020 will amount to 11,150 million kWh (including 3500 million kWh will be produced at hydroelectric power plants), and consumption will reach 10377 million kWh. Whereas in 2015 9951 million kWh were produced, and 9015 million kWh were consumed. In this regard, the demand for energy resources will also increase. Coal (37%) and gas (26%) will remain the priority power sources, while electricity will be only 15% [15, 16].

Two scenarios for further development of the branches of the fuel and energy complex of the Republic have been developed: strategic (more severe) and moderate. According to the strategic scenario for the development of the Republic economy, electricity production by 2030 will increase by 1.76 compared to its level in 2018 and will amount to 17.1 billion kWh. Useful consumption for the period from 2018 to 2030 will increase to 12.9 billion kWh, i.e. by 1.75 times.

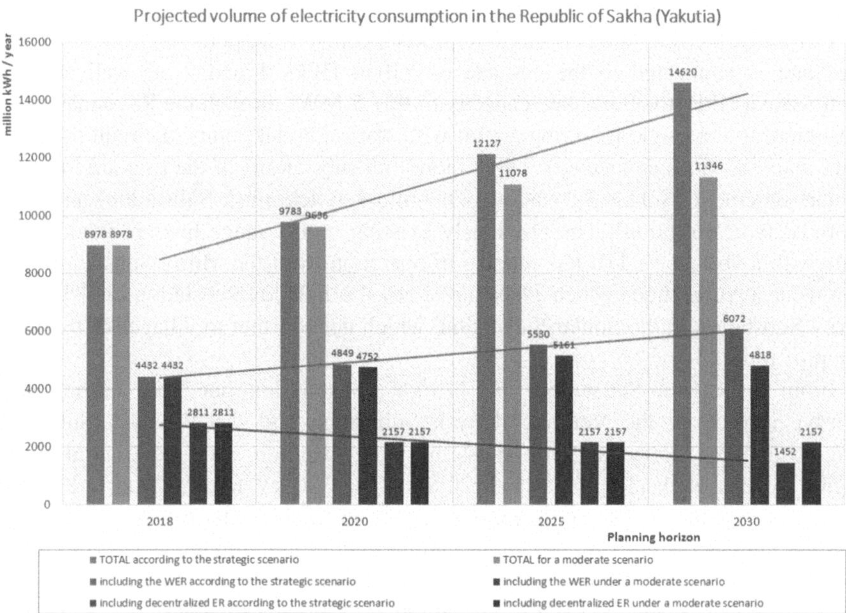

Fig. 3. Electricity consumption levels in the Republic according to the two scenarios.

The volume of electricity consumption in production will consist of energy savings, growth in production electricity consumption and growth in electricity losses. At the same time, an increase in electricity consumption for own needs is not expected. The moderate and strategic (higher in terms of electricity consumption) electricity consumption scenarios for the entire territory of the Republic are presented in Fig. 3.

The graphs show that in the Western Energy Region the projected electricity consumption under the strategic scenario will not grow as intensively as throughout the Republic as a whole, and under the moderate scenario by 2030 a decrease in this level is observed. At the same time, WER currently consumes 49.4% of the total consumption in Yakutia. This is due to the fact that by 2030 the underground development of the Internacionalnaya kimberlite pipe will be completed, as it was decided to develop the remaining under-pit reserves in an open pit [17]. This fact will undoubtedly reduce power consumption, since the electrical equipment of the underground complex of the Internacionalnaya mine has a total capacity of more than 30 MW [18].

The quarantine measures caused by the Covid-19 virus will make certain adjustments to the forecasts for the development of the electric power industry in Yakutia and WER. Most likely, the forecast for 2020 will not be confirmed even for the moderate version, but by 2025 the situation should level out [19].

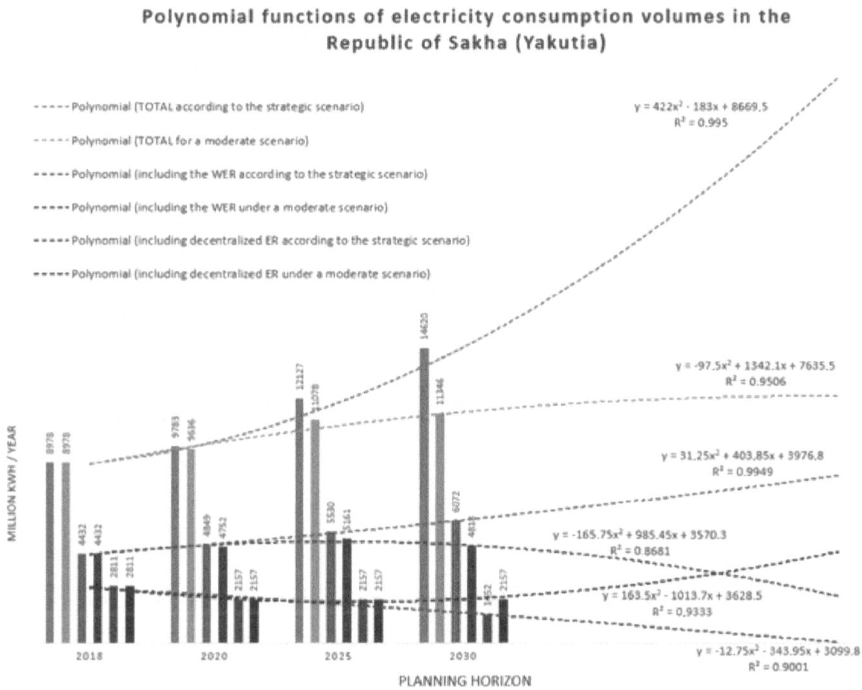

Fig. 4. Projected levels of electricity consumption in the Republic, obtained using polynomial analysis.

We carried out an additional analysis of the energy consumption indicators of the Republic for two scenarios for a longer term and built polynomial functions in the form:

$$f(x) = a_0 + a_1 \cdot x + a_2 \cdot x^2 + \cdots + a_d \cdot x^d, \quad (1)$$

where a are the calculated parameters, d is the order of the polynomial.

A 2nd order polynomial was used, since there is only one extremum on the line graphs (Fig. 3). In our case, no polynomial approximation is required. We use the least squares method, which is based on minimizing the sum of squares of deviations of functions from the variables to be found. The obtained characteristics and equations are shown in Fig. 4.

After analyzing the obtained results, we made the following conclusions: the polynomial characteristic of the strategic scenario of electricity consumption in the

Republic shows the highest reliability with an index of R2 = 0.995. The least reliability is obtained under the moderate scenario of electricity consumption in the Western Energy Region with an index of R2 = 0.868. So, the strategic scenario is too harsh, since even with launching new oil and gas production enterprises, electricity consumption in the Republic is unlikely to exceed 20 billion kWh by 2035. This is due to the fact that in the modern structure of electricity consumption in the Republic, mining accounts for only 32.2%. At the same time, the moderate scenario more accurately predicts the level of electricity consumption both throughout the Republic and in the Western Energy Region, where electricity consumption will continue to decline due to the gradual closure of diamond mining enterprises [20]. One may hope for the work resumption of the Mir diamond mine (by about 2040), which was closed due to an accident in the summer of 2017. As for decentralized energy regions, forecasts for a twofold decrease in electricity consumption according to the strategic scenario seem unrealistic. This is explained by launching the GDK Verkhne-Munskoye in 2018 and by active works on gold and diamond mining by PJSC Polyus and JSC Almazy Anabara, respectively. Under a long-term forecast of a moderate scenario on the polynomial characteristic of electricity consumption in decentralized energy regions, a reverse increase to the levels of 2018 is observed, which can be justified by plans to use unique floating nuclear thermal power plants to supply power to the Northern and Arctic territories [21, 22].

4 Conclusion

After analyzing the current state of the electric power complex and assessing the scenarios for the development of the energy sector of the Republic, the authors made the following conclusions:

1. Development of the power supply system of the Western Energy Region according to the ring principle with connection to the grids of the United Energy System of the East provides ample opportunities for transferring excess electricity to other regions of the country, or, if necessary, obtaining missing capacities from other systems.
2. Useful electricity consumption significantly increases in the period 2025–2030, the average annual growth rate will be 4.8%, while the rate of electricity production during this period will be 1.1%. So, during this period there will be a decrease in the supply of electricity to the UES of the East.
3. The average annual growth rate of electricity production as a whole for the period from 2018 to 2032 will amount to 1.6%, and useful consumption will be 2.4%. In this regard, it is possible to increase the supply of electricity to the Republic both from the UES of the East and the UES of Siberia.
4. According to the moderate scenario, electricity generation will grow at a significantly lower rate on average 0.4% per year and will reach 11.6 billion kWh.

5. The structure of electricity production will change: the share of hydroelectric power plants will decrease, while the share of thermal power plants will increase. Electricity generation by renewable sources will be small, as it was, up to 2–3 million kWh.
6. It is planned to use floating nuclear power plants of low capacity, which will cover the needs of new projects in the Arctic zone.

Acknowledgments. The study was completed within the framework of the research project "Energy Strategy of the Sakha Republic (Yakutia) for the period up to 2032 with goal projection until 2050" under the state contract No. 3217 of 17.12.2019.

References

1. Proskuryakova, L., Starodubtseva, A., Bianco, V.: Modelling a household tariff for reducing sectoral cross-subsidies in the Russian power market. Energy **213**, 118725 (2020)
2. Zemtsov, S.P., Barinova, V.A., Kidyaeva, V.M., Lanshina, T.A.: Ecological efficiency and sustainable regional development in Russia during the 20 years of resource-based growth. Ekonomicheskaya Politika **2020**(2), 18–47 (2020)
3. Proskuryakova, L.N., Ermolenko, G.V.: The future of Russia's renewable energy sector: trends, scenarios and policies. Renew. Energy **143**, 1670–1686 (2019)
4. Igaliyeva, L., Niyazbekova, S., Serikova, M., Kenzhegaliyeva, Z., Mussirov, G., Zueva, A., Tyurina, Y., Maisigova, L.: Towards environmental security via energy efficiency: a case study. Entrepreneurship Sustain. Issues **7**(4), 3488–3499 (2020)
5. Gardiner, R., Hajek, P.: Municipal waste generation, R&D intensity, and economic growth nexus – a case of EU regions. Waste Manag. **114**, 124–135 (2020)
6. Turkova, V., Arkhi, A.: Development of digital economy in Irkutsk region. IOP Conf. Ser. Mater. Sci. Eng. **667**(1), 012104 (2019)
7. Gouveia, J.P., Palma, P.: Harvesting big data from residential building energy performance certificates: retrofitting and climate change mitigation insights at a regional scale. Environ. Res. Lett. **14**(9), 095007 (2019)
8. Markusova, V., Mindeli, L., Libkind, A., Zolotova, A., Akoev, M.: Comparative analysis of Russian and industrialized countries performance on energy and fuels, WoS, 2008–2017. Scientometrics **123**(3), 1281–1300 (2020)
9. Schoenefeld, J.J., Knodt, M.: Softening the surface but hardening the core? Governing renewable energy in the EU. West Eur. Politics **44**(1), 49–71 (2021)
10. Volotkovskaya, Y., Volotkovskaya, N., Semenov, A., Semkova, A., Fedorov, O., Vladimirov, O.: Analysis of the global energy resource market. In: E3S Web of Conferences, vol. 178, p. 01058 (2020)
11. Alyunov, A., Vyatkina, O., Smirnov, I., Nemirovskiy, A., Gracheva, E.: Assessment of efficiency of diesel generators use in distributed energy industry. In: E3S Web of Conferences, vol. 178, p. 01086 (2020)
12. Elyakova, I.D., Elyakov, A.L., Karataeva, T.A., Khristoforov, A.A., Buryanina, N.S., Fedorov, A.A.: Conditions for effective long-term energy supply to the arctic areas of the Republic of Sakha (Yakutia). Int. J. Energy Econ. Policy **8**(6), 361–370 (2018)
13. Khasanov, S.R., Gracheva, E.I., Toshkhodzhaeva, M.I., Dadabaev, S.T., Mirkhalikova, D.S.: Reliability modeling of high-voltage power lines in a sharply continental climate. In: E3S Web of Conferences, vol. 178, p. 01051 (2020)

14. Gerasimov, D.O., Serdyukova, E.V., Suslov, K.V., Buryanina, N.S., Korolyuk, Y.F.: Energy hub component models for multi-energy system. J. Phys. Conf. Ser. **1582**(1), 012033 (2020)
15. Zhang, L., Song, G., Ma, X., Zhan, C., Zhang, S.: Decarbonising residential building energy towards achieving the intended nationally determined contribution at subnational level under uncertainties. J. Clean. Prod. **272**, 122760 (2020)
16. Nevzorova, T.: Biogas production in the Russian federation: current status, potential, and barriers. Energies **13**(14), 3620 (2020)
17. Shevyreva, N.Y., Shevyrev, Y.V., Pichuyev, A.V.: Electricity quality assurance in open-pit mining: considerations of today. In: 2020 International Conference on Industrial Engineering, Applications and Manufacturing (ICIEAM), Sochi, Russia, pp. 1–5 (2020)
18. Kubrin, S.S., Reshetnyak, S.N.: Calculation of specific norms of power consumption for the dredging areas of coal mines. Gornyi Zhurnal **2020**(4), 62–64 (2020)
19. Volotkovskaya, N.S., Semenov, A.S., Fedorov, O.V.: Feasibility of energy sector development and optimization program of the Republic of Sakha (Yakutia). IOP Conf. Ser. Mater. Sci. Eng. **753**(8), 082033 (2020)
20. Bebikhov, Y.V., Semenov, A.S., Egorov, A.N., Glazun, M.A., Fedorov, O.V.: The peculiarities of power conservation and energy efficiency in power supply systems of industrial enterprises. J. Phys. Conf. Ser. **1614**(1), 012027 (2020)
21. Tarasova, O.V., Sokolova, A.A.: Spatial organization of the Chukotka's economy. IOP Conf. Ser. Earth Environ. Sci. **302**(1), 012113 (2019)
22. Handrlica, J.: Underground repositories, reprocessing facilities and floating nuclear power plants: liability issues revisited. J. Energy Nat. Resour. Law **37**(3), 263–288 (2019)

Estimation of Time Intervals Between Flood Events to Increase Durability of Infrastructures

Charith Dushyantha(✉), Irina Ptuhina, Adheesha Fernando, and Yslamnazar Mammetmyradov

Peter the Great St. Petersburg Polytechnic University, Polytechnicheskaya 29, 195251 Saint Petersburg, Russian Federation
wa.charith@gmail.com

Abstract. Distribution functions of flood events in Colombo district of Sri Lanka were developed to estimate the time intervals between particular flood events. National roads in Colombo district were damaged during 2017 mostly due to flood events. The study is composed of records of three hydrometric stations (Nagalagam, Hanwella and Glencourse) located along the river "Kelani Ganga". All required data and information sources were provided by the Ministry of Disaster Management, Department of Metrology and Irrigation Department of Sri Lanka. Current approaches to estimating the time interval function were improved by using the Erlang distribution. Hyper-Erlang distribution method was used to formulate the mathematical model to estimate time intervals between flood events in order to increase the accuracy of the results. Using the developed model, the time interval between flood events at a particular water level can be estimated. Among the above mentioned stations, the Hanwella station is at 90.24% risk of being inundated with the water level rising up to 9.4 m in a time period of 12 years. These results can be used to forecast flood risk in an area and take required measures in order to increase durability of infrastructures.

Keywords: Hyper-Erlang distribution · Flood time intervals · Flood modeling · Kelani Ganga Sri Lanka

1 Introduction

The flood events in May 2017 significantly affected regular land transport operations except rail transport in Colombo and Ratnapura districts causing damages to their infrastructure. The total assessed damages and losses to the transport sector were calculated to be LKR 13,076.2 million [1, 2]. Therefore understanding the time intervals between flood events in an area, where flood risk is high, is crucial in decision making on sustainable development, planning and designing of necessary involvements for funds of flood risk reduction. Analyzing disaster risks is an important task when taking measures to minimize damages and build policies in infrastructure development to reduce disasters and mitigate hazards while improving the living standards and opportunities of the society.

A. Sinitsyn (Ed.): SES 2020, LNCE 141, pp. 241–252, 2021.
https://doi.org/10.1007/978-3-030-67654-4_27

The Japan International Cooperation Agency has reported in their data collection survey, that from 1962 to 2016 river "Kelani Ganga" has exceeded its threshold water level by 35 times flooding the Nagalagam Street, from 1978 to 2016 flooding Hanwella by 22 times and from 1960 to 2016 flooding Glencourse by 28 times [3]. In 2016, the total costs of damages to national road infrastructures of Sri Lanka and Colombo district were 2418 million rupees and 707.79 million rupees respectively [4].

Flood risk management for infrastructures in Sri Lanka can be characterized as modification of susceptibility to flood damages which means actions can be taken before flooding and modifying the impact of flooding during and after the disaster [5]. This paper mainly focused on modifying susceptibility to flood damage while the actions to be taken during and after flooding are taken by the responsible authorities of the country.

In 2017, damage curves for Colombo District were developed by P. Dias by plotting depth vs. Damage curves for building fabrics, building contents of distributed infrastructures and vehicles [6]. In 2017, another paper discussed the qualitative flood risk assessment for the Western province of Sri Lanka. A mathematical model, an economic vulnerability map, social flood risk and economic flood risk maps were developed for the Western Province in that study [7]. However, these papers mainly focused on material damages after a flood event in an area. By analyzing relevant research papers [8–18], studying flood risk modeling and management in different countries such as France, Belgium, Netherland, Poland, Spain and Russia for various scenarios, an approach to flood risk modeling and management was obtained.

C. Dushyantha has discussed the risk of exceeding the threshold water level of Kelani river [19] and this paper will carry out further investigations to estimate the time intervals between flood events at three gauging stations (Nagalagam, Hanwella and Glencourse) along river "Kelani Ganga" in Sri Lanka to increase the durability and reliability of roads, bridges and culverts and decrease erosion of river embankments and shoulders in above mentioned areas.

The aims of this paper are:

1. To develop a mathematical model by analyzing the collected data of river Kelani Ganga at three gauging stations;
2. To calculate and plot time intervals between flood events of each gauging station at particular critical water levels.

2 Methods

Let h be the water level of a river, deviations from the norm of which characterize the flood strength. This relationship can be written as a function of time $h(t)$. Many factors affect the specific implementations of this function over time. Therefore, this process will be considered as an implementation of some quasi-deterministic random process $H(t)$ [20–22].

This process unfolds not only in time, but also in space, so in general it is necessary to study random fields $H(t,x)$, where x - coordinate vector. Discharge from the process of changing of the river level at a particular point $H(t/x)$ is called a flood, for which the criterion for the occurrence is the fulfillment of the condition $h \geq h_H$, where h_H - critical level.

For a possible flood, let the effective load be represented by a random value H_{max}, of which the full probabilistic characteristic is a distribution function [23, 24].

$$F_{max}(h) = P(H_{max} < h) \tag{1}$$

Flood streams (random process outliers/discharge of a random process), considered as random events, are characterized by frequency $\lambda_{max}(h_H)$ [21].

Let the time period for a flood event to occur be T. The relation for the distribution of T until a flood event can be written as formula (2) for a particular point that is impacted by the natural process factors mentioned above:

$$F(t) = P(T < t) \tag{2}$$

Consider the distribution law of time between events at a particular point of the Erlang stream with a $(k-1)$ rate parameter. Let T_m be the time interval between natural phenomena in the simplest stream, which is represented by independent random variables that have an exponential distribution with the parameter λ. Then the expectant time for a k event to occur is represented as [21]:

$$\mathrm{T} = \sum_{\mathrm{m}=0}^{\mathrm{k}-1} \mathrm{T_m} \tag{3}$$

If the conditional probability of flood events occurring at a particular point, provided that a natural phenomenon has occurred, is equal to q, then the stream of the flood event is a stream of Erlang with rate parameter:

$$k - 1 = \frac{1}{q} - 1 \tag{4}$$

The distribution function (2) must consider the sum of exponents, which correspond to the Poisson distribution of events [21]:

$$F(t) = 1 - \sum_{m=0}^{k-1} \frac{(\lambda \cdot t)^m}{m!} \exp(-\lambda \cdot t) \tag{5}$$

By differentiating this formula by t, the probability density function can be derived in the form of [21]:

$$f(t) = \frac{\lambda(\lambda t)^{k-1}}{(k-1)!} \exp(-\lambda t), t \tag{6}$$

3 Results

Let $h_H = 4.8$ m. Based on the criterion of occurrence of a natural phenomenon (a flood event occurs when the water level $h \geq h_H$) an estimate of the frequency of floods was obtained.

$$\lambda(h_H) = \frac{35}{51} = 0.69 \tag{7}$$

Next, a situation which would be considered a flood event was determined. The Nagalagam case was considered for this: exceeding a water level of 5 m during a flood. In this case, the probability of a flood event occurring is equal to:

$$q = P(h \geq 500 | h \geq h_H) = \frac{24}{51} = \frac{1}{2.125} = 0.47 \tag{8}$$

The frequency of occurrence of flood events (average time between floods, which leads to flood events) is equal to:

$$\overline{t_{yc}} = \frac{1}{\lambda \cdot q} = \frac{1}{0.69 \cdot 0.47} = \frac{1}{0.352} = 3.08 \tag{9}$$

The approximate value of the probability of a flood event was taken as $q = 1/3$ from the (8) equation. If $q = 1/3$, then the flow of flood events occurring is equal to the flow of Erlang with an order of 2 (consider Eq. (4)). For the Nagalagam case, the Eq. (4) can be written as:

$$F(t) = 1 - \sum_{m=0}^{2} \frac{(0.69 \cdot t)^m}{m!} \exp(-0.69 \cdot t) \tag{10}$$

Hyper-Erlang distribution:

The Erlang distribution is described only for integers, but in general an arbitrary q of order $(k - 1)$ is not an integer.

The time distribution model between events in this case is a Hyper-Erlang distribution (a combination of Erlang distributions) with a distribution function of the following type:

$$F(t) = 1 - \sum_{k=1}^{n} \left(C_k \cdot \sum_{m=0}^{k-1} \frac{(\lambda \cdot t)^m}{m!} \exp(-\lambda \cdot t) \right) \tag{11}$$

$$n = Ent(k) + 1 \tag{12}$$

$$\sum_{k=1}^{n} C_k = 1 \tag{13}$$

$$C_1 = 0, \ldots, C_{n-2} = 0 \tag{14}$$

$$C_{n-1} = 1 - (k - Ent(k)) \tag{15}$$

$$C_n = 1 - C_{n-1} \tag{16}$$

In Nagalagam case, Eqs. (8), (4), (12), (13), (14), (15) and (16) can be written as:

$$q = \tfrac{24}{51} \approx \tfrac{1}{2.125} \approx 0.47 \quad k = 2.125 \quad n = Ent(2.125) + 1 = 3$$

$$C_1 = 0, C_2 = 0.875, C_3 = 0.125$$

$$F(t) = 1 - \left(0.875 \sum_{m=0}^{1} \frac{(0.69 \cdot t)^m}{m!} \exp(-0.69 \cdot t) + 0.125 \sum_{m=0}^{2} \frac{(0.69 \cdot t)^m}{m!} \exp(-0.69 \cdot t)\right) \tag{17}$$

The exponential distribution function can be written as [23, 24]:

$$F(t) = 1 - \exp(-\lambda_0 \cdot t) \tag{18}$$

Thus, in the Nagalagam case:

$$F(t) = 1 - \exp(-0.324 \cdot t) \tag{19}$$

Similarly, mathematical models were developed to estimate time intervals between flood events with the water level reaching 7 m at Nagalagam station.

$$F(t) = 1 - \sum_{m=0}^{10} \frac{(0.69 \cdot t)^m}{m!} \exp(-0.69 \cdot t) \tag{20}$$

$$F(t) = 1 - \left(0.8 \sum_{m=0}^{9} \frac{(0.69 \cdot t)^m}{m!} \exp(-0.69 \cdot t) + 0.2 \sum_{m=0}^{10} \frac{(0.69 \cdot t)^m}{m!} \exp(-0.69 \cdot t)\right) \tag{21}$$

$$F(t) = 1 - \exp(-0.068 \cdot t) \tag{22}$$

Hanwella gauging station case:

Erlang, Hyper-Erlang and exponential distribution mathematical models were developed to estimate time intervals between 8.6 m water level flood events.

$$F(t) = 1 - \sum_{m=0}^{1} \frac{(0.56 \cdot t)^m}{m!} \exp(-0.56 \cdot t) \tag{23}$$

$$F(t) = 1 - \left(0,14\sum_{m=0}^{0}\frac{(0.56\cdot t)^m}{m!}\exp(-0.56\cdot t) + 0,86\sum_{m=0}^{1}\frac{(0.56\cdot t)^m}{m!}\exp(-0.56\cdot t)\right) \tag{24}$$

$$F(t) = 1 - \exp(-0.3024\cdot t) \tag{25}$$

Erlang, Hyper-Erlang and exponential distribution mathematical models were developed to estimate time intervals between 9.4 m water level flood events.

$$F(t) = 1 - \sum_{m=0}^{3}\frac{(0.56\cdot t)^m}{m!}\exp(-0.56\cdot t) \tag{26}$$

$$F(t) = 1 - \left(0.1\sum_{m=0}^{2}\frac{(0.56\cdot t)^m}{m!}\exp(-0.56\cdot t) + 0.9\sum_{m=0}^{3}\frac{(0.56\cdot t)^m}{m!}\exp(-0.56\cdot t)\right) \tag{27}$$

$$F(t) = 1 - \exp(-0.1456\cdot t) \tag{28}$$

Glencourse gauging station case:

Erlang, Hyper-Erlang and exponential distribution mathematical models were developed to estimate time intervals between 17.1 m water level flood events.

$$F(t) = 1 - \sum_{m=0}^{2}\frac{(0.49\cdot t)^m}{m!}\exp(-0.49\cdot t) \tag{29}$$

$$F(t) = 1 - \left(0,72\sum_{m=0}^{1}\frac{(0.49\cdot t)^m}{m!}\exp(-0.49\cdot t) + 0,28\sum_{m=0}^{2}\frac{(0.49\cdot t)^m}{m!}\exp(-0.49\cdot t)\right) \tag{30}$$

$$F(t) = 1 - \exp(-0.2156\cdot t) \tag{31}$$

Erlang, Hyper-Erlang and exponential distribution mathematical models were developed to estimate time intervals between 19.2 m water level flood events.

$$F(t) = 1 - \sum_{m=0}^{6}\frac{(0.49\cdot t)^m}{m!}\exp(-0.49\cdot t) \tag{32}$$

$$F(t) = 1 - \left(0.67\sum_{m=0}^{5}\frac{(0.49\cdot t)^m}{m!}\exp(-0.49\cdot t) + 0.33\sum_{m=0}^{6}\frac{(0.49\cdot t)^m}{m!}\exp(-0.49\cdot t)\right) \tag{33}$$

$$F(t) = 1 - \exp(-0.0784\cdot t) \tag{34}$$

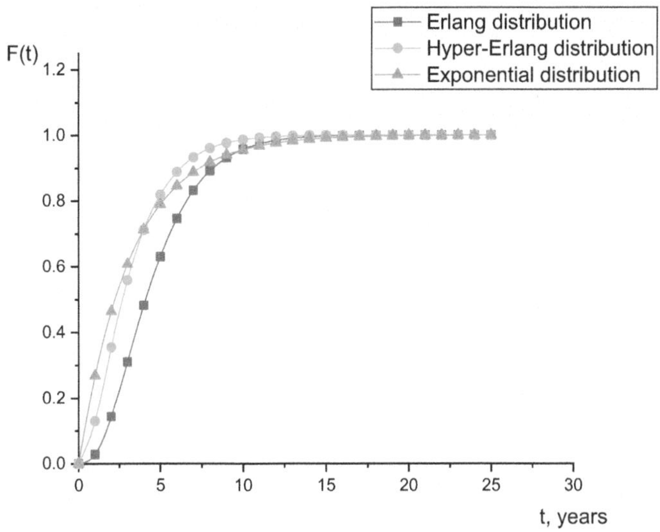

Fig. 1. Distribution functions of flood events at different time intervals at Nagalagam gauging station at $H_{critical} = 5\,\text{m}$

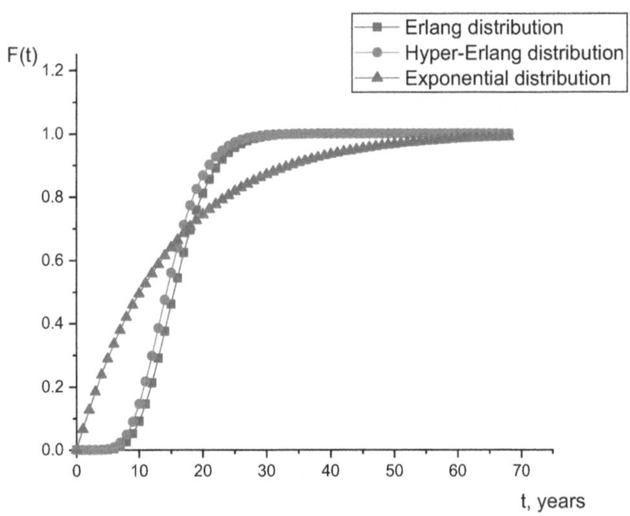

Fig. 2. Distribution functions of flood events at different time intervals at Nagalagam gauging station at $H_{critical} = 7\,\text{m}$

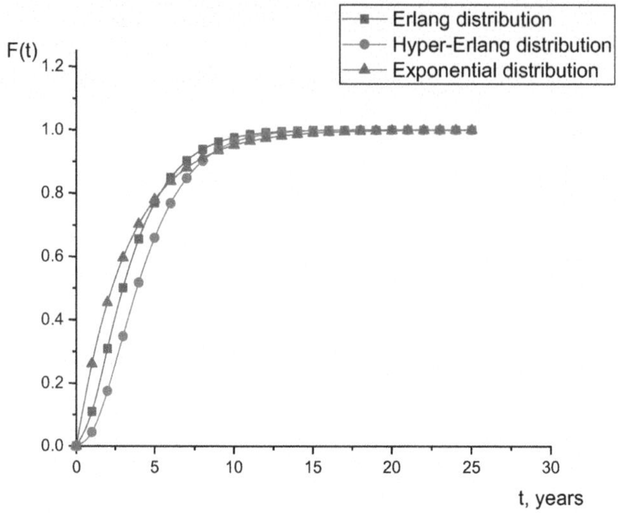

Fig. 3. Distribution functions of flood events at different time intervals at Hanwella gauging station at $H_{critical} = 8.6\,\mathrm{m}$

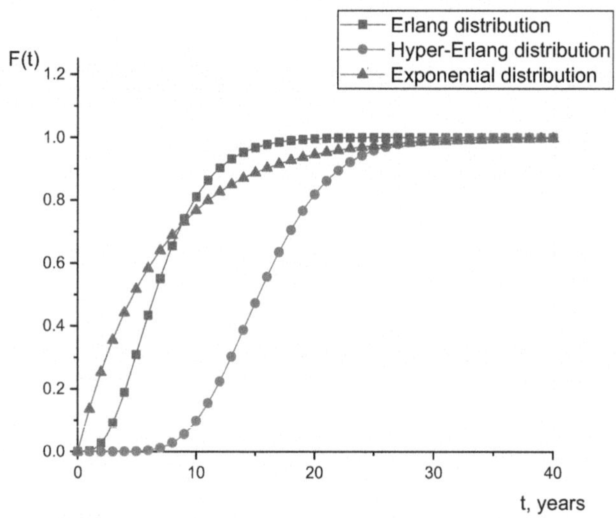

Fig. 4. Distribution functions of flood events at different time intervals at Hanwella gauging station at $H_{critical} = 9.4\,\mathrm{m}$

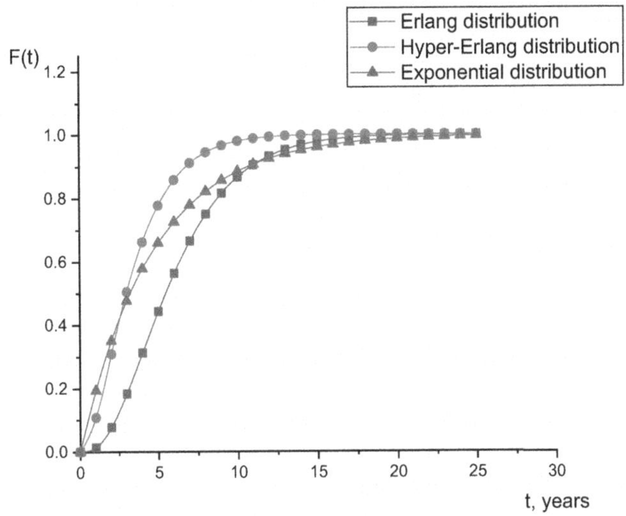

Fig. 5. Distribution functions of flood events at different time intervals at Glencourse gauging station at $H_{critical} = 17.1\,\mathrm{m}$

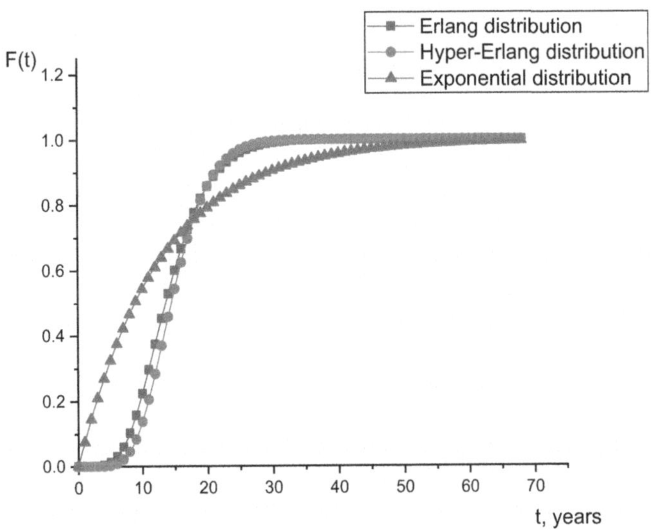

Fig. 6. Distribution functions of flood events at different time intervals at Glencourse gauging station at $H_{critical} = 19.2\,\mathrm{m}$

4 Discussion

The following results were obtained by this research.

1. A mathematical model was developed to estimate the time intervals between occurrences of flood events of river Kelani Ganga in Sri Lanka.
2. Using the obtained graphs, time intervals between flood events at particular critical water levels were calculated and plotted.

Figures 1, 2, 3, 4, 5 and 6 show time distribution graphs between flood events which can lead to catastrophic situations at three gauging stations at its minor and major flood levels. For short periods of time, the exponential distribution model gives the highest estimation of the time period for a flood to occur, while the Erlang and Hyper-Erlang distribution models gives minimum estimations. But for longer time periods, the Hyper-Erlang distribution model gives the highest estimation.

Figures 1 and 2 show distribution functions of flood events occurring at specific time intervals at Nagalagam gauging station, where the reaching of water level up to 5 m is considered as a minor flood event and 7 m is considered as a major flood event. For minor flood events, the exponential distribution model can be used for short time periods such as 4 years and the Hyper-Erlang distribution can be used for periods of time longer than 4 years. For major flood events, the exponential distribution model can be used until a time period of 16 years and the Hyper-Erlang distribution can be used for time periods of 17 years and more.

Figures 3 and 4 show distribution functions of flood events occurring at specific time intervals at Hanwella gauging station, where the reaching of water level up to 8.6 m is considered as a minor flood event and 9.4 m is considered as a major flood event. For minor flood events, the exponential distribution model can be used for short time periods such as 5 years and the Hyper-Erlang distribution can be used for periods of time longer than 5 years. For major flood events, the exponential distribution model can be used until a time period of 8 years and the Hyper-Erlang distribution can be used for time periods of 9 years and more.

Figures 5 and 6 show distribution functions of flood events occurring at specific time intervals at Glencourse gauging station, where the reaching of water level up to 17.1 m is considered as a minor flood event and 19.2 m is considered as a major flood event. For minor flood events, the exponential distribution model can be used for short time periods such as 2 years and the Hyper-Erlang distribution can be used for periods of time longer than 2 years. For major flood events, the exponential distribution model can be used until a time period of 17 years and the Hyper-Erlang distribution can be used for time periods of 18 years and more.

5 Conclusions

Models to estimate time intervals between flood events at river Kelani Ganga were developed by using data from three different stations. The probability of a minor flood event occurring at the three gauging stations is more than 90% in a time period of

7 years. The probability of occurrence of a major flood event is more than 90% for the three gauging stations at different time intervals. Nagalagam station shows a probability of 90.04% of being inundated in a time period of 21 years, Hanwella shows a probability of 90.24% in a time period of 12 years and Glencourse shows a probability of 92.07% in a time period 22 years. Among these three stations, the Hanwella station has the highest risk of being inundated. It is at a 90.24% risk of being inundated with the water level rising up to 9.4 m in a time period of 12 years.

These results can be used to develop hazard maps for relevant areas as one of the criteria that must be taken into consideration when developing infrastructures such as roads, bridges and culverts and decreasing erosion of river embankments and shoulders. The main advantage of having these hazard maps is that people with less technical knowledge can easily understand flood vulnerability of that area and contribute to minimizing flood damages.

References

1. Sri Lanka Post-Disaster Needs Assessment – May 2017, Ministry of Policies and Economic Affairs and Ministry of Disaster Management. https://www.lk.undp.org/content/srilanka/en/home/library/environment_energy/publication5.html, Accessed 29 June 2020
2. Hazard Profiles in Sri Lanka, Ministry of Disaster Management. https://app.adpc.net/sites/default/files/public/publications/attachments/Impacts%20of%20Disasters%20in%20Sri%20Lanka%202016.pdf, Accessed 29 June 2020
3. Data Collection Survey on Disaster Risk Reduction Sector in Sri Lanka, Japan International Cooperation Agency (JICA), Colombo (2017). https://openjicareport.jica.go.jp/pdf/12111894.pdf, Accessed 29 June 2020
4. Sri Lanka Post-Disaster Needs Assessment – May 2016, Ministry of Policies and Economic Affairs and Ministry of Disaster Management. https://www.lk.undp.org/content/srilanka/en/home/library/environment_energy/publication4.html, Accessed 29 June 2020
5. Nandalal, K.D.W.: Use of a hydrodynamic model to forecast floods of Kalu River in Sri Lanka. J. Flood Risk Manag. **2**, 151–158 (2009)
6. Dias, P., Arambepola, N.M.S.I., Weerasinghe, K., Weerasinghe, K.D.N., Wagenaar, D., Bouwer, L.M., Gehrels, H.: Development of damage function for flood risk assessment in the city of Colombo (Sri Lanka). In: Amaratunga, D., Haigh, R. (eds.) ICBR 2017, Procedia Engineering, vol. 212, pp. 332–339. Elsevier B.V, Amsterdam (2018)
7. Weerasinghe, K.M., Gehrels, H., Arambepola, N.M.S.I., Vajja, H.P., Herath, J.M.K., Atapattu, K.B.: Qualitative flood risk assessment for the western province of Sri Lanka. In: Amaratunga, D., Haigh, R. (eds.) ICBR 2017, Procedia Engineering, vol. 212, pp. 503–510. Elsevier B.V, Amsterdam (2018)
8. Juan, A., Gori, A., Sebastian, A.: Comparing floodplain evolution in channelized and unchannelized urban watersheds in Houston. Texas. J. Flood Risk Manag. **13**, e12604 (2020)
9. Gebremedhin, E.T., Basco-Carrera, L., Jonoski, A., Iliffe, M., Winsemius, H.: Crowdsourcing and interactive modelling for urban flood management. J. Flood Risk Manag. **13**, e12602 (2020)
10. Clement, M.A., Kilsby, C.G., Moore, P.: Multi-temporal synthetic aperture radar flood mapping using change detection. J. Flood Risk Manag. **11**, 152–168 (2018)

11. Smith, L., Liang, Q., James, P., Lin, W.: Assessing the utility of social media as a data source for flood risk management using a real-time modelling framework. J. Flood Risk Manag. **10**, 370–380 (2017)
12. Percival, S., Gaterell, M., Teeuw, R.: Urban neighbourhood flood vulnerability and risk assessments at different diurnal levels. J. Flood Risk Manag. **12**, e12466 (2019)
13. Vesuviano, G., Miller, J.D.: Design flood estimation and utility of high-resolution calibration data in small, heavily urbanised catchments. J. Flood Risk Manag. **12**, e12464 (2019)
14. Liefferink, D., Wiering, M., Crabbé, A., Hegger, D.: Explaining stability and change. Comparing flood risk governance in Belgium, France, The Netherlands, and Poland. J. Flood Risk Manag. **11**, 281–290 (2018)
15. Guardiola-Albert, C., Díez-Herrero, A., Amerigo Cuervo-Arango, M., Bodoque, J.M., García, J.A., Naranjo-Fernández, N., Aroca-Jiménez, E.: Analysing flash flood risk perception through a geostatistical approach in the village of Navaluenga. Central Spain. J. Flood Risk Manag. **13**, e12590 (2020)
16. Antonov, V.I., Badenko, V.L., Davydov, R.V., Maslikov, V.I., Molodtsov, D.V.: Modeling parameters of the flood control facilities adapted to climate change. In: Makarov, S.B., Karaseov, P.A., Velichko, E.N., Vinnichenko, M.Y., Zhurikhina, V.V. (eds.) ETACP 2019, Journal of Physics: Conference Series, vol. 1236, p. 012049. Institute of Physics Publishing, Bristol (2019).
17. Antonov, V.I., Davydov, R.V., Maslikov, V.I., Molodtsov, D.V., Badenko, V.L.: Simulation of water flow management by the flood control facilities in the adjacent river basins. In: Averkiev, N.S., Poniaev, S.A., Sokolovskii, G.S. (eds.) PhysicA.SPb 2019, Journal of Physics: Conference Series, vol. 1400, p. 077049. Institute of Physics Publishing, Bristol (2019)
18. Antonov, V.I., Davydov, R.V., Maslikov, V.I., Molodtsov, D.V., Chusov, A.N.: Mathematical models of operating regimes of flood control facility's system. In: Kudryashov, D.V., Kozlova, E.S. (eds.) ITNT 2019, Journal of Physics: Conference Series, vol. 1368, p. 042076. Institute of Physics Publishing, Bristol (2019)
19. Dushyantha, C., Ptuhina, I.: Flood risk assessment of river "Kelani Ganga" exceeding its threshold water level. In: Kudriavtcev, S., Murgul, V. (eds.) KTTI 2019, E3S Web of Conferences, vol. 157, p. 02006. EDP Sciences, Les Ulis (2020)
20. Akimov, V.A., Lesnykh, V.V., Radaev, P.N.: Fundamentals of Risk Analysis and Management in Natural and Techno-Genic Spheres. ZAO FID Business Express, Moscow (2004)
21. Akimov, V.A., Novikov, V.D., Radaev, P.P.: Natural and Man-Made Emergencies: Dangers, Threats, and Risks. ZAO FID Business Express, Moscow (2001)
22. Radaev, N.N.: Assessment of the repeatability of catastrophic floods based on statistical analysis of river level dynamics. Meteorol. Hydrol. **3**, 79–89 (2002)
23. Rodkin, M. V.: Earthquakes and other types of disasters: typical distribution laws and processes of disaster development. The author's abstract of Dis. doktor. Phy-mat. Nauk, Moscow (2004)
24. Rodkin, M.V.: Model of development of synergetic effect at strong catastrophes. Geoecol. Eng. Geol. Hydrogeol. Geocryol. **1**, 81–87 (2005)

Determination of Surface Temperature and Moisture Fields of Structural Elements of Buildings by Thermal Imaging

Denis Karpov[1(✉)], Mikhail Pavlov[1], and Regina Salikhova[2]

[1] Vologda State University, Lenina Street 15, Vologda 160000, Russia
karpov_denis_85@mail.ru

[2] Kazan State Power Engineering University, Str. Krasnoselskaya, 51, Kazan 420066, Russia

Abstract. This paper analyses the topical issues of excess heat losses by construction objects. It was established that building heat losses depend on the technical qualities (density, moisture, thermal conductivity) of external building structures made of various materials. In this regard, practical methods of identifying the temperature and moisture fields of the surfaces of building materials of buildings and their structural elements in operation are of particular importance. The technical capabilities of modern thermal imaging equipment are presented. A practical example of thermographic determination and analysis of the temperature and moisture fields of a translucent building enclosure - a window is given. It was confirmed that thermal imaging diagnostics can identify not only the distribution of heat losses in the control zone, but also the distribution of moisture fields on the surface of the controlled object. Basing on this knowledge, measures are developed to reduce (equalize) them and subsequently evaluate the effectiveness of their implementation.

Keywords: Buildings · Structural elements · Thermal imaging · Heat losses

1 Introduction

The moisture regime of external building enclosures is inextricably linked with the heat engineering regime. An increase in the moisture content of building materials and products increases their thermal conductivity and indoor air moisture. The normal moisture regime has an important sanitary-hygienic and thermal-technical value, as it determines the standard parameters of the microclimate and the durability of building structures. For example, the use of insufficient moisture resistant materials and products in external enclosing structures subject to moisture can also cause their premature wear [1, 2].

Porosity, density, moisture, thermal conductivity, heat capacity, and heat radiation are the main thermal properties of building materials [1, 3–7]. However, the entire set of complex phenomena involved in the transfer of heat in the thickness of materials is usually described by thermal conductivity. The thermal conductivity of building materials is an integrating equivalent coefficient that takes into account all physical phenomena occurring in the material and associated with the heat transfer [1, 8–13].

A. Sinitsyn (Ed.): SES 2020, LNCE 141, pp. 253–258, 2021.
https://doi.org/10.1007/978-3-030-67654-4_28

The material thermal conductivity is not a constant value, it can vary depending on its density, moisture, temperature, heat flow direction, structure (Fig. 1).

It is not possible to establish a general mathematical relationship between the material thermal conductivity and its moisture content, which will be the same for all building materials, since the shape and location of material pores have a significant effect. Therefore, practical methods for determining the temperature and moisture fields of the surfaces of building materials of buildings and their structural elements in operation are becoming especially relevant.

Actualization and justification of the problem being solved. The issues of saving and improving the efficiency of thermal energy use are the priority areas in various industries: construction, energy, mechanical engineering, transport, etc. [1, 2, 9]. Studies show that up to 40% of the energy spent on heating construction objects for various purposes is lost in excess of industry norms [1, 2].

Accounting and control of the rational use of thermal energy can be performed using thermal control and determination of the actual heat engineering characteristics of building structures, construction and buildings during their operation. Moreover, special attention should be paid to thermography of translucent enclosing building structures, since through them most of the thermal energy is lost [1, 8, 14, 15]. It is also necessary to take into account the dynamics of changes in the moisture state of the enclosing structures during full-scale thermal imaging control during the operation of a construction facility.

Economic costs from excess heat losses are determined on the basis of the calculated amount of heat losses taking into account the actual energy tariffs [13, 16, 17]. Obviously, thermal imaging diagnostics itself is only able to ascertain the distribution of heat losses in the control zone, on the basis of which it is possible to develop measures to reduce them (equalize) and subsequently assess their effectiveness. For example, in residential and industrial buildings, the thermal imaging method allows one to evaluate the efficiency of optimization of the heat supply and ventilation system.

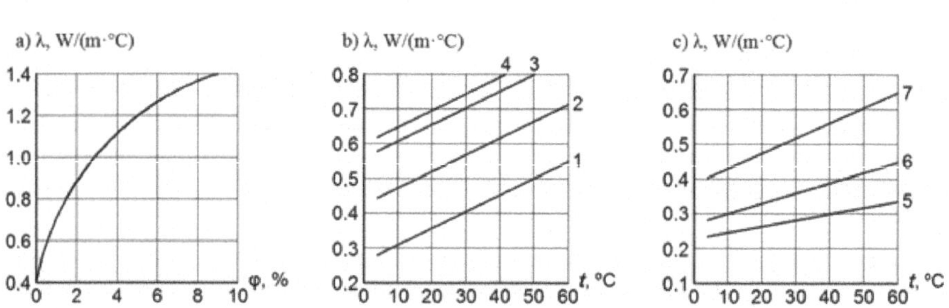

Fig. 1. Relationship between the thermal conductivity: a) of a brickwork and the brick moisture; b) aerated concrete; c) foam concrete and temperature for various moistures: $\varphi_1 = 4.8\%$; $\varphi_2 = 17\%$; $\varphi_3 = 33\%$; $\varphi_4 = 38\%$; $\varphi_5 = 1.4\%$; $\varphi_6 = 6\%$; $\varphi_7 = 28\%$.

2 Materials and Methods

The main technical capabilities of modern thermal imagers include:

1) Digital imaging technology. A digital camera built into the thermal imager performs automatic simultaneous recording of infrared and visible images;
2) Super Resolution technology. It enables to increase the current resolution of the thermal image several times without replacing the detector;
3) Moisture measurement technology. In addition to the usual thermograms, it allows receiving information on the moisture level of the investigated surfaces of the controlled object. This function is important when detecting areas of possible condensation and biocorrosion. To implement this technical possibility, the temperature and relative moisture of the ambient air are pre-measured. Further, the specialized application software supplied with the thermal imager automatically determines the dew point temperature according to the specified air parameters and projects the obtained value onto the temperature field of the surface of the investigated area of the test object. The image areas in which the actual surface temperature is below the dew point are highlighted by the thermal imager in red ("high risk area"), and areas undamaged by moisture are highlighted in green (Fig. 2).

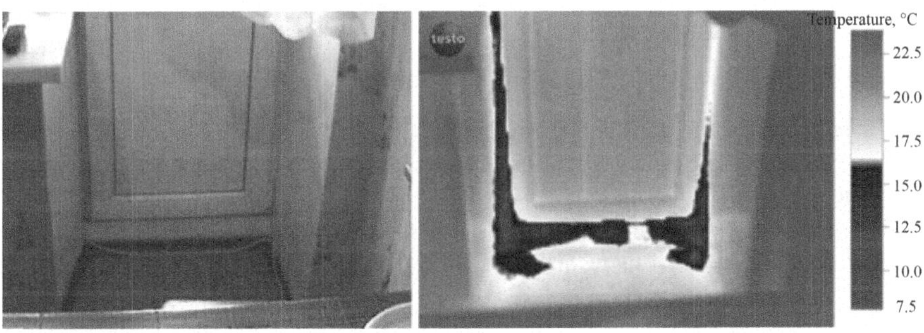

Fig. 2. The moisture field of the surface of the controlled object (door to the balcony) at $t_a = 20$ °C and $\varphi_a = 50\%$ (green color: $\varphi_{sur,a} = 0 \div 64\%$, is not critical; yellow/orange color: $\varphi_{sur,a} = 65 \div 80\%$, is potentially critical; red color: $\varphi_{sur,a} > 80\%$, is critical)

3 Results and Discussion

Consider a practical example of thermal imaging of temperature and moisture fields of a translucent building envelope. As an object of thermal control, consider a translucent enclosing building structure in the form of a window in the window opening of an apartment in a residential building. Figure 3 shows the visible (digital) and thermal images of the test object.

Fig. 3. The visible (digital) and thermal image of a window in a window opening

Table 1 shows the parameters and conditions for the internal thermal imaging of a window in a window opening.

Table 1. Initial data of thermal imaging of a window in a window opening.

Object	Environmental parameters			Parameters of radiation surface		*l*, m
	$t_{int,a}$, °C	$\varphi_{int,a}$, %	$t_{ext,a}$, °C	Material	ε	
Window in a window opening	20	50	–1	Glass/wood	0.94/0.90	3.5

Table 2 shows the results of processing and analysis of the controlled object thermogram using the specialized application software environment IRSoft (for the Testo thermal imager, Germany).

The performed processing and analysis of the thermogram of the window in the window opening, allowed us to make the following conclusions:

1. Excessive heat losses through the joints of window frames are present.
2. There are areas of high moisture that can lead to condensation. The physical phenomenon of moisture condensation is associated with leaks in the butt joints, as a result of which cold outside air penetrates into the room. The internal air bordering with the joint is cooled to the dew point temperature (td = 9.3 °C), which leads to condensation of water vapour.

The main problem areas of the thermal control object are the vertical joint of window frames and the horizontal joint of the window block with the sill. To eliminate these thermal defects, it is recommended to seal the leaky areas with insulation and/or silicone sealant.

Table 2. Results of thermogram processing and analysis for the window in the window opening.

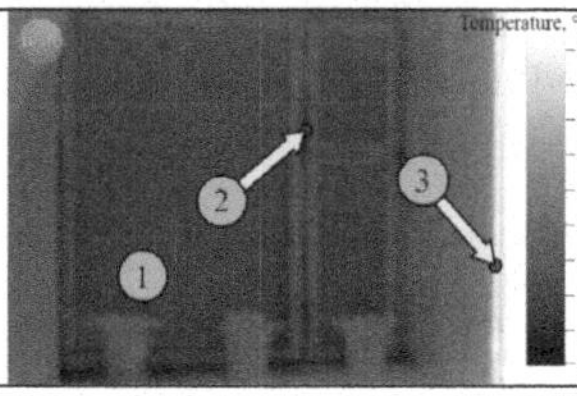	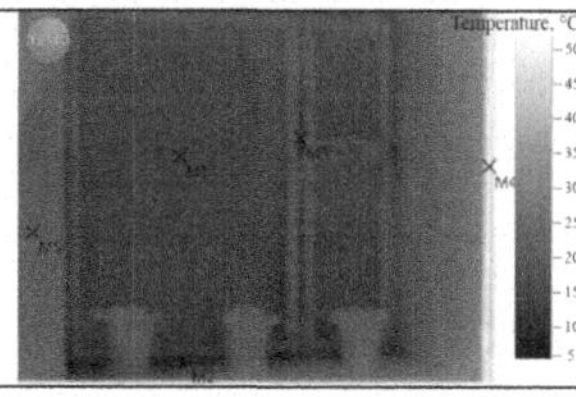
Temperature anomalies: 1, 2 - excessive heat losses through the joints of window frames; 3 - riser of the heating system	Temperatures at reference points: M1: +16.0 °C; M2: +8.3 °C; M3: +9.9 °C; M4: +50.4 °C; M5: +20.5 °C
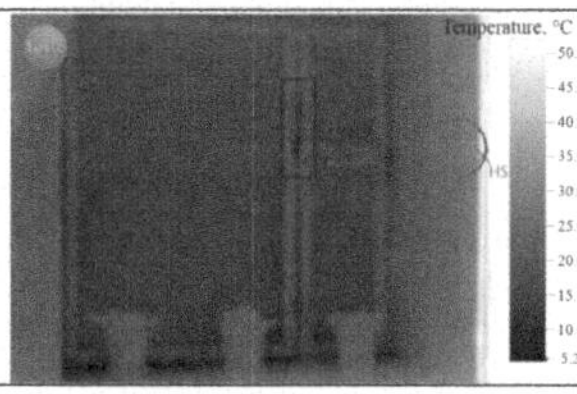	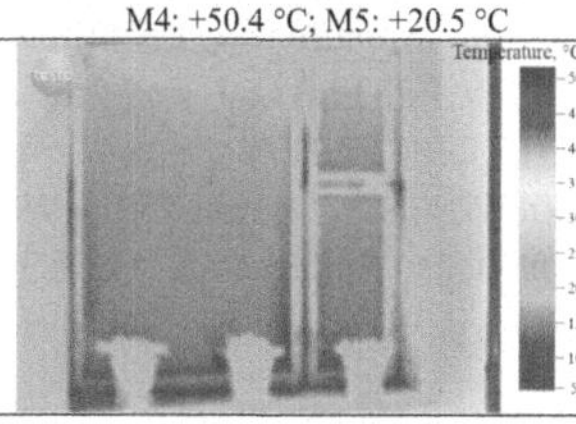
“Hot” and “cold” points: HS1: +50.9 °C; CS1: +9.7 °C	Color palette "Rainbow"
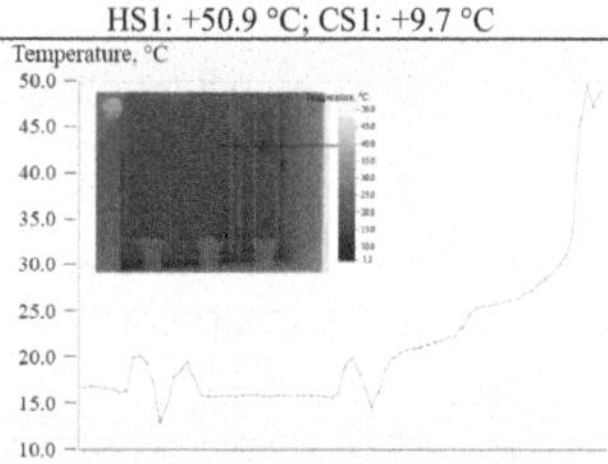	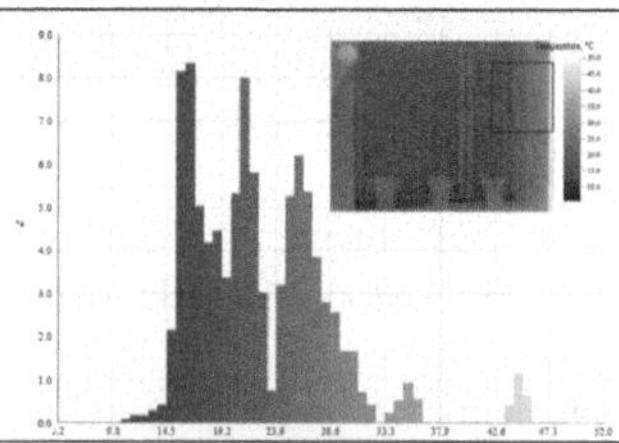
Thermal profile of a fragment of a thermal image	Temperature histogram of a thermal image fragment
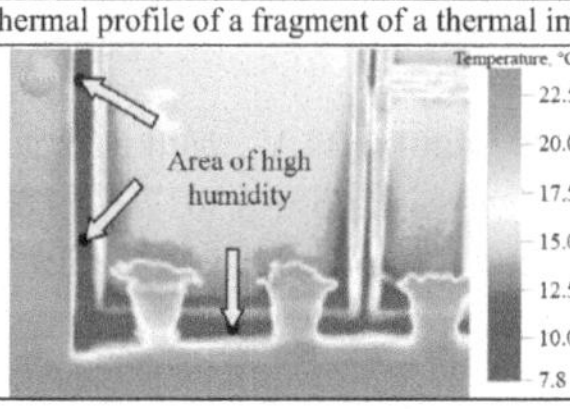	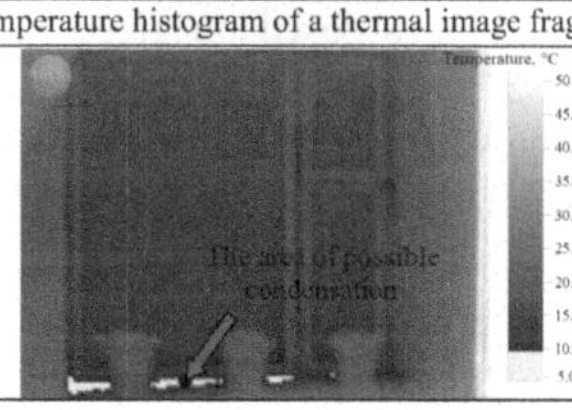
The moisture field of the thermal control object	The area of possible condensation. The section in yellow is for

4 Conclusions

Thermal bridges are the cause of thermal energy leaks into the environment. They contribute to the condensation of moisture from the indoor air, caused by a decrease in its temperature at the border of the thermal defect. Thermal imaging for each individual point of the controlled object allows one to determine the relative surface moisture given the temperature and relative moisture of the environment, and the temperature of the controlled object surface.

Modern thermal imaging technology, and in particular thermal imagers, makes it possible to identify such thermal bridges. Thermal imagers required for conducting thermal imaging of building objects, individual building structures and their elements are portable optical-mechanical devices. They reproduce an image of heated objects and identify the temperature field of the controlled object. The result of thermography is a thermal image (thermogram), which is processed and analyzed in the RAM of the thermal imager and/or with the help of specialized software.

References

1. Sinitsyn, A.A., Karpov, D.F., Pavlov, M.V.: Basic Thermal Imaging Diagnostics of Heat-Consuming Construction Facilities. VoSU (2014)
2. Vavilov, V.P.: Thermal Imaging and Thermal Control for Engineers. Publishing House "SPECTRUM" (2017)
3. Cieślikiewicz, Ł., Łapka, P., Wasik, M., Kubiś, M., Pietrak, K., Wiśniewski, T.S., Furmański, P., Seredyński, M.: E3S Web Conf. vol. 70 (2018)
4. Kovalov, A., Konoval, V., Khmyrova, A., Dudko, K.: E3S Web Conf. vol. 123, p. 01022 (2019)
5. Zhukov, A., Ter-Zakaryan, A., Bobrova, E., Bessonov, I., Medvedev, A., Mukhametzyanov, V., Poserenin, A.: E3S Web Conf. vol. 91, p. 02047 (2019)
6. Alioua, T., Agoudjil, B., Boudenne, A.: E3S Web Conf. vol. 85, p. 02009 (2019)
7. Surikova, A., Sytsyanko, E., Solov'Eva, E., Mukhametova, L.: E3S Web Conf. vol. 178 (2020)
8. Karpov, D., Sinitsyn, A.: E3S Web Conf. vol. 161 (2020)
9. Pavlov, M., Karpov, D., Akhmetova, I., Monarkin, N.: E3S Web Conf. vol. 178 (2020)
10. Karpov, D., Sinitsyn, A.: E3S Web Conf. vol. 161 (2020)
11. Sultanguzin, I.A., et al.: Development strategy of the energy supply system for MPEI campus blocks based on green building. In: IOP Conference Series: Earth and Environmental Science, vol. 288, no. 1, p. 012071 (2019)
12. Akhmetova, I.G., Kalyutik, A.A., Fedukhin, A.V., Derevianko, O.V., Mukhametova, L.R.: IOP Conference Series Earth and Environmental Science, vol. 288, p. 012076 (2019)
13. Akhmetova, I., Chichirova, N., Derevianko, O.: Revisiting heat losses calculation at district heating network. Int. J. Civ. Eng. Technol. **8**, 694 (2017)
14. Vijayaraghavan, G.K., Majumder, M.C., Ramachandran, K.P.: Insight non-destructive test. Cond. Monit. **52**, 481 (2010)
15. George, S., Goravar, S., Mishra, D., Shyamsunder, M.T., Sharma, P., Padmashree, G.K., Kumar, P.S., Bremond, P., Mukherjee, K.: Insight non-destructive test. Cond. Monit. **52**, 470 (2010)
16. Sovetova, N., et al.: Regional management of district heating. In: IOP Conference Series: Earth and Environmental Science, vol. 288, p. 012121 (2019)
17. Ponomareva, N.V., Zvereva, A.O., Golubtsova, E.V., Ilyashenko, S.B.: Tax incentives for use of alternative energy sources in the Russian Federation. Int. J. Energy Econ. Policy **9**, 144 (2019)

Thermal Stressed State at Temperature Variations While Concreting for Hydraulic Structures

Kirill Semenov[1], Maxim Kukolev[1], Ernest Ivanov[1], Aleksandra Makeeva[1](✉), Sergej Manovitskij[2], and Aleksey Kuleshin[2]

[1] Peter the Great St. Petersburg Polytechnic University, Polytechnicheskaya Street 29, 195251 St. Petersburg, Russia
bushmanova.av@edu.spbstu.ru
[2] PB "Luch", Akademika Pavlova Street 14A, 197376 Saint-Petersburg, Russia

Abstract. Massive concrete structures are the most widespread type of constructions in hydraulic engineering and the energy sector. The exothermal reaction of hardening of the concrete mixture leads to the uneven distribution of temperatures in the block and the temperature cracks formation. The crack formation cannot be avoided without special measures to regulate the hardening temperature conditions of a concrete mixture. Previously, a constant structure footings temperature was used in studies of the temperature field distribution and the thermal stressed state. The purpose of the article is to identify the effect of changes in the base temperature on massive concrete structures during their construction and to formulate appropriate recommendations on the method of calculating massive concrete structures during the construction period. The methodology takes into account the dependence of heat release and concrete characteristics on the "temperature" history of hardening, using the lock side slab of the Nizhny Novgorod water-development project as an example. The solution of the one-dimensional problem was implemented in the TERM program. It was found that taking into account the influence of the initial temperature gradient of the base and concrete mixture at the time of concreting qualitatively affects the distribution of temperature fields and stresses in the concrete block and should be taken into consideration when substantiating massive concrete structures.

Keywords: Crack initiation · Concrete construction · Hydraulic structure · Thermal stressed state · Temperature variations

1 Introduction

Massive concrete structures are the most widespread in hydraulic engineering and nuclear construction. Experience has shown that thermal deformations are one of the reasons for the formation of cracks during the construction period in massive concrete structures [1]. Deformations are caused by the release of heat during cement hydration [2], leading to an increase in temperature in the central part of the mass [3].

A. Sinitsyn (Ed.): SES 2020, LNCE 141, pp. 259–267, 2021.
https://doi.org/10.1007/978-3-030-67654-4_29

Temperature effects on the structure are caused by the features of external and internal heat exchanges in massive reinforced concrete structures during the construction phase [4–6]. Another reason for the appearance of cracks is a complex geometry and large dimension, implying a one-time concreting of the supporting structure [7–10]. The availability of temperature cracks in structures of a higher level of responsibility, such as power hydraulic structures, is not permissible [11–16]. Therefore, the prevention of the formation of thermal cracks is a key task to ensure the consumer properties declared in the project.

Earlier, in studies of the distribution of thermal stressed state a constant structure footings temperature was set [17–21]. However, there is a diurnal variation [22].

Temperature fluctuation is one of the most significant factors for a large number of processes [22–26]. It is difficult to take into account the influence of the variable temperature of the base (soil, etc.), because is a complex dynamic process containing both explicit components described by mathematical formulas, and random one. However, the distribution of temperatures inside the block significantly depends on the temperature of the base at the moment of pouring the concrete mixture. Thus, the purpose of this paper is to identify the effect of changes in the base temperature on massive concrete structures during their construction and to formulate appropriate recommendations on the method of calculating massive concrete structures during the construction period.

2 Materials and Methods

2.1 General Data

The object of the study is the lock side slab of the Nizhny Novgorod low-pressure water scheme (Fig. 1). The construction of the Nizhny Novgorod water scheme is provided in the Strategy for the Development of Inland Water Transport of the Russian Federation for the period up to 2030 and includes a scheme for territorial planning of the Russian Federation in the field of federal transport (Fig. 2). The main structures: an earth dam, a spillway dam, a two-line lock with a chamber size of 300 × 30 m.

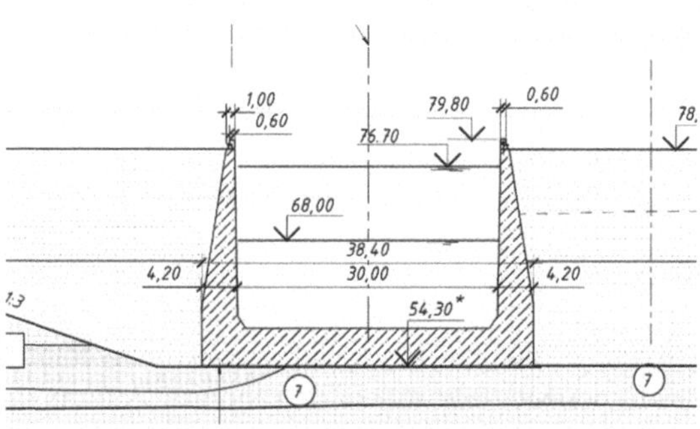

Fig. 1. Fragment of a cross section of the gateway

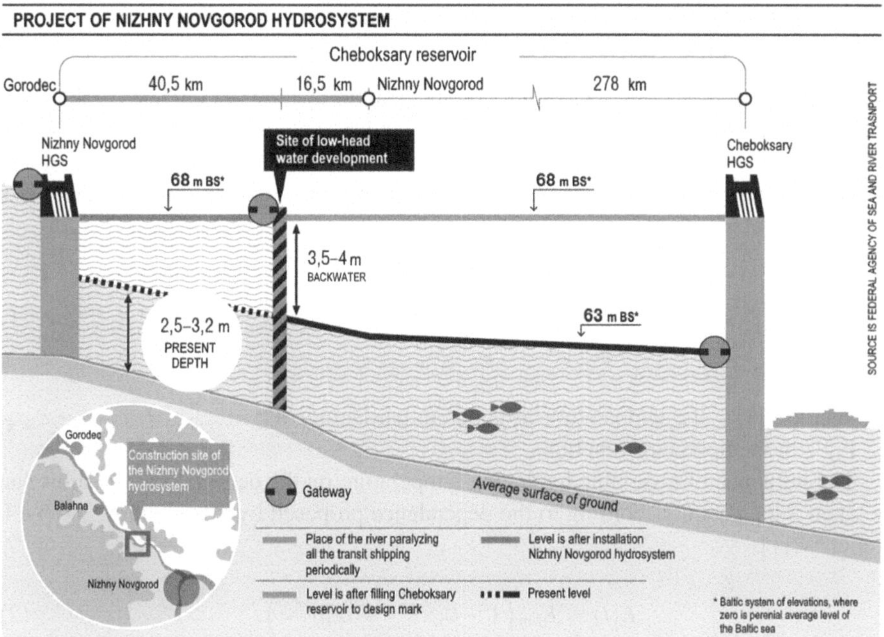

Fig. 2. Project of Nizhny Novgorod Hydrosystem

The Design scheme was adopted as a monolithic section of a lock gate plate with a thickness of 4.2 m made of B30 concrete without technological holes. The design scheme is set as a single block. The one-dimensional problem is solved by the finite difference method. As the results, the values of temperatures, thermal stresses, and deformations at the nodes of the MCR mesh will be obtained. In the paper, we consider that the lock gate rests on a layer of concrete preparation on a rocky foundation.

Strength and deformation characteristics of concrete are adopted according to SP 41.13330.2012 “Concrete and reinforced concrete structures of hydraulic structures”. The heat release parameters, thermophysical characteristics of concrete were adopted based on the results of experimental studies the Polytech-SKiM-Test laboratory. The solution of the one-dimensional problem was implemented in the TERM program, intended for calculating the thermal stress state of concrete massifs during their construction.

As an investigated method of combating temperature cracks and influencing the thermal stress state of a foundation slab array, to achieve the objectives of the work, we chose to cover the concrete surface with a heat-insulating material - extruded polystyrene foam with a thermal conductivity of λ = 0.033 W/(m × °C).

2.2 Effect of Hardening Temperature on Heat Generation, Deformation Modulus and Concrete Creep

The influence of hardening temperature on the heat dissipation of concrete is taken into account using the adjusted time hypothesis, which makes provision for the entire "temperature history" and allows one to determine heat emission at any time and any temperature regime by using the curve of exothermic heat dissipation of concrete at a temperature of 20 °C [1]:

$$t_{adj} = \int_0^t F_Q[T(\tau)]d\tau, \tag{1}$$

Where T(τ) – the temperature of concrete hardening, °C;

F_Q – the temperature function.

The influence of a variable hardening temperature on the deformation modulus can be taken into account according to the dependence proposed by P.I. Vasiliev and Yu.I. Kononov [1]:

$$E(t) = E_{lim}\left(1 - 0.5^{\alpha_1 t_{adj}} - 0.5^{\alpha_2 t^p_{adj}}\right), \tag{2}$$

Where α_1, α_2, p – the empirical coefficients,

t_{adj} - "adjusted" time.

The influence of temperature conditions on the magnitude of creep deformations can be taken into account by replacing the actual time with "adjusted" [27]:

$$t_{adj} = \int_0^t F_E[T(\tau)]d\tau. \tag{3}$$

It is proposed to take the temperature function F_E in the form:

$$F_E[T(\tau)] = \left[\frac{T(\tau) + 30}{50}\right]^{3.2}. \tag{4}$$

2.3 Temperature Variations

Temperature is a stable characteristic of the terrain. The amplitude of annual fluctuations A is in the range: 12 ÷ 22 °C, and depends on the climatic conditions of the area. The lower the heat capacity and thermal conductivity of the soil, the greater its amplitude.

Taking into account the influence of variable temperature is difficult because of complex dynamic process containing both explicit, described by mathematical formulas and random components [22–26, 28–30]. Most models are based on long-term meteorological observations with daily resolution [28, 29] and for a given period and populated locality is characterized by three parameters - the average temperature T_0, as well as the amplitude of the sinusoidal function A and its maximum point t_0.

The research was carried out on two principal analytical models:

- The first analytical model implied a simplified approach, which means it did not take into account the temperature variations: the structure footings temperature is constant. The temperature for warm and cold periods is accepted according to SP 131.13330.2018 Construction climatology for Nizhny Novgorod.
- The second analytical model implied the changes in the structure footings temperature over a certain period:

$$T(t) = A \cdot Sin(2\pi t \cdot v + \varphi) + c, \tag{5}$$

Where A – the amplitude, ° C;
ν – the frequency;
φ – the phase;
c – the rise, °C

3 Results

The results of application of methodology for calculating thermo-stressed state in a massive concrete structure, with one-dimensional setting, during construction is presented in Table 1.

A typical graph of the temperature distribution in the block depending on age for the case of concreting No. 1 (Table 1) in the warm period of the year is shown on Fig. 3, Fig. 4.

The table and graphs show that the most unfavorable conditions, from the point of view of ensuring crack resistance, is when concreting in the warm period for the case of increased base temperature (the thickness of the thermal insulation is 28% higher relative to the absence of a temperature difference between the concrete mixture and the base at the time of laying). For the cold period, a more heated base compared to a concrete mix also leads to an increased thickness of thermal insulation (by 5%). In the case of low base temperatures (within the range of daily fluctuations), both periods of concreting require a lower insulation thickness (by 14% and 10%, respectively).

Table 1. Results of selection of the required thickness of thermal insulation depending on the boundary temperature conditions at the time of concreting

No	Warm period				Cold period			
	T_{base}, °C	T_{air}, °C	T_{conc}, °C	$t_{th.is.}$, sm	T_{base}, °C	T_{air}, °C	T_{conc}, °C	$t_{th.is.}$, sm
1	18	18	18	2.1	−12	−12	10	2.0
2	27	18	18	2.7	−6	−12	10	2.1
3	9	18	18	1.7	−18	−12	10	1.8

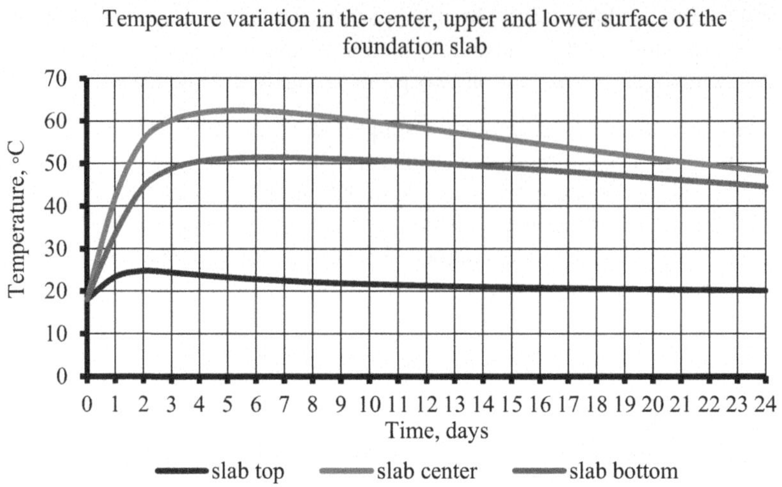

Fig. 3. Temperature distribution in the block depending on the age of concrete in the warm period of the year (without a thermal insulation)

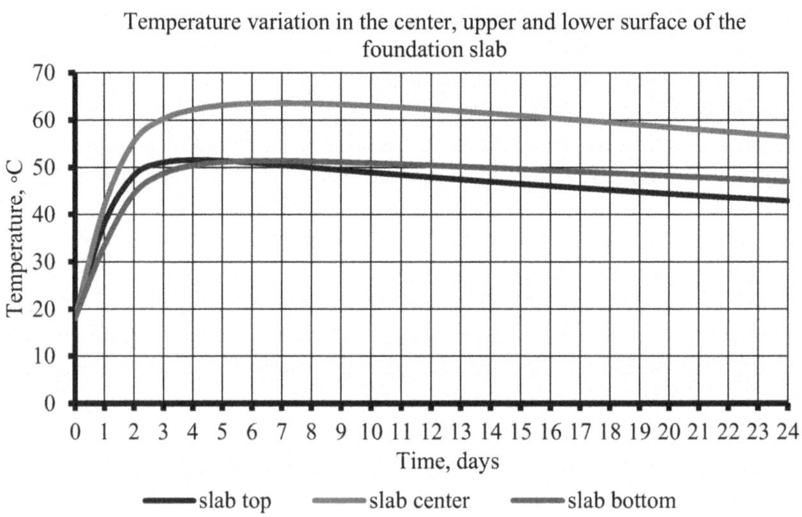

Fig. 4. Temperature distribution in the block depending on the age of concrete in the warm period of the year (with a thermal insulation)

4 Discussion

Fluctuations in temperature are one of the most significant factors for a large number of processes [22–26]. For example, in article [25], based on the obtained experimental results of the study, for the first time it was possible to make direct statements about the deformation conditions arising in the asphalt base layer due to temperature. Taking into account the influence of variable temperature is difficult because of complex dynamic process containing both explicit, described by mathematical formulas and random components [22–26, 28–30].

In calculating the thermal stress state and thermal crack resistance of massive concrete structures during the building period, base temperature along with the initial data on the geometric shape and characteristics of concrete is a fundamental factor [7–10]. Earlier, in studies of the distribution of temperature fields and the thermal stressed state a constant structure footings temperature was set [7–10]. However, there is a variation. The influence of the initial temperature gradient of the base and concrete mixture at the time of concreting qualitatively affects the distribution of temperature fields and stresses in the concrete block and should be taken into consideration when substantiating massive concrete structures.

5 Conclusion

Based on the analysis of the results of all calculations, the following conclusions can be proposed and appropriate recommendations formulated:

1. Taking into account the influence of the initial temperature gradient of the base and concrete mixture at the time of concreting qualitatively affects the distribution of temperature fields and stresses in the concrete block;
2. In quantitative terms, the thickness of the required thermal insulation can be greater from 5% (for cold) to 28% (for warm periods of concreting) relative to the basic calculation (without taking into account the temperature gradient of the base-concrete mixture);
3. The colder base leads to a lower thickness of thermal insulation on the block surface (by 10–14%), which can be used as a measure to combat temperature cracks in massive concrete and reinforced concrete structures during the construction period.

References

1. Semenov, K., Barabanshchikov, Y.: Thermal cracking resistance in massive concrete structures in the winter building period. Appl. Mech. Mater. **725–726**, 431–441 (2015)
2. Taweel, T.J.B., Sokolova, E., Sergeev, V., Solovev, D.B.: Energy and exergy analysis of clinker cooler in the cement industry. In: IOP Conference Series: Materials Science and Engineering, vol. 463, p. 032101. IOP Publishing, Bristol (2018)
3. Schackow, A., Effting, C., Gomes, I.R., Patruni, I.Z., Vicenzi, F., Kramel, C.: Temperature variation in concrete samples due to cement hydration. Appl. Therm. Eng. **103**, 1362–1369 (2016)
4. Zreiki, J., Bouchelaghem, F., Chaouche, M.: Early-age behavior of concrete in massive structures, experimentation and modelling. Nucl. Eng. Des. **240**(10), 2643–2654 (2010)
5. Maruyama, I., Lura, P.: Properties of early-age concrete relevant to cracking in massive concrete. Cem. Conc. Res. **123**, 105770 (2019)

6. Benin, A.V., Semenov, A.S., Semenov, S.G.: Fracture analysis of reinforced concrete bridge structures with account of concrete cracking under steel corrosion. Adv. Mater. Res. **831**, 364–369 (2014)
7. Korsun, V., Vatin, N., Korsun, A., Nemova, D.: Physical-mechanical properties of the modified fine-grained concrete subjected to thermal effects up to 2000C. Appl. Mech. Mater. **633–634**, 1013–1017 (2014)
8. Nguyen, T.C., Luu, X.B.: Reducing temperature difference in mass concrete by surface insulation. Mag. Civ. Eng. **88**(4), 70–79 (2019)
9. Won, J.Y., Lee, S.H., Park, T.W., Nam, K.Y.: Basic applicability of an insulated gang form for concrete building construction in cold weather. Constr. Build. Mater. **125**, 458–464 (2016)
10. Korsun, V., Korsun, A.: The influence of precompression on strength and strain properties of concrete under the effect of elevated temperatures. Appl. Mech. Mater. **725–726**, 469–474 (2015)
11. Kuriakose, B., Rao, B.N., Dodagoudar, G.R.: Early-age temperature distribution in a massive concrete foundation. Procedia Technol. **25**, 107–114 (2016)
12. Zhang, J.: Control measures of temperature cracks in mass concrete for bridge base slab. In: 2010 International Conference on E-Product E-Service and E-Entertainment, ICEEE2010, pp. 5305–5309. IEEE, New York (2010)
13. Zhang, Z., Zhang, X., Wang, X., Zhang, T., Zhang, X.: Merge concreting and crack control analysis of mass concrete base slab of nuclear power plant. Appl. Mech. Mater. **94–96**, 2107–2110 (2011)
14. Kolosova, G.S., Lalin, V.V., Kolosova, A.V.: The effect of construction joints and cracks on the stress-strain state of the arch-gravity dam. Mag. Civ. Eng. **40**(5), 76–85 (2013)
15. Orekhov, V.G., Tolstikov, V.V.: Stress-strain state of the concrete dam at the Kankunskii hydroproject. Power Technol. Eng. **46**, 121–128 (2012)
16. Aniskin, N.A., Chuc, N.T.: Temperature regime and thermal stress in a concrete massif with pipe cooling. Power Technol. Eng. **52**, 638–643 (2019)
17. Abdulrazeg, A.A., Khanehzaei, P., Noorzaei, J., Jaafar, M.S., Mohammed, T.A., Thanoon, W.A.: Cracking safety evaluation of massive concrete structures. Key Eng. Mater. **462–463**, 1403–1408 (2011)
18. Aniskin, N., Nguyen Trong, C.: The thermal stress of roller-compacted concrete dams during construction. In: MATEC Web of Conferences, vol. 196, p. 04059. EDP Sciences, Les Ulis (2018)
19. Shaibakova, A., Semenov, K., Barabanshcikov, Y.: Thermal cracking resistance of stacking concrete blocks. In: MATEC Web of Conferences, vol. 245, p. 08005. EDP Sciences, Les Ulis (2018)
20. Wu, S., Huang, D., Lin, F.B., Zhao, H., Wang, P.: Estimation of cracking risk of concrete at early age based on thermal stress analysis. J. Therm. Anal. Calorim. **105**, 171–186 (2011)
21. Benboudjema, F., Darquennes, A., Hilaire, A., Briffaut, M.: Prediction of cracking in massive concrete structures by numerical simulations. In: 4th International fib Congress 2014: Improving Performance of Concrete Structures, FIB 2014 – Proceedings, pp. 50–51 (2014)
22. Huang, Y.Y., Chen, C.T., Liu, W.T.: Thermal performance of extensive green roofs in a subtropical metropolitan area. Energy Build. **159**, 39–53 (2018)
23. Thiis, T.K., Burud, I., Flø, A., Kraniotis, D., Charisia, S., Stefanssona, P.: Monitoring and simulation of diurnal surface conditions of a wooden facade. Procedia Environ. Sci. **38**, 331–339 (2017)

24. Wang, B.L., Tian, Z.H.: Application of finite element-finite difference method to the determination of transient temperature field in functionally graded material. Finite Elem. Anal. Des. **41**(4), 335–349 (2005)
25. Otto, F., Liu, P., Zhang, Z., Wang, D., Oeser, M.: Influence of temperature on the cracking behavior of asphalt base courses with structural weaknesses. Int. J. Transp. Sci. Technol. **3** (5), 208–216 (2018)
26. Gorshkov, A.S., Rymkevich, P.P., Vatin, N.I.: Simulation of non-stationary heat transfer processesin autoclaved aerated concrete-walls. Mag. Civ. Eng. **8**(52), 38–48 (2014)
27. Korotchenko, I.A., Ivanov, E.N., Manovitsky, S.S., Borisova, V.A., Semenov, K.V., Barabanshchikov, Y.G.: Deformation of concrete creep in the thermal stress state calculation of massive concrete and reinforced concrete structures. Mag. Civ. Eng. **1**(69), 56–63 (2017)
28. Semenov, S.M., Gelver, E.S.: Sine approximation of the annual course of daily mean air temperature in Russia in the twentieth century. Meteorol. Hydrol. **11**, 25–30 (2002)
29. Qu, M., Wan, J., Hao, X.: Analysis of diurnal air temperature range change in the continental United States. Weather Clim. Extrem. **4**, 86–95 (2014)
30. Khrapunov, E.F., Potechin, I.V., Chumakov, Y.S.: Structure of a free convective flow over a horizontal heated surface under conditions of conjugate heat transfer. In: IOP Conference Series: Journal of Physics: Conference Series, vol. 891, p. 012081. IOP Publishing, Bristol (2017)

Study of the Impact of Ultrasonic Treatment on the Technological Properties of Cationic Bitumen Emulsion

V. A. Shorin[1(✉)], A. Y. Velsovskij[1], T. R. Akhmetov[2], and C. A. Myasnikova[1]

[1] Department of Highways, Vologda State University, Lenina Street 15, 160000 Vologda, Russia
v2u@yandex. ru
[2] Kazan State Power Engineering University, Street Krasnoselskaya, 51, 420066 Kazan, Russia

Abstract. This work is devoted to the study of the influence of ultrasonic treatment on the properties of cationic bitumen emulsion for patching repair of roads. For the first time, the influence of high-frequency sound treatment on the technological properties of road bitumen emulsion EBK-2 was studied. It was shown that treatment of a bitumen emulsion with ultrasound increases its homogeneity by 25%. Due to this, during the emulsion breaking, bitumen is distributed more evenly over the surface of the stone material. It results in a better formation of contact between crushed stone (slag) and bitumen emulsion. The ultrasonicated emulsion EBK-2A is a promising material for improving the quality of patching, its viscosity decreases by 1.6 times. With a decrease in viscosity, the emulsion becomes more mobile and better envelops the grains of stone material. This is especially effective for crushed stone materials based on blast furnace or steelmaking slag with their highly developed rough surface.

Keywords: Cationic bitumen emulsion · Ultrasonic treatment · Repair of roads

1 Introduction

Nowadays a vast variety of road construction techniques can be used to achieve the set tasks depending on the season, weather and other factors [1–7]. Currently, patching repair or corrugation repair of the top layer of the asphalt concrete pavement are performed annually. Patching repair of roads by jet-injection method is of high interest now [8–12]. The use of emulsions instead of hot bitumen simplifies the repair technology, allows one to conduct works in early spring, using a pre-prepared mixture, and wet stone materials. It is most advisable to use a bitumen emulsion for repairs during the wet season, as it ensures quick formation and good quality of work.

At present, the search for bitumen emulsions with improved technological properties is of particular interest. In terms of solving this problem, we were the first to study the influence of high-frequency sound treatment on the properties of bitumen emulsion. The obtained results open up undoubted prospects for using this technology to improve the properties of bitumen emulsions.

A. Sinitsyn (Ed.): SES 2020, LNCE 141, pp. 268–275, 2021.
https://doi.org/10.1007/978-3-030-67654-4_30

2 Materials and Methods

In our study we used the ultrasonic installation shown in Fig. 1. The installation consists of a reactor, the walls of which are made of stainless steel. Two emitters are mounted in it so that the walls themselves become emitters and provide a uniform cavitation effect on both viscous media and liquid media of low viscosity. Ultrasonic installation has the following characteristics:

- Total power of emitters: 200 W;
- Chamber volume: 2L;
- Total number of cavitation according to IC-4D cavitometer: 1000–1200;
- Reactor material: stainless steel 12X18H10T;
- Material of emitters: piezoceramic;

A source of high-frequency electrical vibrations.

Fig. 1. The ultrasonic installation.

An electromechanical device is used to generate ultrasonic vibrations (Fig. 2), in which ultrasonic energy is obtained by converting electrical energy. Energy is converted by the inverse piezoelectric effect. If alternating electric voltage is applied to a quartz plate or piezo-materials based on barium titanium and titanium lead zirconate, the crystal begins to contract and expand (change its geometric dimensions) with the frequency of the applied voltage. Piezoelectric elements are part of the piezoelectric transducer. The converter provides the conversion of electrical vibration energy into mechanical vibration energy.

Fig. 2. Device for converting electrical energy into ultrasonic.

In our study we used the cationic bitumen emulsion EBK-2. The review of scientific and technical information unambiguously showed that despite a vast variety of ultrasound applications [13–16], it was previously used only for bitumen, dispersed systems, but was not used for bitumen emulsions. Ultrasonic treatment of bitumen causes an increase in its adhesion properties in comparison with conventional bitumen by 30–70% and reduces its viscosity. Due to these prerequisites, when treating the bitumen emulsion with ultrasound, we were right to expect an improvement in the bitumen properties in this emulsion, as well as the properties of the bitumen emulsion itself, such as uniformity, storage stability, relative viscosity.

3 Results and Discussion

During experiments we examined emulsions before and after ultrasound treatment (Fig. 3).

Fig. 3. Emulsions before and after ultrasound treatment.

3.1 The Homogeneity of the Bitumen Emulsion

The homogeneity of the bitumen emulsion is the identity of the composition and structure of the material throughout its volume. The homogeneity of the bitumen emulsion was determined by the sieve residue. The "sieve residue" is assessed in order to determine the mass of the binder remaining on a sieve with a mesh No. 014 after filtering the emulsion through a sieve. The smaller the sieve residue is, the more homogeneous the emulsion is.

The results of experimental studies are shown in Table 1 and Fig. 4. When the bitumen emulsion is treated by ultrasound, the residue on the sieve decreases, which indicates that its homogeneity increases. The optimum processing time for bitumen emulsion is 5 min at a cavitation number of 400–500 and a temperature of 30–42 °C. At the same time, the sieve residue of the ultrasonicated emulsion (0.18%) decreased by 0.06% compared to the original emulsion (0.24%). The emulsion has become more homogeneous in its composition. The emulsion quality has improved.

Table 1. Influence of ultrasound on the homogeneity of cationic bitumen emulsion.

Emulsion type	Cavitation number	Time (min)	t, °C	Residue on sieve, %
EBK -2	0	0	–	0.24
EBK -2	400–500	5	30–42	0.18
EBK -2	500–700	5	48–54	0.18

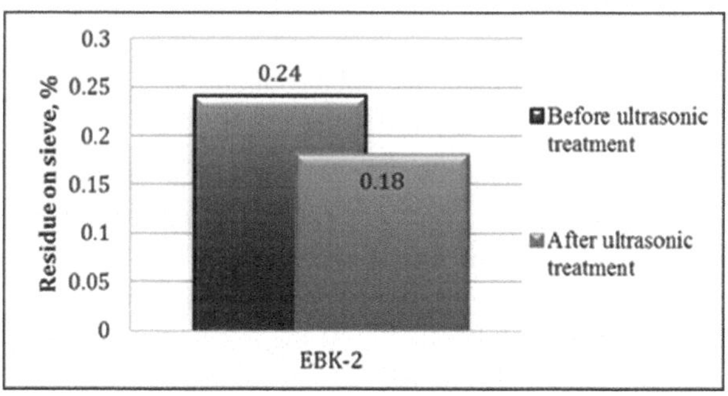

Fig. 4. The influence of ultrasound on the homogeneity of the cationic emulsion.

According to [17], emulsions must be homogeneous. On a glass rod, immersed for 1 s in the emulsion, there should be no droplets or sparkles of water visible to the eye, the emulsion should not delaminate within 7 days at a positive temperature.

3.2 Storage Stability of Bitumen Cationic Emulsion

Due to the difference in density between water and bitumen, the bitumen droplets will move either downward or upward. Storage stability is the ability of the emulsion to resist delamination (not separating into bitumen and water). The storage stability is assessed in order to determine the residue on a sieve with a mesh No. 014 after filtering through it the emulsion, stored at room temperature for 7 and 30 days. The lower the residue on the sieve is, the longer the storage stability is.

The results of experimental studies are presented in Table 2 and Fig. 5.

Table 2. Influence of ultrasound on cationic bitumen emulsion.

Emulsion type	Cavitation number	Time (min)	t, °C	Storage stability, % (after 3 months)
EBK -2	0	0	–	0.5
EBK -2	400–500	5	30–42	0.24
EBK -2	500–700	5	48–54	0.27

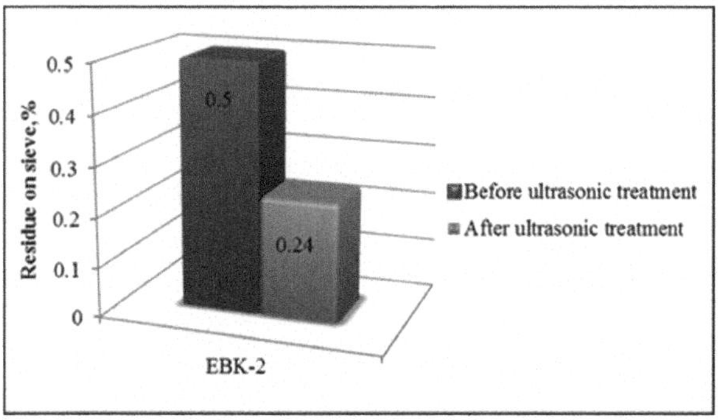

Fig. 5. The influence of ultrasound on the stability during storage the cationic emulsion.

When the bitumen emulsion is treated by ultrasound after 3 months, the sieve residue decreases, which indicates that the storage stability of the treated bitumen emulsion increased compared to the initial one.

The optimum processing time for bitumen emulsion is 5 min at a cavitation number of 400–500 and a temperature of 30–42 °C. At the same time, the residue on the sieve of the emulsion treated with ultrasound (0.24%) in comparison with the initial emulsion (0.5%) decreased by 0.26%. The emulsion does not delaminate within three months. The emulsion quality has improved.

3.3 The Relative Viscosity of the Bitumen Cationic Emulsion

Viscosity (internal friction) is one of the transport phenomena, the property of fluid bodies to resist the movement of one part of them relative to another. Relative viscosity is a value that indirectly characterizes the hydraulic resistance to flow, measured by the time of expiration of a given volume of solution through a vertical tube (of a certain diameter) [18].

The relative viscosity is assessed in order to determine the time of the flow of 50 ml of the emulsion from the viscometer through the drain hole with a diameter of 3 mm. The low viscosity of the bitumen emulsion provides a good ability to process stone materials without drying and heating them.

The results of the experiment can be seen in Table 3 and Fig. 6.

Table 3. Influence of ultrasound on the relative viscosity of cationic bitumen emulsion.

Emulsion type	Cavitation number	Time (min)	t, °C	Relative viscosity, s
EBK -2	0	0	–	24
EBK -2	600–700	2	28–30	15
EBK -2	400–500	5	30–42	15–17

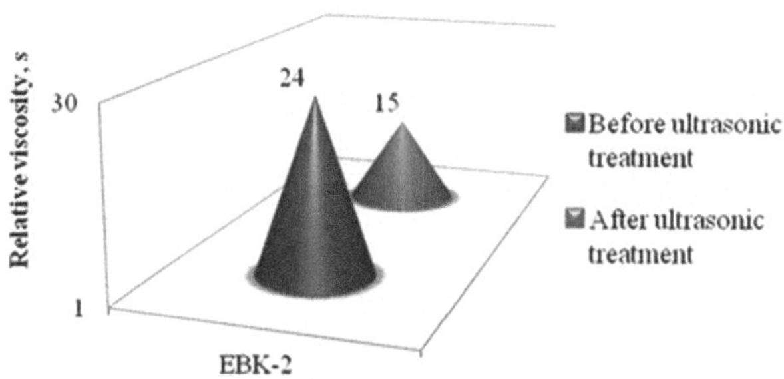

Fig. 6. Influence of ultrasound on the relative viscosity of the cationic bitumen emulsion.

When processing a bitumen emulsion with ultrasound, the relative viscosity decreases by 37.5%. The optimal processing time for bitumen emulsion is 2 min at a cavitation number of 600–700 and a temperature of 28–30 °C. The emulsion becomes less viscous, and therefore it becomes possible to use it with less heating, the BCM-24 pipeline is clogged less.

Repeating this experiment with the inverter turned on, made it possible to stabilize the capacitor voltage at a lower level (see Fig. 5, b).

4 Conclusions

When treating the bitumen emulsion with ultrasound, its homogeneity increases (the residue on the sieve decreases by 25%). When interacting with stone material, the ultrasonicated bitumen emulsion disintegrates. When decaying, bitumen is distributed more evenly over the surface of the crushed stone by increasing the homogeneity of the emulsion. Consequently, during patchwork repair, when the stone material is treated with an ultrasonicated bitumen emulsion, a better formation of contact between crushed stone and emulsion occurs.

When the bitumen emulsion is treated with ultrasound, its storage stability is doubled. Therefore, its longer storage can be achieved. It is convenient for such areas as Veliky Ustyug, Vytegra and others, where there is no own base for the production of bitumen emulsion, because the costs of transporting bitumen emulsion are reduced.

The ultrasonicated bitumen emulsion EBK-2A has a good prospect for improving the quality of patchwork repair, since its viscosity decreases in 1.6 times. With a decrease in viscosity, the emulsion becomes more mobile, envelops the grains of stone material better, which is good for blast furnace or steel-making slags with their especially developed rough surface.

Taken together, all three characteristics of the bitumen emulsion (homogeneity, storage stability, relative viscosity) obtained as a result of ultrasonic treatment increase the quality of road patching by the jet injection method.

References

1. Appiah, J.K., Berko-Boateng, V.N., Tagbor, T.A.: Use of waste plastic materials for road construction in Ghana. Case Stud. Constr. Mater. **6**, 1 (2017)
2. Kamanga, M.J., Steyn, W.J.V.D.M.: J. S. Afr. Inst. Civ. Eng. **55**, 79 (2013)
3. Laurance, W.F., Clements, G.R., Sloan, S., O'Connell, C.S., Mueller, N.D., Goosem, M., Venter, O., Edwards, D.P., Phalan, B., Balmford, A., Van Der Ree, R., Arrea, I.B.: A global strategy for road building. Nature **513**, 229 (2014)
4. Huang, Y., Bird, R., Heidrich, O.: Development of a life cycle assessment tool for construction and maintenance of asphalt pavements. J. Clean. Prod. **17**, 283 (2009)
5. Kagan, G.L., Shorin, V.A., Velsovskij, A.Y.: E3S Web Conf. vol. 161, p. 1038 (2020)
6. Shorin, V.A., Kagan, G., Velesovskii, A.Y.: A new diagnostic instrument and method for stabilization of heaving soil in the beds of structures. Soil Mech. Found. Eng. **45**, 144 (2008)
7. Kagan, G.L., Mukhametova, L.R., Velsovskij, A.Y.: E3S Web Conf. vol. 178, p. 1064 (2020)
8. Vel'sovskij, A., Karpov, B., Smirnova, E.: Development of a new method for checking frost heave in roads. In: Proceedings of the Institution of Civil Engineers-Civil Engineering, vol. 168, p. 49 (2015)
9. Rokugo, K., Kunieda, M., Lim, S.C.: in ConMat'05, pp. 04–4680 (2005)
10. Chen, D.H., Zhou, W., Kun, L.: Fiber reinforced polymer patching binder for concrete pavement rehabilitation and repair. Constr. Build. Mater. **48**, 325 (2013)
11. Byzyka, J., Rahman, M., Chamberlain, D.A.: Thermal segregation of asphalt material in road repair. Traffic J. Transp. Eng. (Engl. Ed.) **4**, 360 (2017)
12. Nener-Plante, D.J., Zofka, A.: Long-term study on asphalt mixture segregation in connecticut: preliminary results on use of MTV. In: Proceedings of the 8th International Conference on the Bearing Capacity of Roads, Railways and Airfields, pp. 397–406 (2009)
13. Lebedev, O., Gavrilov, Y., Timkovskii, A.L., Bezrukova, A.G., Lezhnev, E.I., Polyanskii, V.A., Timofeev, A.N., Akhmetova, I.G.: Non-chemical water treatment in unified light and sound field and testing on waste water decontamination. In: IOP Conference Series: Earth and Environmental Science, vol. 337, p. 012074 (2019)
14. Smyatskaya, Y., Politaeva, N., Toumi, A., Olshanskaya, L.: in MATEC Web Conf. (2018)
15. Pilli, S., Bhunia, P., Yan, S., LeBlanc, R.J., Tyagi, R.D., Surampalli, R.Y.: Ultrasonic pretreatment of sludge: a review. Ultrason. Sonochem. **18**, 1 (2011)

16. Santos, H.M., Capelo, J.L.: Trends in ultrasonic-based equipment for analytical sample treatment. Talanta **73**, 795 (2007)
17. Soyuzdor NII, A.: Guide for the Preparation and Use of Bituminous Road Emulsions (to the Construction Norms & Regulations SNiP 3.06.03-85) (Stroyizdat, Moscow) (1989)
18. Budnik, V.A., Evdokimova, N.G., Zhirnov, B.S.: Oil Gas Business Electronic Science Journal 1 (2006)

Expertise and Testing of Pedestrian Overpasses

Valery Eremeev[2], Gennady Shmelev[1], Pavel Eremeev[1(✉)], Daniil Eremeev[1], and Arthur Ippolitov[1]

[1] Kazan State University of Architecture and Engineering, 1 Zelenaya Street, Kazan 420043, Russian Federation
pavil-66@inbox.ru

[2] JSC Spetsremproekt, 5 Zaslonova Street, Kazan 420097, Russian Federation

Abstract. Pedestrian overpasses traditionally have a low weight and high flexibility, and are subject to dynamic loading from the crowd of people, which makes it very likely, that resonant phenomena will occur. The article discusses the technical condition of three pedestrian overpasses built in the city of Kazan. Calculation models based on the finite element method were created. Dynamic characteristics of the structure, such as the natural vibration frequency and the logarithmic decrement of attenuation, were determined. Conclusions were made on the static and dynamic state of structures. The design was determined to meet the requirements for safety and comfort of pedestrians in both national and international documentation. 3 tests were carried out with loading from people. In the first case, during dynamic tests, the excitation of natural vibrations of the structure was performed by swinging by a group of 5 people, which was located in the middle of the span. In the second case, the excitation of natural vibrations of the structure was performed by swinging by a group of 40 people, which was located in the middle of the span. In the third case, the recording was performed without exciting vibrations. The vibrations were recorded using vertical and horizontal vibrosensors, the docking station, as well as the "Vibran-3" device installed in the middle of the span. The test results were compared with the calculated data. Based on the results of the work, a pedestrian overpass that is vulnerable to dynamic impacts was found and recommendations for further operation were given.

Keywords: Pedestrian overpasses · Dynamics · Oscillation · Dynamic loading · Construction · Structures

1 Introduction

Traditionally, the main materials that were used and continue to be used in the construction of overpasses are wood, stone, concrete and steel. Recently, polymer and composite materials have been increasingly used for bridge construction [1].

The improvement of pedestrian overpasses has led to a reduction in their weight and an increase in the length of spans, which creates the risk of getting the period of natural oscillation of structures in the range corresponding to the period of human step. This can

A. Sinitsyn (Ed.): SES 2020, LNCE 141, pp. 276–289, 2021.
https://doi.org/10.1007/978-3-030-67654-4_31

lead to a decrease in the dynamic comfort of pedestrians, as well as collapse. Because of this, the issues of vibration damping are actively investigated [2–4].

There are many studies devoted to finding the optimal solution for setting the load from the group, taking into account its dynamic influence [5–8].

In Russian standards, the calculation of structures for dynamic effects is carried out, taking them as quasi-static, that is, the static component of the load is multiplied by the coefficient of dynamism. And in international standards, such a load is expressed in the form of Fourier series, taking into account the dynamic characteristics of structures and the impact itself. Special attention is paid to preventing resonance at transitions with low horizontal and vertical natural frequencies.

The dynamic state of pedestrian overpasses is usually determined by two criteria – the period of natural vibrations of structures, the limitation of which makes it possible to avoid resonant phenomena, and the acceleration of structural changes, which ensures pedestrian comfort [9–11].

According to clause 5.48 of SP 35.13330.2011 Bridges and pipes [12], the periods of self-oscillation of the pedestrian bridge in the vertical plane for the two lowest forms should not fall within the range of 0.45–0.6 s., while the load from the crowd is set by a load of 0.5 kPa.

Permissible values of acceleration of vibrations of pedestrian bridges, at which people do not experience significant discomfort, can be found in foreign literature.

For example, Fig. 1, 2 show the permissible values of acceleration of vibrations of structures in the vertical and horizontal directions depending on the frequency of vibrations according to various standards and studies. The figures were taken from the article [13].

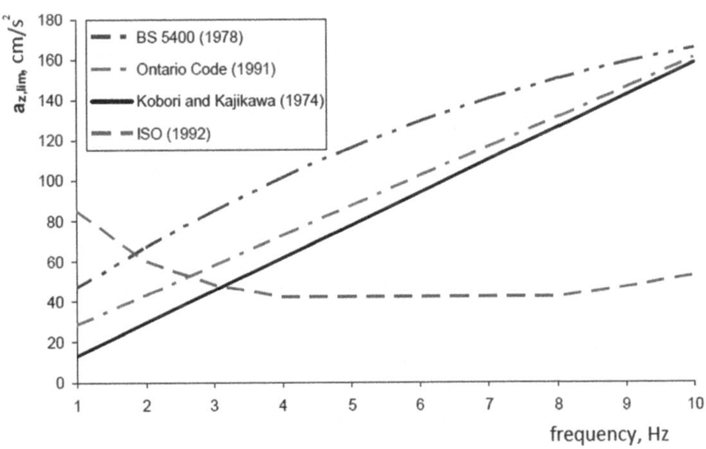

Fig. 1. Permissible values of vertical oscillation accelerations

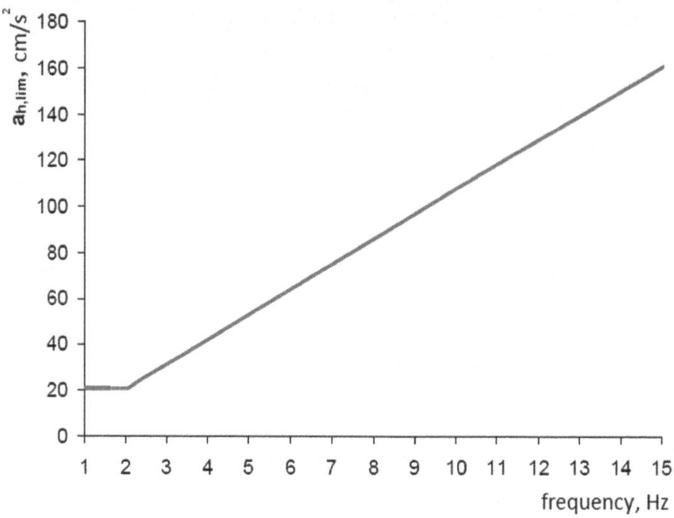

Fig. 2. Permissible values of horizontal oscillation accelerations

According to the international standards ISO 10137:2007 (E) [14], the limit value of the acceleration of oscillation is determined by the basis curve (see Fig. 3), multiplying the value obtained from this graph by a multiplier equal to 60 for pedestrian overpasses.

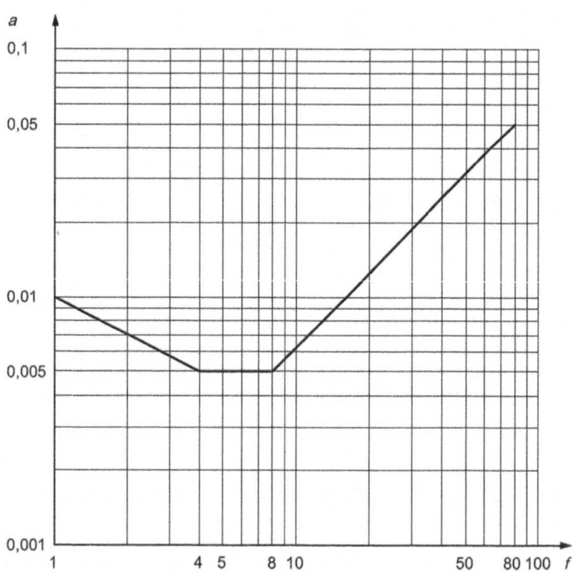

Fig. 3. Graph of the dependence of the mean-square acceleration a (m/s^2) on the frequency in Hz.

2 Methods

2.1 Static Calculation

The article considers 3 pedestrian overpasses, with spans of 20, 30 and 44.6 m, built in the city of Kazan. Calculation schemes including one-dimensional and two-dimensional finite elements were made (Figs. 4, 5 and 6).

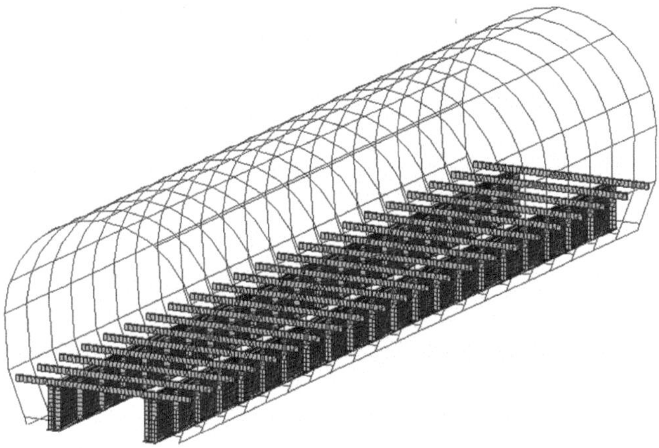

Fig. 4. The design scheme of the overpass (span of 20 m).

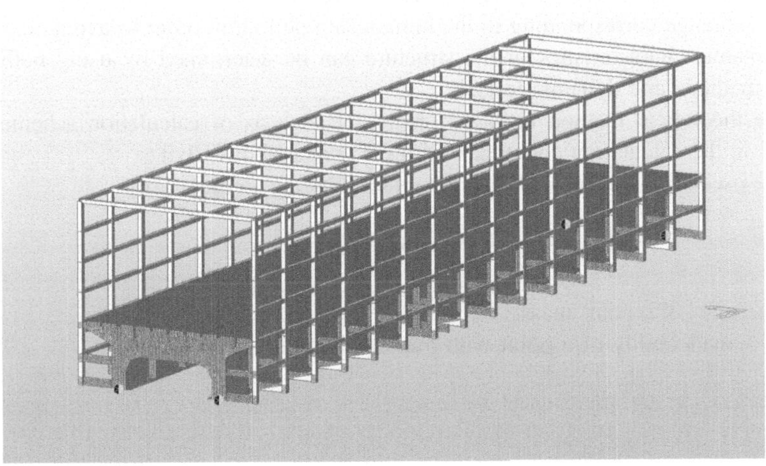

Fig. 5. The design scheme of the overpass (span of 30 m).

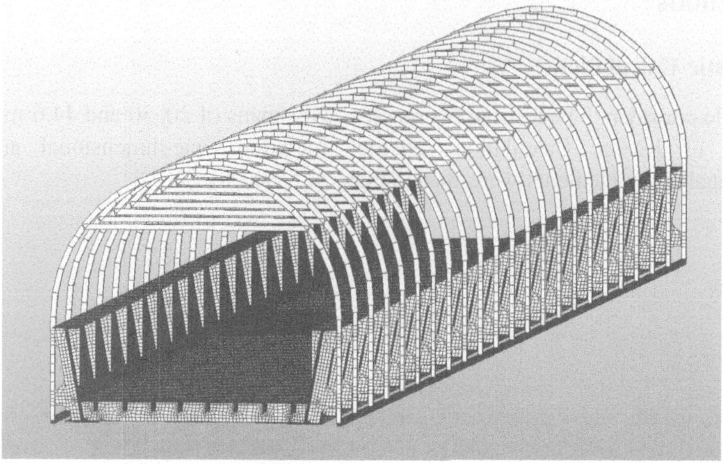

Fig. 6. The design scheme of the overpass (span of 44.6 m).

In the design scheme, the following loads were performed: own weight, weight of enclosing structures, snow load, pedestrian load, wind load. The net weight was calculated automatically by the virtual system, based on the assigned materials and cross sections. The remaining loads were set in accordance with Russian standards [15].

2.2 Determination of Dynamic Characteristics

According to the normative documents, the natural oscillation period should not fall within the range corresponding to the human step period, in order to avoid resonance. The dynamic characteristics of the structure can be determined by using both theoretical methods and dynamic tests.

The theoretical method is based on modal analysis of calculation schemes performed in the calculation complexes Lira CAD and Midas Civil.

The oscillation period was also determined by the formula [16]:

$$T = 2\pi\sqrt{m\delta_{11}} \tag{1}$$

Where m - the point mass;
δ_{11} – malleability of a point with mass in a single state (Fig. 7).

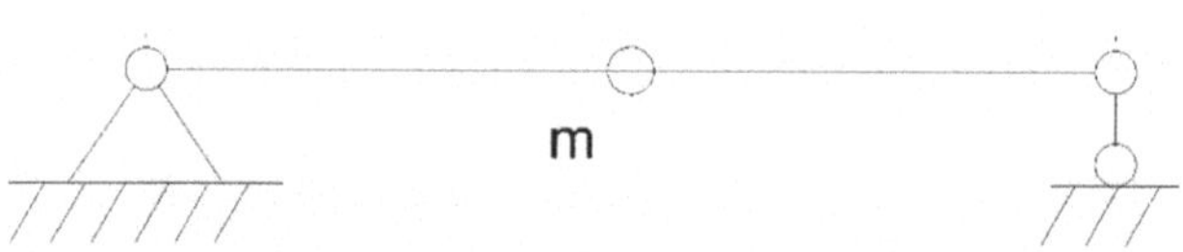

Fig. 7. The design dynamic model

Dynamic testing consists in measuring oscillations with vibration sensors. Questions arise when determining the state in which it is necessary to conduct the measurements. Therefore, within the framework of this article, the initiation of the investigation was carried out in different ways.

In the first case, for the overpass of a span of 44.6 m, the excitation of natural oscillations of the structure was developed by swinging of a group of 5 people, which was located in the middle of the span. Then the swinging stopped and measurements of oscillation damping were made.

In the second case, for the overpass of a span of 20 m, the excitation of natural oscillations of the structure was developed by swinging of a group of 40 people, which was located in the middle of the span. Then the swinging stopped and measurements of oscillation damping were made.

In the third case, the overpass of a span of 30 m, the recording was performed without exciting oscillations.

The vibrations were recorded using vertical and horizontal vibrosensors "VEG" and the Ni SCXI-1000 docking station with the Ni 1317 module, as well as the device "Vibran-3" installed in the middle of the span of the pedestrian bridge (Figs. 8, 9 and 10).

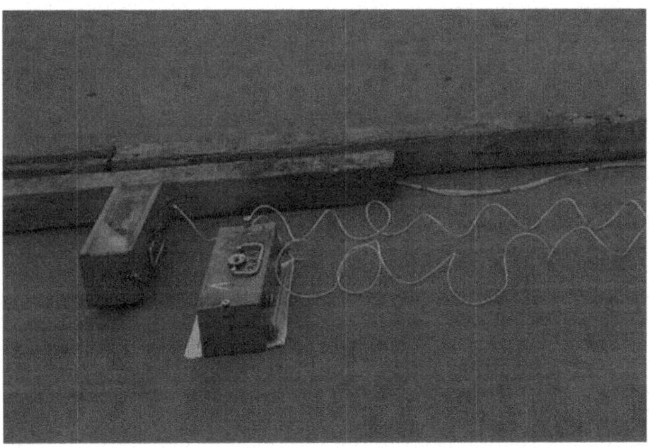

Fig. 8. General view of vertical and horizontal vibration sensors

Fig. 9. General view of the Ni SCXI-1000 docking station with the Ni 1317 module

Fig. 10. Vibration measurements with the "Vibran-3" device

3 Results and Discussion

3.1 The Results of Static Calculation

The calculation results are presented in graphical form. The stress values in the plate elements are equivalent (according to von Mises) (Figs. 11, 12, 13, 14, 15 and 16).

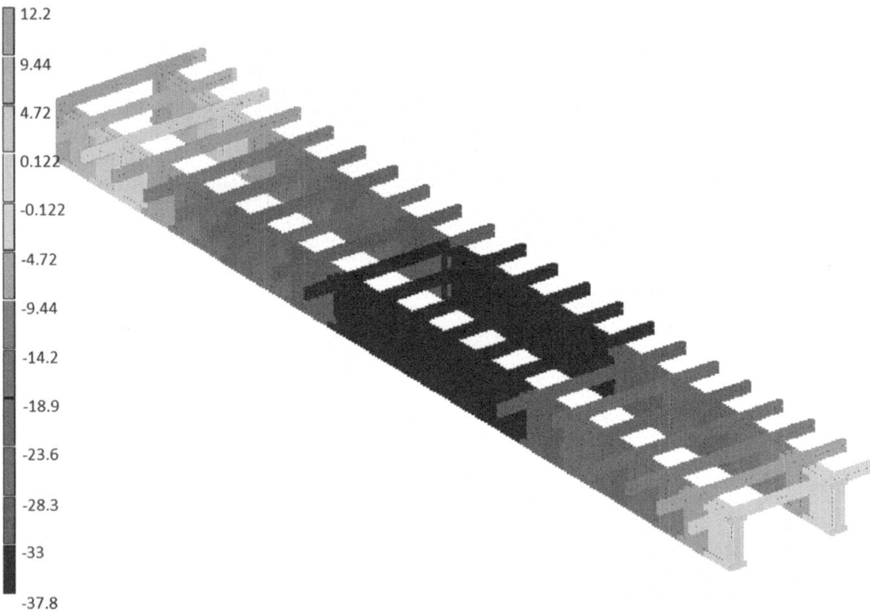

Fig. 11. Vertical movement from the action of the standard load on the overpass (span = 20 m), mm

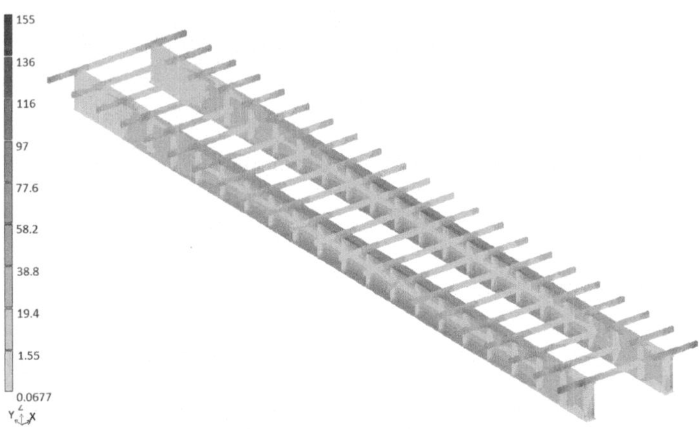

Fig. 12. Isofield of stresses from the action of a combination of design loads on the overpass (span = 20 m), MPa

Fig. 13. Vertical movement from the action of the standard load on the overpass (span = 30 m), m

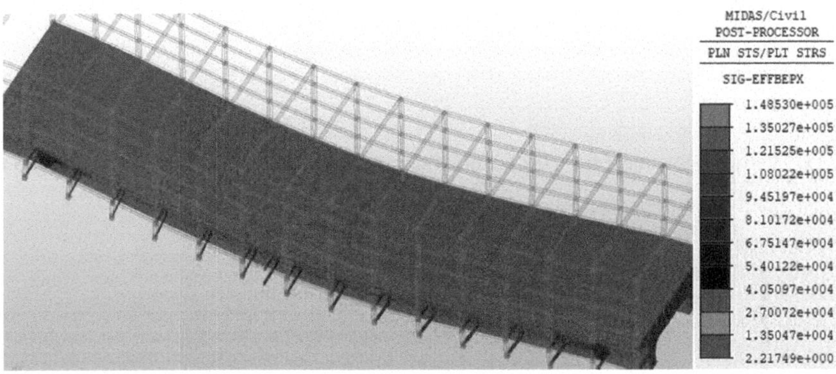

Fig. 14. Isofield of stresses from the action of a combination of design loads on the overpass (span = 20 m), kPa

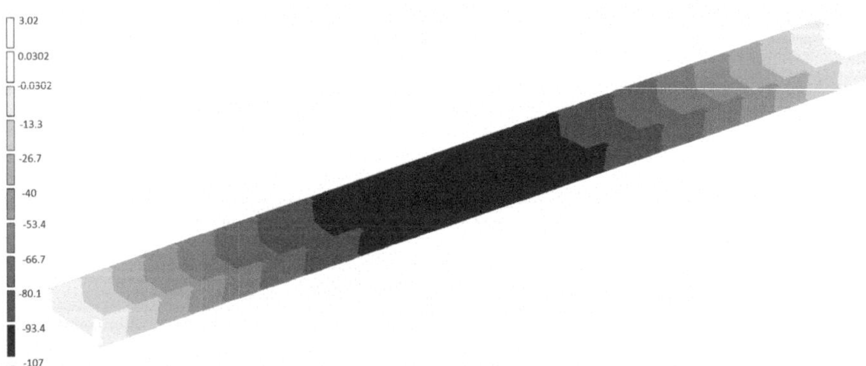

Fig. 15. Vertical movement from the action of the standard load on the overpass (span = 44.6 m), mm

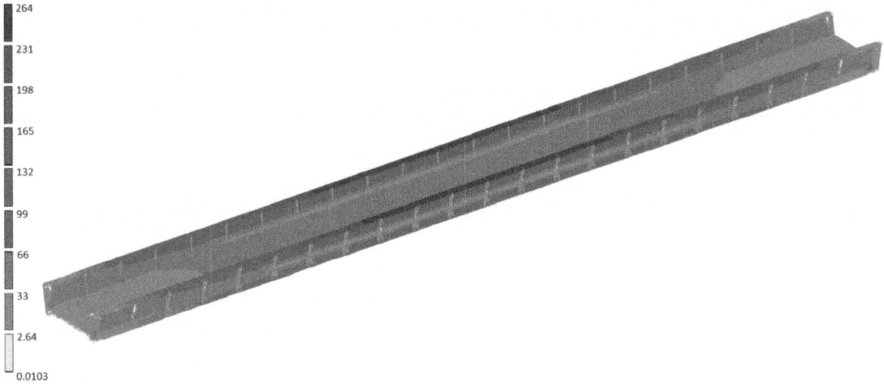

Fig. 16. Isofield of stresses from the action of a combination of design loads on the overpass (span = 44.6 m), MPa

Based on the results of static calculation, it is determined that all 3 overpasses meet the requirements of Russian regulatory documents for groups 1 and 2 of limit states.

3.2 Results of Numerical Studies and Dynamic Tests to Determine Dynamic Characteristics

The results of dynamic tests are represented by oscillation diagrams obtained in single-channel and four-channel modes (Figs. 17, 18, 19, 20, 21 and 22).

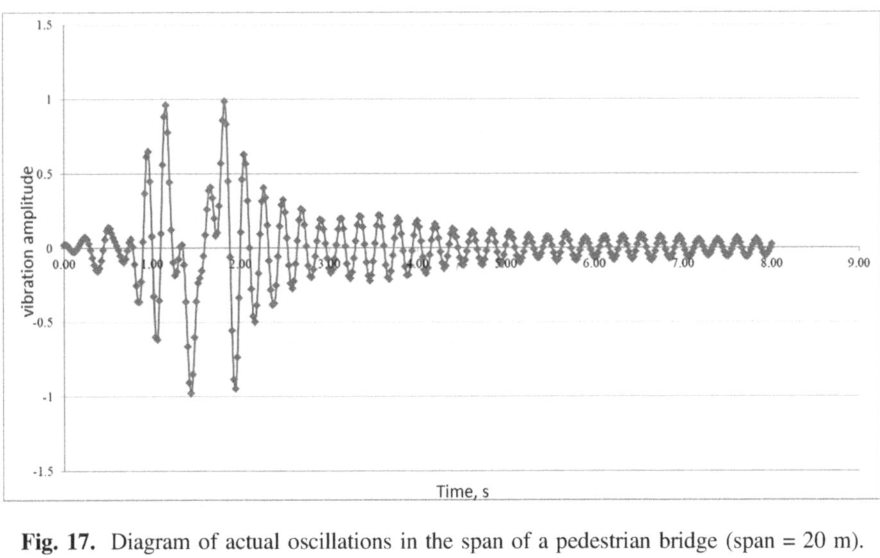

Fig. 17. Diagram of actual oscillations in the span of a pedestrian bridge (span = 20 m).

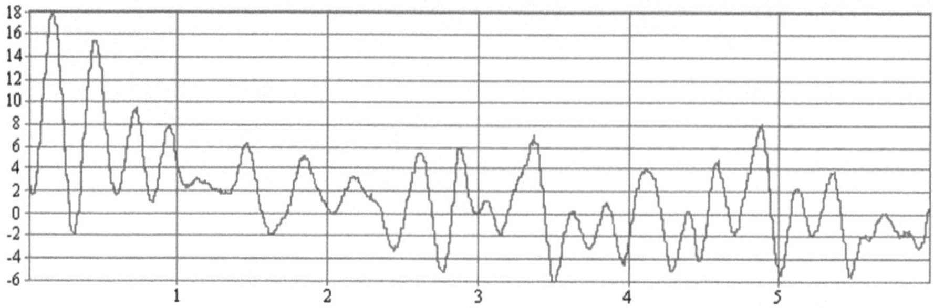

Fig. 18. Diagram of actual oscillations in the span of a pedestrian bridge (span = 30 m, the first channel).

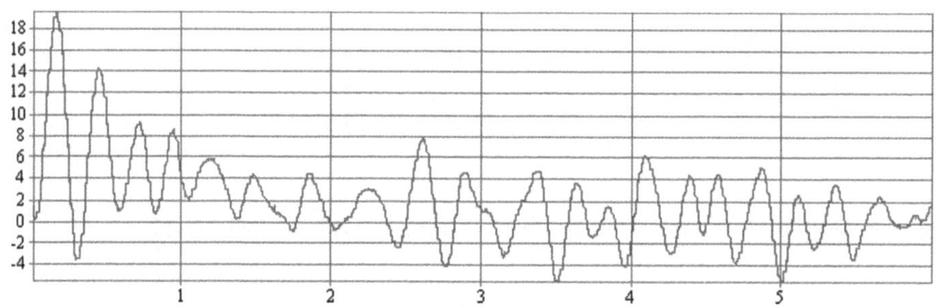

Fig. 19. Diagram of actual oscillations in the span of a pedestrian bridge (span = 30 m, the second channel).

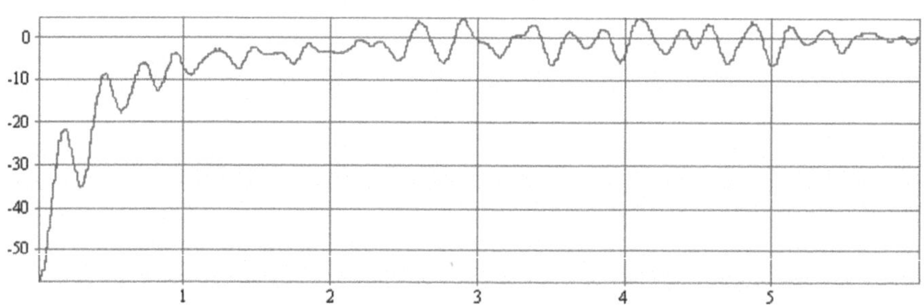

Fig. 20. Diagram of actual oscillations in the span of a pedestrian bridge (span = 30 m, the third channel).

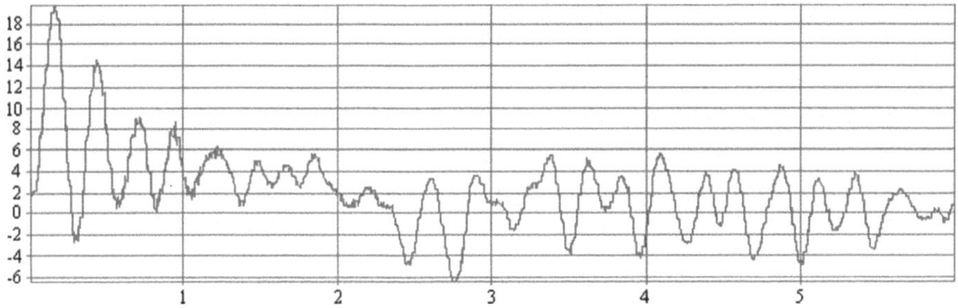

Fig. 21. Diagram of actual oscillations in the span of a pedestrian bridge (span = 30 m, the fourth channel).

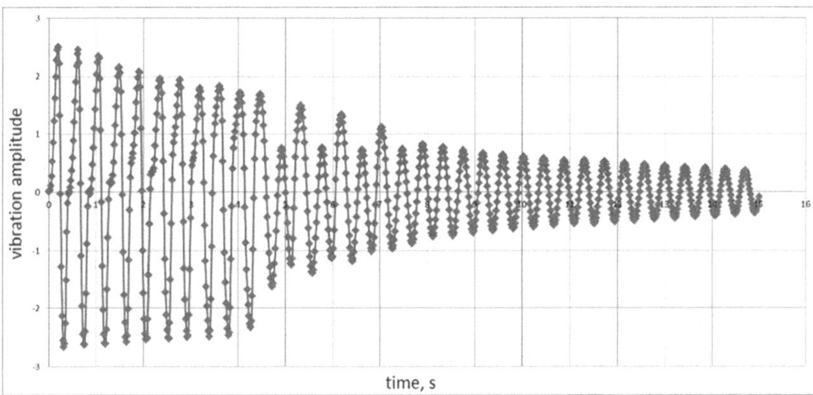

Fig. 22. Diagram of oscillations in the span of a pedestrian bridge (span = 44.6 m).

The actual frequency of oscillations is calculated from 10 oscillations using the formula:

$$\vartheta = \frac{N}{t},$$

where N – number of oscillations;

t – the time during which the span structure makes N oscillations.

The calculation results are presented in table form (Table 1).

Table 1. Results of determining the period of natural oscillations

№	The span of the overpass, m	The period of oscillations in the 1st form, s	The period of oscillations in the 2nd form, s	The period of oscillations obtained by the calculation method, s	The actual period of the oscillations, s
1	20	0,214	0,185	0,178	0,21
2	30	0,3	0,22	0,294	0,3125
3	44,6	0,42	0,369	0,436	0,424

All three pedestrian overpasses have confirmed compliance with the requirements of the regulatory documentation [12] for non-falling in the range of 0.45–0.6 s, but the overpass with a span of 44.6 m can be considered dangerous.

Tests and calculations, according to SP 35.13330.2011 Bridges and pipes, were carried out in an unloaded state, when the additional mass increases the period of natural oscillations of this structure. Also, this transition has a relatively high flexibility, which creates risks of the phenomenon of unconscious synchronization [17–19], when a person walking along a vibrating structure adjusts the frequency of his step to the vibrations. Thus, seemingly harmless vibrations can develop into resonant ones, which is unacceptable. The situation is complicated by the absence of a fundamental theory [20–22] describing the interaction of two dynamic systems such as a person and the structure excited by it. Although in most models the load from a pedestrian is considered as an external force, the movement of the structure can actually affect human behavior, thereby changing the forces of dynamic interaction [23–26].

4 Conclusions

Pedestrian overpass structures that include glazing must be carefully checked for maximum movement, as there are risks of damage to it. There are known cases when the glazing "flew" out of the design position due to significant deformations.

The numerical method for determining the period of natural oscillations showed high convergence with experimental results, less than 5%. Whereas, the calculated method showed low convergence, more than 15%. According to the results of the study, it is concluded that it is recommended to determine the period of natural oscillations by means of modal analysis in virtual complexes.

The pedestrian overpasses considered in the article meet the normative requirements [12], but for the reasons described earlier, it is recommended to carry out vibration damping measures for overpass with a span of 44.6 m.

References

1. Ovchinnikov, I.: Pedestrian bridges: design, construction, architecture. Saratov state technical University, Saratov (2005)
2. Bedon, C.: Diagnostic analysis and dynamic identification of a glass suspension footbridge via on-site vibration experiments and FE numerical modelling. Compos. Struct. **216**, 366–378 (2019)
3. Chróścielewski, J.: Modal properties identification of a novel sandwich footbridge. Comparison of measured dynamic response and FEA. Compos. Part B Eng. **151**, 245–255 (2018)
4. Bedon, C., Bergamo, E.: Vibration experiments for diagnostic investigations on a glass suspension footbridge. Vibroengineering Procedia **24**, 41–46 (2019)
5. Venuti, F., Racic, V., Corbetta, A.: Modelling framework for dynamic interaction between multiple pedestrians and vertical vibrations of footbridges. J. Sound Vib. **379**, 245–263 (2016)

6. Lai, E., Gentile, C., Mulas, M.G.: Experimental and numerical serviceability assessment of a steel suspension footbridge. J. Constr. Steel Res. **132**(5), 16–28 (2017)
7. Demartino, C., Avossa, A.M., Ricciardelli, F.: Deterministic and probabilistic serviceability assessment of footbridge vibrations due to a single walker crossing. Shock Vib. **1**, 1–26 (2018)
8. Tubino, F.: Human-structure interaction in pedestrian bridges: a probabilistic approach. Procedia Eng. **199**, 2883–2888 (2017)
9. Qin, S., Zhou, Y.L., Kang, J.: Footbridge serviceability analysis: from system identification to tuned mass damper implementation. KSCE J. Civ. Eng. **23**(2), 754–762 (2019)
10. Wang, D., Wu, C., Zhang, Y., Li, S.: Study on vertical vibration control of long-span steel footbridge with tuned mass dampers under pedestrian excitation. J. Constr. Steel Res. **154**, 84–98 (2019)
11. Miguel, L.F.F., Fadel Miguel, L.F., Lopez, R.H.: A firefly algorithm for the design of force and placement of friction dampers for control of man-induced vibrations in footbridges. Optim. Eng. **16**(3), 633–661 (2015)
12. SP 35.13330.2011 "Bridges and pipes. Updated version of SNiP 2.05.03-84* (with change №1)" Russia (2011)
13. Nazarov, U., Symbirkin V.: Analysis and limitation of vibrations of structures under the influence of people. J. Cent. Res. Inst. V. A. Kucherenko. 10–18 (2009). Research on the theory of structures
14. ISO 10137:2007(E) International standard, Bases for design of structures — Serviceability of buildings and walkways against vibrations (2007)
15. SP 20.13330.2016 Loads and impacts. Updated version of SNiP 2.01.07-85* Russia (2016)
16. Shakirzyanov, R.: Dynamics and stability of structures. KSUAE, Kazan (2015)
17. Bachmann, H., Ammann, W.: Vibrations in structures induced by man and machines. Structural Engineering Document 3e, International Association for Bridge and Structural Engineering, Zürich, 2(A) (1987)
18. Newland, D.E.: Pedestrian excitation of bridges. Proc. Inst. Mech. Eng. Part C J. Mech. Eng. Sci. **218**(5), 477–492 (2004)
19. Fujino, Y., Pacheco, B.M., Nakamura, S.-I., Warnitchai, P.: Synchronization of human walking observed during lateral vibration of a congested pedestrian bridge. Earthquake Eng. Struct. Dyn. **22**, 741–758 (1993)
20. Bocian, M., Macdonald, J.H.G., Burn, J.F.: Biomechanically inspired modeling of pedestrian-induced vertical self-excited forces. J. Bridg. Eng. **18**(12), 1336–1346 (2013)
21. Carroll, S.P., Owen, J.S., Hussein, M.F.M.: Modelling crowd-bridge dynamic interaction with a discretely defined crowd. J. Sound Vib. **331**(11), 2685–2709 (2012)
22. Qin, J.W., Law, S.S., Yang, Q.S., Yang, N.: Pedestrian-bridge dynamic interaction, including human participation. J. Sound Vib. **332**(4), 1107–1124 (2013)
23. Hu, J., Bian, X., Jiang, J.: Critical velocity of high-speed train running on soft soil and induced dynamic soil response. Procedia Eng. **143**, 1034–1042 (2016)
24. Ntotsios, E.: Modelling of train induced vibration. In: Stephenson Conference Research for Railways 2015, April 2015, pp. 153–162 (2015)
25. Ahmadi, E., Caprani, C., Živanović, S., Heidarpour, A.: Vertical ground reaction forces on rigid and vibrating surfaces for vibration serviceability assessment of structures. Eng. Struct. **172**, 723–738 (2018)
26. Van Nimmen, K., et al.: Identification and modelling of vertical human-structure interaction. Conference Proceedings of the Society for Experimental Mechanics Series, vol. 2, pp. 319–330 (2015)

Investigation of the Recovery Mode with Different Transformation Coefficients of the Autotransformer Reverse Rectifier and Harmonic Components of the Network with an Optimally Selected Transformation Coefficient

Tatyana Sinyukova[1(✉)], Alexey Sinyukov[1], Vladimir Dotsenko[1], Alfred Safin[2], and Viktor Meshcheryakov[1]

[1] Lipetsk State Technical University, 30 Moskovskaya Street, Lipetsk, Russia
stw0411@maik.ru
[2] Kazan State Power Engineering University, Str. Krasnoselskaya, 51, 420066 Kazan, Russia

Abstract. The loading machine is designed for feeding slabs coming through the loading roller to the methodic furnace. Slabs are loaded by walking beams driven by a hydraulic system, the beams themselves are located on the carriage moving by means of an asynchronous electric drive. In this work the study of the processes of start-up, braking performance of the engine carriage of the machine is a furnace room, and three roller machine loading – K5 shop hot rolling public joint stock company "Novolipetsk steel". Special attention is paid to the recovery mode in the electric drive. The analysis of other methods of limiting overvoltage is made. The recovery mode with different transformation coefficient of the reverse rectifier autotransformer is investigated. The analysis of harmonic components is made. The optimal mode for limiting overvoltage in the mode of energy recovery in the network is determined. The transformation coefficient of the reverse rectifier autotransformer is selected for maximum use of the regenerative properties of the "Asynchronous motor – frequency Converter – Rectifier – Network – link.

Keywords: Transformation coefficient · Asynchronous motor · Frequency converter

1 Basic Information About the Object

Many studies pay active attention to the study of industrial facilities and energy saving modes [1–21].

Step furnaces of mill 2000 of Novolipetsk metallurgical combine are the largest furnaces of this type and are designed for heating up to 1250 °C slabs with a thickness of 150..300 mm, a width of 700…1850 mm, a length of 4500…10500 mm. The productivity of the furnace at a cold setting is 260 t/h, at a hot setting-350 t/h.

A. Sinitsyn (Ed.): SES 2020, LNCE 141, pp. 290–299, 2021.
https://doi.org/10.1007/978-3-030-67654-4_32

2 Description of the RRU-Based Drive System Model

2.1 Complete Model Structure and Accepted Assumptions

Figure 1 shows a model of the complete auxiliary drive system for furnace No. 3. Recuperative rectifier on thyristors, RRU control system adopted taking into account the delay for opening thyristors to avoid bridge short circuit, autotransformers selected taking into account the properties of the rectifier to avoid its overturning during overvoltages and drawdowns, roller roller inverter and loading machine, vector and scalar control systems adopted for analysis and comparison based on operating conditions, vector control system is based on a relay current controller, scalar control system is built taking into account the flow controller.

The designations used in Fig. 1 are: 1 – autotransformers (1) of the forward and (2) reverse rectifier bridge; 2.1 – straight (upper) rectifier bridge; 2.2-reverse (lower) rectifier bridge; 3 – RRU type control system for rectifier bridges; 4.1 – loading machine inverter; 4.2 – roller group inverter; 5.1 – vector control system for the loading machine; 5.2 – scalar control system for the roller group; 6.1 – asynchronous motor of the loading machine carriage; 6.2 – asynchronous roller roller motor with total parameters of 8 real drives (to simplify the model); 7 – block of the sinusoidal voltage source.

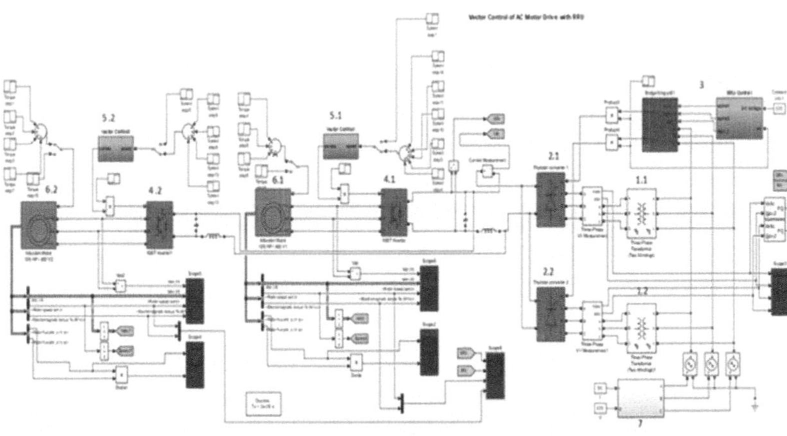

Fig. 1. Complete model of the drive system.

2.2 Analysis of Engine Start-Up and Braking Schedules at Idle in the RRU System

These graphs, shown in Fig. 2, 3 and 4, show the characteristics of the network, rectifier and motor, which show the occurrence of overvoltage at the time of regenerative braking working out of current and voltage regulators for a given perturbation. When the recuperative (reverse) bridge is connected, the voltage in the DC circuit increases to 650 V.

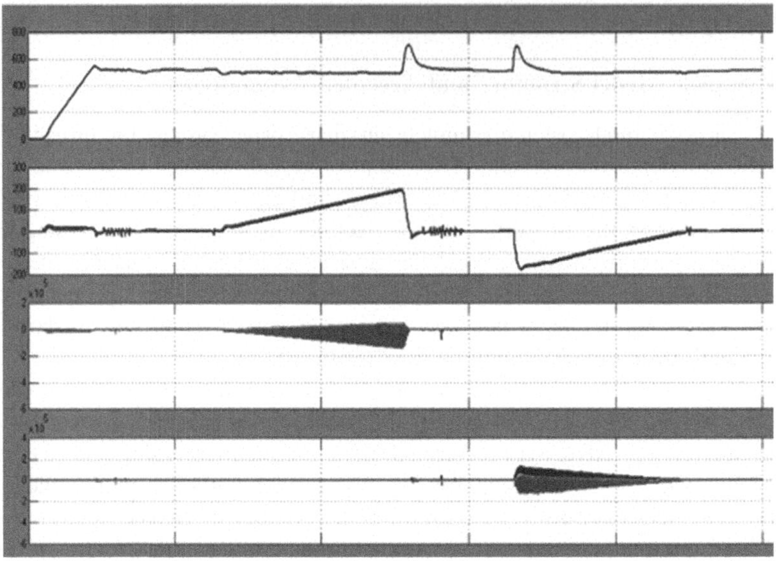

Fig. 2. Voltage and current of the DC link, power consumed and given to the network: graph 1 (axes: x-time; y-constant voltage); graph 2 (axes: x-time; y-current); graphs 3, 4 (axes: x-time; y-power (active power is indicated in blue, reactive power is indicated in red)).

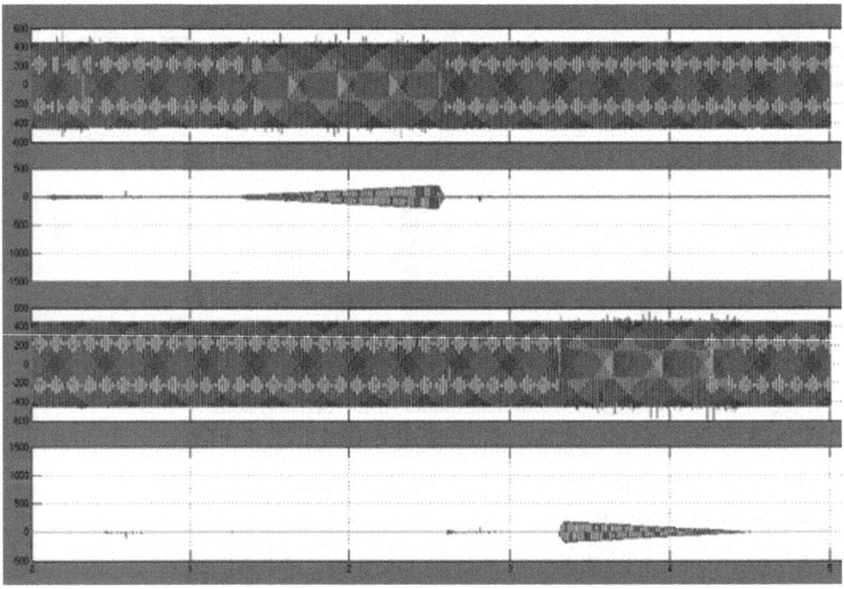

Fig. 3. The voltage and current of the network (in front of the master rectifier output rectifier): the graphs 1 and 3 on the x-axis – time, y – axis AC voltage; graphs 2 and 4 on the x-axis – time, y – axis current.

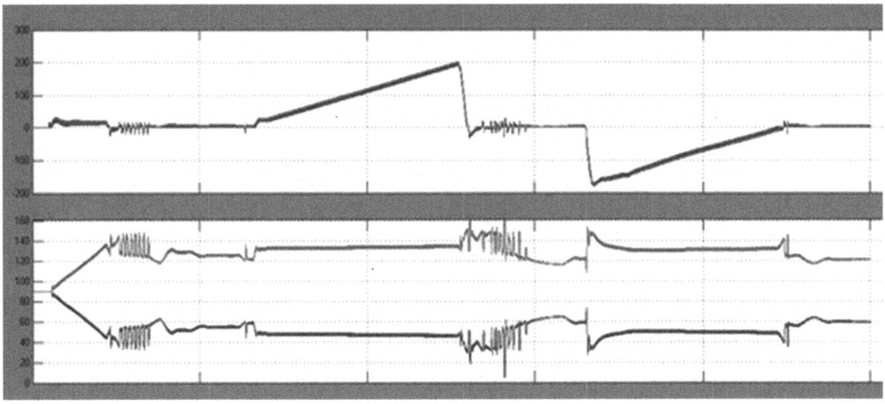

Fig. 4. A The operation of the current controller and the value of the opening angle on the master and slave bridges: graph 1 x-axis – time, y axis – current; graph 2 x-axis – time, y-axis – the angle in degrees.

Based on these graphs, we can conclude that the timing of testing the voltage regulator.

3 Selection of the Transformation Coefficient to Optimize the Return of Energy to the Network

3.1 Analysis of Used Autotransformers

Simence uses step-up autotransformers to optimize the power output to the grid and protect the rectifier from over-voltage or drawdown. The transformation coefficient of these transformers ranges from 0.5 to 0.85, depending on the characteristics of the drives and the network. Figures 5, 6, 7 and 8 show graphs of the dependence of the power given to the network and the current in the DC circuit at different transformation coefficients.

In Figs. 5, 6, 7 and 8: graph 1 x-axis – time, y-axis – DC voltage; Fig. 2 x-axis – time, y axis – current; graphs 3 and 4 on the x-axis – time, y-axis – power (blue color denotes the active power, reactive red power).

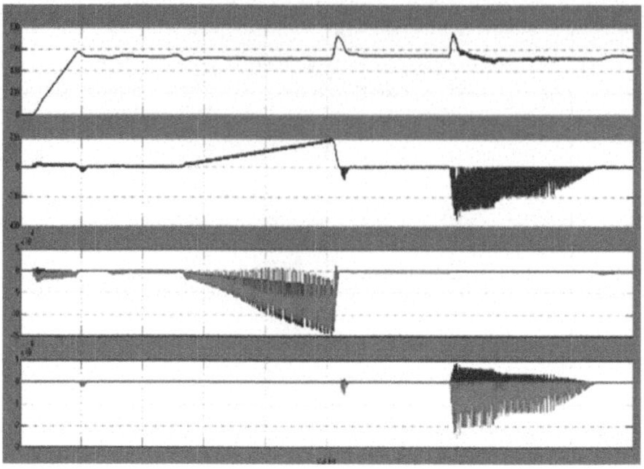

Fig. 5. Network characteristics with a coefficient of 0.08.

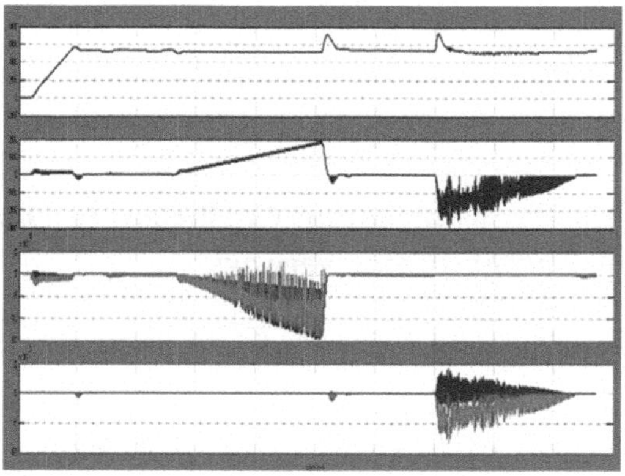

Fig. 6. Network characteristics with a coefficient of 0.2.

Based on these graphs, we can conclude that with a decrease in the transformation coefficient, the amount of energy given off increases, but its quality significantly deteriorates, as well as fluctuations in the DC voltage circuit occur.

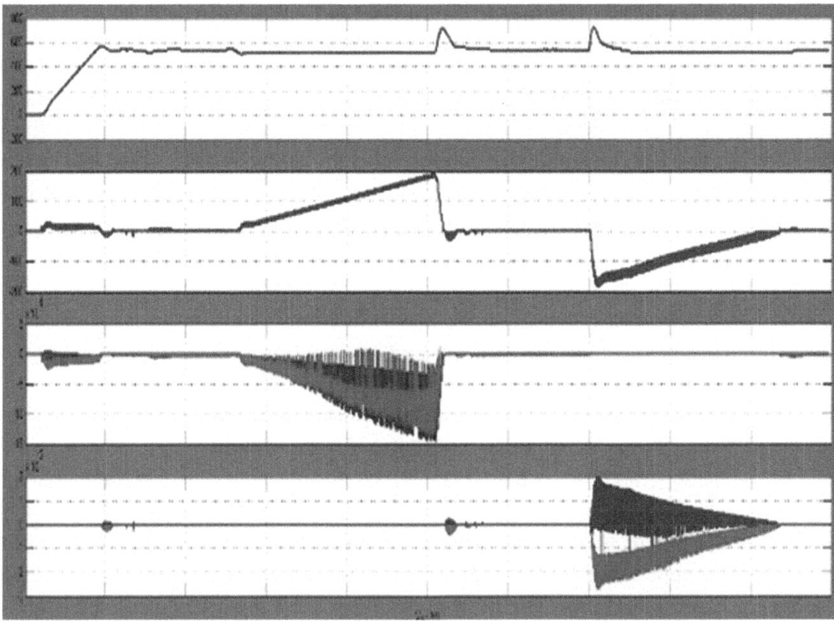

Fig. 7. Network characteristics with a coefficient of 0.5.

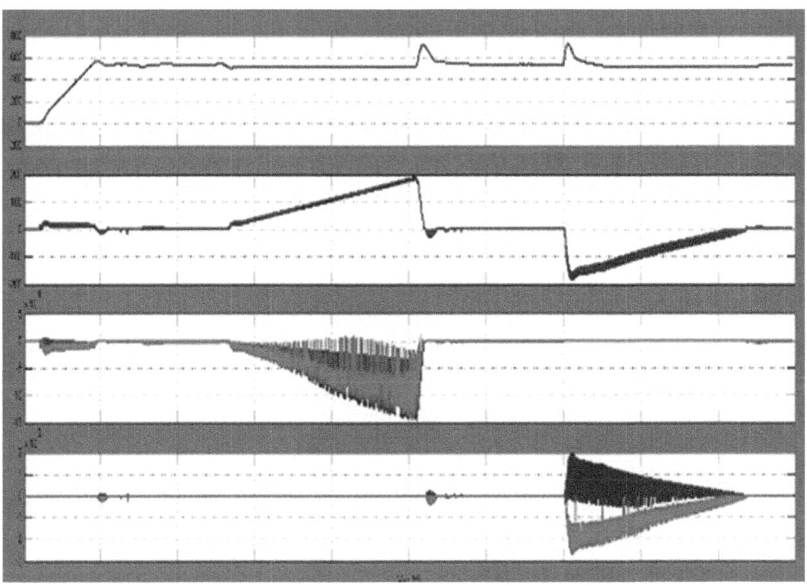

Fig. 8. Network characteristics at a coefficient of 0.85.

3.2 Investigation of the Harmonic Components of the Network with an Optimally Selected Transformation Coefficient

When using a six-pulse circuit, the key harmonics will be the fifth harmonic and the seventh harmonic. They will have a predominant influence on the network. Getting rid of these low-frequency harmonics is possible only by using twelve pulse and twenty-four pulse circuits. Reducing the influence of these harmonics on the network is the task of filters, but it is worth considering that the installation of filters must be carried out in pairs, the filter of the fifth harmonic must be a filter of the seventh harmonic. The harmonic composition of a single system can include harmonics of higher orders, and their total percentage can reach 30–35% in installations with a large number of converters. The main thing is that the total number of harmonics on the power transformer does not exceed 3–5%.

Figures 9 and 10 show an analysis of the harmonic components. Figure 11 shows the shape of voltages and currents in different operating modes.

Based on these graphs shown in Figs. 9, 10 and 11, we can conclude that it is necessary to reduce the harmonics of high orders, as well as the use of high-frequency harmonics filters. There is no need to use twelve-pulse and twenty-four pulse circuits due to the percentage of harmonics that does not exceed the maximum. The necessary and sufficient condition is met. The need to use sine filters must be calculated based on the total number of harmonics on the supply transformer.

Data harmonic composition, the total percentage of harmonics and the magnitude of the output energy at different transformation ratios:

- if the transformation ratio of 0,08 magnitude of the output of energy equal to 1000000, with significant harmonics 5, 7, 11, 13, and the percentage of influence of harmonics on the network is 45%;
- with a transformation coefficient of 0.2, the order of magnitude of the energy output is 500,000, with significant harmonics of 5, 7, 11, 13, and the percentage of harmonics affecting the network is 35%;

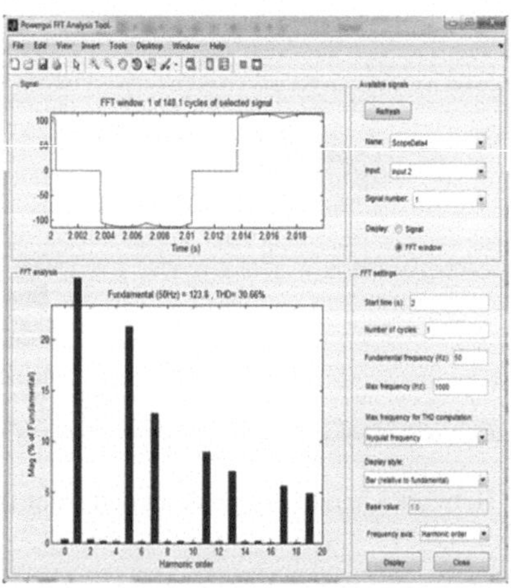

Fig. 9. Analysis of harmonic components behind the lead rectifier.

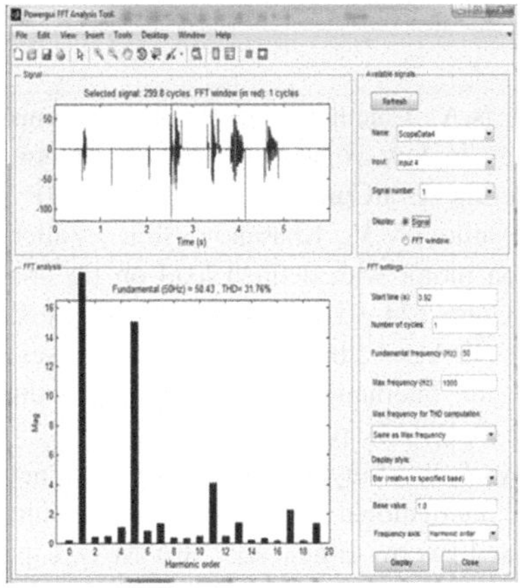

Fig. 10. Analysis of the harmonic components for the output rectifier.

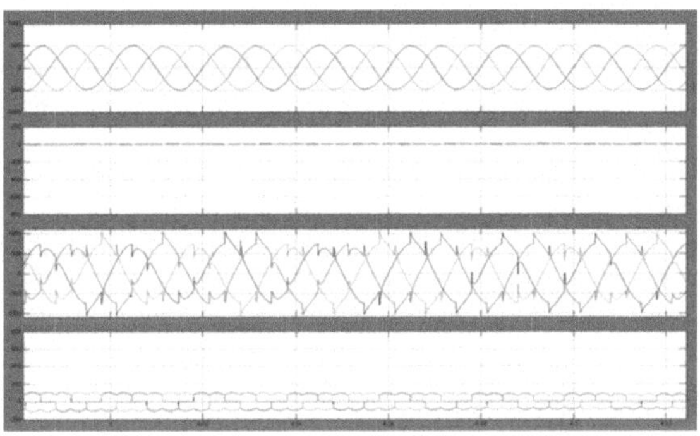

Fig. 11. Form of voltages and currents at idle and under load.

- with a transformation coefficient of 0.4, the order of magnitude of the energy output is 200,000, with significant harmonics of 5, 7, 13, 19, and the percentage of harmonics affecting the network is 32%;
- with a transformation coefficient of 0.5, the order of magnitude of the energy output is equal to 150,000, while the significant harmonics are 5, 7, 13, 19, and the percentage of harmonics ' influence on the network is 31.8%;
- with a transformation coefficient of 0.8, the order of magnitude of the energy output is 75,000, with significant harmonics of 5, 7, 13, 21, and the percentage of harmonics affecting the network is 28.7%).

References

1. Gracheva, E.I., Alimova, N.A.: Calculating methods and comparative analysis of losses of active and electric energy in low voltage devices. In: International Ural Conference on Electrical Power Engineering (UralCon), pp. 361–367 (2019)
2. Meshcheryakov, V.N., Danilov, V.V., Khasanov, Sh.R., Valtchev, S.: Minimization of the stator current in induction motor with defined load on the shaft by maintaining optimum absolute slip. E3S Web Conf. **124**, 01036 (2019). Kazan, SES 2019
3. Gracheva, E.I., Naumov, O.V.: Estimation of power losses in electric devices of the electrotechnical complex. In: International Conference on Industrial Engineering. Applications and Manufacturing (ICIEAM), p. 6 (2019)
4. Sinykova, T.V., Sentsov, E.V., Sinyukov, A.V.: Neural network speed observers. In: Proceedings 2019 1st International Conference on Control Systems, Mathematical Modelling, Automation and Energy Efficiency (SUMMA), Lipetsk, Russia, p. 288 (2019)
5. Pugachev, A.: Efficiency increasing of induction motor scalar control systems. In: International Conference on Industrial Engineering, Applications and Manufacturing (ICIEM), St. Petersburg, Russia, 16–19 May 2017, pp. 1–5 (2017)
6. Thien, R.T.Y., Kim, Y.: Decentralized formation flight via PID and integral sliding mode control. Aerosp. Sci. Technol. **81**, 322–332 (2018)
7. Hu, Z., Bodyanskiy, Y.V., Tyshchenko, O.K.: A Multidimensional adaptive growing neuro-fuzzy system and its online learning procedure. Advances in Intelligent System and Computing, vol. 689, pp. 186–203 (2018)
8. Kukishev, D.Yu., Meshcheryakov, V.N., Voyekov, V.N., Ivshin, I.V.: Asynchronous electric drive with an additional switch in the DC link of the frequency converter. In: High Speed Turbomachines and Electrical Dreves Conference 2020 (HSTED 2020), Prague, Czech Republic, 09 July 2020, p. 01043 (2020)
9. Sinykova, T.V., Gladyshev, V.E., Sinyukov, A.V.: Methods for reducing electromechanical oscillations in conveyor control systems. In: Proceedings 2019 1st International Conference on Control Systems, Mathematical Modelling, Automation and Energy Efficiency (SUMMA), Lipetsk, Russia, p. 435 (2019)
10. Fedotov, E.A., Gracheva, E.I., Fedotov, A.I., Chernova, N.V.: Local fourier transformation application for mathematic modeling of synchronous machine valve actuator. J. Eng. Appl. Sci. **11**(13), 2939–2945 (2016)
11. Fiser, J., Sitek, P., Vyhlidal, T.: Neutral PID control loop investigated in terms of similarity theory. IFAC-Papers Online **51**(14), 212–217 (2018)
12. Fiser, J., Sitek, P., Vyhlidal, T.: Robust PID controller design for plants with delay using similarity theory. IFAC-Papers Online **51**(14), 236–241 (2018)
13. Coelho, L.D.S., Corlho, A.A.R.: Automatic tuning of PID and gain scheduling PID controllers by a derandomized evolution strategy. AI EDEM **13**(5), 341–349 (1999)
14. Meshcheryakov, V., Sinykova, T., Sinyukov, A., Boikov, A., Mukhametzhanov, R.: Modeling and analysis of vector control systems for asynchronous motor. In: High Speed Turbomachines and Electrical Drives Conference 2020 (HSTED 2020), Prague, Czech Republic, 09 July 2020, p. 01001 (2020)
15. Meshcheryakov, V.N., Sibirtsev, D.S.: Frequency asynchronous electric drive with correction of phase shift between moment-forming vectors. Control Syst. Inf. Technol. **2** (68), 48–57 (2017)
16. Gracheva, E., Alimova, A.: Calculating probability of faultless work of shop nets with the help of coefficients of ratio. In: 2019 International Russian Automation Conference, RusAutoCon 2019, p. 8867727 (2019)

17. Gracheva, E.I., Naumov, O.V.: Evaluation criteria of contact group technical state concerning electrical appliances. Int. J. Pharm. Technol. **8**(4), 26763–26770 (2016)
18. Meshcheryakov, V., Sibirtsev, D., Mikhailova, E.: Mathematical simulation of synchronized asynchronous electric drive. In: High Speed Turbomachines and Electrical Drives Conference 2020 (HSTED 2020), Prague, Czech Republic, 09 July 2020, p. 01021 (2020)
19. Abakumov, A.M., Randin, D.G.: Research of dual-mass oscillation system with linear motor. In: International Conference on Industrial Engineering, Applications and Manufacturing (ICIEM), Sochi, Russia, 25–27 March 2019, pp. 1–5 (2019)
20. Meshcheryakov, V.N., Lastochkin, D.V., Shakurova, Z.M., Valtchev, S.: Energy saving system of cascade variable frequency induction electric drive. E3S Web Conf. **124**, 01037 (2019). Kazan, SES 2019
21. Verma, P., Saxena, R., Chitra, A., Sultana, R.: Implementing fuzzy PI scalar control of induction motor. In: IEEE International Conference on Power, Control, Signals and Instrumentation Engineering (ICPCSI), Chennai, India, 21–22 September 2017, pp. 1674–1678 (2017)

Effect of Metakaolin and MC Adhesive Additives on the Mechanical Properties of Concrete

Yuri Barabanshikov, The Hien Pham(✉), and Stanislav Akimov

Peter the Great St. Petersburg Polytechnic University, St. Petersburg, Russian Federation
Phamthehien710@gmail.com

Abstract. This article discusses the influence of two types of additives on the compressive and flexural strength and elastic modulus of concrete. Two different types of additives (Metakaolin and MC Adhesive) were used to produce concrete with target strengths of 30 to 45 MPa and a reduced modulus of elasticity. With a partial replacement of cement with metakaolin, as well as in the case of the use of a liquid additive MC Adhesive, a decrease in both the bending and compression strength and the modulus of elasticity is observed. At the same time, the addition of metakaolin, replacing 17.5% of cement, reduces these characteristics to the greatest extent. However, the tensile strength in the bending test of concrete with metakaolin decreases slightly. It was found that the addition of metakaolin increases the crack resistance of concrete as a result of a significant decrease in the modulus of elasticity and a slight decrease in the tensile strength of concrete.

Keywords: Crack resistance · Modulus of elasticity · Compressive strength · Metakaolin · MC adhesive

1 Introduction

Currently, the problem of crack resistance of concrete buildings and structures has acquired significant relevance. Since concrete cracking occurs everywhere as in the case of massive and thin-walled structures. Crack resistance is a complex property of concrete. Along with the strength characteristics, one of the most important properties that determine the crack resistance of concrete is its deformability.

One of the most popular and environmentally friendly methods for regulating concrete properties, including mechanical characteristics, is the introduction of chemical and mineral additives into the concrete mixture [1–13].

In recent years, metakaolin has begun to be studied as an active dispersed additive for high-quality concrete [14–18]. Metakaolin in its activity can be compared with silica fume [19–24]. Usually metakaolin is taken as a partial replacement of cement in the amount of 5–15% by weight of cement [25–27]. The number of studies on the effect of metakaolin on the modulus of elasticity of concrete is relatively small. In this

A. Sinitsyn (Ed.): SES 2020, LNCE 141, pp. 300–309, 2021.
https://doi.org/10.1007/978-3-030-67654-4_33

connection, the aim of this study is to investigate the influence of metakaolin on the deformative properties of concrete with partial replacement of cement. Also, at the same time, the study was conducted of the effect on the properties of concrete of a polymer additive "MC Adhesive" manufactured by LLC "MC Bauchemie".

2 Materials and Methods

2.1 Materials

- Cement: in this work, a non-additive Portland cement - Eurocement M500 D0 CEM I 42.5, manufactured by Eurocement Group JSC was used as a binder.
- Fine aggregate: sand quartz I-st class with a fineness modulus $M_f = 2.0$.
- Coarse aggregate: crushed granite fraction 5 - 10 mm, density $\rho^g = 1350$ kg/m^3.
- Additives:

- MC Adhesive: polymer additive manufactured by LLC "MC Bauchemie" according to TU 5745–079-51552155–2009 is a liquid of milky white color. Density ρ^a - 1015 kg/m^3.
- Metakaolin: highly active VMK-45 manufactured by TD ArmMix LLC, density $\rho^m = 480$ kg/m^3.

2.2 Methods of Experiment

The production and testing of concrete specimens were carried out at the Polytech-SKiM-Test laboratory of Peter the Great St. Petersburg Polytechnic University.

In order to study the influence of the studied additives on the mechanical characteristics of concrete, 3 concrete mixtures were prepared with the same grade of mobility - S2. In the control composition, as a rule, additives are not added. The second composition with the addition of a polymer additive MC Adhesive to the concrete mixture with a dose of 5% by weight of cement. And in the third composition, metakaolin was used as a replacement for cement in the amount of 17.5% by weight of cement. The compositions and properties of concrete mixtures are presented in Table 1.

Table 1. The composition and mobility of concrete mixtures

№	Additive	Mix proportion, kg/m^3				Additive, % by weight of cement	Slump, cm
		Cement	Water	Fine aggregate	Coarse aggregate		
1	Control	408	225	852	916	0	6.5
2	MC Adhesive	408	209	852	916	5	6.6
3	Metakaolin	337	273	852	916	17.5	6.8

From each composition, 6 cubes with dimensions of 70 × 70 × 70 mm were prepared for determining compressive strength, 6 prisms with dimensions of 70 × 70 × 280 mm for determining the modulus of elasticity, and 3 prisms with dimensions of 40 × 40 × 160 mm for determining flexural strength. After molding, the specimens were kept in molds for 24 h and then stored in air in a climatic chamber at a temperature t = 20 ± 2 °C and relative humidity φ = 60 ± 5%.

Compression tests of concrete cubes with dimensions of 70 × 70 × 70 mm were carried out in accordance with GOST 10180–2012 on a PSU-50A hydraulic press. Tests for the elastic modulus of concrete prisms with dimensions of 70 × 70 × 280 mm were carried out on a PSU-10A hydraulic press in accordance with GOST 24452 - 80. Longitudinal force strains were measured using four dial gauges ICh-1 with a graduation interval of 0.001 mm attached to the prisms using special framework devices (Fig. 1). The longitudinal strain measurement base was 150 mm. The bending tests of concrete prisms with dimensions of 40 × 40 × 160 mm were carried out using the MII-100 testing machine equipped with an ICh-1 dial gauges with a graduation interval of 0.001 mm (Fig. 2) to measure bending strain in the middle of the test specimens. The distance between the supports of the prisms was 100 mm.

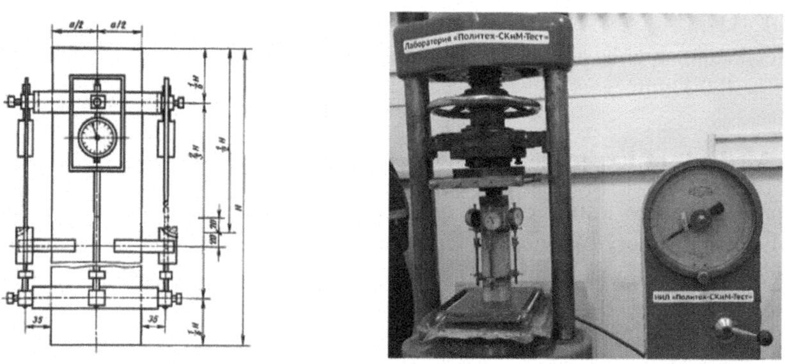

Fig. 1. The scheme and the general view of the test on modulus of elasticity

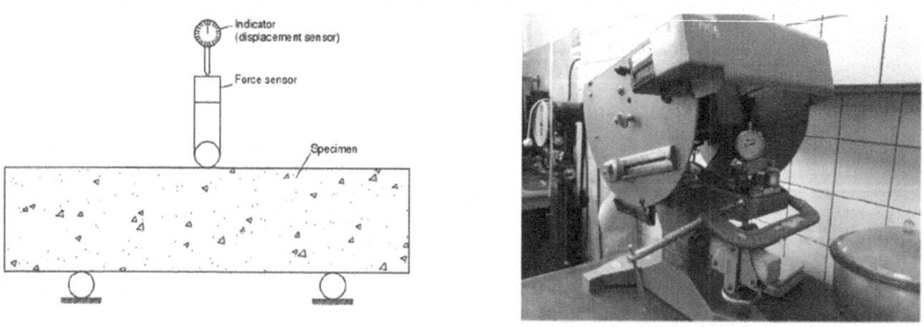

Fig. 2. The scheme and the general view of the bending test

3 Results and Discussion

3.1 Compressive Strength

Concrete cubes measuring 70 × 70 × 70 mm were tested for strength at the age of 7 and 28 days. The compressive strength of the control specimens (without additives) is greatest and amounts to 37.0 MPa at the age of 7 days and 42.2 MPa at the age of 28 days (Table 2). When replacing 17.5% of cement with metakaolin, the compressive strength at the age of 7 and 28 days decreased, respectively, by 28.8 and 32.3%. If 5% of the mass of cement was used in the concrete mixture, the MC Adhesive liquid polymer additive, the compressive strength of concrete decreased by 11.7% and 9.4%, respectively, at the age of 7 and 28 days.

Table 2. Results of testing concrete for strength and deformability.

№	Additive	Compressive strength, MPa, age, days		Flexural strength, MPa, at the age of 28 days	Modulus of elasticity, MPa, age, days	
		7	28		7	28
1	Control	37.0	42.2	6.5	28720	30300
2	MC Adhesive	32.6	38.2	6.2	27980	28670
3	Metakaolin	26.3	28.6	6.1	24180	24790

3.2 Tensile Strength in Bending

The results of the bending test of prism specimens with dimensions of 40 × 40 × 160 mm at the age of 28 days are presented in Table 2 and in the form of graphs of the dependence "stress - deflection" in Fig. 3. It can be seen that both the flexural strength of the tested three concrete compositions and their deformation differ little from each other. The Flexural strength of specimens with the addition of "MC Adhesive" and specimens with metakaolin is lower than the flexural strength of control specimens by only 4.5 and 6.2%, respectively, which may lie within the measurement error. The arrangement of the curves on the graph shows that the deformative properties of concrete with the addition of MC Adhesive are slightly higher than that of the other compositions and it can be expected that the modulus of elasticity in compression test of this composition will be lower than the control composition and the composition with metakaolin. However, the results of this test are not entirely correct, since, in the accepted methodology, the measured deflection includes the deformation of the crushing of the material by the support rollers, as well as the deformation of the supports themselves. These results are considered indicative.

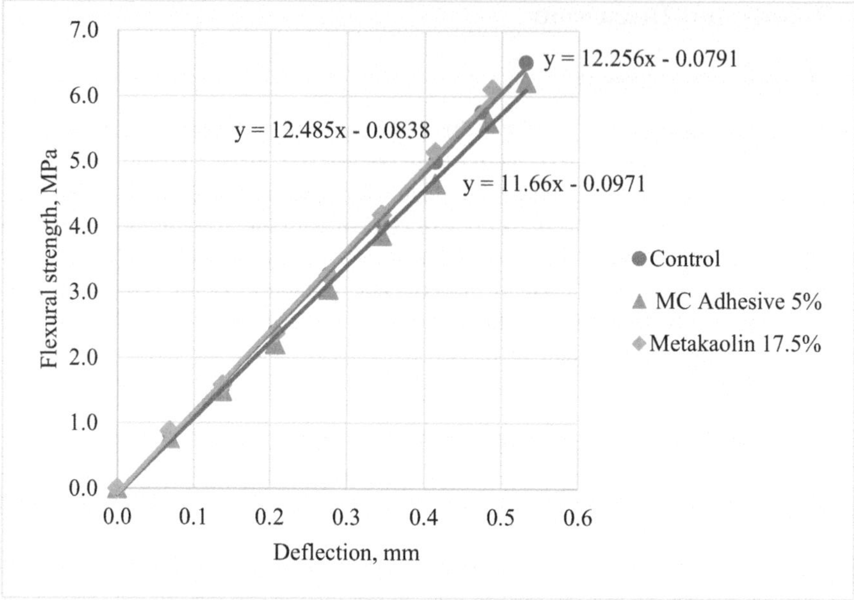

Fig. 3. The dependence "Stress - strain" in bending

3.3 Modulus of Elasticity

Prism specimens with dimensions of 70 × 70 × 280 mm were loaded in steps to a level equal to 40% of the expected breaking load. The results of the stress – strain relationship at the corresponding age are presented in Figs. 4 and 5.

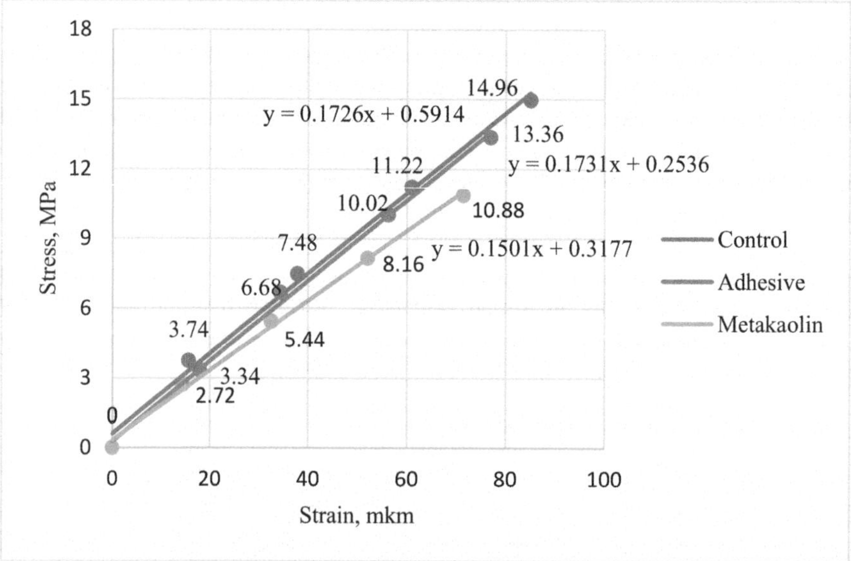

Fig. 4. The dependence "stress - strain" of specimens at the age of 7 days

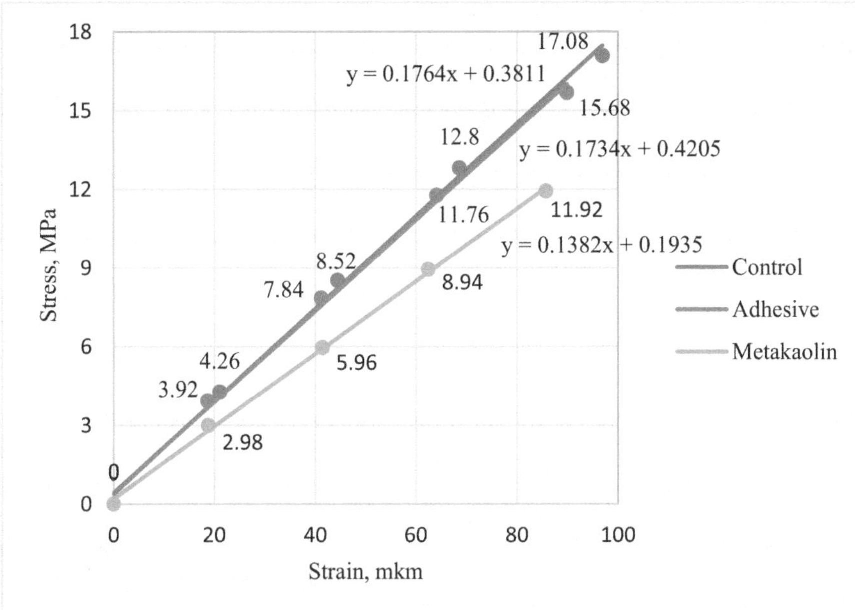

Fig. 5. The dependence "stress - strain" of specimens at the age of 28 days

The results of determining the modulus of elasticity of concrete with various additives are presented in Table 2 and in Fig. 6. The values of the elastic modulus of the control specimens are the highest and at the age of 7 days - 28720, 28 days - 30300 MPa. When using additives, a decrease in the elastic modulus is observed. In the case of using metakaolin, the decrease is 15.8 and 18.2%, respectively, at the age of 7 and 28 days, and in the case of concrete with a polymer additive MC - 2.6 and 5.4% at the same time.

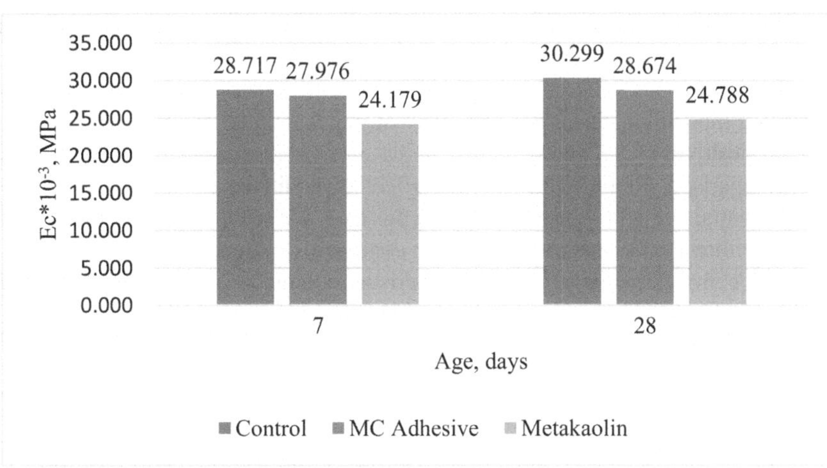

Fig. 6. The effect of additives on the modulus of elasticity of test specimens

The criterion for crack resistance of concrete according to P.I. Vasiliev is the ultimate tensile strength of concrete ε_{red}, this value is calculated by the formula (1):

$$\varepsilon_{red} = f_c^t / E_t + \varepsilon_{pl} + \varepsilon_{creep} \tag{1}$$

where f_c^t is the tensile strength; E_t is the tensile modulus of elasticity; ε_{pl} – the plastic strain; ε_{creep}- the creep strain.

Concrete is considered crack resistant if the actual deformation ε_a does not exceed the ultimate tensile strength (condition (2)):

$$\varepsilon_a < \varepsilon_{red} \tag{2}$$

The f_c/ E_c ratio is the ultimate elastic strain. In our case, the modulus of elasticity was determined in compression test, however, assuming that the modulus of elasticity both in compression and in tension is the same $E_t = E_c$ (as is commonly believed) and using the results of determining the tensile strength in bending $f_c{}^t$, we can calculate the ratio $f_c{}^t/E_c$. Since elements in real reinforced concrete structures perceive mainly bending loads, this ratio can to some extent characterize the crack resistance of concrete. The results of calculating the ultimate elastic strain in bending $f_c{}^t/E_c$ are given in Table 3.

Table 3. The values of the ultimate elastic strain in bending $f_c{}^t/E_c$ for the tested mixtures.

№	Additive	Tensile strength, MPa, at the age of 28 days	The modulus of elasticity, MPa, at the age of 28 days	$f_c{}^t/E_c \times 10^{-4}$
1	Control	6.5	30300	2.15
2	MC Adhesive	6.2	28670	2.16
3	Metakaolin	6.1	24790	2.46

As can be seen from the Table 3, the “MC Adhesive” additive had practically no effect on the ultimate elastic tensile strain in bending. In concrete with the addition of metakaolin, replacing 17.5% of cement, the value of $f_c{}^t/E_c$ is 15% higher than that of the control composition. Therefore, the effect of metakaolin on crack resistance can be considered positive.

For comparison, in Fig. 7 shows the effect of additives on the ultimate elastic compression strain equal to the ratio f_c/E_c. In this case, we see the opposite picture. This is due to the direct relationship between the compressive strength and the modulus of elasticity in compression test. This characteristic cannot be used to judge the crack resistance of concrete, since cracks occur under the action of tensile stresses.

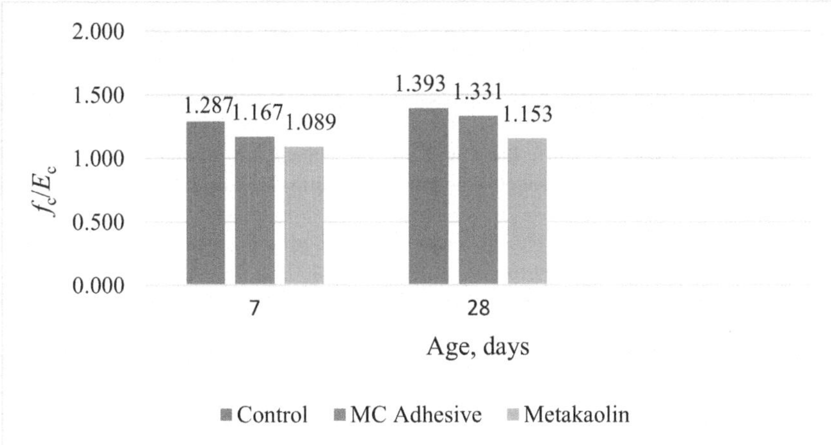

Fig. 7. The effect of additives on the ultimate elastic compression strain equal to the ratio f_c/E_c

4 Conclusions

1. The effect of metakaolin and MC Adhesive additives on the compressive strength and elastic modulus of concrete was investigated.
2. Both tested additives reduce the strength and modulus of elasticity of concrete. At the same time, the addition of metakaolin, replacing 17.5% of cement, reduces these characteristics to the greatest extent. However, the flexural strength of concrete specimen in bending with metakaolin decreases slightly.
3. It was found that the addition of metakaolin increases the crack resistance of concrete as a result of a significant decrease in the modulus of elasticity and a slight decrease in the flexural strength of concrete.

References

1. Bharatkumaranaik, T., Srishaila, J.M., Rajendra Kumar, K.M., Veerangouda, R.: Effect of fly ash on mechanical properties of high strength concrete. Int. Res. J. Eng. Technol. **4**, 1455–1461 (2017)
2. Felekoğlu, B., Sarikahya, H.: Effect of chemical structure of polycarboxylate-based superplasticizers on workability retention of self-compacting concrete. Constr. Build. Mater. **22**, 1972–1980 (2008). https://doi.org/10.1016/j.conbuildmat.2007.07.005
3. Ismail, M., Muhammad, B., Yatim, J.M., Noruzman, A.H., Soon, Y.W.: Behavior of concrete with polymer additive at fresh and hardened states (2011). https://doi.org/10.1016/j.proeng.2011.07.281
4. Mazloom, M., Ramezanianpour, A.A., Brooks, J.J.: Effect of silica fume on mechanical properties of high-strength concrete (2004). https://doi.org/10.1016/S0958-9465(03)00017-9
5. Memphis: Fly ash, Slag, silica fume, and natural Pozzolans (1996)

6. Onuaguluchi, O., Panesar, D.K.: Hardened properties of concrete mixtures containing pre-coated crumb rubber and silica fume. J. Clean. Prod. **82**, 125–131 (2014). https://doi.org/10.1016/j.jclepro.2014.06.068
7. Rassokhin, A.S., Ponomarev, A.N., Figovsky, O.L.: Silica fumes of different types for high-performance fine-grained concrete (2018). https://doi.org/10.18720/MCE.78.12
8. Saridemir, M.: Effect of silica fume and ground pumice on compressive strength and modulus of elasticity of high strength concrete (2013). https://doi.org/10.1016/j.conbuildmat.2013.08.091
9. Sakthivel, T., Gettu, R., Pillai, R.G.: Compressive strength and elastic modulus of concretes with fly ash and slag (2019). https://doi.org/10.1007/s40030-019-00376-w
10. Barabanshikov, Y.G.: Komarinskiy: superplasticizer S-3 and its effect on the technological properties of concrete mixtures. Constr. Unique Build. Struct. **6**, 58–69 (2014)
11. Kolosova, N.B., Kulish, I.V., Ershov, V.V.: Study properties of road surface for foot pavements and courtyard areas. In: Materials Science Forum (2020). https://doi.org/10.4028/www.scientific.net/MSF.974.325
12. Grigoryan, E., Babanina, A., Eropkin, A.: Concrete mixture grout for printing extruder. In: E3S Web of Conferences (2019). https://doi.org/10.1051/e3sconf/201913503072
13. Makaronichev, A., Belyaeva, S., Pakrastins, L.: The impact of aggregates type to abrasion resistance of mortar. Lecture Notes in Civil Engineering (2020). https://doi.org/10.1007/978-3-030-42351-3_12
14. Ahmed, A.L.: Effects of reactivity metakaolin on properties of high strength concrete. Int. J. Eng. Technol. **7**, 2132–2136 (2018). https://doi.org/10.14419/ijet.v7i4.13584
15. Tafraoui, A., Escadeillas, G., Vidal, T.: Durability of the ultra high performances concrete containing metakaolin. Constr. Build. Mater. **112**, 980–987 (2016)
16. Ashfi Rahman, H.S.: Effect of mineral admixtures on characteristics of high strength concrete. Int. J. Res. Eng. Technol. **04**, 610–614 (2015). https://doi.org/10.15623/ijret.2015.0402085
17. Kim, H.K., Hwang, E.A., Lee, H.K.: Impacts of metakaolin on lightweight concrete by type of fine aggregate. Constr. Build. Mater. **36**, 719–726 (2016)
18. Salimi, J., Ramezanianpour, A.M., Moradi, M.J.: Studying the effect of low reactivity metakaolin on free and restrained shrinkage of high performance concrete. J. Build. Eng. **28**, 101053 (2020)
19. Ding, J.T., Li, Z.: Effects of metakaolin and silica fume on properties of concrete (2002). https://doi.org/10.14359/12222
20. Güneyisi, E., Gesoglu, M., Karaoglu, S., Mermerdaş, K.: Strength, permeability and shrinkage cracking of silica fume and metakaolin concretes. Constr. Build. Mater. **34**, 120–130 (2012)
21. Seelapureddy, J., Bommisetty, J., Seshagiri Rao, M.V.: Effect of metakaolin and micro silica on strength characteristics of standard grades of self-compacting concrete. Mater. Today Proc. J. (2020)
22. Saboo, N., Shivhare, S., Kori, K.K., Chandrappa, A.K.: Effect of fly ash and metakaolin on pervious concrete properties. Constr. Build. Mater. J. **223**, 322–328 (2019)
23. Mohammadi, M., Moghtadaei, R.M., Samani, N.A.: Influence of silica fume and metakaolin with two different types of interfacial adhesives on the bond strength of repaired concrete. Constr. Build. Mater. J. **51**, 141–150 (2014)
24. Roy, D.M., Arjunan, P., Silsbee, M.R.: Effect of Silica fume, metakaolin, and low-calcium fly ash on chemical resistance of concrete. Cem. Concr. Res. **31**, 1809–1813 (2001)
25. Murthy, N.K.: The influence of metakaolin on the modulus of elasticity of concrete. IOSR J. Eng. **02**, 18–23 (2012). https://doi.org/10.9790/3021-021131823

26. Rajkumar, R., Umamaheswari, N., Kumar, A., Mrinal Kumar, L.R., Vineeth Gupta, R.P.: Flexural behaviour of reinforced concrete beams with partial replacements of metakaolin and marble powder. Mater. Today (2020)
27. Rakesh Muduli, B.B.M.: Effect of incorporation of metakaolin and recycled coarse aggregate on properties of concrete. J. Clean. Prod. J. **209**, 398–414 (2019)

Predicting Single-Phase Damages in 10 kV Electric Networks Taking into Account the Seasonal Influence of Weather Factors

Alexander Kachanov[1(✉)], Vadim Chernyshov[1], Artem Revkov[2], Maxim Borodin[3], and Oleg Vladimirov[4]

[1] Department "Electrical Equipment and Energy Saving", FSBE HE «OSU name by I. S. Turgenev», Oryol, Russia
kan@ostu.ru
[2] Department "Mechatronics, Mechanics and Robotics", FSBE HE «OSU name by I. S. Turgenev», Oryol, Russia
[3] Department of Electric Power Supply, FSBEE HE Oryol State Agrarian University named after N. I. Parakhin, Oryol, Russia
[4] Department "Power Supply of Industrial Enterprises", Kazan State Power Engineering University, Kazan, Russia

Abstract. The paper considers a technique for predicting a number of single-phase ground faults in 10 kV overhead distribution networks for a ten-day period for warm and cold seasons and the software that implements it. The technique is based on the developed regression equations that take into account the influence of weather and climatic factors on the process of occurrence of single-phase ground faults in overhead power lines with a voltage of 10 kV, differing in the length and degree of their physical wear, normalized by the approved indicators of the technical condition of power facilities. To obtain the regression equations, statistical data of three-year observations on the occurrence of single-phase ground faults in the overhead distribution networks of the Oryol region were used. The developed engineering technique and the proposed software can be useful to specialists engaged in the operation of 10 kV overhead distribution networks. The methodology and software will make it possible to use a ten-day weather forecast, quickly to draw up confidence schedule for the predictive and preventive maintenance of lines with different levels of wear, which will increase the level of their reliability and safety.

Keywords: Damages · Electric networks

1 Introduction

For energy enterprises operating 10 kV overhead distribution networks, one of the urgent problems is to ensure their reliable and safe operation [1]. The most common type of damage in 10 kV overhead distribution networks is single-phase earth faults (SPEF), which in some cases are the precursors of serious accidents, as well as sources of industrial and domestic electrical injuries [2].

A. Sinitsyn (Ed.): SES 2020, LNCE 141, pp. 310–320, 2021.
https://doi.org/10.1007/978-3-030-67654-4_34

According to statistical data, the accident rate in 10 kV overhead distribution networks in terms of 100 km of overhead power transmission lines (OPTL) is 6–7 SPEF per year for temperate regions and 20–30 - for regions with difficult climatic conditions [3]. It should be noted that almost any 10 kV distribution network consists of separate sections that differ from each other in length and technical condition. This leads to the fact that the likelihood of the occurrence of SPGF in a 10 kV overhead transmission line during the year is distributed unevenly, which is also due to the influence of weather and climatic factors on them. In recent years, there has been a steady trend of abnormal climate changes. Therefore, the accumulated data on accidents are unsuitable for the development of the predictive and preventive maintenance schedules for the repair and maintenance of sections of 10 kV overhead transmission lines, and also do not allow to optimize the amount of spare parts and the number of personnel.

From the above, it follows that scientific research in the field of predicting the SPGF in distribution electrical networks of 10 kV with overhead transmission lines, differing in length and degree of wear, as well as taking into account the seasonal effects of weather and climatic factors, are relevant, providing further development of theory and practice in the field of improving the efficiency of functioning distribution networks.

2 Aims and Tasks

To develop a mathematical model that makes it possible to reliably predict the number of SPGF in 10 kV overhead transmission lines for a ten-day period, it is necessary to establish the relationship their number and the most significant average indicators of weather and climatic factors of a ten-day period.

Table 1 presents the results of three-year statistical observations of the occurrence of SPGF in Oryol electrical networks with a voltage of 10 kV, taking into account their length and technical condition.

Table 1. Results of three-year observations of SPGF in Oryol electrical networks with a voltage of 10 kV.

Year observations	Length 10 kV lines, km		Technical condition 10 kV lines, km	Number of single-phase ground faults on the 10 kV line	Number of single-phase ground faults per 100 km of 10 kV line
First year	General up to 25 km	12705.491	With no regard to condition	764	6.015
		6064.604	With no regard to condition	337	5.556
		3007.642	Good	101	3.358
		1854.962	Satisfactory	115	6.199
		1201.998	Unsatisfactory	121	10.066
	From 25 km up to50 km	5155.836	With no regard to condition	316	6.128
		2573.698	Good	86	3.341

(*continued*)

Table 1. (*continued*)

Year observations	Length 10 kV lines, km		Technical condition 10 kV lines, km	Number of single-phase ground faults on the 10 kV line	Number of single-phase ground faults per 100 km of 10 kV line
		1540.547	Satisfactory	113	7.335
		1041.591	Unsatisfactory	117	11.232
	From 50 km and higher	1485.051	With no regard to condition	111	7.474
		724.879	Good	37	5.104
		464.512	Satisfactory	37	7.965
		295.660	Unsatisfactory	37	12.514
Second year	General up to 25 km	12695.741	With no regard to condition	776	6.112
		6148.863	With no regard to condition	343	5.578
		3163.733	Good	108	3.413
		1826.389	Satisfactory	116	6.351
		1158.741	Unsatisfactory	119	10.269
	From 25 km up to 50 km	5090.449	With no regard to condition	324	6.364
		2459.269	Good	84	3.415
		1540.225	Satisfactory	116	7.531
		1090.955	Unsatisfactory	124	11.366
	From 50 km and higher	1456.429	With no regard to condition	109	7.484
		712.911	Good	36	5.049
		452.739	Satisfactory	37	8.172
		290.779	Unsatisfactory	36	12.380
Third year	General up to 25 km	13068.073	With no regard to condition	767	5.869
		6260.692	With no regard to condition	337	5.382
		3204.978	Good	106	3.307
		1804.84	Satisfactory	107	5.928
		1250.874	Unsatisfactory	124	9.913
	From 25 km up to 50 km	5273.926	With no regard to condition	319	6.048
		2603.618	Good	86	3.303
		1534.472	Satisfactory	108	7.038
		1135.836	Unsatisfactory	125	11.005
	From 50 km and higher	1533.455	With no regard to condition	111	7.238
		744.822	Good	37	4.967
		484.841	Satisfactory	36	7.631
		303.792	Unsatisfactory	37	12.179

When choosing indicators of the technical condition of 10 kV overhead transmission lines, the authors were guided by the well-known method [4], according to which 5 levels of physical wear are accepted: "Critical" (wear is more than 75%); "Unsatisfactory" (wear from 50 to 75%); "Satisfactory" (wear from 30 to 50%); "Good" (wear from 15 to 30%) and "Very good" (wear less than 15%). Considering the fact that there is virtually no 10 kV overhead transmission line with a critical level of wear in the Oryol region, this category is not discussed. It was also assumed that two categories of the level of physical wear of 10 kV overhead transmission lines, namely, "Good" and "Very good", are considered the same – "Good".

It should be noted that 10 kV overhead transmission lines, the level of physical wear of which is recognized as unsatisfactory, are being operated today, despite the fact that the individual elements and nodes included in them are inoperative [4].

Taking into account the accepted assumptions, an engineering methodology will be developed for predicting the number of SPGF for a ten-day period, with regard to the influence of weather and climatic factors, for overhead transmission lines with different levels of physical (good, satisfactory and unsatisfactory), presented in Table 1.

3 Conditions, Materials and Methods

The process of step-by-step development of mathematical models describing the dependence of the number of SPGF of a ten-day period per 100 km of 10 kV OPTL on the most significant weather and climatic factors will be considered using the example of 10 kV OPTL, up to 25 km long, since their total length in the Oryol distribution networks is more than 50% (6 thousand km.).

According to the instruction [5], the facts of an accident at the 10 kV overhead transmission line are recorded in the operational log, which is kept for 3 years. At the same time, the values of the average indicators of a ten-day period concerning the characteristics of the environment (air temperature, wind speed, amount and type of precipitation, etc.) are also recorded by the services of the Hydrometeorological Center.

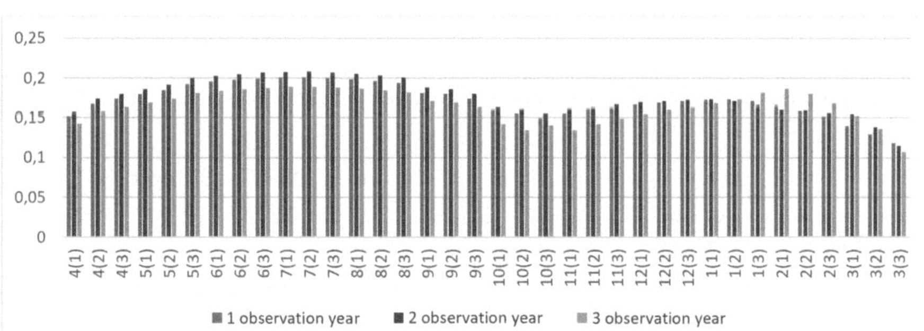

Fig. 1. Histogram of the ten-day distribution of the specific amount of SPGF in 10 kV overhead transmission lines with good technical condition

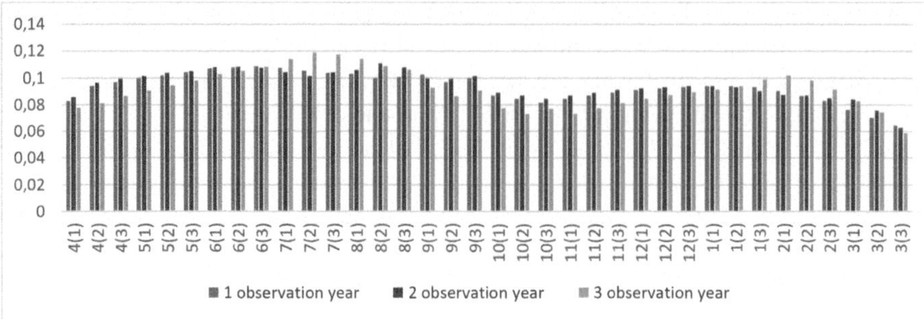

Fig. 2. Histogram of the ten-day distribution of SPGF in 10 kV overhead transmission lines, having a satisfactory technical condition

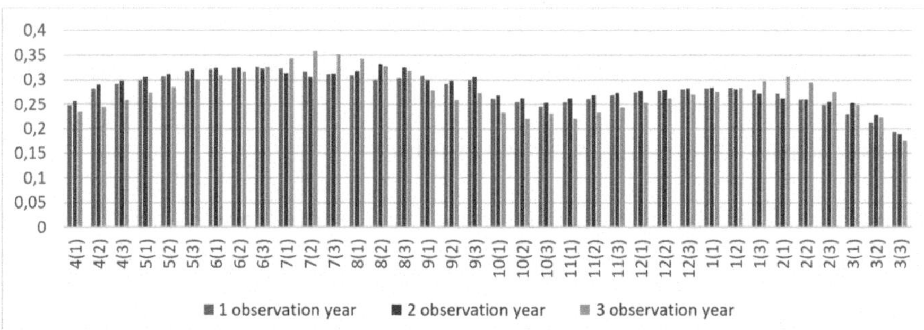

Fig. 3. Histogram of the ten-day distribution of SPGF in 10 kV overhead transmission lines, having an unsatisfactory technical condition

Figures 1, 2 and 3 show the histograms of the ten-day distribution of the specific amount of SPGF for three years of observations in 10 kV networks with overhead transmission lines up to 25 km long, having a different level of technical condition.

Table 2. Levels and intervals of variation of weather and climatic factors for the warm (X1, X2, X3, X4) and cold (X'1, X'2, X'3, X'4) period of the year kV.

Factors	Notation	Variation levels			Variation interval
		Lower	Main	Upper	
Average surface temperature of a ten-day period /(air temperature), °C	X_1/ X'_1	20/−15	30/7,5	40/0	10/7,5
Average wind speed of a ten-day period, m/s	X_2/ X'_2	3/3	4,5/4,5	6/6	1,5/1,5
Precipitation of a ten-day period, mm	X_3/ X'_3	5/5	17,5/17,5	30/30	12,5/12,5
Number of hours of thunderstorm activity / (days of icy load on the wires per ten days)	X_4/ X'_4	0/0	10,5/4,5	21/9	10,5/4,5

When analyzing the histograms presented in Figs. 1, 2 and 3, the following pattern was revealed, namely, the seasonal cyclicity of the coincidence of the curves of smooth growth and decline in the specific amount of SPGF in two areas.

The first section is determined by the time interval from the first ten-day period of April to the second ten-day period of October inclusive, and the second section, respectively, from the third ten-day period of October to the third ten-day period of March inclusive. The revealed regularity is due to the different influence of weather and climatic factors on the reliability of the operation of 10 kV overhead transmission lines in warm and cold seasons.

The following factors were chosen as the main weather and climatic factors that have the greatest impact on the likelihood of the occurrence of SPGF in the 10 kV OPTL:

for the warm season: X1 - average soil surface temperature of a ten-day period (°C); X2 - average wind speed of a ten-day period (m/s); X3 - the amount of precipitation per ten days (mm); X4 - the duration of thunderstorm activity per ten days (h).

for the cold season: X'1 – average air temperature of a ten-day period (°C); X'2 - average wind speed of a ten-day period (m/s); X'3 - the amount of precipitation per ten days (mm); X'4 - ice load for a ten-day period (days).

Table 2 shows the selected levels and intervals of variation of weather and climatic factors [6, 7].

To construct a standard experiment planning matrix, the natural values of the factor levels were converted into dimensionless code values according to the formula [6]:

$$X_i = \frac{x_i - x_{io}}{\Delta x_i}, \qquad (1)$$

where X_i – code value of the i-th factor; x_i – natural current value of the i-th factor; x_{io} – the main level of the factor; Δx_i – interval of variation of the i-th factor.

$$\Delta x_i = \frac{x_{iMAX} - x_{iMIN}}{2}, \qquad (2)$$

where x_{iMAX} and x_{iMIN} are the natural values of the upper and lower levels of variation of the i-th factor, respectively.

Calculated by formula (1) coded values of factors for the lower and upper levels of variation, respectively, are equal: $X_i^H = -1$; $X_i = 1$.

The experiment design matrix is presented in Table 3.

Table 3. Experiment planning matrix for warm/cold seasons.

Experiment number	(°C), X_1	Average wind speed of a ten-day period (m/s), X_2	The amount of precipitation per ten-day period (mm), X_3	Number of hours of thunderstorm activity per ten-day period (h), X_4
1	−1,0/−1,0	1,0/1,0	−1,0/−1,0	−1,0/−1,0
2	−1,0/−1,0	1,0/1,0	1,0/1,0	−1,0/−1,0
3	−1,0/−1,0	−1,0/1,0	1,0/−1,0	−1,0/1,0
4	−1,0/−1,0	−1,0/1,0	−1,0/1,0	−1,0/1,0
5	1,0/−1,0	1,0/−1,0	1,0/−1,0	−1,0/1,0
6	1,0/1,0	−1,0/−1,0	−1,0/−1,0	−1,0/−1,0
7	1,0/1,0	−1,0/−1,0	1,0/1,0	−1,0/−1,0
8	1,0/1,0	−1,0/−1,0	−1,0/−1,0	1,0/1,0
9	1,0/1,0	1,0/1,0	−1,0/−1,0	1,0/1,0
10	1,0/1,0	1,0/1,0	1,0/1,0	1,0/1,0
11	1,0/1,0	−1,0/−1,0	1,0/1,0	1,0/1,0
12	1,0/1,0	1,0/1,0	−1,0/−1,0	−1,0/−1,0
13	−1,0/1,0	−1,0/1,0	−1,0/1,0	1,0/−1,0
14	−1,0/−1,0	−1,0/−1,0	1,0/1,0	1,0/1,0
15	−1,0/−1,0	1,0/−1,0	−1,0/1,0	1,0/−1,0
16	−1,0/−1,0	1,0/−1,0	1,0/−1,0	1,0/−1,0

The regression equation characterizing the dependence of the number of ground faults (Y) on four influencing factors (X1, X2, X3, X4) has the form of a first-order mathematical equation [6]:

$$\begin{aligned} Y = {} & b_0 + b_1X_1 + b_2X_2 + b_3X_3 + b_4X_4 + b_{12}X_1X_2 \\ & + b_{13}X_1X_3 + b_{14}X_1X_4 + b_{23}X_2X_3 + b_{24}X_2X_4 \\ & + b_{34}X_3X_4 + b_{123}X_1X_2X_3 + b_{124}X_1X_2X_4 \\ & + b_{134}X_1X_3X_4 + b_{234}X_2X_3X_4 + b_{1234}X_1X_2X_3X_4 \end{aligned} \quad (3)$$

The coefficients of the regression Eq. (3) are determined by using the dependencies [8, 9]:

$$b_0 = \frac{\sum_{U=1}^{N} X_0 \cdot y_{UC}}{N}, \quad (4)$$

$$b_i = \frac{\sum_{U=1}^{N} X_i \cdot y_{UC}}{N} \quad (5)$$

$$b_{ij} = \frac{\sum_{U=1}^{N} X_i \cdot X_j \cdot y_{UC}}{N} \tag{6}$$

$$b_{ijk} = \frac{\sum_{U=1}^{N} X_i \cdot X_j \cdot X_k \cdot y_{UC}}{N} \tag{7}$$

$$b_{ijkl} = \frac{\sum_{U=1}^{N} X_i \cdot X_j \cdot X_k \cdot X_l \cdot y_{UC}}{N} \tag{8}$$

After finding the coefficients of the equations for all three categories of 10 kV OPTL with a length of up to 25 km and checking their static significance according to the Student's criterion [8, 10], the obtained equations take the following form.

In particular, consider the regression equations for 10 kV overhead transmission lines for the following possible options:

- good technical condition for the warm season:

$$\begin{aligned} Y = 0,1017438 &+ (0,0042429X_1) + (0,0018000X_2) \\ &+ (0,0042858X_4) + (-0,0018964X_1X_2) \\ &+ (-0,0023036X_1X_4) + (0,0017464X_1X_2X_3X_4) \end{aligned} \tag{9}$$

- good technical condition for the cold season:

$$\begin{aligned} Y = 0,0857563 &+ (0,0059305X_1) + (0,0015076X_2) \\ &+ (0,0036476X_4) + (-0,0020935X_1X_2) \\ &+ (-0,0018399X_1X_4) + (0,0020971X_1X_2X_4) \\ &+ (0,0017005X_2X_3X_4) \end{aligned} \tag{10}$$

- satisfactory technical condition for the warm season:

$$\begin{aligned} Y = 0,1866788 &+ (0,0080094X_1) + (0,0035292X_2) \\ &+ (0,0076427X_4) + (-0,0032656X_1X_2) \\ &+ (-0,0044687X_1X_4) + (0,0029906X_1X_2X_3X_4) \end{aligned} \tag{11}$$

– satisfactory technical condition for the cold season:

$$\begin{aligned} Y = {} & 0,1572770 + (0,0108740X_1) + (0,0027156X_2) \\ & + (0,0067146X_4) + (-0,0037812X_1X_2) \\ & + (-0,0033802X_1X_4) + (0,0038500X_1X_2X_4) \\ & + (0,0031052X_2X_3X_4) \end{aligned} \tag{12}$$

– unsatisfactory technical condition for the warm season:

$$\begin{aligned} Y = {} & 0,3062605 + (0,0131211X_1) + (0,0057671X_2) \\ & + (0,0125408X_4) + (-0,0053526X_1X_2) \\ & + (-0,0073184X_1X_4) + (0,0048908X_1X_2X_3X_4) \end{aligned} \tag{13}$$

– unsatisfactory technical condition for the cold season:

$$\begin{aligned} Y = {} & 0,2579980 + (0,0178401X_1) + (0,0044704X_2) \\ & + (0,0110072X_4) + (-0,0062349X_1X_2) \\ & + (-0,0055480X_1X_4) + (0,0063296X_1X_2X_4) \\ & + (0,0050980X_2X_3X_4) \end{aligned} \tag{14}$$

4 Results and Discussion

Checking according to Fisher's criterion [6, 9] confirmed that the regression Eqs. (9–14) are adequate and suitable for predicting the SPGF.

Using the above technique, similar regression equations were obtained for other categories of 10 kV over-head transmission lines, with a length of 25 to 50 km and from 50 km and more, having good, satisfactory and unsatisfactory technical condition.

To speed up the process of processing a large array of information data, the SIZAM software (synoptic of closures) [11] was developed, Fig. 4 shows the working window of the program.

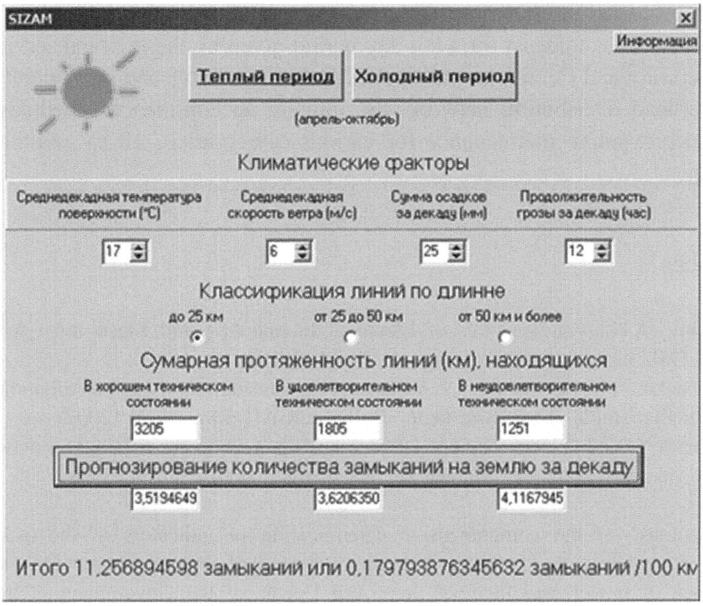

Fig. 4. Working window of the SIZAM program.

5 Conclusion

The obtained results of predicting the number of SPGF for various categories of 10 kV overhead transmission lines, taking into account the seasons, allow us to draw the following conclusions.

1. For the warm season, weather and climatic factors are ranked as follows according to the degree of their independent influence on the occurrence of SPGF in 10 kV OPTL: average wind speed of a ten-day period (45.26%); average soil surface temperature of a ten-day period (24.13%); the number of hours of thunderstorm activity per ten days (16.12%), the amount of precipitation per ten days (14.51%). It should be noted that the influence of the average thunderstorm activity and the amount of precipitation of a ten-day period naturally increases with an increase in the average wind speed and soil surface temperature of a ten-day period.

2. For the cold season, the weather and climatic factors, according to the degree of their independent influence on the occurrence of SPGF in the 10 kV OPTL, are ranked as follows: average wind speed of a ten-day period (49.71%), average air temperature of a ten-day period (24.05%); number of days with ice load per ten days (14.81%); the amount of precipitation per ten days (11.43%). It should be noted that the separate influence of the average number of days with ice load and the amount of precipitation of a ten-day period increases with an increase in the average wind speed and air temperature of a ten-day period.

3. The proposed engineering methodology will allow energy enterprises, on the basis of the weather forecast for a ten-day period given by the regional services of the Hydrometeorological Center, to ensure effectively the reliable and safe operation of the 10 kV overhead distribution network, by drawing up confidence schedules for predictive and preventive maintenance for various categories of 10 kV overhead transmission lines.

References

1. Kachanov, A.N.: Fundamentals of Electrical Engineering and Electronics: A Textbook, p. 340. OSU named after I.S. Turgenev, Oryol (2019)
2. Gavrichenko, A.I., Chernyshov, V.A.: Improving the safety of 10 kV distribution networks by predicting single-phase earth faults. Bull. OrelGAU **5**(8), 19–21 (2007)
3. Artsishevsky, Ya.L.: Determination of places of damage to power lines in networks with isolated neutral: Textbook. Manual for vocational schools, p. 87. Higher. shk., Moscow (1989)
4. Methodology for the comprehensive determination of indicators of the technical and economic state of power facilities, including indicators of physical wear and tear and energy efficiency of power grid facilities, approved Decree of the Government of the Russian Federation 1401, 19 December 2016
5. Instructional letter of MINTOPENERGO of the Russian Federation, on maintaining an operational log 42-6/35-ET, 9 December 1995
6. Ivobotenko, B.A., Ilyinsky, N.F., Kopylov, I.P.: Experiment Planning in Electromechanics, p. 184. Energiya Publishing House, Moscow (1975)
7. GOST 16350-80, The climate of the SSR, Zoning and statistical parameters of climatic factors for technical purposes, p. 140. Publishing house of standards, Moscow (1986)
8. Rumshinsky, Ya.Z.: Mathematical processing of experimental results, p. 192. Nauka, Moscow (1971)
9. Gmurman, V.E.: Probability theory and mathematical statistics: Textbook. Manual for universities, p. 479. Higher. shk., Moscow (2003)
10. Kachanov, A.N., Korenkov, D.A.: Mathematical model of the distribution of the electromagnetic field in long flat capacitors with a dielectric made of wood. Ind. energy **9**, 23–26 (2016)
11. Chernyshov, V.A., Gavrichenko, A.I.: Calculation of the expected number of earth faults in 10 kV distribution networks with issolated neutral SIZAM (short-circuit forecaster): certificate of state registration of computer programs, 2008610196, Ros. Federation, applicant Oryol State Agrarian University, 200761427, declared 29.10.2007, publ. 09.01.2008

Comparative Analysis of the Calculation of Composite Reinforced Concrete Slabs with Steel Decking According to Russian and European Standards

V. I. Korsun(✉), N. A. Vinogradova, and G. A. Shvets

Peter the Great St. Petersburg Polytechnic University, Polytechnicheskaya, 29, St. Petersburg 195251, Russia
korsun_vi@mail.ru

Abstract. Reinforced concrete slabs with steel profiled decking as permanent formwork are a modern, technologically advanced and cost-effective type of structure. Their calculation is performed for two stages: construction and operation. At the stage of floor construction, the main supporting structure is the formwork, which also plays the role of a working platform. At the stage of operation, the steel flooring is included in the work of the floor slab due to the ties (punchings, profile type, anchors), which prevent the steel from slipping relative to the concrete. This type of structure has become widespread, including in the territory of the Russian Federation. The authors carried out a comparative analysis of Russian (STO 0047–2005 "Steel-reinforced concrete slabs with a monolithic slab on steel profiled flooring. Calculation and design") and European (Eurocode-4 "Design of composite steel and concrete structures") calculation methods. Revealed significant differences in the considered methods. The conclusion is made about the impossibility of applying the European normative document in the Russian design practice without its adaptation, as well as the need for further research in this area.

Keywords: Steel-reinforced concrete floor · Calculation · Russian regulations · Eurocode

1 Introduction

Reinforced concrete slabs are efficient structures both in technical and economic aspects. However, due to the insufficient development of the relevant regulatory framework in Russia, when designing them, it is usually not taken into account that the profiled flooring acts not only as a convenient permanent formwork, but also can serve as a working external reinforcement, the effect of which can be taken into account by an appropriate calculation.

The issue of ensuring the joint work of the steel and reinforced concrete parts of the composite structure is important and actively studied [2–8]. The results of the scientists' work formed the basis of the current regulatory documents for the design of steel-reinforced concrete structures.

A. Sinitsyn (Ed.): SES 2020, LNCE 141, pp. 321–330, 2021.
https://doi.org/10.1007/978-3-030-67654-4_35

The paper describes two methods for calculating steel-reinforced concrete floors, which take into account the steel profiled flooring as a working reinforcement:

- according to the manual for STO 0047–2005 “Steel-reinforced concrete slabs with a monolithic slab on a steel profiled flooring. Calculation and design” [9];
- according to the manual for Eurocode-4 “Design of composite steel and concrete structures” [10].

The purpose of this study is to compare the two methods used in Europe and in the Russian Federation. Including an assessment of the possibility of using Eurocodes and their adaptation in our country.

2 Materials and Methods

STO 0047–2005, as well as Eurocode-4, provides for the calculation of the bearing capacity of structures at the stage of construction and operation.

At the stage of construction, the supporting structure is a steel profiled flooring. When calculating, its strength and stiffness are determined as for a thin-walled bending element operating on a load from its own weight of the flooring, the mass of freshly laid concrete and the assembly load, including the mass of equipment and people in the process of erecting floors.

In the operational stage, the supporting structure is a reinforced concrete slab, in which the steel profiled flooring is used as external working reinforcement. When calculating the slab, the full design load on it, including its own weight, should be taken into account (Fig. 1).

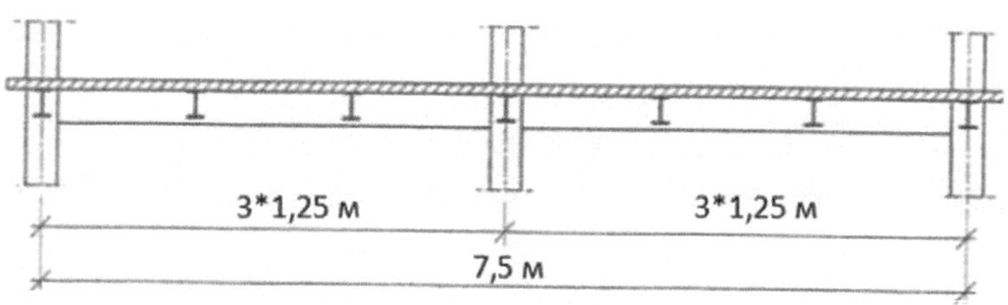

Fig. 1. Overlap design scheme

2.1 STO 0047–2005

Calculation for the Construction Stage. The strength of the profiled sheet at the construction stage is checked for supporting and span sections according to the conditions:

$$\frac{M}{W_x} \leq R_n; \tag{1}$$

$$\frac{Q}{\sum th_n} \leq R_{ns}; \tag{2}$$

where

M – maximum bending moment from design loads per 1 m of corrugated board width, N•m;

W_x – design moment of resistance per 1 m of corrugated board width, $(sm)^3$;

R_n – design tensile strength of steel profiled decking, MPa;

Q – transverse force from design loads per 1 m width of corrugated board, N;

$\sum t$ – the total number of sheeting walls that absorb the shear force across the width 1 m, mm;

h_n – height of the steel profiled deck, mm;

R_{ns} – design shear resistance of profiled steel decking, MPa.

Before calculating the strength of the steel profiled flooring on the support, the stability condition of the compressed wide flange is checked (Fig. 2):

$$\sigma_n = \frac{M_{sup}}{W_{x2}} \leq 34,3 \cdot 10^4 (tb^i)^2; \tag{3}$$

where $b_i = 9{,}3 - 2 \cdot r$.

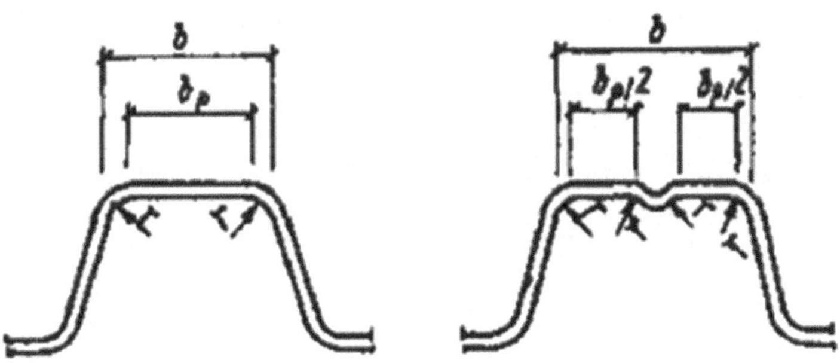

Fig. 2. To determine the width of flat sections of compressed flanges

The rigidity of the steel profiled decking at the construction stage is checked according to the formula:

$$f_n = \frac{k_n (q^n i_n^4)}{E_n I_x} + a \leq \frac{i_n}{200}; \tag{4}$$

where

f_n – deflection of the deck under load in the middle of the outer span, sm;

k_n – coefficient determined depending on the layout of the flooring (for flooring with a span of three or more 0,0088);

q^n – total regulatory load, N/sm;
i_n– calculated deck span, sm;
I_x – calculated moment of inertia of the considered section of the deck, sm^4;
a – empirical value equal to 2 mm for multi-span decks.

Calculation for the Operational Stage. The calculation of the strength of normal sections to the longitudinal axis of a bent element is performed according to the formulas:

$$M_{span} \leq R_b \cdot b_f \cdot x \cdot (h_0 - 0,5x) + R_{sc} \cdot A_s^{'}\left(h_0 - a^{'}\right); \tag{5}$$

$$M_{span} - M_n \leq R_b \cdot S_{bx} + \gamma_c \cdot R_y \cdot \left[S_n^{'} + t(x - h_f)^2 + t(h_n - x)^2 + S_n^{''}\right] + R_s(6) \cdot S_{sx} + R_{sc} \cdot S_{sx}; \tag{6}$$

$$M_{span} - M_n \leq 0,5 \cdot R_b \cdot b_f \cdot h_f^2 + \gamma_c \cdot R_s\left(A_n^{''} \cdot h_n + t \cdot h_n^2\right) + R_s \cdot A_s(h_n - a) + R_{sc} \cdot A_s\left(h_f - a^{'}\right); \tag{7}$$

where:

M_{span} – bending moment in the considered section of the slab, N•m;
R_b – design resistance of concrete to axial compression, MPa;
b_f – width of one corrugation of steel profiled decking, sm;
x – height of the compressed zone of concrete, sm;
$h0$ – working section height, sm;
R_{sc} – design resistance of rebars to compression, MPa;
A'_s – cross-sectional area of rebars located in the compressed zone, sm^2;
a' – distance from the resultant forces in the bar reinforcement in the compressed zone to the nearest face of the section, sm;
A''_n – cross-sectional area of the lower flange of one corrugation, sm^2;
S_{bx} – static moment of the sectional area of compressed concrete relative to the axis X-X, sm^3;
A'_n – cross-sectional area of the upper flange of one corrugation, sm^2;
S'_n – static moment of the area of the upper flange of the profiled flooring relative to the axis X-X, sm^3;
S''_n – static moment of the area of the bottom flange of the profiled flooring relative to the axis X-X, sm^3;
S_{sx}, S'_{sx} – static moments of the area of the tensile and compressed bar reinforcement, respectively, relative to the axis X-X, sm^3.

If the neutral axis is within the thickness of the slab flange and does not cross the walls of the profiled flooring, then formula (5) is used, if it does, then formula (6), and if it is at the level of the upper flange of the profiled flooring, then formula (7).

Strength calculation of inclined sections to the longitudinal axis is performed according to the formulas:

$$Q \leq 0,17R_n h_n 2t + \sum R_{sw} A_{sw} + Q_b; \tag{8}$$

$$Q \leq 0,3\varphi_{w1}\varphi_{b1}R_b\left(\frac{b+b'}{2}\right)\cdot h_0; \tag{9}$$

where:

Q – lateral force, N;
$0{,}17R_n h_n 2t$ – transverse force absorbed by the sheeting walls in one corrugation;
$\sum R_{sw}A_{sw}$ – the sum of the shear forces taken by the transverse rods crossing the inclined section;
Q_b – transverse force absorbed by concrete, N;
φ_{w1} and φ_{b1} – coefficients taken by SNiP 2.03.01–84.

The calculation of the floor slab for deformations is performed according to the formula:

$$f_m = f_{rc} + f_{add} \leq \frac{i}{150}; \tag{10}$$

where:

f_m – full deflection, sm;
f_{rc} – deflection of a reinforced concrete slab due to loads during operation (excluding its own weight), sm;
f_{add} – additional deflection of a reinforced concrete slab due to the compliance of anchor ties, sm.

2.2 Eurocode-4

Calculation for the Construction Stage. The verification of the fulfillment of the conditions of the bearing capacity in the span and on the support during bending is carried out according to the formulas:

$$\frac{\mathrm{M}_{\pi \mathrm{p,d}}}{\mathrm{M}_{\mathrm{Rd,1}}} < 1; \tag{11}$$

$$\frac{\mathrm{M}_{\pi \mathrm{p,d}}}{\mathrm{M}_{\mathrm{Rd,2}}} < 1; \tag{12}$$

where:
$M_{np,d}$ – maximum bending moment in span;
$M_{on,d}$ – maximum bending moment of supports;

$M_{Rd,1}$, $M_{Rd,2}$ – standard values of the bearing capacity of the steel-reinforced concrete section for the bending moment in the span and on the support.

The maximum deflection of a profiled sheet during concreting should not exceed the limit value:

$$w_{max} < \frac{L}{180}. \tag{13}$$

Calculation for the Operational Stage. Condition to ensure the bearing capacity in bending:

$$\frac{M_{Ed}}{M_{pl,Rd}} < 1; \tag{14}$$

where

M_{Ed} – bending moment from the calculated combinations of actions in the middle of the span;

$M_{pl,Rd}$ – bearing capacity of the section in bending.

The condition of the bearing capacity of the slab in longitudinal shear:

$$\frac{V_{Ed}}{V_{1,Rd}} < 1. \tag{15}$$

where

V_{Ed} – lateral force from design actions;

$V_{1,Rd}$ – bearing capacity of a slab in longitudinal shear, determined by the m-k method.

The condition for the bearing capacity of the slab at transverse shear:

$$\frac{V_{Ed}}{V_{v,Rd}} < 1. \tag{16}$$

where

$V_{v,Rd}$ – load-bearing capacity of a slab in transverse shear, determined without taking into account the work of the profiled sheet.

Plate stiffness condition:

$$w_{fin} = w_p + w < L/250 \tag{17}$$

where

w_{fin} – full slab deflection;

w – deflection of the slab due to characteristic actions.

3 Results

When calculating the overlap in accordance with STO 0047–2005, the following materials were adopted:
– profiled sheets of grade H80A-674–0.9;
– concrete of class B20 (Table 1).

Table 1. Determination of floor loads according to STO 0047–2005

Description	Normative load on 1 m^2, N	Overload factor	Design load on 1 m^2, N
Construction stage			
Self-weight load of the profiled sheet	150	1,05	157,5
Load from the mass of freshly laid concrete mixture	2290	1,2	2748
Installation load: when unloading concrete mix from buckets with a capacity of up to 0,8 m^3	2500	1,3	3250
Total	**4940**		**6155,5**
Operation stage			
Self-weight load of the profiled sheet	150	1,3	195
Concrete mass load	2290	1,3	2975
Tile + glue (t = 20 mm)	390	1,3	505
Cement-sand screed (t = 40 mm)	710	1,3	925
Suspended ceiling	500	1,3	650
Payload	4000	1,2	4800
Total	**8040**		**10050**

Based on the calculation of a multi-span continuous beam, we find that at the construction stage, the bending moment in the extreme span from the design loads is M_{span} = 770 N•m, on the middle support - M_{sup} = 962 N•m, lateral force from design load - Q = 4617 N.

At the stage of operation, the bending moment in the extreme span from the design loads is equal to M = 1256 N•m, lateral force from design load - Q = 7538 N.

When calculating the overlap according to Eurocode-4, the following materials were used:

– profiled sheets of grade H80A-674–0.9;
– concrete of class C20/25 in accordance with EN 206:2013 [11].

Table 2. Determination of floor loads according to Eurocode-4

Impact type	Impact characteristic value (kN/m^2)	Partial safety factor	Estimated Impact Value (kN/m^2)
Construction stage			
Self-weight load of profiled sheet	0,15	1,1	0,165
Load from the mass of freshly laid concrete mixture	2,29	1,5	3,44
Total permanent load p	2,44		3,6
Installation load q	1,0	1,5	1,5
Total	**3,44**		**5,1**
Operation stage			
Concrete slab mass load	2,29	1,1	2,52
Weight load of profiled sheets	0,15	1,1	0,165
Total permanent load p	4,04	1,1	4,44
Variable payload q	4,0	1,5	6
Combination of influences p + q	**8,04**		**10,44**

The joint work of a profiled sheet with concrete is carried out due to their mechanical adhesion, the parameters of which, according to experimental data, have the following values: m = 184 N/mm^2, k = 0,053 N/mm^2.

Table 3. Calculation results

Index	STO 0047–2005	Eurocode-4
Construction stage		
Strength of support sections	5,35 MPa < 140 MPa	0,84 kNm < 10,1 kNm
Strength of spans	20 MPa < 220 MPa	0,69 kNm < 8,5 kNm
Compressed wide flange stability	21 MPa < 40,3 MPa	–
Deflection of steel deck	0,23 sm < 0,63 sm	0,03 sm < 0,69 sm
Operation stage		
Strength of normal section in span	1256 N·m $\leq$ 3795 N·m	2040 N·m < 4560 N·m
Cross-sectional shear strength	7538 N $\leq$ 13450 N	6530 N < 27300 N
Full deflection	0,2 sm $\leq$ 0,83 sm	0,14 sm < 0,5 sm

4 Conclusion

From a comparison of the results of calculations of steel-reinforced concrete floors according to the norms of the Russian Federation and according to Eurocodes, as well as on the basis of an analysis of the research results of other authors, the following conclusions can be formulated:

- The basic principles for assigning the characteristics of concrete and reinforcement in the Eurocode and in Russian standards are close to each other (strength classes, standard resistances, safety factors for materials, deformation diagrams, deformation characteristics). However, the quantitative values of the characteristics of concrete and reinforcement are different. Methods of conformity assessment differ.
- The Eurocode includes design characteristics for concrete in terms of cylindrical strength, and in Russia - in terms of prismatic strength.
- Methods for calculating the strength of reinforced concrete elements to the action of bending moments and longitudinal forces, as the main calculation of reinforced concrete elements, which determines the number of longitudinal reinforcement, in the Eurocode and in Russian standards are based on uniform prerequisites for the application of equilibrium equations (hypotheses of flat sections and deformation diagrams of concrete and reinforcement) and differ only in some details.
- Methods for calculating the strength of reinforced concrete elements against the action of transverse forces and torques in Russian standards and in the Eurocode are fundamentally different. In Russian standards, these methods are based on the technique of inclined and spatial sections, in the Eurocode - on conditional bar models.
- The general principles for calculating reinforced concrete elements for the second limit state (for deformations) in the Russian standards and in the Eurocode are close to each other. The differences lie in individual design parameters. In the Eurocode, approximate methods for assessing deformations are used, and in Russian standards, the calculated dependences are given in a more complete and detailed form.
- The requirements for the design of steel-reinforced concrete structures given in Russian standards are generally close to the Eurocode. This applies, first of all, to the anchoring of the reinforcement, the thickness of the concrete cover, the distance between the reinforcing bars.
- Russian normative documents contain more complete requirements for the calculation of structural systems (calculation for stability, determination of horizontal and vertical displacements, calculation for progressive failure, determination of forces in the elements of a structural system, etc.), as well as guidelines for determining the stiffness characteristics of reinforced concrete elements for various design situations.
- Many design provisions in the Eurocode are given in the form of textual formulations, while in Russian regulatory documents the design material is present in the form of a sequential system of design dependencies.
- The Eurocode does not contain documents on the design of prefabricated and precast-monolithic floor structures and large-span roofs, taking into account the joint work of elements (slabs with slabs, slabs with girders and walls).

- The Eurocode provides more detailed recommendations for ensuring adhesion and anchoring of reinforcement with concrete, as well as requirements for ensuring endurance, standards of variability, etc.
- A number of basic assessments of the degree of aggressive effects of environments, given in the EN and in domestic standards, practically coincide, but Russian standards are distinguished by greater detail.

Further research in this area, improving the regulatory framework for the calculation and design of steel-reinforced concrete floors will contribute to their wider application in construction and reconstruction and will make it possible to fully use the advantages of steel-reinforced concrete.

References

1. Ahmed, I.M., Tsavdaridis, K.D.: The evolution of composite flooring systems: applications, testing, modelling and eurocode design approaches. J. Constr. Steel Res. **155**, 286–300 (2019)
2. Medvedev, V.N., Semeniuk, S.D.: Durability and deformability of braced bending elements with external sheet reinforcement. Mag. Civil Eng. **63**(3), 3–15 (2016)
3. Holomek, J., Bajera, M., Vilda, M.: Test arrangement of small-scale shear tests of composite slabs. Procedia Eng. **161**, 716–721 (2016)
4. Derysz, J., Lewiński, P.M., Więch, P.P.: New concept of composite steel-reinforced concrete floor slab in the light of computational model and experimental research. Procedia Eng. **193**, 168–175 (2017)
5. Holomek, J., Bajer, M., Vild, M.: Cast screws as shear anchors for composite slabs. Procedia Eng. **195**, 114–119 (2017)
6. Rybakov, V.A., Ananeva, I.A., Rodicheva, A.O., Ogidan, O.T.: Stress-strain state of composite reinforced concrete slab elements under fire activity. Mag. Civil Eng. **74**(6), 161–174 (2017)
7. Gholamhoseini, A., Gilbert, R.I., Bradford, M.A.: Creep and shrinkage effects on the bond-slip characteristics and ultimate strength of composite slabs. J. Civil Eng. Architecture **82**, 1085–1097 (2014)
8. Gholamhoseini, A., Gilbert, R.I., Bradford, M.: Long-term behavior of continuous composite concrete slabs with steel decking. ACI Struct. J. **115**(2), 439–449 (2018)
9. STO 0047–2005 Steel-reinforced concrete floors with a monolithic slab on profiled steel flooring. Calculation and design. Homepage. https://docs.cntd.ru/document/1200080354.
10. EN 1994 Eurocode 4: Design of composite steel and concrete structures.
11. GOST R 57345–2016 / EN 206–1: 2013 Concrete. General specifications. Homepage. https://docs.cntd.ru/document/1200142845.
12. SP 63.13330.2018 Concrete and reinforced concrete structures. The main provisions. Homepage. https://docs.cntd.ru/document/554403082.
13. Hicks S.J.: Eurocode 4: Design of Composite Steel and Concrete Structures. In: The essential guide to Eurocodes transition, pp. 105–122. BSI (2010).
14. Tam, C.T., Babu, D.S., Li, W.: EN 206 conformity testing for concrete strength in compression. Procedia Eng. **171**, 227–237 (2017)

Some Issues of the Functioning of Electric Power Systems with Distributed Generation Sources

Muhayo Islomovna Toshkhodzhaeva[1(✉)], Elena Ivanovna Gracheva[2], Akram Akhmadovich Mirzoakhmedev[1], and Shakhboz Tolibjonovich Dadabaev[1]

[1] Department of Power Supply and Automation, Khujand Polytechnic Institute of the Tajik Technical University Named After Academician M.S. Osimi, Khujand 735700, Tajikistan
shukronal4_01_2011@mail.ru

[2] Department Power Supply of Industrial Enterprises, Kazan State Power Engineering University, Kazan 420066, Russian Federation

Abstract. The article provides an overview of regulatory legal acts, incentive mechanisms, the share of implementation of distributed generation systems, considers the features of a centralized and decentralized power supply system in terms of changing operating conditions. The possibilities of using systems with distributed generation sources are considered. Ways of integrating distributed generation systems that form several local sources of electrical energy with a centralized source. As an example, the power supply system of industrial enterprises of the Konibodom district of the Sughd region is considered. The existing single-line diagram of power supply of industrial enterprises is presented. The perspective loads of industrial enterprises are analyzed. The perspective consumed active, reactive and full power of enterprises, power factor taking into account the length of the distribution network are given. The calculated currents are determined in all branches of the circuit with a centralized power supply. Taking into account the power consumption and geographical location, a design diagram was drawn up with three power sources. When calculating the electrical circuit, the method of two nodes was applied. The total, active, reactance of the circuit with three power supplies has been determined. The currents in individual branches of the power supply system of industrial enterprises are determined.

Keywords: Electric power systems · Distributed generation sources

1 Introduction

In developed countries of the world (USA, Denmark, Norway, Germany, Canada, etc.), along with the development of centralized power supply, systems with local energy sources are being intensively introduced [1]. The purpose of the presented article is to review and analyze the implementation of distributed generation systems in various countries of the world (Russian Federation and China) and the management of the electric power system, when several local energy sources are connected.

A. Sinitsyn (Ed.): SES 2020, LNCE 141, pp. 331–338, 2021.
https://doi.org/10.1007/978-3-030-67654-4_36

Table 1. Prospective design active and reactive load taking into account the length of the supply lines of the above industrial enterprises.

№	Name enterprise	P_H kWt	L km	$cos\varphi$	$tg\varphi$	I, A	P_P kWt	Q_p kVAr	S_p kVA
1	Konibodom cement plant	2037.5	0.8	0.8	0.56	117.7	1628.9	912.2	1866.9
2	JSC Bodom	13071.9	1850	0.75	0.58	705.2	9149.9	5306.9	1057.5
3	Orien Cotton Company	1250	1500	0.8	0.56	72.25	999.9	559.9	1145.9
4	Naftrason Company	2973.68	3000	0.76	0.75	171.8	2318.3	1738.7	2897.8
5	Asphalt plant	818.2	3000	0.77	0.69	79.7	638	439.5	773.9

With a centralized scheme of the electric power system, the distribution of the power flow is carried out from one conventional source to the consumers and the selectivity of the relay protection settings is adjusted according to their response time. However, in the presence of several sources in one system, ensuring the selectivity of relay protection and automation devices is a complex mechanism for a number of reasons:

1. The consumer of electricity can simultaneously function as a source of electricity, therefore, a strict detuning of the relay protection and control devices is required, which is coordinated with the setting maps of the operational dispatch service;
2. In case of violation of synchronization in one of the sources, there is a probability of strong disturbances in the power system, i.e. cascade development of the accident occur.
3. In case of failure of a local energy source, the power system must provide the required power for this consumer;
4. The need to maintain the permissible network frequency parameters during emergency and post-emergency power system modes.

In the process of reforming the electric power system, i.e. in the transition from a traditional system to a decentralized system, the management of the system becomes difficult, in view of the fact that not only the distribution of the power flow changes, but also the problem of timely elimination of emergency modes arises, in completely different conditions [2].

Depending on the economic, social, climatic characteristics, various mechanisms for managing and stimulating distributed generation (DG) systems are used.

For example, in Russia there are a number of regulatory and legal acts that stimulate and regulate the implementation of systems DG:

Russia's energy strategy for the period up to 2030 (approved by the order of the Government of the Russian Federation of November 2009 No. 1715);

Decree of the Government of the Russian Federation on the promotion of the use of renewable energy sources in the retail electricity markets;

Resolution of the Government of the Russian Federation of May 28, 2013 No. 449 (as amended on February 28, 2017) "On the Mechanism for Encouraging the Use of Renewable Energy Sources in the Wholesale Electricity and Power Market;

The draft energy strategy of the Russian Federation for the period up to 2035;

The rule of technological connection of power receivers of electricity consumers, power generation facilities, as well as power grid facilities belonging to grid organizations and other persons, etc. [3, 4].

The analysis of the above normative legal acts showed that currently in the Russian Federation there are no normative legal documents that regulate the technical requirements for DG facilities, taking into account their technological features, connection and functioning as part of the Energy System of the Russian Federation.

The share of DG systems implementation in the Russian Federation is 8% [5]. In the Russian Federation, the following sources of DG are mainly used: solar, wind cogeneration (gas turbine, gas piston) sources of various capacities.

In the Russian Federation, a mechanism has been introduced to stimulate the production of electricity and capacity through contracts for the provision of capacity by qualified generating facilities [6]. At the state level, support is provided for small generations based on renewable energy sources. The Government of the Russian Federation has developed regulations, according to which individuals can sell electricity from micro generation plants up to 15 kW, operating on the basis of renewable energy sources, to the general network. At the same time, the guaranteeing suppliers (the power grid) will be obliged to purchase the volumes of electricity created by private generation, and from this income will not be subject to taxation [5, 8–10].

The regulatory framework for the introduction of DG sources in China is regulated by the law on renewable energy sources and White paper on Energy [7]. China is the world leader in terms of the volume of renewable energy investments and in terms of the total capacity of renewable energy equipment and by 2020 according to the share of DG sources in total energy consumption, it was 18%. The use of renewable energy sources is among the priorities of the Chinese energy policy. To ensure this process, a flexible regulatory and legal framework has been created.

The mechanisms for stimulating the introduction of DG sources are:

- Priority to energy saving;
- Reliance on internal, including renewable and non-traditional resources;
- Stimulating the development of new energy technologies;
- Encouraging different development models and deepening reforms in the energy sector;
- Improvement of the fuel and energy pricing system;
- Environmental protection;
- International cooperation and cooperation based on mutual benefit.

The possibility of functioning of distributed generation sources both for a common network with centralized sources and for individual consumers creates good conditions for the integration of the sectors of centralized and decentralized electricity production. In the decentralized sector, sources of distributed generation have an advantage, although local power supply systems can also be formed here, combining several distributed sources for joint work on a single network [11].

2 Methodology

As an example, consider the power supply system of industrial enterprises from the substation “Kanibadam 220/110/10 kV” located in the Sughd region (Fig. 1). As can be seen from the figure, the power supply circuit has a radial shape and is powered from a single source.

From this substation, such enterprises as the Konibodom cement plant, Bodom OJSC, Orien Cotton company, Naftrason company and an asphalt plant receive power. The prospective calculated active and reactive load, taking into account the length of the supply lines of the above industrial enterprises, is given in Table 1. The actual load is taken as the prospective load, taking into account its growth by 20% [12, 13].

The rated current is determined by the formula:

$$I = \frac{S_p}{\sqrt{3} U_H cos\varphi} \tag{1}$$

where, S_p is the total design capacity of the enterprise, U_H - rated voltage of the supply network, we take 10 kV; *cos* φ - power factor of the enterprise.

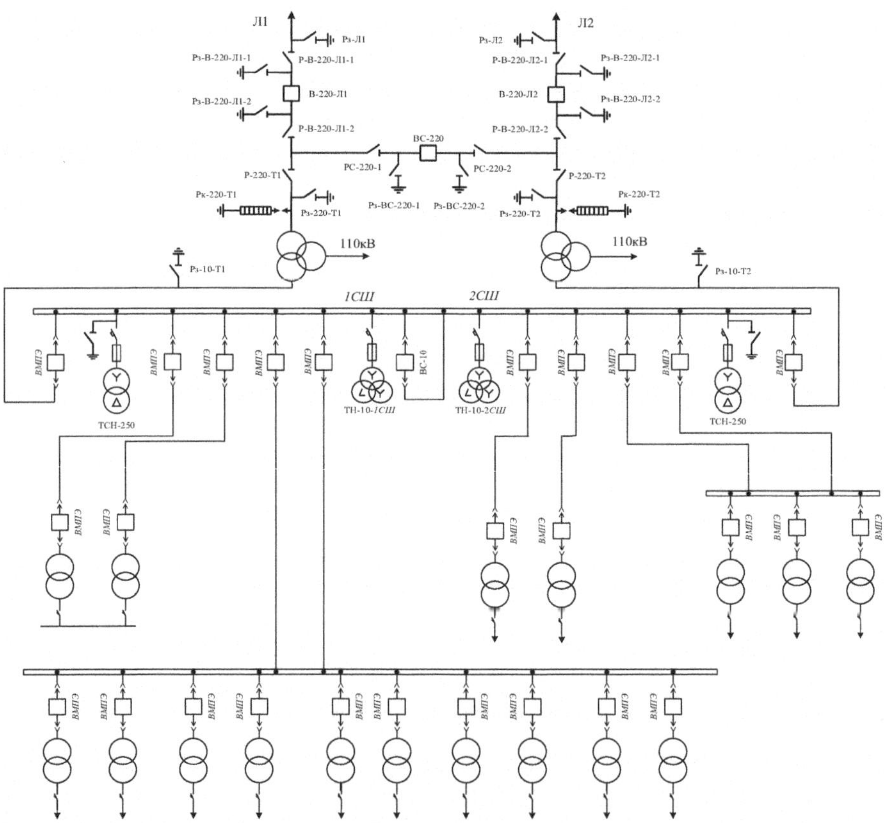

Fig. 1. Power supply system of industrial enterprises from the substation “Konibodom 220/110/10 kV”

3 Experimental Part

When introducing distributed generation systems, it is planned to install at each enterprise its own source of electricity, equal to the prospective load [14].

EMF E and load Z are taken as sources. To calculate this scheme, the method of two nodes is used - a method for calculating electrical circuits, in which the voltage between the two nodes of the circuit is taken as the desired one (with its help, the currents of the branches are then determined) [15]. Formula for calculating the voltage between two nodes (Fig. 1):

Based on the geographical location and capacity of enterprises, the power supply scheme of the distribution network changes, i.e. a network with a common load node with three sources is created between the above-mentioned enterprises. The calculated scheme is shown in Fig. 2.

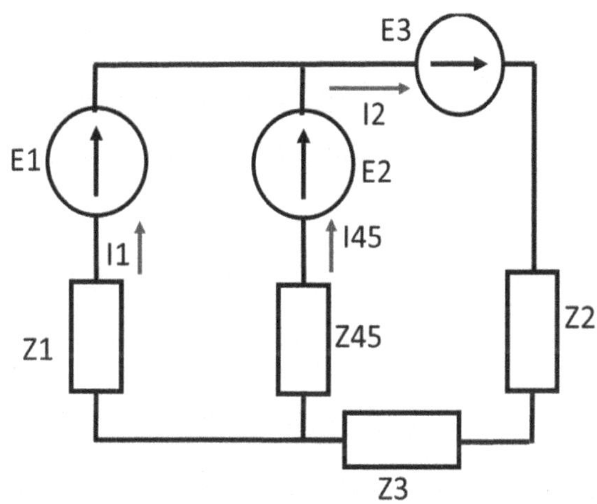

Fig. 2. Design scheme of enterprises, taking into account sources distributed generation.

$$U_{ab} = \frac{\sum E_k y_k}{\sum y_k} \tag{2}$$

where, E_k is the voltage of the EMF source of the k-th branch; y_k - conductivity of the k-th branch.

The current in a separate branch is determined from the expression:

$$I_n = (E_n - U_{ab})y_k \tag{3}$$

The design impedance is determined by the formula:

$$z_{\Pi} = \frac{S_p}{\sqrt{3} I_p^2} \quad (4)$$

where, S_p is the total power of the enterprise, kVA; I_p- rated current at the enterprise, A.

Active and reactance is defined as the geometric sum of the impedance Z, Ohm:

$$z_{\Pi} = \sqrt{(r^2 + x^2)} \quad (5)$$

where, r is the active resistance of the circuit; x is the reactance of the circuit.

To determine the active and reactive components of the circuit, a given power factor is used according to Table 1:

$$r = z_{\Pi} cos\varphi \quad (6)$$

The conductivity of individual branches is determined by the formula:

$$y = \frac{1}{z_{\Pi}} \quad (7)$$

The calculated data of the presented scheme are shown in Table 2.

Table 2. Calculated data of the presented scheme.

№	Name enterprise	z_{Π} Ohm	r Ohm	x Ohm	y	I A
1	Konibodom cement plant	47,6	38,08	28,56	0,021	37,8
2	Bodom OJSC, Orien Cotton Company	12,8	9,6	8,46	0,078	140,4
3	Naftrason company, Asphalt plant	77,5	62	46,5	0,013	161,9

As can be seen from the calculation results, with the introduction of distributed generation sources and separation of electricity consumers from the main source, the current distribution, and, therefore, the power flow change its direction can be used as sources of distributed generation, based on the capabilities of enterprises.

4 Conclusion

Thus, to stimulate DG systems, three main conditions must be met:

- Adoption of a national strategy in the field of renewable energy source and DG (task setting);
- Adoption of a regulatory and legal framework (establishment of structure and rules in the market);
- Increased transparency and fairness of competition from the traditional energy sector.

Thus, a distributed generation system in low voltage nodes of industrial enterprises is acceptable in conditions of power limitation. In the future, in the event of excess capacity, industrial enterprises can export electricity to the power system. Renewable energy sources can be used as sources of distributed generation, based on the capabilities of enterprises.

References

1. Hazra, J., Sinha, A.K.: Congestion management using multi objective particle swarm optimization. IEEE Trans. Power Syst. **22**(4), 1726–1734 (2007)
2. Huang, W., et al.: A review on micro grid technology containing distributed generation system. Power Syst. Technol. **9**, 006 (2009)
3. Popadyuk, T.G., Kupreyev, D.A.: Stimulirovaniye innovatsionnogo razvitiy araspredelonnoy energetiki (Stimulation of innovative development of distributed energy). SRRM 3, 108. (in Russian)
4. Chernyshev, A.V.: Stimulirovaniye ispol'zovaniya vozobnovlyayemykh istochnikov energii i faktory, prepyatstvuyushchiye razvitiyu netraditsionnoy energetiki v Rossii (Stimulation of the use of renewable energy sources and factors hindering the development of non-traditional energy in Russia). IzvestiyaSPbGEU **5**, 83–87 (2013). (in Russian)
5. Kharitonov, S.V., Churikova, P.A.: Vyrabotka metodov poderzhki i stimulirovaniya innovatsionnoy deyatel'nosti v elektroenergetike Rossii v ramkakh « Tsifrovogoperekhoda » (Development of methods to support and stimulate innovation in the electric power industry of Russia within the framework of the Digital transition). BIZNES – PARK UNIKUM-M **2**, 77-88 (2019). (in Russian)
6. White Paper on Energy, China's energy conditions and politicians (2007)
7. Tolibjonovich, D.S., Islomovna, T.M., Saidulloevna, M.D.: Modeling of starting transition processes of asynchronous motors with reduced voltage of the supply network. Eur. J. Electr. Eng. **22**(1), 23–28 (2020). https://doi.org/10.18280/ejee.220103
8. Gracheva, E.I., Naumov, O.V.: Vliyanie rezhima raboty na veroyatnostnye kharakteristiki elektricheskikh ustroystv (Operating Mode Influence on Probability Characteristics of Electric Devices). J. Eng. Appl. Sci. **11**(1), 2934–2938 (2016)
9. Gracheva, E.I., Naumov, O.V.: Estimation of power losses in electric devices of the electrotechnical complex. In: International Conference on Industrial Engineering, Applications and Manufacturing (ICIEAM), p. 6 (2019)
10. Khasanov, S.R., Gracheva, E.I., Toshkhodzhaeva, M.I., Dadabaev, S.T., Mirkhalikova, D.S.: Reliability modeling of high-voltage power lines in a sharply continental climate. In: E3S Web of Conferences, vol. 178, p. 01051 (2020). https://doi.org/10.1051/e3sconf/202017801051
11. Bessonov, L.A.: Teoreticheskiye osnovy elektrotekhniki, Elektricheskiye tsepi (Theoretical foundations of electrical engineering, Electrical circuits. Moscow, Vyssh. shk. (1996). (in Russian)
12. Voropay, N.I., Keyko, A.V., Kler, A.M., Stennikov, V.A.: Tekhniko-ekonomicheskiye problemy ispol'zovaniya netraditsionnoy energetiki (Technical and economic problems of the use of non-traditional energy). Problemyne traditsionnoy energetiki, Mat-lynauch. sessiiPrez., SO RAN, 13.12.2005, Novosibirsk: Izd-vo SO RAN, pp. 32–54 (2006). (in Russian)
13. Ackermann, T., Anderson, G., Soeder, L.: Distributed generation: a definition. Electric. Power Syst. Res. **57**, 195–204 (2001)

14. Decentralized generation technologies: potentials, success factors and impacts in the liberalized EU energy markets, Final report, DECENT 234, October 2002
15. World survey of decentralized energy 2005 [Electronic resource], WADE, Edinburgh, 45 (2004). http://www.localpower.org/documents_pub/report_worldsurvey05.pdf

The Qualimetric Method for Optimising the Selection of Components for Making High Quality Asphalt-Concrete Mixes

Nikolay Ermoshin and Aleksei Ismailov(✉)

Peter the Great St. Petersburg Polytechnic University, Saint-Petersburg, Russia
ismailov-aleksei@mail.ru

Abstract. The article describes the qualimetric method for optimising the selection of components for the preparation of high quality asphalt mixtures. This method can be used to calculate a complex integrated quality indicator of the basic components of an asphalt-concrete mix and for completing various tasks. These include analysing the quality of crushed stone, analysing the qualitative characteristics of crushing screenings; crushing screenings and mineral powder; analysing the qualitative characteristics of cementing media and making a choice of the best quality basic components for asphalt-concrete mixes. The laboratory tests as an experimental study were carried out to make precise selection of the high quality components for asphalt-concrete mixes using special method. Experimental data was obtained and analysed in an accredited testing laboratory. The efficiency of applying the qualimetric method was substantiated through these experiments. The possibility of reducing costs at the design stage by automatic control was also proven, as was, the possibility of reducing or eliminating hard laboratory operations for the selection of asphalt-concrete mixes. Through using such a method, the production of high-quality asphalt-concrete mixes may be maximised.

Keywords: Routes · Roads · Bitumen · Increased durability · Technological processes · Technological solutions · Asphalt-concrete mixes · Asphalt-concrete granulated material · Segregation · Design · Quality control

1 Introduction

The problem selecting the basic components for making asphalt-concrete mixes of the a high quality gains more and more importance because of an increase in the overhaul life of asphalt concrete coatings of up to 12 years [1]. Many factors have an impact on the durability and the quality of asphalt-concrete road paving. Firstly, the quality of mixture design, production and placing is key. Among all the factors that have an impact on the durability of asphalt-concrete paving, determining the asphalt-concrete composition is very important [2]. It is necessary to ensure the use of only high-quality components even at the design stage of the asphalt-concrete mixture.

The most important step of quality control in asphalt-concrete mixture production is the provision of the necessary parameters of the initial inert aggregates (crushed stone, gravel and screenings) and bituminous binders. The existing technical documents for

A. Sinitsyn (Ed.): SES 2020, LNCE 141, pp. 339–350, 2021.
https://doi.org/10.1007/978-3-030-67654-4_37

asphalt-concrete mixture production [3–5] do not define the proper quality components needed for the preparation of asphalt-concrete mixtures. Moreover, the issue of proper quality components for the preparation of asphalt-concrete mixtures (ACM) have not been studied in detail. Therefore, there is a contradiction between the requirements for the design quality of asphalt-concrete mixture and the status of scientific and methodological knowledge to solve this problem [6–8].

The solution to may be found in the application of the "optimization-qualimetric approach", which is a synthesis of linear programming methods and qualimetric study of asphalt-concrete mixture quality using QFD analysis procedures [9] (Quality Function Deployment), FCA [10–12] (Functional Cost Analysis) and FMEA analysis [13] (Failure Mode and Effects Analysis).

The selection of basic components for making asphalt-concrete mixes of the necessary quality consists in optimizing available solutions taking into account the actual characteristics of road building materials (see Fig. 1).

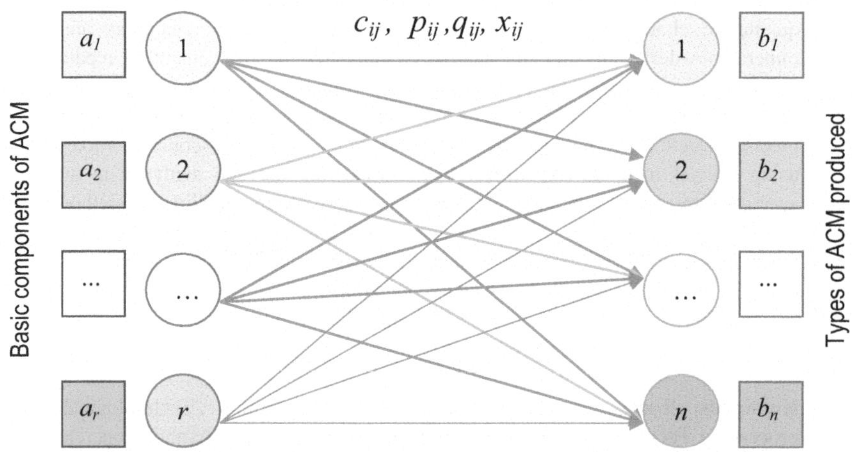

Fig. 1. A graphic model of selecting basic components for making asphalt-concrete mixes.

Such optimization shall be carried out not only for composition of asphalt-concrete mixes, but also generally for the processes at the asphalt-concrete plant (ACP). In other words, ACP shall provide construction projects with the needed types of asphalt-concrete mixes in the required volume and with the necessary quality at minimum cost

$$\sum_{i=1}^{n}\sum_{j=1}^{m}\left(\frac{c^{e}_{ij}/q^{e}_{ij}}{c_{ij}/q_{ij}}\right)\left(c^{e}_{ij,}/c_{ij,}\right)c_{ij}x_{ij} \rightarrow \min \tag{1}$$

taking into account the following limitations:

$$\sum_{i=1}^{n} x_{ij} = b_j \quad (2)$$

$$\sum_{j=1}^{r} x_{ij} \leq a_i \quad (3)$$

$$0 \leq x_{ij} \leq p_{ij} \quad (4)$$

where c_{ij}^{e} and c_{ij} – accordingly are the reference and actual total costs of the purchase of basic components and ACM production upon purchase i – type raw materials for production of j – type (brand and type) ACM, rubble;

q_{ij}^{e} and q_{ij} – accordingly, reference (normative) and actual values of the key or integral quality coefficient of ACM using i – type raw materials for production of j – type ACM (brand and type);

$c_{ij,}^{e}$ and $c_{ij,}$ – accordingly, reference and actual production costs of j – type ACM (brand and type) using i – type raw materials, rubbles;

x_{ij} – amount of j – type (brand and type) ACM produced using i - type raw materials, t;

b_j – construction demand for j – type ACM, t;

p_{ij} – ACP production capacity of j – type (brand and type) ACM using i – type raw materials, t;

a_i – volume of i – type raw materials that can be used at ACP, t;

r – number of raw materials types that can be used at ACP.

Objective function and task limitations are interpreted as follows:

objective function (1) minimizes production costs of ACM of necessary quality;

limitation (2) requires meeting the construction demand of the asphalt-concrete mixture;

limitation (3) requires that the total volume of ACM produced does not exceed the available basic components of the company;

limitation (4) takes into account production capacity of ACP p_{ij} and non-negativity requirements for x_{ij}.

As the main topic of study and a fundamentally important task for the model concerned is the determination of the method for formalization of the particular and key (integral, generalized) quality coefficient of basic components and ACM in general.

The analysis of technological regulations [14] in production quality management of the asphalt-concrete mixtures revealed the key features (Table 1), determining integral coefficient and reference (normative) values of K_i^e

$$K_i^e - \sigma \leq K_i^e \leq K_i^e + \sigma \quad (5)$$

where $\sigma_{K_i^e}$-standard deviation of random K_i^e.

Basic components selection for making asphalt-concrete mixes of the necessary quality (see Fig. 2) is based on the evaluation of laboratory results, the analysis of

Table 1. Key features determining the integral coefficient of basic components for making asphalt-concrete mixes.

Basic components for making asphalt-concrete mixes	Quality indicators	Name	Reference value, K_i^e	Unit	Convergence of the function
Crushed stone	Fraction gradation	K_1	100	%	max
	True density	K_2	2.7	g/cm^3	max
	Durability	K_3	1400	%	max
	Content of thin flat chips	K_4	10	%	min
Screening crushing	Fraction gradation	K_5	100	%	max
	True density	K_6	2.7	g/cm^3	max
	Content of dust and clay particles	K_7	0.1	%	min
Mineral powder	Fraction gradation	K_8	100	%	max
	Dilatation	K_9	2.5	%	min
	True density	K_{10}	2.82	g/cm^3	max
Bitumen	Depth of indentation	K_{11}	63	mm	min
	Melting point	K_{12}	52	°C	max
	Dynamic viscosity	K_{13}	270	Pa*c	max

physical and mechanical parameters of the basic components, as well as a comparison of their actual values of quality (taking into account the significance coefficient) with reference values.

It is used the method of multi-additive convolution of Quality Indicators to have the comprehensive quality rating of basic components for making asphalt-concrete mixes.

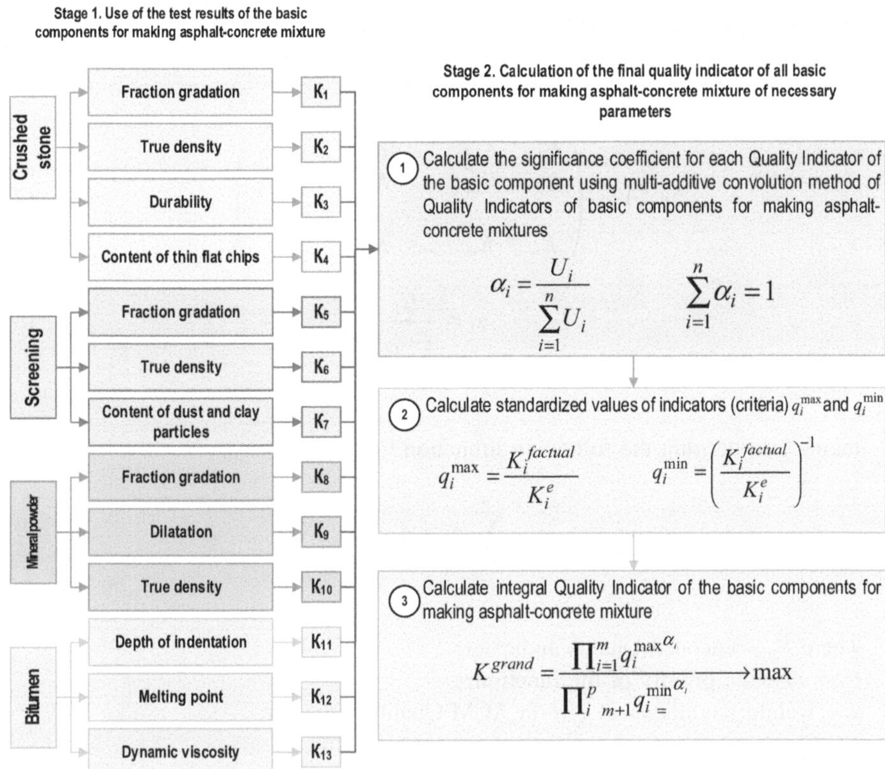

Fig. 2. Structure of the Qualimetric Method of basic component selection for making asphalt-concrete mixes of necessary quality

In this case, the calculation of a generalized or integral criterion (Quality Indicator) K^{grand} is carried out by the multiplicative convolution of Quality Indicators defined as (6)

$$K^{grand} = \frac{\prod_{i=1}^{m} q_i^{\max \alpha_i}}{\prod_{i=m+1}^{p} q_i^{\min \alpha_i}} \longrightarrow \max \tag{6}$$

where q_i^{max} – standardized value of *i*–th Quality Indicator of ACM while the function goes to the maximum;

q_i^{min} – standardized value of *i*–th Quality Indicator of ACM while the function goes to the minimum;

a_i – weight (significance) of *i*–th Quality Indicator of ACM;

The calculation of quality parameters for method (1) is carried out as follows.

1. The calculation of the significance coefficient of each Indicator is carried out by the method of pairwise comparisons – the sum of Ui assigned preference coefficients is

determined for each row of the matrix. Then the absolute priorities of the Ui function are calculated, so each row in the matrix is multiplied by a Ui column-vector (7), and using the Eq. (8), we calculate α_i, for which the matrix is presented in the Table 2.

$$A = \begin{pmatrix} K_{11} & K_{12} & \ldots & K_n \\ K_{21} & K_{22} & \ldots & K_n \\ \ldots & \ldots & \ldots & \ldots \\ K_n & K_n & \ldots & K_n \end{pmatrix} * \begin{pmatrix} U_1 \\ U_2 \\ \ldots \\ U_n \end{pmatrix} \tag{7}$$

$$\alpha_i = \frac{U_i}{\sum_{i=1}^{n} U_i} \tag{8}$$

taking into account the following limitation:

$$\sum_{i=1}^{n} \alpha_i = 1 \tag{9}$$

where K_n – encoded Quality Indicator;
U_i – absolute priority of the function;
α_i – weight (significance) of i–th ACM Quality Indicator.

Table 2. Quality comparison matrix of indicators of basic components for making asphalt-concrete mixes, where Kn is the encoded Quality Indicator, U_i is absolute priorities of the functions and α_i is a weight coefficient of Quality Indicator (1-Equal weight; 2-Some prevalence of the significance of one indicator over another, the weak dependence; 3-Strong dependence)

No	K_1	K_2	K_3	K_4	K_5	K_6	K_7	K_8	K_9	K_{10}	K_{11}	K_{12}	K_{13}	$\sum_{i=1}^{n} U_i$	U_i	α_i
K_1		3	3	3	2	3	3	2	3	3	3	3	3	34	775	0.111
K_2	1		3	3	1	2	3	3	3	2	3	3	3	30	664	0.095
K_3	1	1		2	1	1	2	1	3	1	3	3	3	22	462	0.066
K_4	1	1	2		1	1	2	1	3	1	3	3	3	22	462	0.066
K_5	2	3	3	3		3	3	2	3	3	3	3	3	34	775	0.111
K_6	1	2	3	3	1		3	1	3	2	3	3	3	28	600	0.086
K_7	1	1	2	2	1	1		1	2	1	3	3	3	21	443	0.064
K_8	2	3	3	3	2	3	3		3	3	3	3	3	34	775	0.111
K_9	1	1	2	2	1	1	2	1		1	3	3	3	21	443	0.064
K_{10}	1	1	3	3	1	2	3	1	3		3	3	3	27	572	0.082
K_{11}	1	1	1	1	1	1	1	1	1	1		2	2	14	329	0.047
K_{12}	1	1	1	1	1	1	1	1	1	1	2		2	14	329	0.047
K_{13}	1	1	1	1	1	1	1	1	1	1	2	2		14	329	0.047

The normalized values of indicators going to the maximum $q_i^{\max}$ shall be calculated using Eq. (10) and those going to the minimum $q_i^{\min}$ shall be calculated using Eq. (11)

$$q_i^{\max} = \frac{K_i^{factual}}{K_i^e} \tag{10}$$

$$q_i^{\min} = \left(\frac{K_i^{factual}}{K_i^e}\right)^{-1} \tag{11}$$

where $K_i^{factual}$ and K_i^e accordingly, actual and needed (reference) values of i–th Quality Indicator.

2 Method Validation

Method validation was done during a laboratory study and consisted of the following steps:

1) Basic component selection for making asphalt-concrete mixtures using the method.
2) Making two projected asphalt-concrete mixtures in the laboratory:

- Mixture No. 1 contains the basic components of the highest quality;
- Mixture No. 2 contains the basic components of the lowest quality;

3) Laboratory tests of the asphalt-concrete mixtures;
4) Analysis and comparison of research results.

3 Results and Discussion

The following type of asphalt concrete was chosen as laboratory samples for comparing the characteristics of "Hot dense fine grained mixture, type A, grade I" according to GOST 9128-2013, made in the laboratory of "StroyEXPERT" Ltd. in Saint Petersburg.

Quantitative and qualitative characteristics of the basic mineral components for making asphalt-concrete mixture No. 1 are in Table 3 and for making asphalt-concrete mixture No. 2 are in Table 4.

Quantitative and qualitative characteristics of the basic organic binders for making asphalt-concrete mixture No. 1 and asphalt-concrete mixture No. 2 are in Table 5.

Table 3. Characteristics of basic mineral components for making asphalt-concrete mixture No. 1

	Grain composition (through a sieve with holes, mm) % weight											True density	Durability	Thin flat chips	Content of dust and clay particles	Dilatation
	40	20	15	10	5	2.5	1.25	0.63	0.315	0.16	0.071					
Crushed stone fr. 10–20 mm	100	89.7	52.5	2.5	0.8	0.5	0.5	0.5	0.5	0.5	0.5	2.72	1200	8.0	–	–
Crushed stone fr. 5–10 mm	100	100	100	84.5	4.7	2.4	1.2	0.4	0.4	0.4	0.4	2.72	1200	8.0	–	–
Sand from crushing screenings	100	100	100	100	86.5	71.6	46.1	28.8	18.7	9.6	4.5	2.71	–	–	0.5	–
Mineral powder	100	100	100	100	100	100	100	99.6	95.5	89.6	74	2.82	–	–	–	0.7

Table 4. Characteristics of basic mineral components for making asphalt–concrete mixture No. 2

	Grain composition (through a sieve with holes, mm) % weight											True density	Durability	Thin flat chips	Content of dust and clay particles	Dilatation
	40	20	15	10	5	2.5	1.25	0.63	0.315	0.16	0.071					
Crushed stone fr. 10–20 mm	100	100	69.0	8.6	0.6	0.4	0.4	0.4	0.5	0.4	0.4	2.67	1200	7.0	–	–
Crushed stone fr. 5–10 mm	100	100	100	79.0	5.3	2.0	1.1	0.6	0.6	0.6	0.6	2.67	1200	12.0	–	–
Sand from crushing screenings	100	100	100	100	91.4	73.4	44.9	29.3	18.9	11.8	5.2	2.69	–	–	0.7	–
Mineral powder	100	100	100	100	100	100	100	99.6	95.5	89.6	74.0	2.82	–	–	–	1.3

Table 5. Characteristics of the basic organic binders for making asphalt-concrete mixtures

Binder name	Depth of indentation at 25 °C, 0,1 mm	Melting point, °C	Dynamic viscosity at 60 °C, Pa*c
Binder BND 60/90 JSC "GazpromNeft-Moscow NPZ"	69	49	178
Binder BND 60/90 LUKOIL-NizhnyNovgorodnefte-orgsintez LLC	71	50	236

Analysis data of a set of basic components using the method are in Table 6 for mixture No. 1, are in Table 7 for mixture No. 2.

Table 6. Analysis data of quality characteristics of the basic components for making asphalt-concrete mixture No. 1 according to the method.

K_n	Characteristic	Convergence of the function	K_i^e	$K_i^{factual}$	q_i^{max}/q_i^{min}	α_i	K^{grand}
K_1	Fraction gradation	max	100	87.5	0.8750	0.111	0.93
K_2	True density	max	2.7	2.72	1.0074	0.095	
K_3	Durability	max	1400	1200	0.8571	0.066	
K_4	Content of thin flat chips	min	10	8	1.2500	0.066	
K_5	Gradation	max	100	86.5	0.8650	0.111	
K_6	True density	max	2.7	2.71	1.0037	0.086	
K_7	Content of dust and clay particles	min	0.1	0.5	0.2000	0.064	
K_8	Gradation	max	100	74	0.7400	0.111	
K_9	Dilatation	min	2.5	0.7	3.5714	0.064	
K_{10}	True density	max	2.82	2.82	1.0000	0.082	
K_{11}	Depth of indentation	min	63	71	0.8873	0.047	
K_{12}	Melting point	max	52	50	0.9615	0.047	
K_{13}	Dynamic viscosity	max	270	236	0.8741	0.047	

Table 7. Analysis data of quality characteristics of the basic components for making asphalt-concrete mixture No. 2 according to the method

K_n	Characteristic	Convergence of the function	K_i^e	$K_i^{factual}$	q_i^{max}/q_i^{min}	α_i	K^{grand}
K_1	Fraction gradation	max	100	74	0.7400	0.111	0.91
K_2	True density	max	2.7	2.67	0.9889	0.095	
K_3	Durability	max	1400	1200	0.8571	0.066	
K_4	Content of thin flat chips	min	10	7	1.4286	0.066	
K_5	Gradation	max	100	91.5	0.9150	0.111	
K_6	True density	max	2.7	2.69	0.9963	0.086	
K_7	Content of dust and clay particles	min	0.1	0.7	0.1429	0.064	

(*continued*)

Table 7. (*continued*)

K_n	Characteristic	Convergence of the function	K_i^e	$K_i^{factual}$	q_i^{max}/q_i^{min}	α_i	K^{grand}
K_8	Gradation	max	100	74	0.7400	0.111	
K_9	Dilatation	min	2.5	0.7	3.5714	0.064	
K_{10}	True density	max	2.82	2.82	1.0000	0.082	
K_{11}	Depth of indentation	min	63	69	0.9130	0.047	
K_{12}	Melting point	max	52	49	0.9423	0.047	
K_{13}	Dynamic viscosity	max	270	156	0.5778	0.047	

Samples of tested asphalt-concrete mixtures were made and tested according to GOST 12801-98 "Materials based on organic binders for road and airfield construction. Test Methods" (with amendments no. 1) The results are shown in Tables 8 and 9.

Table 8. Acceptance test results of asphalt-concrete mixture No. 1

Acceptance tests	Requirements of GOST 9128-2013	Factual data
Water saturation, % vol	from 2.0 to 5.0	2.00
Water resistance	not less than 0.90	0.99
Compressive strength at 20 °C, MPa	not less than 2.5	3.90
Compressive strength at 50 °C, MPa	not less than 1.0	1.34

Table 9. Acceptance test results of asphalt-concrete mixture No. 2

Acceptance tests	Requirements of GOST 9128-2013	Factual data
Water saturation, % vol	From 2.0 to 5.0	3.10
Water resistance	Not less than 0.90	0.92
Compressive strength at 20 °C, MPa	Not less than 2.5	3.36
Compressive strength at 50 °C, MPa	Not less than 1.0	1.26

4 Conclusion

The proposed method shall primarily solve the problems of managing the production of asphalt concrete mixtures while selecting their compositions.

The results obtained allow us to draw the following conclusions:

1. The method allows a choice of materials at the design stage of composition to obtain the asphalt mixture with the best characteristics through taking into account the quality variability of the basic components for making asphalt-concrete mixtures.
2. The Qualimetric method for optimising the selection of basic components for making asphalt-concrete mixes of the necessary quality allows:
a. the analysis of the quality characteristics of crushed stone;
b. the analysis of the quality characteristics of screenings;
c. the analysis of the quality characteristics of aggregate;
d. the analysis of the quality characteristics of binder;
e. the choice of the best quality basic components.
3. The method was tested in the laboratories "Stroy EXPERT" LLC and JSC ABZ "Magistral".

References

1. On the standards of financial costs and the Rules for calculating the size of the federal budget appropriations for overhaul, repair and maintenance of highways of federal significance. Decree of the Government of the Russian Federation of May 30, 2017 N [Electronic resource]
2. Ismailov, A.M.: Optimization-qualimetric model of the process of designing asphalt-concrete mixes of required quality. Mod. High Technol. **11**, 270–279 (2019)
3. GOST R 54401–2011. General automobile roads. Hot asphalt concrete road. Technical requirements. Standartinform, p. 19 (2012). Accessed 01 May 2012
4. STO NOSTROY 2.25.39–2011. The device of asphalt concrete pavement made of cast asphalt concrete. Publishing House "BST", p. 28 (2011). Accessed 05 Dec 2011
5. STO NOSTROY 2.25.40–2011. The device of asphalt concrete coatings from cold asphalt concrete. Publishing House "BST", p. 26 (2011). Accessed 05 Dec 2011
6. Anishchenko, D, Novik, A, Boytsov, A, Lazarev, Y., Trubina, D.: Underground streets in residential area: aspects of design on the example of Russia. In: EECE-2019 International, International Scientific Conference on Energy, Environmental and Construction Engineering; Lecture Notes in Civil Engineering, vol. 70, pp. 433–440 (2019)
7. Makovetskaya-Abramova, O., Lazarev, Y., Gravit, M., Silla, S., Shakhova, M.: Multiplicative method for creating the traffic monitoring base in a megapolis. In: E3S Web of Conferences, vol.164, p. 03023, Section 2. Energy Efficiency in Transportation (2020).
8. Makovetskaya-Abramova, O., Ivanov, A., Lazarev, Y., Shakhova, M., Rozov, A.: Economic assessment of construction of the roadside service facilities. In: E3S Web of Conferences, Section Environmental and Energy Management, Key Trends in Transportation Innovation KTTI-2019, vol. 157, Article Number 04035 (2020)
9. Kumar, A., Antony, J., Dhakar, T.S.: Integrating quality function deployment and benchmarking to achieve greater profitability. Benchmarking (2006)
10. Yoshikawa, T., Innes, J.: Functional analysis. The Routledge Companion to Cost Management (2013)

11. Shephard, R.W.: Theory of Cost and Production Functions/Theory of Cost and Production Functions (2015)
12. Daly, A., Sanko, N., Wardman, M.: Cost and time damping: evidence from aggregate rail direct demand models. Transportation (Amst) **4**, 1499–1517 (2017)
13. Chin, K.S., et al.: Failure mode and effects analysis by data envelopment analysis. Decis. Supp. Syst. **48**, 246–256 (2009)
14. The technical regulation of JSC ABZ "Magistral" "Production of asphalt mixes at the Vibau and Amomatik plants No. TP 03218337.003–2012. Accessed 12 Dec 2012

Investigation of Defects in Flat Conductors Using Electron-Optical Moire

Evgeny Aleksandrovich Pechagin[1(✉)], Vadim Alekseevich Chernyshov[2], Roman Pavlovich Belikov[3], Evgeniy Andreevich Mironov[4], and Nailya Kamilevna Miftakhova[5]

[1] Department "Power Industry", FSBE HE «TSTU», Tambov, Russia
pechagin_ea@mail.ru
[2] Department "Electrical Equipment and Energy Saving", FSBE HE «OSU Name By I. S. Turgenev», Oryol, Russia
[3] Department "Power Supply", FSBE HE «OSAU Named After N. V. Parakhin», Oryol, Russia
[4] Department "Electrical Equipment and Energy Saving", FSBE HE «OSU Name By I. S. Turgenev», Oryol, Russia
[5] Department "Power Supply of Industrial Enterprises", Kazan State Power Engineering University, Kazan, Russia

Abstract. The article describes investigation on the observation of scattering magnetic fields created by flat conductors using an electron-optical moire method. The defects behavior in conductive materials under the influence of an electromagnetic field is studied by electron-optical moire method in order to identify their interaction and to obtain informative assessing criteria of electromagnetic situation around holes and cracks that trigger material destruction mechanisms. The possibility of sorting flat conductors using a measuring and computing system, according to the distortions of the magnetic field introduced by the edge crack, is ascertained.

Keywords: Flat conductors · Defects · Electron-optical moire method

1 Introduction

Electronics reliable operation is closely related to the electromagnetic fields generated around its elements and is characterized by the electromagnetic compatibility of these elements. Electromagnetic and thermal energy, localized around the defect, create favorable conditions for its destruction. The main conductor discontinuity factors, apart from its physical constants, are the crack size and the enveloping electric current. For example, in thin strip conductors there are always mounting, technical, geometric and structural defects appearing in their production and operation process. As already noted, the electromagnetic energy concentration around such defects can lead to equipment failure. Therefore, the observation of electromagnetic fields around defects, the strength measurement and their interaction is an urgent task.

A. Sinitsyn (Ed.): SES 2020, LNCE 141, pp. 351–359, 2021.
https://doi.org/10.1007/978-3-030-67654-4_38

2 Work Objective

The main work objective is investigation of the defects detection possibility in conductors by electron microscopy using an electron-optical moire method.

3 Research Methods

It is difficult to obtain an experimental estimate of the magnetic field strength around a defect by traditional measurement methods, and in some cases it is impossible due to great methodological difficulties associated with its low values and field boundary [1, 2]. Therefore, the use of electron microscopy in order to obtain electron-optical moire is more validated.

During calculating and modeling short-range magnetic fields, the geometric and physical object features associated with its physico-chemical background are often not taken in considering [3, 4]. For example, the compression, tension and bending consequences of a flat conductor, as well as the consequences of its mechanical processing and heat treatment, significantly affect the magnetic fields distribution pattern.

Let us consider the observation and calculation research of local magnetic fields in flat conductors with defects using the electron-optical moire method. A copper plate with an edge notch (80 mm × 80 mm × 1 mm) was used as a prototype. It was placed in the column of the EG-100 A electronograph so that the electron beam slid along its surface and was as close as possible to the defect (Fig. 1).

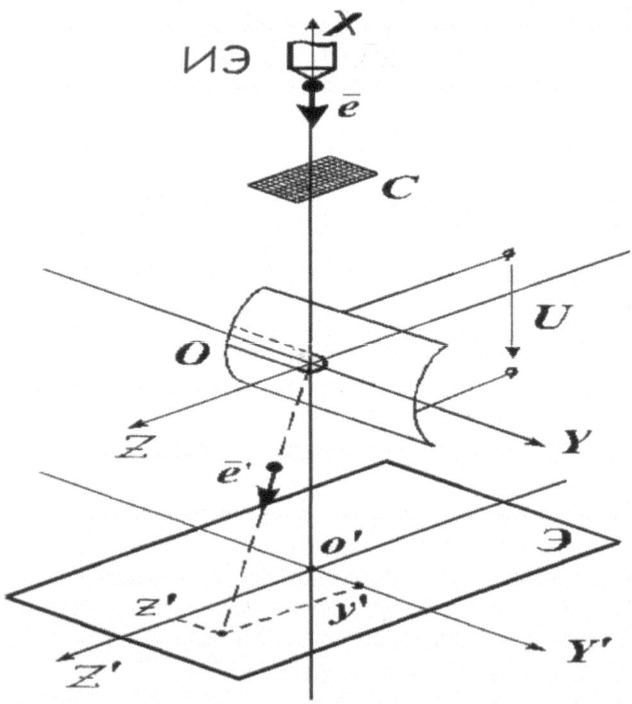

Fig. 1. Experimental scheme: *O*– object with a crack(curved plate with a crack); *IE*– electron source; *C* – grid; *E* – electronograph screen; $\bar{e}$ – beam of electrons; *U* – sourcevoltage.

The electron flux created by the electronograph electron gun is forming into a diverging electron beam using a focusing magnetic lens. Then this beam passes through a metal mesh and forms its shadow image on the electronograph screen. If an electromagnetic field is introduced in the electron beam path, the grid image is distorted accordingly. In the case with a diverging beam, the object that creates the magnetic field and the grid are observed in the shadow projection. The magnification factor is determined by the focus, grid, object, screen location. This version of the electron-optical scheme can be implemented on almost any design of the electron microscope, practically without alterations, by adding devices to set the grid and align the object that creates the field [5–7]. The magnetic field was set up by passing rectangular current pulses along the sample with different amplitudes and a duty cycle of two. To form such pulses, the contacts of a polarized relay were connected in series with the current source, which winding was powered by a audio-frequency generator at a frequency of 30–50 Hz. The magnitude of the accelerating voltage was selected from the condition for obtaining moire patterns on the screen containing the number of bands necessary for machine processing, and was 40 kV.

The moire pattern on the electronograph screen arose as a result of an overlay undistorted and distorted field grid image.

The photometric result of combining grid images is shown on the chart (Fig. 2).

Let a grid with a period α along the X axis and with an opaque section width d to be projected by a parallel electron beam on the screen. The screen glow brightness along the X axis is modulated with the grid period (curve A, Fig. 2). The grid, which image is distorted by researched field by an amount $\Delta\alpha$, gives a similar screen brightness modulation (curve B, Fig. 2).

If we combine, for example, by sequentially photographing images A and B on one plate, then with a linear reaction of the emulsion to the blackening distribution brightness in the total image it will correspond (curve $(A + B)$, Fig. 2). A black image is obtained where there is no obstacle in the electron beam path (II), gray - if the obstacle occurs once (I), and white when the obstacle occurs twice (III). Contrast on photographic prints will be the opposite.

Three types of features are observed in the image of the moire pattern: coincidence of horizontal lines, coincidence of vertical lines, as well as nodes in which completely cells of the distorted and undistorted grid image coincide.

The horizontal lines segments centers give the geometric points place at which the displacements in the vertical direction are equal and, accordingly, the horizontal field components are equal. Similarly, the vertical strokes centers give the geometric points place at which the vertical field components are equal. In places where the cells completely coincide, the component fields on the X and Y axis are treated as integers.

4 Experiments and Results

The experimental results showed a significant magnetic field distortion of the plate containing the crack.

Figure 3 presents the shadow electron-optical moire images of samples with a crack of different diameters and a sample without a crack.

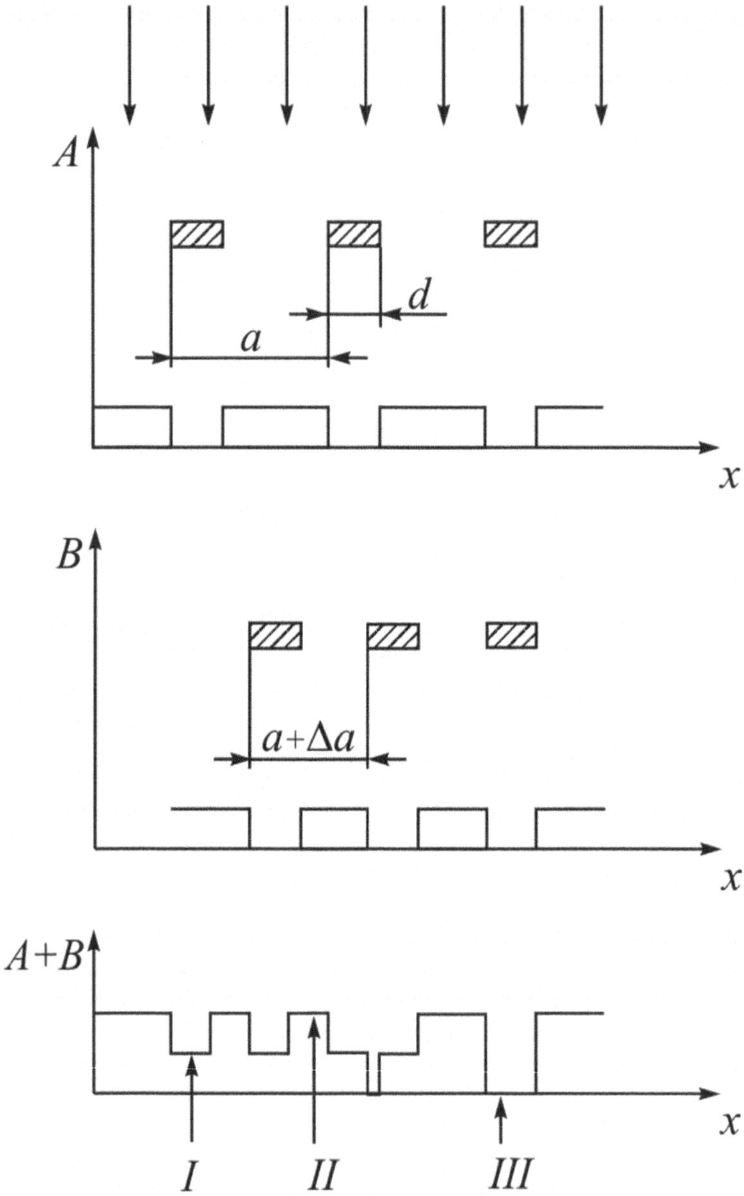

Fig. 2. Scheme for photometric explanation of the moire patterns production.

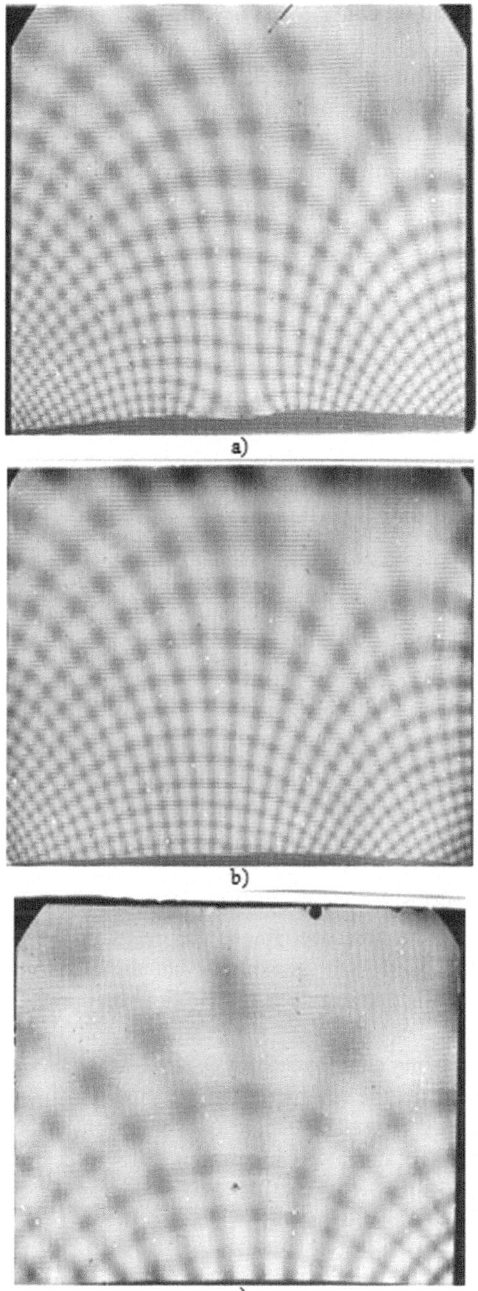

Fig. 3. Flat conductor electron-optical moire (I = 15 A): a) with a crack d = *12* mm, b) with a crack d = *18* mm, c) without crack.

The moire images shown that:

a) The magnetic field of a flat conductor without a crack has axial symmetry, while this symmetry for conductors with a crack is broken;

b) The magnetic field of a flat conductor without a defect has virtually no distortion, and with a crack have obvious distortion;

c) The thickness and length of the magnetic fields force lines of a flat conductor, varies depending on a particular defect existence.

As a result, according to the obtained moire patterns in the measuring and computing system, it is possible to analyze the electrical product quality as a whole.

Observations of scattering fields by electron microscopy can simplify control and reduce the inspector personality factor during visual quality inspection of an electrical product assembly and processing.

In determining the flux density of electromagnetic energy, it is necessary to evaluate the electric field strength of the current E and the magnetic field H around the defect tip when the current flows around it with a density $j\infty$. It is known that the electric field in a cracked conductor is estimated as (1):

$$E = E_{\infty}(1 + \alpha) \tag{1}$$

Where E and $E\infty$ electric field strength at the edge of the defect and electric field strength away from it;

$\alpha = \left({}^{l}\!/_{a}\right)^{{}^{1}\!/_{2}}$, l - crack length, a - radius at its apex.

With the actual defect size, the field strength at its edge is two degree of magnitude greater than the field at infinity. Consequently, the current density at the crack tip j will also exceed the average density in the conductor $j\infty$ by a hundred times, which will affect the local heating of the region around the defect and the magnetic field strength increase in this zone.

A double current effect is manifested here: it locally heats the conductor at the defect edge and at the same time strengthens the magnetic field around it.

A qualitative moire bands analysis shows that the electrons curvature trajectory moving along the X axis is due to two magnetic field components of the plate Hz and Hy (Fig. 4).

Of greatest interest is the component Hz(x), since at the crack top it is much larger than the others and controls the conductor deformation and fracture processes.

Complete with the current electric field concentration at the defect, it creates an electromagnetic energy flux density, which often leads to destruction by thermal and ponderomotive mechanisms.

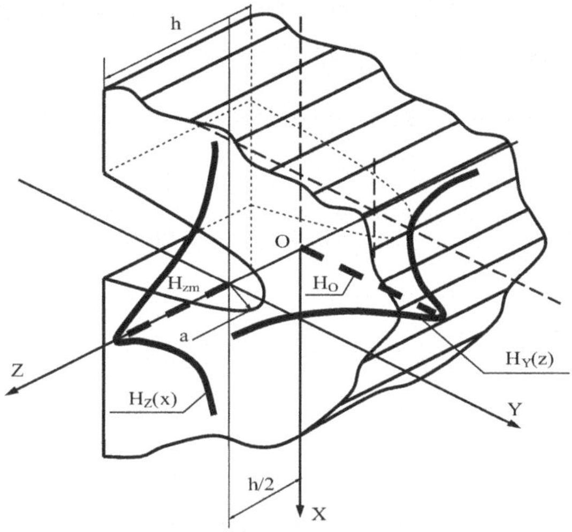

Fig. 4. The magnetic field distribution around the crack tip: h – plate thickness; a – radius at the crack tip; $H_z(x)$ – current direction along the X axis.

For a quantitative analysis of the magnetic field around the defect by the electron beam displacement magnitude on the electronograph screen, we use the charged particle motion equation along the X axis (2):

$$V_x = \frac{e}{m}\mu_0 V_0 \int H_x(z)dz \tag{2}$$

where V_x – electron escape velocity from the field action area; V_0 – particle acceleration velocity; Hz – magnetic field around the hole; e, m – charge and mass of the electron; μ_0 – magnetic constant.

The beam displacement in a magnetic field visible on the screen can be calculated by (3):

$$x' = V_x \frac{L}{V_0} = \frac{e}{m}\mu_0 \frac{L}{V_0} \int H_x(y)dy \tag{3}$$

where L – distance between the object and the screen, equal to 0,4 m.

Equipotentials obtained by grid vertical coincidence lines, which characterize the magnetic field radial component, are processed in a similar way. In this case, the displacement along the Y axis can be calculated by (4):

$$y' = V_y \frac{L}{V_0} = \frac{e}{m}\mu_0 \frac{L}{V_0} \int H_y(x)dx \tag{4}$$

Substituting the numerical data and the corresponding bell-shaped distribution at $d = 5{\cdot}10^{-6}$ m in Eq. (4) and integrating it, we obtain (5):

$$H_{ym} = \frac{y'}{5 \cdot 10^{-6}} \tag{5}$$

On Fig. 5a shown the distance dependence from the hole to the point with the axial component H_{yt} = 1287 A/m, depending on the current flowing through the flat conductor.

Similarly processed the magnetic field radial component results H_{xt} = 940 A/m (Fig. 5b).

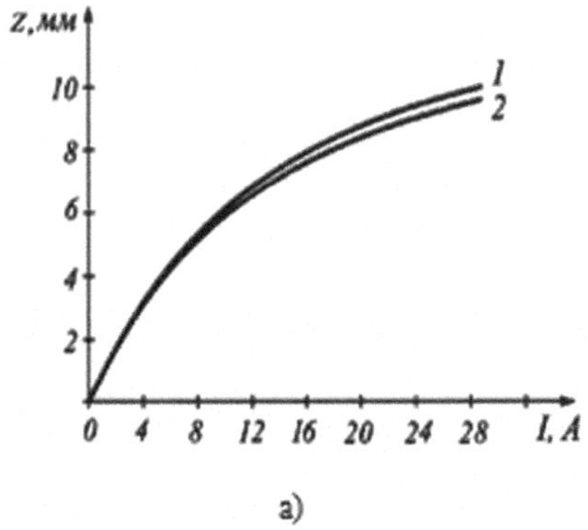

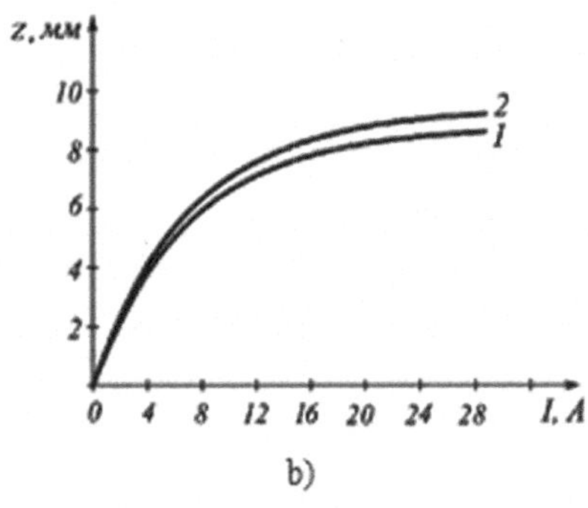

Fig. 5. The equipotential distance dependence on the current for: a) H_{yt} = 1287 A/m; b) H_{xt} = 940 A/m: 1 – experimental; 2 – theoretical.

From a comparison of theoretical and experimental data, it follows that the discrepancy is no more than 5% in the entire range, which proves the adequacy of the moire pattern to the real inhomogeneous magnetic field of the flat conductor.

With the actual defect size, the field strength at its edge is two degree of magnitude greater than the field at infinity. Consequently, the current density at the crack tip will also exceed the average density in the conductor by a hundred times, which will affect the local heating of the region around the defect and the increase in the magnetic field strength in this zone. A double current effect is manifested here: it locally heats the conductor at the defect edge and at the same time strengthens the magnetic field around it.

5 Conclusions

The magnetic field distortions in a flat conductor introduced by an edge crack were investigated, and the experimental data obtained by the electron-optical method ascertained a significant magnetic fields amplification on flat conductors with defects, which makes it possible to sort them.

References

1. Vagin, D.V., Kasatkin, S.I., Polyakov, P.A.: Strip magnetic field concentrators for magnetoresistive current sensors and Hall sensors. Sens. Syst. **12**, 25–29 (2010)
2. Petersen, T.H., Carpenter, K.H., May, C.M.: Comparison of experimental measurements of current distribution in a flat conductor with simulated results from the partial inductance method. IEEE Trans. Electromagn. Compat. **51**(2), 345–350 (2009)
3. Lozanova, S.V., Noykov, S.A., Ivanov, A.J., Roumenin, C.: Functional multisensor for temperature and subsequent 3D magnetic-field measurement. Procedia Eng. **120**, 824–827 (2015)
4. Lai, Y.-S., Kao, C.-L.: Electrothermal coupling analysis of current crowding and Joule heating in flip-chip packages. Microelectron. Reliab. **46**, 1357–1368 (2006)
5. Sinha, B., Ramulu, T.S., Kim, K.W., Venu, R., Lee, J.J., Kim, C.G.: Planar Hall magnetoresistiveaptasensor for thrombin detection. Biosens. Bioelectron. **59**, 140–144 (2014)
6. Pechagin, E.A.: Measuring and computing system for controlling the quality of magnetic products using electron-optical moire effects, Abstract of Ph.D. thesis, Lipetsk (2002)
7. Kalinin, V.F., Ivanov, V.M., Uvarov, A.N., Limonov, D.N., Pechagin, E.A.: Pat. 2292053 Russian Federation, IPC G01R 33/02, A method of measuring magnetic fields by electron-optical method, Applicant and patent holder FGBOU VPO TSTU, 2004129351/28, declared 10.03.2006, publ. 20.01.2007, Bull. 2, 14 (2007)

Support Node for Temporary Spatial Rod Structures of the Layher Type

Ilnur Gimatdinov(✉), Gennady Shmelev, and Khaidarov Lenar

Kazan State University of Architecture and Engineering, 1 Zelenaya Street, Kazan 420043, Russian Federation
akkfok@mail.ru

Abstract. This article discusses the support node (jack) of the Layher type of scaffolding and falsework. There are many ways to calculate such structures, but currently there are no normative documents specifically for designing temporary spatial structures of the Layher type. In this regard, the question of studying the work of the Layher type of scaffolding jack is very relevant to the present. By studying the stress-strain state, a detailed graph of the operation of the studied node was determined. Based on the obtained data, the load-bearing capacity of the jack under study as a compressed-flexible element in the structure of the studied Layher type scaffolding was considered in detail. A graph of the bearing capacity of the rod part of the jacks was also determined, depending on the ratio of the transverse force and the longitudinal compressive force. By considering the node as part of the structure, a design recommendation was proposed to increase the accuracy of calculations during design. The main factor of this recommendation is the consideration of slip in one-way communication when calculating and designing.

Keywords: Jack · Layher · Scaffolding · Falsework · Support node · Construction

1 Introduction

When conducting cultural events, an important issue is the organization of the stage sector. Usually, there is a need to organize an event in a short time. To solve this issue, the technology of modular building structures, which are based on high-strength wedge-type scaffolding, is increasingly often used [1–3]. The purpose of structures is very diverse and they can reach large sizes, both in plan and in height [4, 5].

Scaffolding as temporary structures for cultural events were first used in foreign countries with the addition of special structures such as seats, chairs, railings, etc. to the list of elements [6, 7]. Only in the last decade such temporary structures were used on the territory of the Russian Federation [8, 9].

But Russian regulatory documents do not yet take into account the features of these structures, such as the flexibility of collapsible joints, and the presence of gaps, installation on an unprepared base, etc. [10, 11].

A. Sinitsyn (Ed.): SES 2020, LNCE 141, pp. 360–370, 2021.
https://doi.org/10.1007/978-3-030-67654-4_39

In contrast to traditional scaffolding, these structures are independent and the horizontal loads acting on them are transmitted not by anchor elements to a nearby building, but to the base through support structures [12, 13].

Jacks are steel support structures consisting of a support plate and a vertical rod with a screw thread and a screw element for height adjustment [14–16]. The base plate with the rod part can be connected pivotally or by welding. Jacks are usually installed on square-shaped distribution boards with side dimensions from 0.25 m to 0.4 m made of wooden beams and plywood sheets, and the latter – without sinking on the base, including asphalt, concrete or unprepared soil [17, 18]. There are usually no additional fastening elements in the form of screws and anchors between the jack base plate, the switchboard and the base, and the horizontal load is transmitted only through friction between the touching surfaces [19, 20].

A number of articles indicate errors in the installation of support units and indicate that due to non-compaction of the base soil, very often there is a difference in the precipitation of the base. It is also noted that the elements of the support unit can slip without transmitting a transverse force to the base [21].

There are also several calculation approaches for the angular stiffness of the support node.

According to the first approach, it is assumed that the node works rigidly until the tipping moment is greater than the holding moment, equal to the multiplication of the longitudinal compressive force in the jack and the arm, equal to half the width of the plate (Fig. 1).

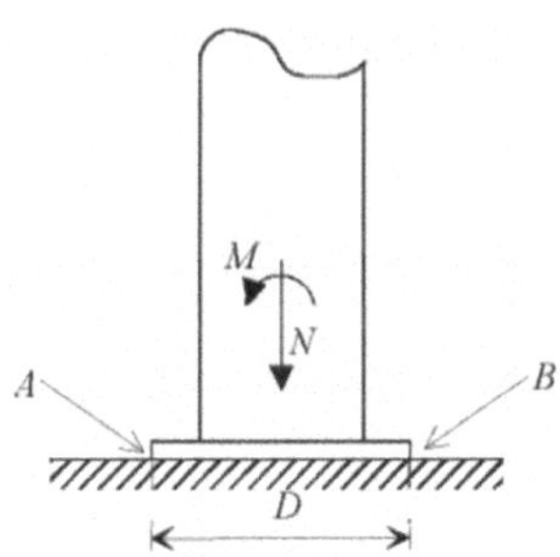

(a) Rigid baseplate

Fig. 1. Design scheme for the first calculation approach

The second approach is proposed by the UK standard document BS EN12811–11 “Temporary structures for construction structures”, according to which the angular stiffness of support units is assumed to be equal to 2 * 10^7 N * mm/rad.

The third approach is proposed by BS EN 1065: plate is attached to the ground at a distance of 0.4 D (D is the length of the base plate) from the center of plate, rod of the jack is attached to the hinge plate, but its rotation relative to “earth” is limited by a spring with a stiffness of 3 * 10^7 N * mm/rad, after 1° turn (Fig. 2).

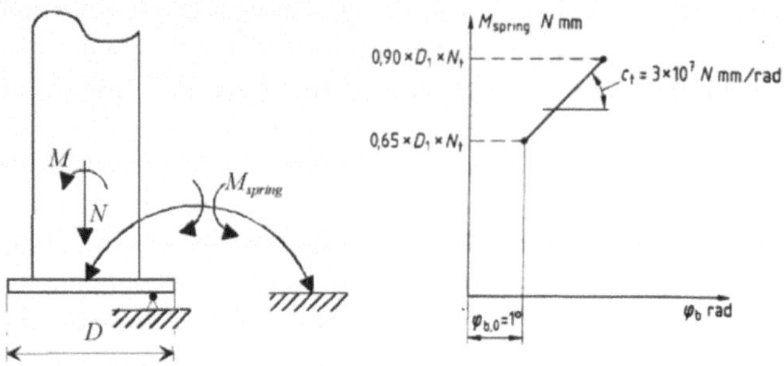

(b) Baseplate model used in BS EN 1065

Fig. 2. Design scheme for the third calculation approach

The fourth approach is suggested by Andre et al. "Experimental study of steel telescopic racks" in 2009, which involves direct modeling of normal and tangential contact between the contacting surfaces of the base plate and the base.

2 Methods

2.1 Stress Strain State

To study the stress-strain state, the base plate of a typical screw jack was modeled using the finite element method in Ansys Mechanical. The base plate is a stamped sheet with a plan size of 150 * 150 mm and a thickness of 5 mm (Fig. 3).

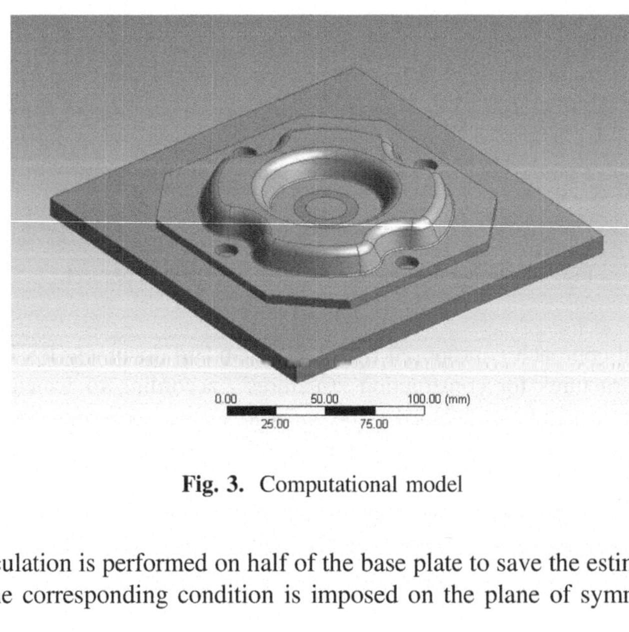

Fig. 3. Computational model

The calculation is performed on half of the base plate to save the estimated time. In this case, the corresponding condition is imposed on the plane of symmetry. On the lower side of the base plate, a one-way connection condition is applied that works on compression.

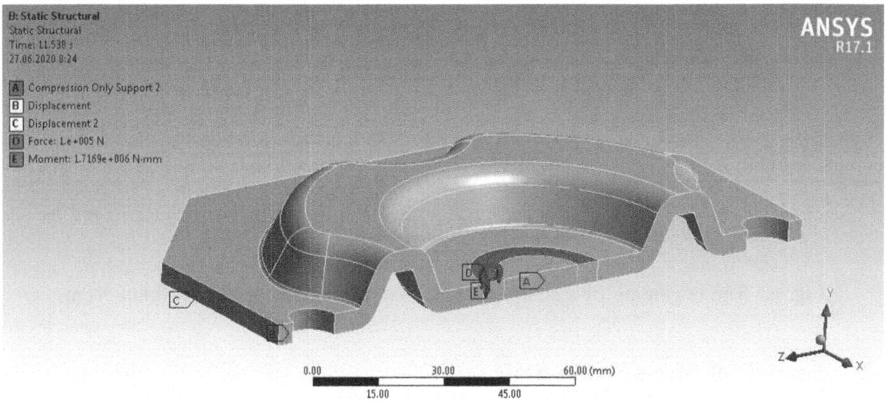

Fig. 4. Boundary conditions

The plate is split by three-dimensional finite elements in the form of a tetrahedron with a size of 2.5 mm from the condition of splitting the plate by thickness into at least two elements. The load in the form of a vertical compressive force and bending moment was transmitted to the surface of the connection of the base plate with the rod part (highlighted in red in the Fig. 4).

Under the action of a bending moment of 60 kN * cm, the plate material begins to work in the plastic stage, which is accompanied by a faster increase in deformations. This moment value is the limit of the bearing capacity of the jack base plate (Fig. 5).

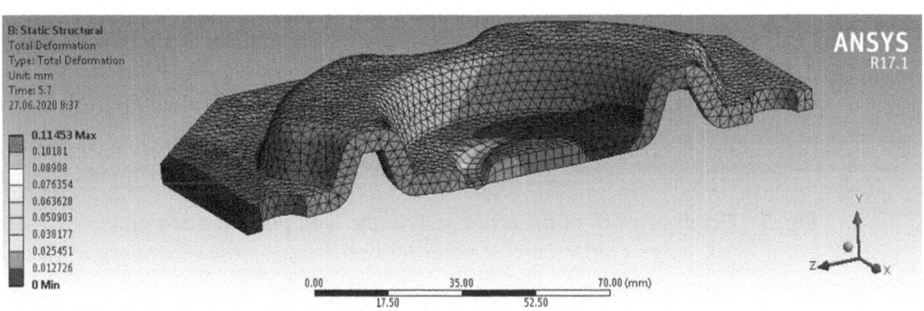

Fig. 5. Deformations at M = 60 kN * cm

With a further increase in the bending moment, the deformations increase and at a bending moment of 100 kN * cm, the node begins to work as an absolute hinge (Fig. 6).

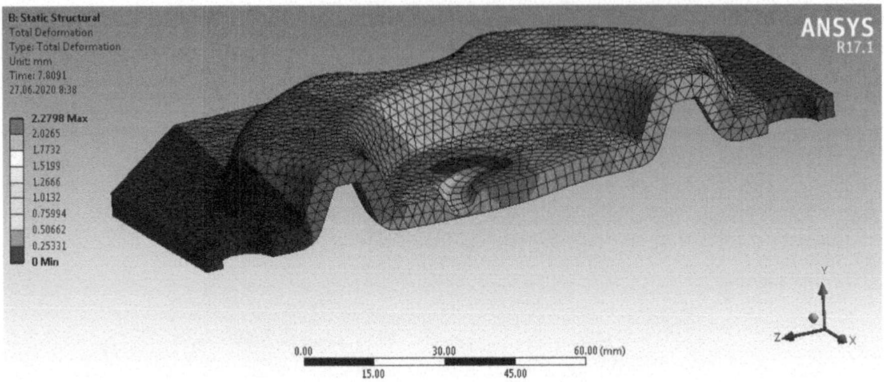

Fig. 6. The beginning of a sharp increase in deformations at M = 100 kN * cm

2.2 Analysis of Stress Strain State

Based on the results of the calculation, the node's work graph was created. The graph shows that when the bending moment is from 0 to 60 kN * cm, the rotation angle of a node is directly proportional to the moment. For this phase the rigidity of the node is defined as the tangent of the inclination of the graph (Fig. 7).

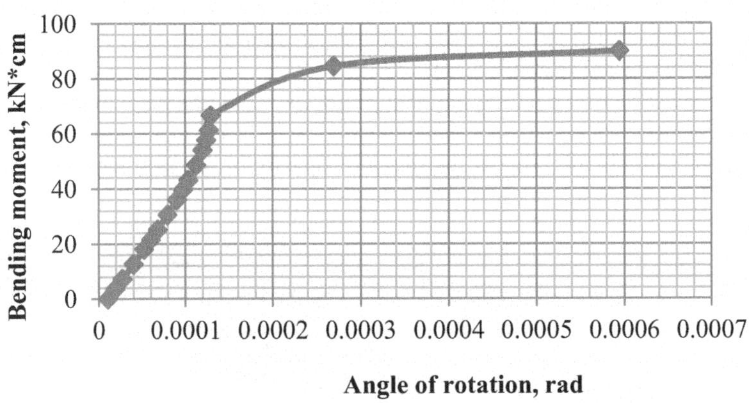

Fig. 7. Graph of work of the screw jack to the base plate junction unit

According to SP 16.13330.2016 "Steel structures", the coefficient of the calculated lengths of the jack screw rod is known when the screw element is extended by 10, 20, 30, 40 cm, taking into account the rigidity of the attachment node to the base plate. Based on this, a graph of the bearing capacity of the jack rod as a compressed bending element was created. The bearing capacity of the rod part of the jack is provided if the combination of forces is between the axes and the corresponding graph (Fig. 8).

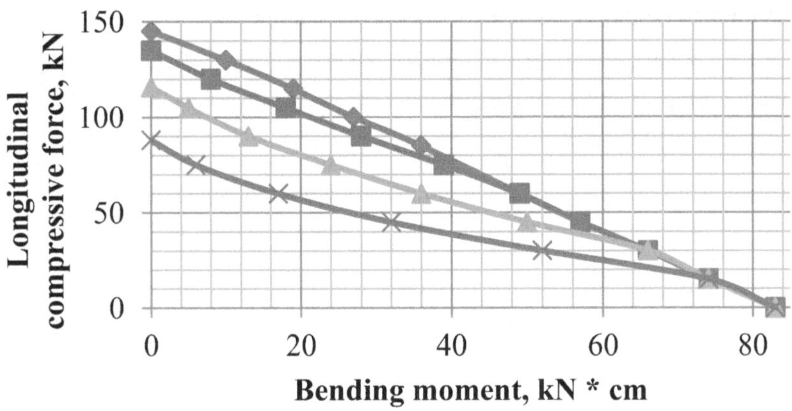

Fig. 8. Graph of the load capacity of the jack as compressed-bent element

Based on this, a graph of the bearing capacity of the rod part of jacks is constructed depending on the ratio of the transverse force and the longitudinal compressive force (Fig. 9).

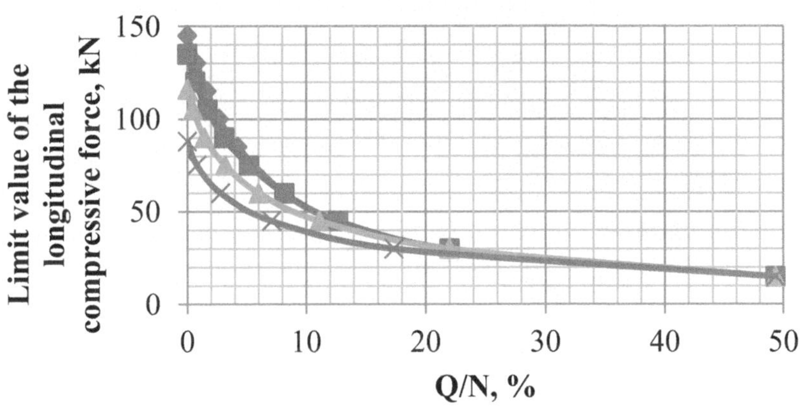

Fig. 9. Graph of the dependence of the maximum compressive force on the ratio of the transverse force to the longitudinal force

The graph shows that the action of the transverse force significantly affects the bearing capacity of the jack, and therefore it is important to correctly determine their design values when creating structures with screw jacks.

2.3 Temporary Spatial Structures Based on Scaffolding

To determine the forces, namely, the transverse forces acting on the jack, calculations were made for the structures of the stage and the podium with a canopy in the SP Lira. The calculation was performed with two models. In the first version, the support nodes of the jacks were fixed in a horizontal plane, as is usually modeled in the calculation of such structures. In the second variant, nonlinear single-node finite elements of type 263 were created in the lower nodes. This type of finite element is intended for modeling sliding in a one-way connection.

A comparative analysis of the transverse forces in jacks in the first and second versions of the simulation was performed.

The first structure under consideration is the stage (Fig. 10).

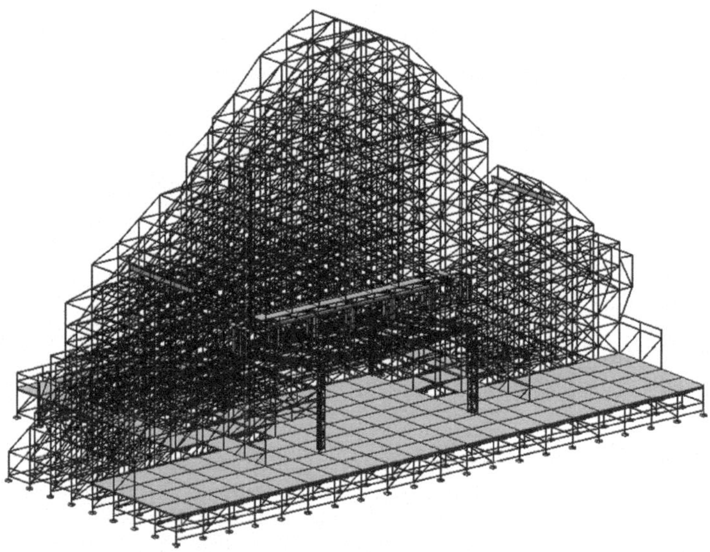

Fig. 10. Spatial model of the stage

The second structure is a grandstand with a canopy (Fig. 11).

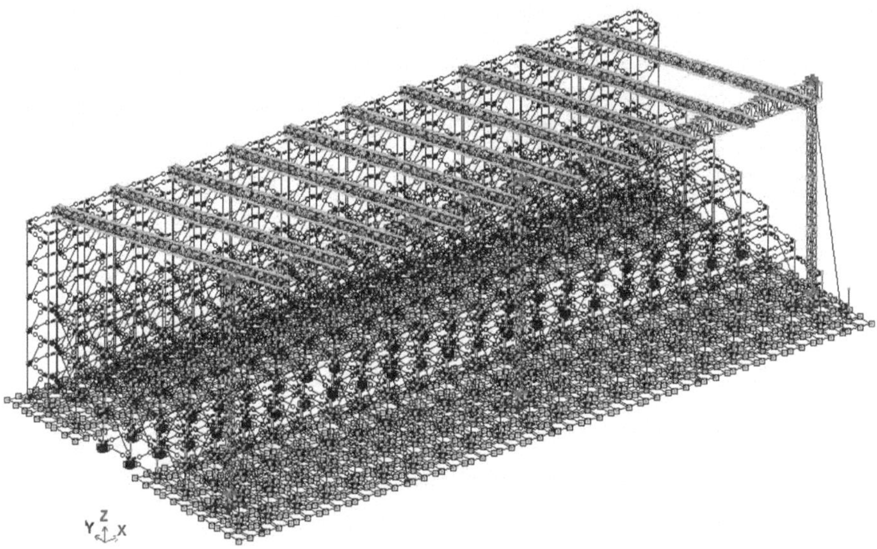

Fig. 11. Design scheme of the grandstand

3 Results and Discussion

3.1 Calculation Results for the Effect of Sliding in a One-Way Connection.

For the structure of the stage, the maximum value of the transverse force when fixing the support nodes was 0.2 t, for modeling with FE263 was 0.233 t. The increase was equal to 16.5%. Modeling with fixing in the horizontal plane underestimates the value of the transverse force of the support jack, as a result, the load-bearing capacity of the node and, consequently, the entire structure may not be provided.

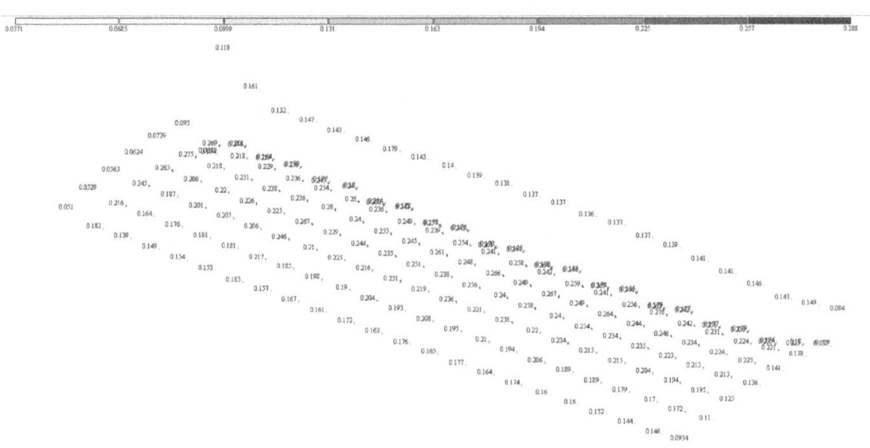

Fig. 12. Mosaic of forces taking into account FE 263

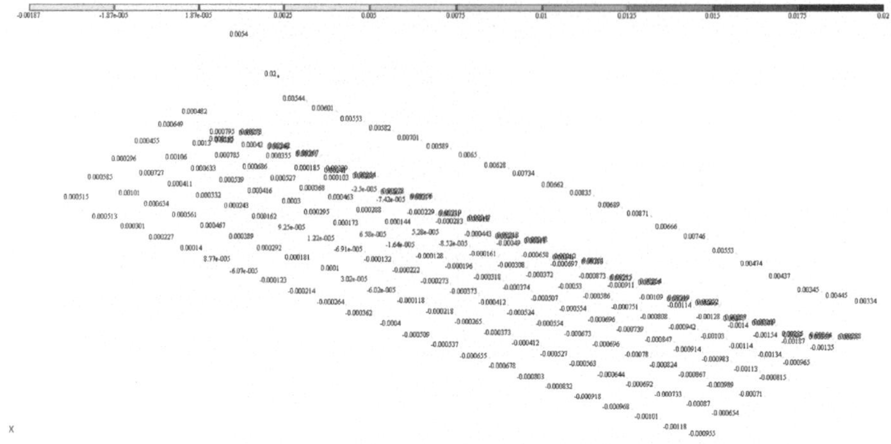

Fig. 13. Mosaic of forces without taking into account FE 263

For the construction of a grandstand with a canopy, the maximum values of transverse forces were 0.288 t and 0.36 t, when modeling for the first and second options, respectively. The difference was 25%. Modeling with horizontal anchoring also underestimates the value of the transversal force of the support Jack.

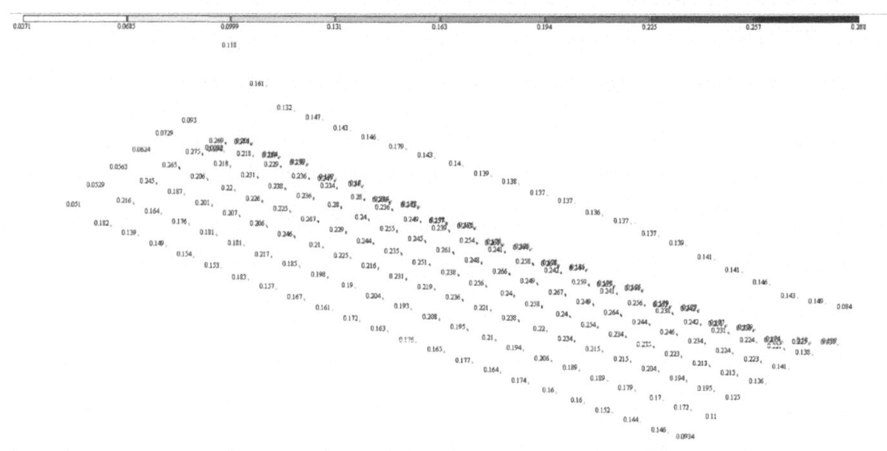

Fig. 14. Mosaic of forces without taking into account FE

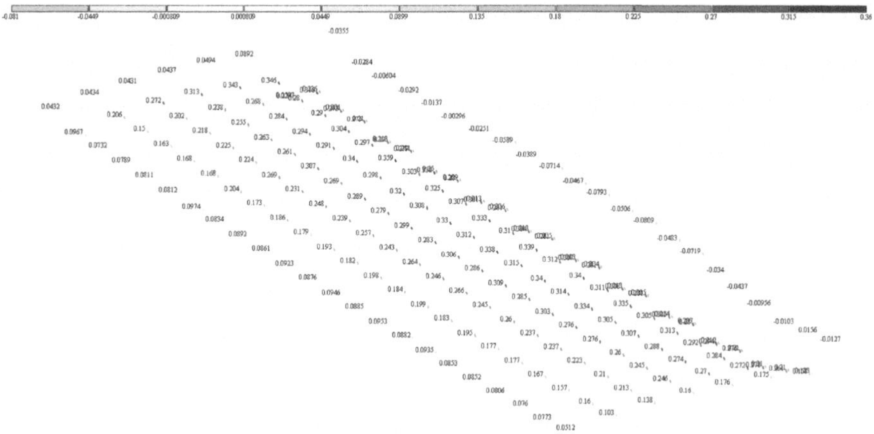

Fig. 15. Mosaic of forces taking into account FE

4 Conclusions

During the non-linear calculation of the base plate, the limit value of the bending moment equal to 60 kN * cm was determined, at which the plate material begins to work in the plastic stage not only under the screw part, but also on the other sections. Up to this point, the rotation of the screw part of the jack is proportional to the bending moment. During the analysis, several graphs were considered, and one of them showed a significant influence of the transverse force on the bearing capacity of the screw Jack.

As can be seen from the presented results (Figs. 12, 13, 14 and 15), the influence of a finite element of type 263 (accounting for slip in a one-way connection) has an essential role. The average percentage of increased effort given this condition is about 20%.

An incorrect assessment of the transverse forces in the jack can lead to a loss of load-bearing capacity of the jack itself and the entire structure as a whole. Therefore, it is recommended to model the support nodes using the type 263 FE, due to which the transverse forces in the jacks are distributed proportionally to the longitudinal forces, or to establish horizontal connections where there will be obviously large longitudinal forces compared to the rest of the racks, for example, where the ballast is installed.

References

1. Layher. Allround scaffolding system, p. 48. Technical brochure (2017)
2. Cimellaro, G.P., Domaneschi, M.: Stability analysis of different types of steel scaffolds. J. Eng. Struct. **152**, 535–548 (2017)
3. Błazik-Borowa, E., Pieńko, M.: Scaffoldings. Monografie – Politechnika Lubelska, p. 115. University of Technology, Lublin (2017)
4. Błazik-Borowa, E., Szer, J.: The analysis of the stages of scaffolding «life» with regard to the decrease in the hazard at building works. Arch. Civ. Mech. Eng. **15**, 516–524 (2015)

5. Hunnebeck company. Product portfolio Modex, p. 66 (2017)
6. Bluff, E.: Safety in machinery design and construction: Knowledge and performance. Saf. Sci. **74**, 59–69 (2015)
7. Carrillo-Castrillo, J.A., Trillo-Romero, A., Rubio-Romero, J.C.: Construction accidents: Identification of the main associations between causes, mechanisms and stages of the construction process. Int. J. Occup. Saf. Ergon. **23**, 240–250 (2017)
8. Hara, T., Hamano, K., Miyake, S.: Advanced design and planning of scaffolding system for construction. In: Proceedings of 17th International Conference on Computing in Civil and Building Engineering, Tampere, Finland (2018)
9. Peng, J., Ho, C., Chan, S., Chen, W.: Stability study on structural systems assembled by system scaffolds. J. Constr. Steel Res. **137**, 135 (2017)
10. Lin, C., He, L., Wu, Z., Yuan, J.: Experimental study on joint stiffness with vision-based system and geometric imperfections of temporary member structure. J. Civ. Eng. Manag. **24**, 43–52 (2018)
11. Layher technische brochure belastbaarheid allround-en systeemvrije onderdelen, p. 33. Kontich, Belgium (2017)
12. Layher allround scaffolding – instructions for assembly and use. Edition 04.2018. Ref. No. 8116.230. Gueglingen-Eibensbach, Layher, Germany 42 (2018)
13. Bakker, A., Smit, J., Wegerif, R.: Scaffolding and dialogic teaching in mathematics education: Introduction and review. ZDM Math. Educ. **47**, 1047–1065 (2015)
14. Yuan, X., Anumba, C.J., Parfitt, M.K.: Cyberphysical systems for temporary structure monitoring. Autom. Constr. **66**, 1–14 (2016)
15. Chen, G., Zhang, H., Rasmussen, K.J.R., Fan, F.: Modeling geometric imperfections for reticulated shell structures using random field theory. Eng. Struct. **126**, 481–489 (2016)
16. Liu, H., Jia, L., Wen, S., Liu, Q., Wang, G., Chen, Z.: Experimental and theoretical studies on the stability of steel tube–coupler scaffolds with different connection joints. Eng. Struct. **106**, 80–95 (2016)
17. Manos, G.C., Nalmpantidou, A., Kourtides, V., Anastasiadis, A.: Cyclic response of a steel beam-to-column connections – an experimental and numerical study, In: COMPDYN 2015, Greece (2015)
18. Nepelski, K., Błazik-Borowa, E.: Determining the bearing capacity of the ground as the part of designing scaffoldings. Rusztowania **10**, 11–13 (2017)
19. Błazik-Borowa, E., Gontarz, J.: The influence of the dimension and configuration of geometric imperfections on the static strength of a typical façade scaffolding. Arch. Civ. Mech. Eng. **16**, 269–281 (2016)
20. Loss, C., Frangi, A.: Experimental investigation on in-plane stiffness and strength of innovative steel-timber hybrid floor diaphragms. Eng. Struct. **138**, 48–98 (2017)
21. Gordeev, V.N., Lantukh-Lyashenco, A.I., Pashinskiy, V.A., Perelmuter, A.B. and Pichugin, S.F.: Loads and impacts on buildings and structures, pp. 195–222. IASW, Moscow (2007)

Analysis of Electric Energy Consumption Depending on Consumer Groups: A Case Study from Dmitrovsky District, Russia

Alexander Vinogradov[1], Alexey Vasiliev[1], Vadim Bolshev[1(✉)], Alina Vinogradova[1], Alexey Bukreev[1], and Leonid Dolomanyuk[2]

[1] Federal Scientific Agroengineering Center VIM, 5 1st Institutsky Proezd, Moscow 109428, Russia
vadimbolshev@gmail.com
[2] Kazan State Power Engineering University, 51 Krasnoselskaya Street, Kazan 420066, Russia

Abstract. Analysis of electricity consumption by consumer groups in the Dmitrovsky district of the Orel region, Russian Federation shows that changes in the consumption structure takes place in the direction of a slight increase in the energy consumption by 'other consumers' whereas there is a decrease in the proportion of the households, industry and budget sector. The reduction in electricity consumption in the budget sector is associated with the closure of a number of budget organizations (primarily rural schools) and the transfer of others to a shorter work schedule. At the same time, there is no noticeable decrease in payments for electricity by this group in the whole district. A slight increase in consumption by 'other consumers' is associated with the transfer of part of population to homeowners' association and cooperatives. The reduction in electricity consumption by industry is also associated with the closure of a number of enterprises and with a change in the load of existing ones. Increased costs for electricity bill, even while reducing its consumption in real terms across all groups of consumers, allows us to conclude that the effort to reduce the tariff burden for all groups of consumers have not yielded significant results.

Keywords: Electric energy consumption · Consumer groups

1 Introduction

Power supply system reliability is one of the main criteria for power supply efficiency [1–4]. In accordance with the PUE [5], all electrical receivers are divided into three categories according to power supply reliability: I, II, III. It also gives the following definition of power supply reliability: "a continuous supply of electricity to consumers in accordance with the schedule of power consumption and according to the scheme that is provided for long-term operation". It is logical to assume that the higher the category of electrical consumers in the power supply system, the more attention should be paid to it in the field of its modernization by replacing equipment with expired service life [6, 7] and equipping electrical networks with technical systems for their control and management [8–11]. In this article we analyze power supply to end

A. Sinitsyn (Ed.): SES 2020, LNCE 141, pp. 371–382, 2021.
https://doi.org/10.1007/978-3-030-67654-4_40

consumers taking into account consumer groups. There can be the following groups: industrial consumers, budget consumers (financed from government budget), ‘other consumers’, including all legal entities that do not belong to the previous two groups, and households. In this paper we analyze energy consumption of such groups for the period 2013…2017 on the example of Dmitrovsky district of the Orel region, Russia Federation. Dmitrovsky district is located in the southwestern part of the Orel region, on the highly rugged terrain of the Central Russian Upland between rivers: the Oka, the Desna and the Seim. The total area is 124,979 hectares [12]. The administrative center is the city Dmitrovsk. The resident population is 10712 people. The population density is 8.6 people per km^2. The economy has a pronounced industrial and agrarian character. The decline in production in the 1990s in the processing industries of the agro-industrial complex has not been overcome. The main branches of industry are food processing and building materials industry.

2 Dynamics of Electricity Consumption Structure

The dynamics of changes in the structure of electricity consumption by the above groups of consumers for the period under consideration is shown in Fig. 1.

Over the five years, the share of consumption in the budget sector has gradually decreased from 17.3% of the total productive supply to 15%. In general, in the district, the shares of the budget sector of different levels are approximately equal: 31% is the share of consumption of municipal budget sector, 35% is the regional budget sector and 34% is the federal budget sector.

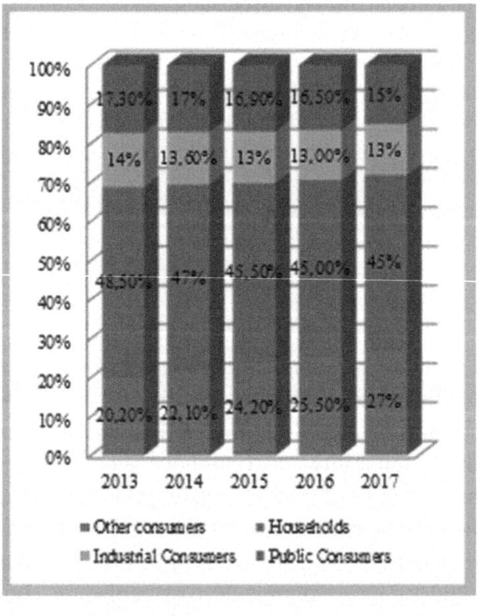

Fig. 1. Dynamics of changes in the structure of electricity consumption by consumer groups of Dmitrovsky district, Orel region, Russia Federation for the period 2013…2017.

Further, considering the dynamics, we have noted that the share of industrial consumers had decreased from 14% in 2013 to 13% in 2017, which indicates a decline in production. There were not many industrial companies in the region and the volume of their products did not increase, hence the consumption of electric energy for their needs did not not tend to grow. An additional influence on the dynamics of industrial production was exerted by cross-subsidization, i. e., the industry bore a more significant price burden compared to the population. In order to stimulate industrial production, the Russian government had taken measures to both freeze the growth of tariffs and eliminate cross-subsidization. However, not all measures undertaken by the Government has yielded positive results. Therefore, according to the Government's plan, the changes in the pricing of industrial consumers, introduced by the Government Decree No. 442 and entered into force on 01.07.2012 [13], should have led to fairer pricing in the retail market. However, due to a number of objective reasons, the result turned out to be the opposite.

Changing the structure of productive power supply in the category of "households" and "others" was mainly due to the entry into force of the Russian Government dated 23.05.2006 N 307 (amended on 12.17.2014) [14] and the introduction of new groups of consumers – homeowners' association (in this case, they are part of a group of "others" along with other legal entities). At the same time, the part of energy consumption from the households group moved to the 'other consumers' group (as part of the homeowners' associations). It was associated with the emergence of the need to calculate the consumption of electricity consumed for general household needs and, accordingly, the need to conclude separate energy supply contracts with them. It should be noted that the majority of homeowners' association were not ready to perform these functions. In this situation, power suppliers protecting their interests were forced to work with the population for taking meeting of owners of apartment buildings and take decisions to pay for electricity consumed under individual metering devices (in residential areas). Collection of payments for electricity consumed in non-residential premises (general building needs) remained with homeowners' associations. However, not all homeowners' associations issued invoices for these services to owners of premises and collected payments. With the most persistent violators, power suppliers terminated energy supply contracts and started working with the population in terms of paying for consumed electricity (both for individual metering devices and for general household needs) directly. The Orel Region has been included in the project on the use of social norms when paying for electricity since 2013 [15]. During this time, analysis of productive supply tariffs within the social norms showed that only about 2% of consumption is higher norms (taking into account the urban and rural areas). At the same time, the rural excessive power supply is very small and is less than 0.2% of the total useful supply.

3 Analysis of Electricity Consumption by Budget Consumers

The budget category of consumers can be attributed to non-production institutions: socio-cultural, government bodies, defense, courts, prosecutors, science, education, health, etc. These institutions receive all the necessary funds from the state budget.

The largest presence of budget sector organizations was observed in the city Dmitrovsky because it has been the largest settlement in the region and is also an administrative center. The share of consumption by the city Dmitrovsky was about 70% of the total productive power supply of this group of consumers. Figure 2 shows the dynamics of electricity consumption by the budget sector of the Dmitrovsky district in physical terms, Fig. 3 do it in monetary terms.

Due to a significant reduction in electricity consumption by budget consumers in 2017 compared to 2016, electricity costs decreased slightly.

Despite the general decrease in the cost of paying for electricity by budget organizations in 2017, the picture was somewhat different. Using the example of one of the district schools, we considered and analyzed the dynamics of electricity costs for the same period.

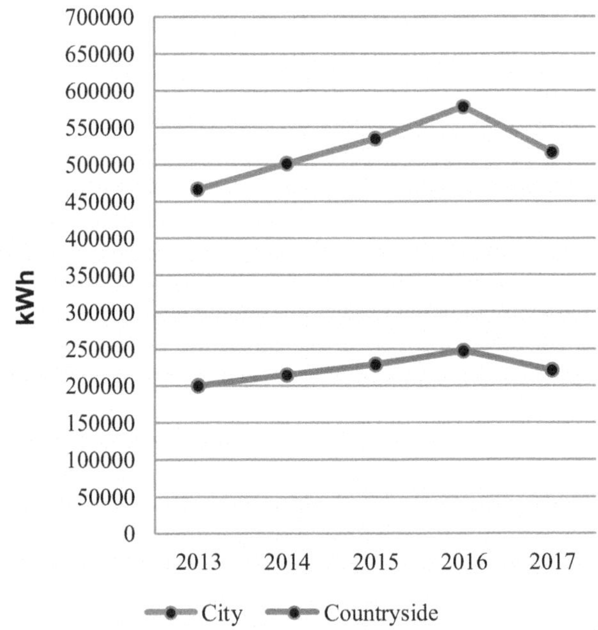

Fig. 2. Dynamics of electricity consumption by the budget sector of the Dmitrovsky district in physical terms.

For the period under review 2013…2017, the amount of electricity consumed by the school went through two stages: growth and decline (Fig. 4). Until 2015, the growth in school consumption had been due to the introduction of new energy-consuming equipment (computer class, new electric stoves, etc.) as well as an additional influx of students from closed rural schools and a new educational program for disabled children. The decline in energy consumption after 2016 was primarily due to the transition of schools to 5-day education instead of 6-day education, the rejection of the education program for children with disabilities and a decrease in the number of applicants.

Regardless of the decrease in energy consumption, utility bills increased by 4.5% due to the annual tariff increase (Fig. 5).

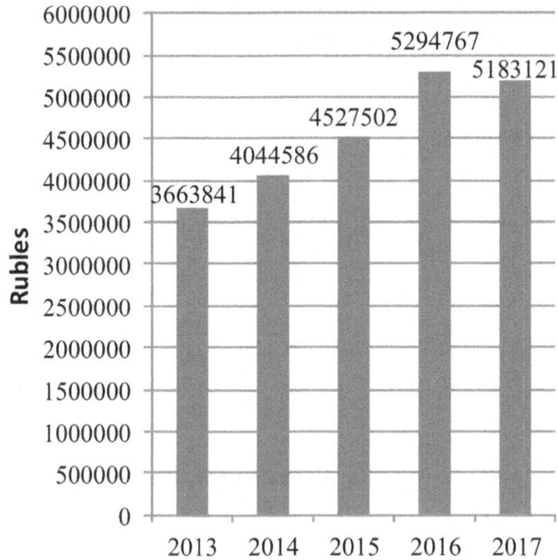

Fig. 3. Dynamics of electricity consumption by the budget sector of the Dmitrovsky district in monetary terms.

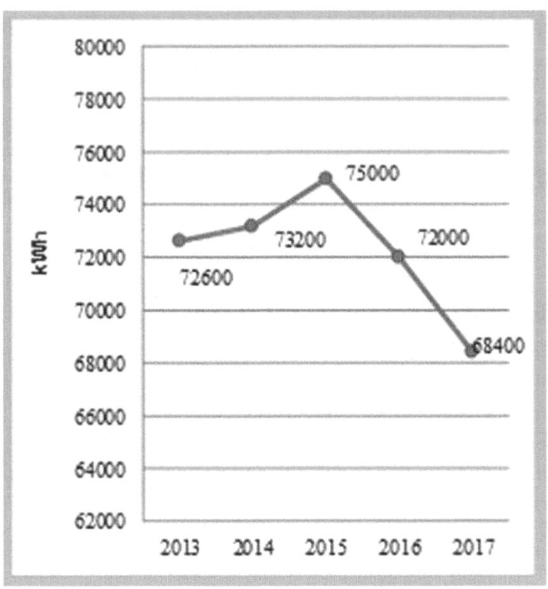

Fig. 4. Dynamics of electricity consumption by one of the schools in the city Dmitrovsk in physical terms.

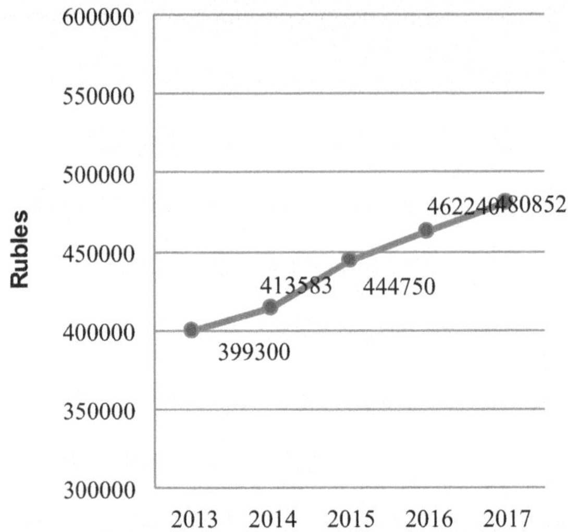

Fig. 5. Dynamics of electricity consumption by one of the schools in the city Dmitrovsk in monetary terms.

The study of specific examples of electricity payment dynamics by budget organizations showed therefore that the total cost reduction was primarily associated with the closure of some organizations, a reduction in their operating time and not with energy savings, although the closure of schools did not always lead to a noticeable decrease in consumption [16]. At the same time, even if there were savings, the growth of the tariff covered it and the costs of specific organizations were growing instead of decreasing.

4 Analysis of Electricity Consumption by Industrial Consumers

Industry in rural areas in the territory of the Dmitrovsky district, Orel region is underdeveloped due to several reasons such as lack of infrastructure, lack of qualified staff, ecologically unfavorable area, the lack of direct investing in the area. Nevertheless, almost until 2017, there had been an increase in electricity consumption by the industrial sector of the district. In 2017, consumption stabilized at the level of 650 thousand kWh of electricity (Fig. 6).

Despite a slight reduction in electricity consumption by industrial enterprises, the cost of paying for electricity increased (Fig. 7). This was due to the annual tariff increase.

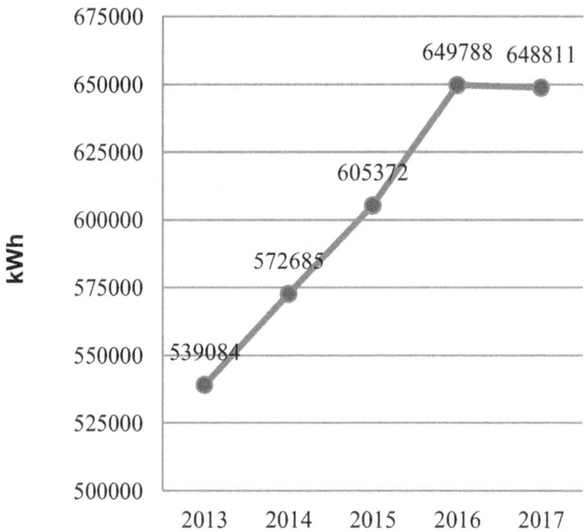

Fig. 6. Dynamics of electricity consumption by the industrial sector of Dmitrovsky district in physical terms.

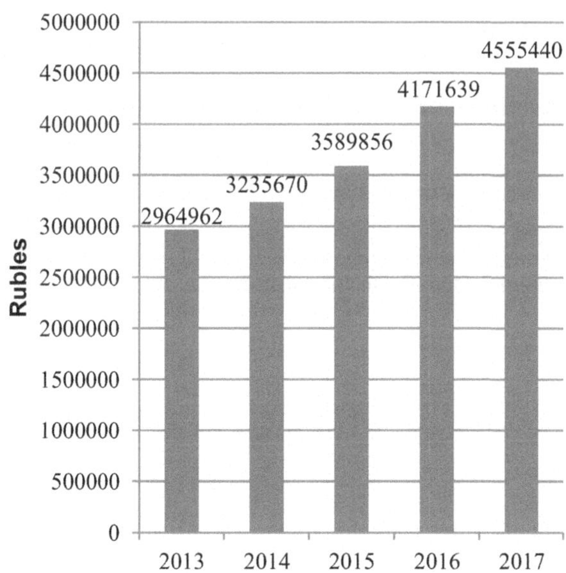

Fig. 7. Dynamics of electricity consumption by the industrial sector of Dmitrovsky district in monetary terms.

Using the example of one of enterprises in the Dmitrovsky district, we considered and analyzed the dynamics of electricity consumption and payment for it. LLC "Dmitrovsky Meat Processing Plant" (LLC "DMK") began its activities in 2013. The gradual development of the enterprise, the opening of additional workshops and shops, new points of sale - all this contributed to the increase in production, and, as a consequence, an increase in electricity consumption. In 2016 LLC "DMK" switched to imported raw materials in connection with the conclusion of a profitable deal. Production stopped working at full capacity, but this trend did not last long. As the enterprise opened up new opportunities for marketing the product, which also affected power consumption (Fig. 8).

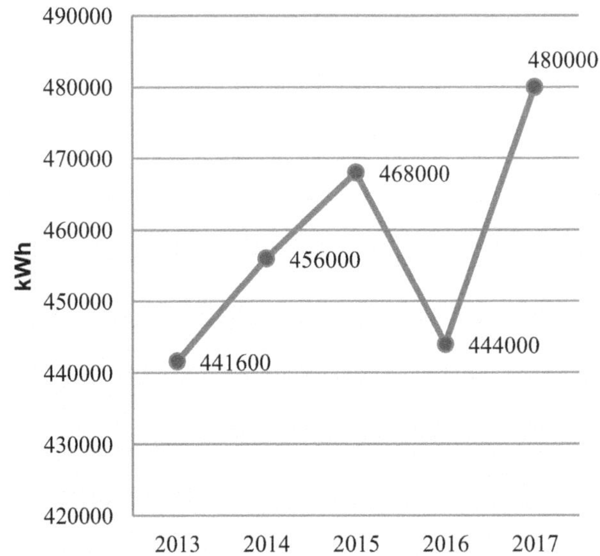

Fig. 8. Dynamics of electricity consumption by LLC "DMK" in physical terms.

Despite the decrease in electricity consumption in 2016, the payment increased by 7.75% due to the annual increase in the tariff and an increase in production (Fig. 9).

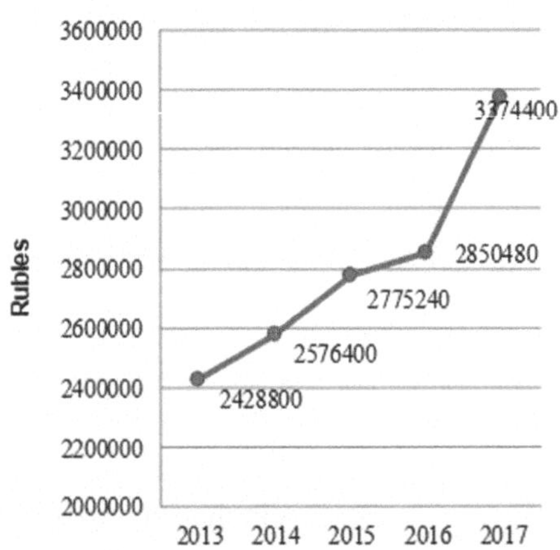

Fig. 9. Dynamics of electricity consumption by LLC "DMK" in monetary terms.

5 Analysis of Electricity Consumption by Other Consumers

The category of ‘other consumers’ includes all legal entities that have concluded energy supply contracts with a sales company except for those included in the categories ‘industry’ and ‘budget’. These are various LLC, OJSC, individual entrepreneur, private enterprise as well as relatively new formations - homeowners’ association and cooperatives. The growth in consumption by the ‘others’ group (Fig. 10) in recent years was due to two factors: the conclusion of new contracts with consumers and the transition of a part of households to homeowners’ associations (Fig. 11).

Using the example of LLC “Remstroyproekt v Dmitrovsk”, we considered and analyzed the dynamics of a specific enterprise from the category of ‘other consumers’ in the Dmitrovsky district, Orel region. Remstroyproekt includes several points of electricity consumption: recreation center “Selkin Prud”, cafe “Nerussa”. The main consumption fell on the recreation center, where there were street lighting, a hotel, individual cottages on a large territory. According to the graph shown in Fig. 12, it can be seen that from 2013 to 2015, electricity consumption had grown, but it significantly decreased after 2015 due to a decrease in effective demand for some types of services (Fig. 13).

Comparison of electricity consumption by the organization in physical and monetary terms showed that even a significant reduction in natural consumption volumes did not lead to a noticeable decrease in payment.

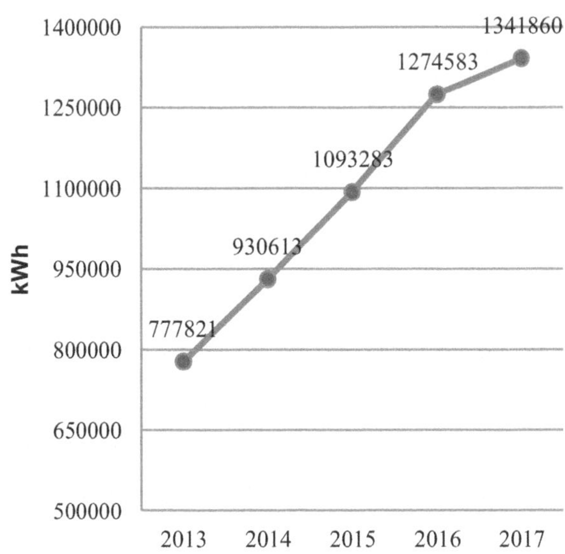

Fig. 10. Dynamics of electricity consumption by the ‘other consumers’ category in Dmitrovsky district in physical terms.

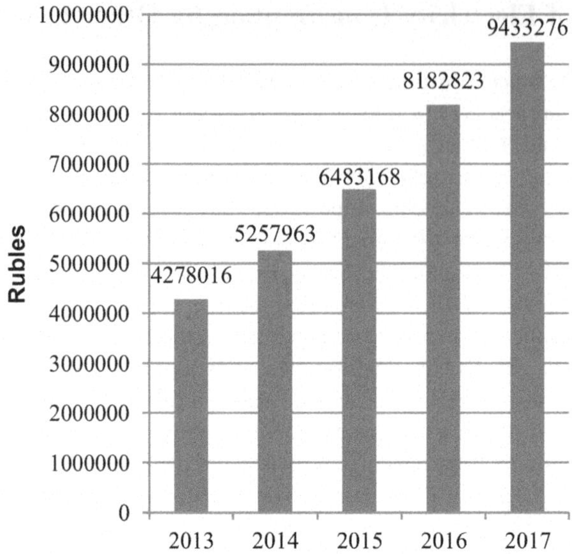

Fig. 11. Dynamics of electricity consumption by the 'other consumers' category in Dmitrovsky district in monetary terms.

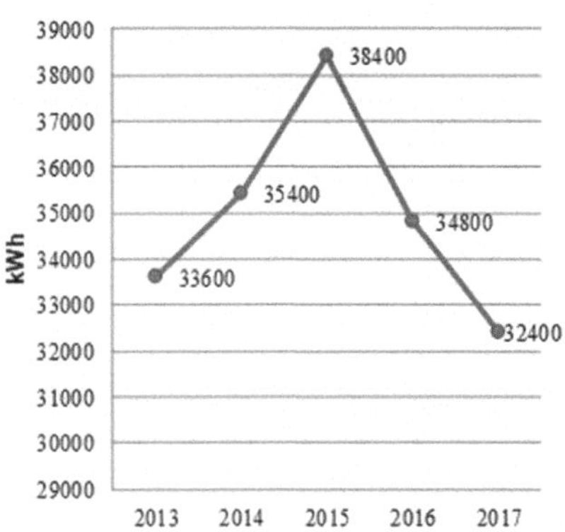

Fig. 12. Dynamics of electricity consumption by LLC "Remstroyproekt v Dmitrovsk" in physical terms.

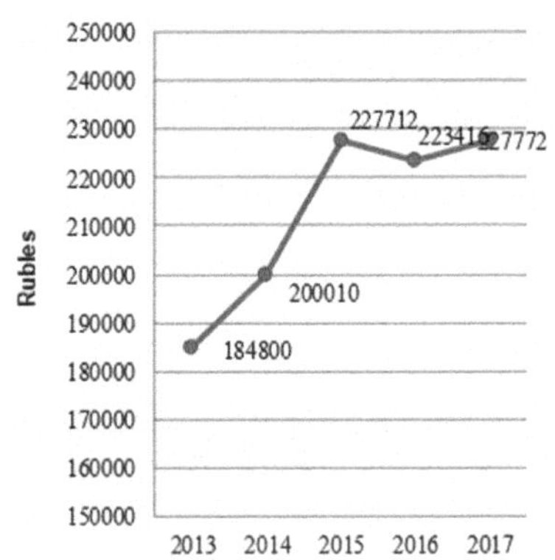

Fig. 13. Dynamics of electricity consumption by LLC "Remstroyproekt v Dmitrovsk" in monetary terms.

6 Conclusions

1. Analysis of electricity consumption by consumer groups in the Dmitrovsky district, Orel region showed that changes in the consumption structure were taking place in the direction of a slight increase in the share of consumption by 'other consumers' with a decrease in the share of households, industry and budget sector.

2. The reduction in electricity consumption in the budget sphere was associated with the closure of a number of budget organizations (primarily rural schools) and the transfer of others to a shorter work schedule. At the same time, there was no a noticeable decrease in the payment for electricity by this group as a whole in the district. An analysis using the example of a particular organization showed that even with a noticeable decrease in electricity consumption, there was an increase in the cost of paying for it. A similar picture was observed in the group of 'other consumers'. Both of these groups as well as the group of household consumers, largely depend on the situation in the industry that exists in the region and provides jobs for population. Since there was no noticeable development of industry, there was a decrease in the number of population and a decrease in demand for services of budget organizations.
3. The reduction in electricity consumption by industry was also associated with the closure of a number of enterprises and with a change in the load of existing ones. There were examples of enterprises showing an increase in electricity consumption with a significant increase in payments for it. But even here (most clearly), an analysis, using the example of a particular enterprise, showed that with a noticeable decrease in electricity consumption, there was an increase in the cost of paying for it.
4. The increase in the cost of paying for electricity, even with a decrease in its consumption in physical terms, allowed us to conclude that efforts to reduce the tariff burden on all areas of consumers had not brought any noticeable result. The financial burden on organizations associated with payment for energy resources had grown, which, in turn, increased the cost of their products, the cost of provided services and thus decreased demand for them. Therefore, it is necessary to revise both the tariff policy and the policy in the field of industrial support.

References

1. Vinogradov, A., Vinogradova, A., Bolshev, V.: Analysis of the major constituents of the power supply system efficiency for rural consumers. IETE J. Res. 1–9 (2020). https://doi.org/10.1080/03772063.2020.1782782
2. Vinogradov. A., et al.: Analysis of the power supply restoration time after failures in power transmission lines. Energies **13**(11), 2736 (2020). https://doi.org/10.3390/EN13112736
3. Kushler, M., York, D.: Energy efficiency and electric system reliability (2002)
4. Zatsepina, V.I., Zatsepin, E.P.: Reliability of power supply systems under the influence of negative factors. In: 2017 International Conference on Industrial Engineering. Applications and Manufacturing (ICIEAM), pp. 1–4 (2017)
5. Ministry of Energy of the Russian Federation, Electrical Installations Code, 7th edition (2003)
6. Vinogradov, A., Vinogradova, A., Bolshev, V.: Analysis of the quantity and causes of outages in LV/MV electrical grids. CSEE J. Power Energy Syst. **6**(3), 537–542 (2020). https://doi.org/10.17775/cseejpes.2019.01920

7. Cliteur, G.J., Wetzer, J.M.: Condition assessment of power transmission and distribution components. In: 16th International Conference and Exhibition on Electricity Distribution, Part 1: Contributions, CIRED, IEE Conference Publication No. 482, vol. 1, p. 5 (2001). https://doi.org/10.1049/cp:20010698
8. Bolshev, V., Vinogradov, A., Jasiński, M., Sikorski, T., Leonowicz, Z., Gono, R.: Monitoring the number and duration of power outages and voltage deviations at both sides of switching devices. IEEE Access **8**, 137174–137184 (2020). https://doi.org/10.1109/ACCESS.2020.3011836
9. Vinogradov, A., et al.: A system for monitoring the number and duration of power outages and power quality in 0.38kV electrical networks. In: Advances in Intelligent Systems and Computing, vol. 866 (2019). https://doi.org/10.1007/978-3-030-00979-3_1
10. Erol-Kantarci, M., Mouftah, H.T.: Wireless multimedia sensor and actor networks for the next generation power grid. Ad Hoc Netw. **9**(4), 542–551 (2011). https://doi.org/10.1016/j.adhoc.2010.08.005
11. Chatterjee, D., et al.: Smart power cut monitoring in distribution lines using sensors. In: 2019 IEEE International WIE Conference on Electrical and Computer Engineering (WIECON-ECE), pp. 1–7 (2019). https://doi.org/10.1109/WIECON-ECE48653.2019.9019912
12. Zhudin, I.M.: Dmitrovsky district (2003)
13. On the functioning of retail electricity markets, full and (or) partial limitation of the mode of consumption of electrical energy, Decree of the Government of the Russian Federation 442
14. On the procedure for providing public services to citizens, Decree of the Government of the Russian Federation 307
15. Bitsukova, N.I., Vinogradov, A.V.: Electricity consumption in the districts of the Oryol region included in the area of responsibility of the guaranteeing supplier, LLC INTER RAO – Orlovsky energosbyt in the period 2008 … 2013. OrelSAU, Orel, Russia (2013)
16. Vinogradov, A.V., Vinogradova A.V.: The result of an energy survey of rural schools and the task of increasing their energy efficiency. Features of technical and technological equipment of modern agricultural production, pp. 330–333 (2013)

Research of the Influence of Wind Flow Direction on Loads on Temporary Structures

Ilshat Galimullin(✉), Arthur Ippolitov, Gennady Shmelev, and Lenar Khaidarov

Kazan State University of Architecture and Engineering, 1 Zelenaya street, Kazan 420043, Russian Federation
alchimik96@mail.ru

Abstract. The article is devoted to the numerical study of the wind load on structures under various common schemes of installation of covering material and various wind directions. Five main schemes were studied when viewed from above – "line", "square", "rectangle", "U-shape" and "right angle" for different angles of wind direction with a step of 15°, while taking into account the symmetry of the structures in question in the plan. When changing the wind direction angle on temporary structures, a twisting moment with an arm of 0.1–0.12 of the size of the largest side of the temporary structure may occur, which is important to consider when designing structures. This resulting twisting moment is not taken into account in the regulatory documents, such as SP "Loads and impacts". This regulatory document provides a limited number of schemes for calculating wind impacts. The schemes considered in the study can also be applied to the stands with a canopy, displaying various stages during their construction on the site for a specific entertainment event. The results of the numerical study are relevant not only for temporary structures, but also for structures in principle. However, the heavier the structure, the smaller the share of its loads is the influence of wind.

Keywords: Structures · Wind load · Particle image velocimetry · Computational fluid dynamics · Grandstand · Wind-induced pressure

1 Introduction

Temporary structures are often used to accommodate spectators at various entertainment events [1, 2]. For temporary structures, wind load is a fairly significant part of all loads on the structure [3–5]. Therefore, it is important to study the influence of the direction of wind flow on the loads on temporary structures [6, 7]. It is worth noting that the results of a detailed numerical study may be relevant not only for temporary structures, but also for structures in principle. However, the heavier the structure, the smaller the share of its loads is the influence of wind.

At the same time, both domestic and foreign regulatory documents provide a limited number of schemes of structures for calculating wind impacts. Alternative schemes are proposed to be considered using tests in specialized large-scale wind tunnels [8–11]. Unfortunately, such tests are very time-consuming, and for the most

A. Sinitsyn (Ed.): SES 2020, LNCE 141, pp. 383–392, 2021.
https://doi.org/10.1007/978-3-030-67654-4_41

large-scale models they have to be performed in special laboratories, such experiments are near impossible for many scientists [12–15].

In modern construction, it is important to be able to build buildings and structures of complex geometric configurations. At the same time, the more complex the geometric configuration of a particular structure, the less suitable it is for using standard schemes of structures from domestic and foreign regulatory documents. The actual values of the aerodynamic coefficients may differ significantly from the values according to the SP "Loads and impacts" [16, 17]. In recent years, there has been an intensive development of computer-based hydro-gas dynamics (CFD) systems. Thus, there is an alternative to calculating temporary structures in specialized wind tunnels. Such methods of modeling wind impacts have sufficient accuracy and significantly reduce the time resources for calculating aerodynamic coefficients for temporary structures [18–24]. In turn, popular and effective programs of the ANSYS complex are selected from the programs that are suitable for calculation. From the complex, the FLUENT and CFX modules are usually used for calculating wind impacts. This is due to the ability to accurately model the wind impact and the breadth of data obtained [25–27].

The ANSYS FLUENT software package was selected for use in this numerical study as the most optimal for solving the set tasks. It is also important to note that the results obtained by this method can be further used for verification of wind tunnels and optimization of experiments conducted in them.

2 Materials and Methods

The study of the influence of air flow is carried out on the example of suspended structures and temporary structures of stands, due to the predominance of wind forces on these structures.

In order to track and establish the dependencies we are interested in, we started with creating and calculating simplified models of temporary structures in the plan. Analysis of existing options for both stands with canopies in particular, and entertainment structures in general, helped to identify the following main variations of the form in the plan:

a) Line (free-standing banners, very narrow and long structures);
b) Square (for temporary structures with equal or almost equal aspect ratio);
c) Rectangle (for temporary structures with an uneven aspect ratio – one side is larger than the other);
d) U-shape (a common form used, for example, for a single structure of the grandstand with the location of the entertainment event inside the letter "U", at the same distance from the sides);
e) Right angle (an unfavorable, but common form for temporary structures, used, for example, for a single structure of the podium with the location of a spectacular event inside an improvised "corner", at the same distance from the sides);

It is worth noting that each of these five basic forms has one or two axes of symmetry in the plan. Accordingly, the wind direction can be considered not for 360°, as for a completely asymmetric structure, but only for a certain range of angular values (Figs. 1, 2, 3, 4 and 5).

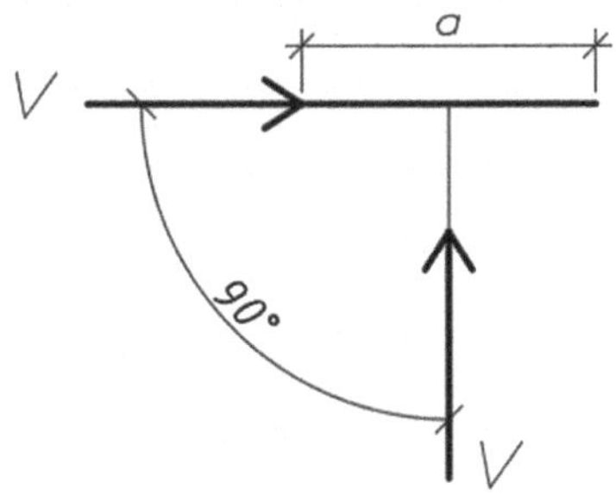

Fig. 1. Form in plan №1 "Line"

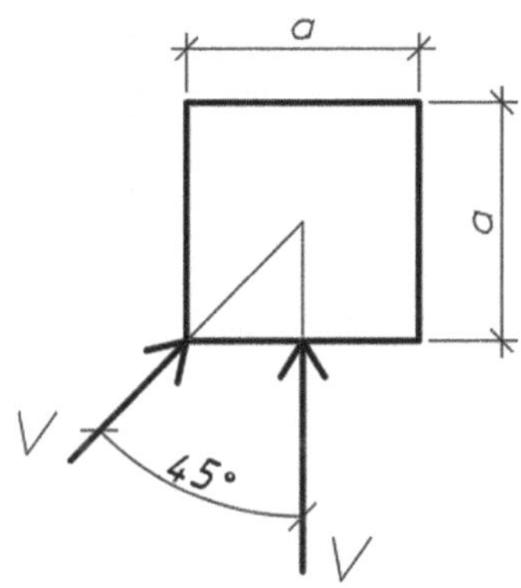

Fig. 2. Form in plan №2 "Square"

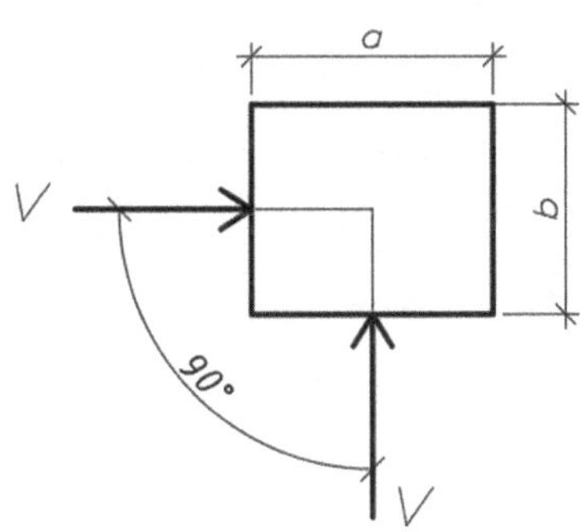

Fig. 3. Form in plan №3 "Rectangle"

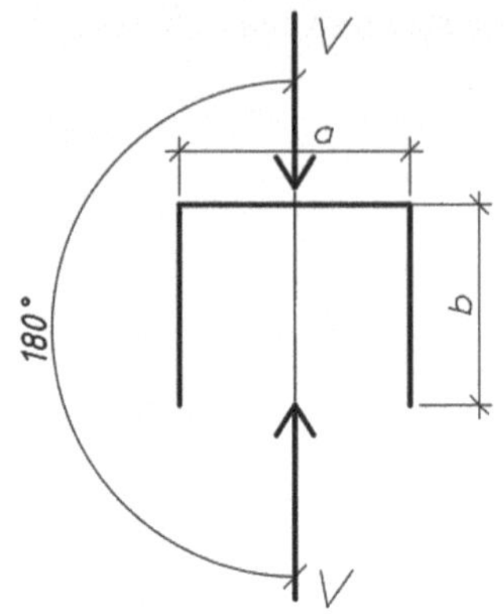

Fig. 4. Form in plan №4 “U-shape”

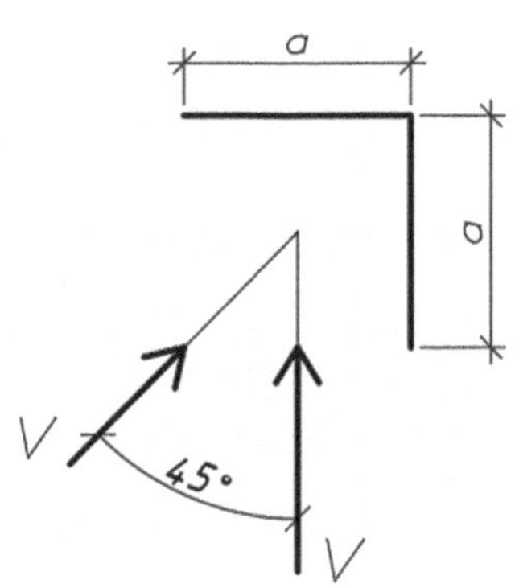

Fig. 5. Form in plan №5 “Right angle”

In the five diagrams of various shapes shown in the plan, “V” is the wind direction, “a” and “b” are the dimensions of the sides of the calculated model of the temporary structure. The wind speed in the numerical study was set to 10 m/s, and the obtained aerodynamic coefficients were rechecked with wind speeds of 5, 7.5, 12.5 and 15 m/s. Numerical research is performed in such a way that both the dimensions of the sides (and their ratio) and the distances to objects are not taken as specific values, but relative to each other. This is done for the universality of the results obtained.

3 Results and Discussion

The illustrations obtained in the course of numerical research in the ANSYS FLUENT software package for the character of wind flow for each of the five simple forms of construction in the plan turned out to be quite realistic, as can be seen from the illustrations given as an example below (Figs. 6 and 7).

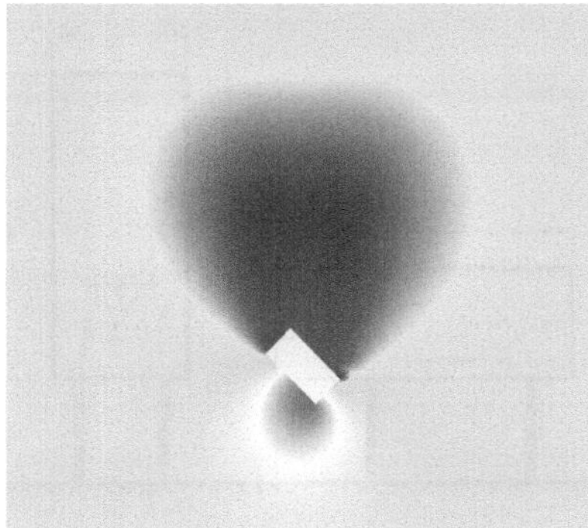

Fig. 6. Wind flow pattern №3 "Rectangle" when the wind direction is at an angle of 45°

Fig. 7. Wind flow pattern №3 "Rectangle" when the wind direction is at an angle of 60°

The nature of actual wind loads on temporary structures in the plan is compared through the aerodynamic coefficient of drag C_f and the shoulder of the total twisting moment acting on the temporary structure. In this case, the moment's arm is output as a fraction of "a" - one of the sides of the temporary structure.

It is worth noting that in the SP "Loads and impacts", twisting moments are mentioned. For example, appendix "B.1.1 Free-standing flat solid structures" provides a diagram for billboards raised above the ground (Fig. 8).

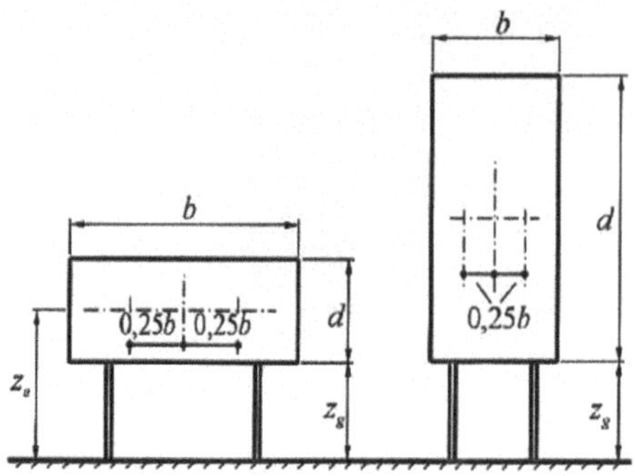

Fig. 8. Figure B.2. From the SP "Loads and impacts", billboards raised above the ground to a height of at least d/4

It is important to note that such scheme (shown in the figure above) from the considered forms of temporary structures in the plan corresponds to the form №1 "Line".

Using the obtained values, we plot the distribution of the shoulder of twisting moment (in fractions of the "a" side) for different wind direction angles (Figs. 9, 10, 11, 12 and 13).

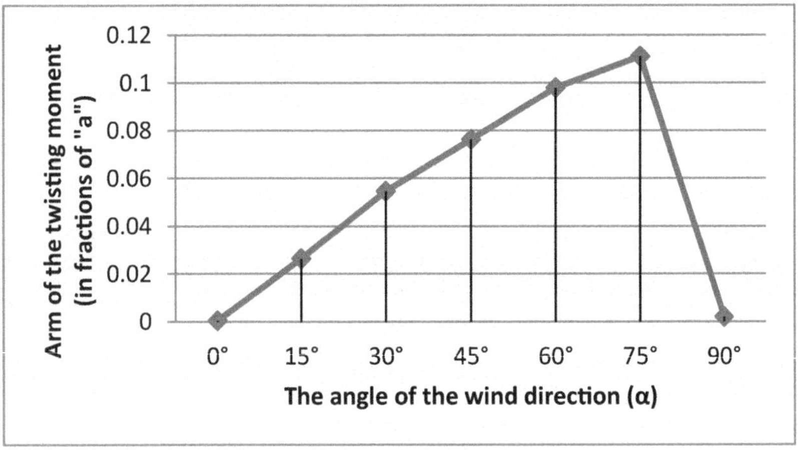

Fig. 9. Graph of the distribution of the twisting moment's arm depending on the angle of the wind direction (Diagram №1 "Line")

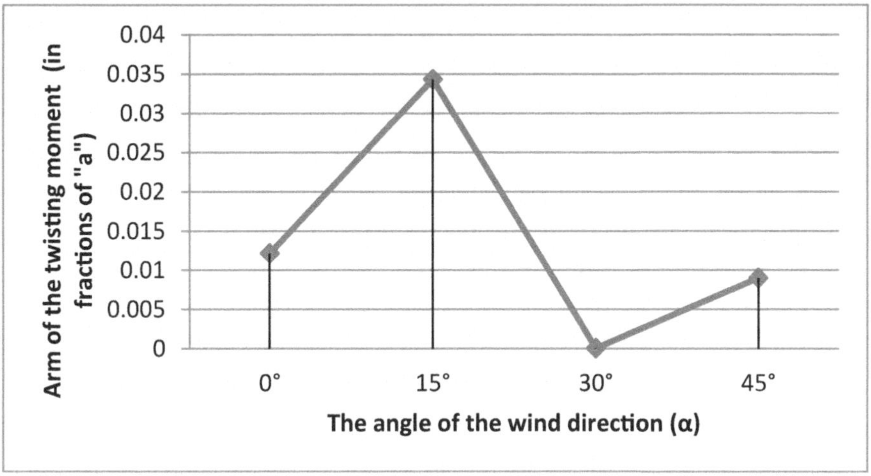

Fig. 10. Graph of the distribution of the twisting moment's arm depending on the angle of the wind direction (Diagram №2 "Square")

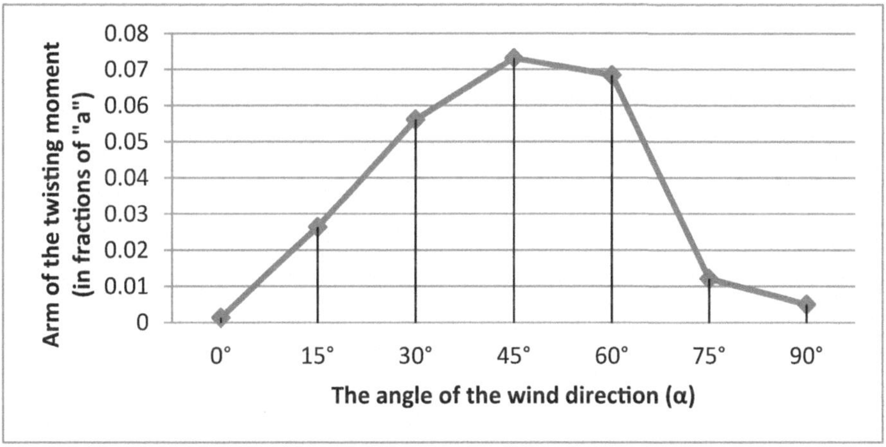

Fig. 11. Graph of the distribution of the twisting moment's arm depending on the angle of the wind direction (Diagram №3 "Rectangle")

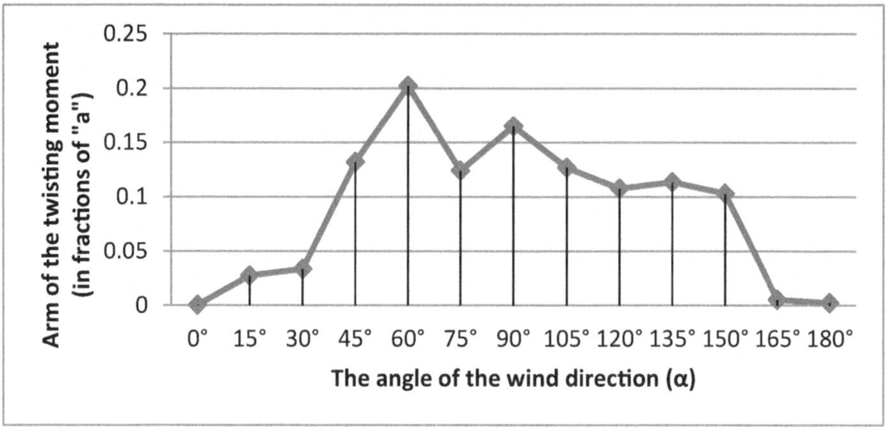

Fig. 12. Graph of the distribution of the twisting moment's arm depending on the angle of the wind direction (Diagram №4 "U-shape")

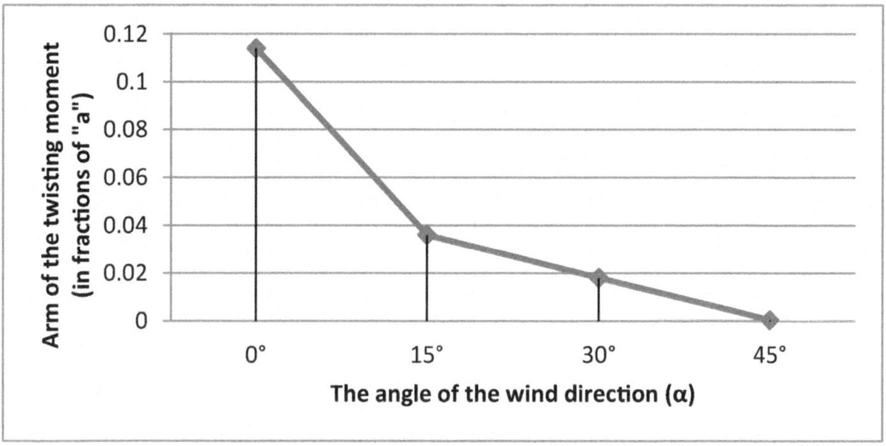

Fig. 13. Graph of the distribution of the twisting moment's arm depending on the angle of the wind direction (Diagram №5 "Right angle")

As it can be seen, the nature of the distribution of the twisting moment's arms is quite realistic. It is noted at which angles the maximum values are reached from the considered ones - 75° for "line", 15° for "square", 45° for "rectangle", 60° for "U-shape" and, finally, 0° for "right angle".

It is important to note that the value of the moment's arm at some angles of wind direction reaches 0.1–0.12 of the size of the largest side ("a"), which is not taken into account by both domestic and foreign regulatory documents on wind impacts on structures. These schemes can also be applied to the stands with a canopy, showing the various stages of their construction on the site for a specific event.

4 Conclusion

Modern entertainment events often feature temporary structures of complex geometric shapes that differ from standard schemes of normative documents. For such structures, in particular, for stands with a canopy, it is important to accurately determine the aerodynamic coefficients that reflect the wind load. Based on the above results obtained in the course of numerical research, it is possible to draw the following main conclusions:

1. The ANSYS FLUENT is used for solving the set tasks of hydro gas dynamics.
2. A numerical study of the wind load on structures under various schemes of installation of cladding material and various wind directions was performed.
3. Five main schemes were studied – "line", "square", "rectangle", "U-shape" and "right angle" for different angles of wind direction in 15° increments, taking into account the symmetry of the considered structures in the plan.
4. The nature of actual wind loads on temporary structures in the plan was compared using the aerodynamic drag coefficient C_f and the arm of the twisting moment acting on the temporary structure. In this case, the moment's arm was output as a fraction of "a" - one of the sides of the temporary structure.
5. For each of the considered forms of temporary structures, a graph of the distribution of the moment's arm is drawn up in the plan, and the largest value of the moment's arm is found from the considered values of the wind flow direction angle.
6. When changing the angle of wind direction to temporary structures, a twisting moment with an arm of 0.1–0.12 of the size of the largest side may occur. This is not taken into account by regulatory documents, but it is important to take into account when designing a structure.

References

1. Gordeev, V.N., Lantukh-Lyashenco, A.I., Pashinskiy, V.A., Perelmuter, A.B., Pichugin, S. F.: Loads and impacts on buildings and structures, pp. 195–222. IASW, Moscow (2007)
2. Overend, M., Zammit, K.: Wind loading on cladding and glazed façades. In: International Symposium on the Application of Architectural Glass, ISAAG, pp. 1–3, Nottingham (2006)
3. Mendis, P., Ngo, T., Haritos, N., Hira, A., Samali, B., Cheung, J.: Wind loading on tall buildings, pp. 41–42. EJSE Special Issue: Loading on Structures, Australia (2007)
4. Ghosh, A.K.: International Workshop on Wind Engineering and Science, pp. 107–112. Elite Publishing House Pvt. Ltd., New Delhi (2004)
5. Yuan, X., Anumba, C.J., Parfitt, M.K.: Cyber-physical systems for temporary structure monitoring. Autom. Constr. **66**, 2–6 (2016)
6. Liu, C., He, L., Wu, Z., Yuan, J.: Experimental and numerical study on lateral stability of temporary structures. Arch. Civ. Mech. Eng. **18**, 1475–1489 (2018)
7. Van Uffelen, G.M.: Wind-induced building interference: increase of wind loads on existing buildings after erection of new high-rises, vol. 5, pp. 1–3. EACWE, Florence (2009)
8. Hoffman, J., Johnson, C.: A new approach to computational turbulence modeling. Comput. Methods Appl. Mech. Eng. **195**(23–24), 2865–2870 (2006)

9. Hu, H., Yang, Z., Sarkar, P.: Dynamic Wind Loads and Wake Characteristics of a Wind Turbine Model in an Atmospheric Boundary Layer Wind, pp. 1277–1279. Springer-Verlag, Ames (2011)
10. Ahmad, S., Muzzammil, M., Zaheer, I.: Numerical prediction of wind loads on low buildings. Aligarh Int. J. Eng. Sci. Technol. **3**, 59–60 (2011)
11. Crick, D., Grondin, G.Y.: Monitoring and analysis of a temporary grandstand structural engineering report No. 275, pp. 48–98. Department of Civil & Environmental Engineering of University of Alberta, Edmonton (2008)
12. Chandrangsu, T., Rasmussen, K.J.: Structural modelling of support scaffold systems. J. Constr. Steel Res. **67**, 866–875 (2011)
13. Sauca, A., Milchis, T., Gobesz, Z.: Wind loading on solar panels. Műszaki Tudományos Közlemények. **10**, 73–78 (2019)
14. Killen, J.P., Letchford, C.W.: A parametric study of wind loads on grandstand roofs. Eng. Struct. **23**, 725–735 (2001)
15. Killen, J.P., Letchford, C.W.: Equivalent static wind loads for cantilevered grandstand roofs. Eng. Struct. **24**, 207–217 (2002)
16. Zhenggang, C., Ning, S., Yue, W., Shitao, P.: Gust response envelope approach to the equivalent static wind load for large-span grandstand roofs. J. Wind Eng. Ind. Aerodyn. **180**, 108–121 (2018)
17. Wang, F., Lam, K.M., Zu, G.B., Cheng, L.: Coherent structures and wind force generation of square-section building model. J. Wind Eng. Ind. Aerodyn. **188**, 177–193 (2019)
18. Zu, G., Lam, K.M.: Simultaneous measurement of wind velocity field and wind forces on a square tall building. Adv. Struct. Eng. **21**, 2241–2258 (2018)
19. Lam, K.M., To, A.P.: Generation of wind loads on a horizontal grandstand roof of large aspect ratio. J. Wind Eng. Ind. Aerodyn. **54–55,** 345–357 (1995)
20. Lam, K.M., Zhao, J.G.: Occurrence of peak lifting actions on a large horizontal cantilevered roof. J. Wind Eng. Ind. Aerodyn. **90**, 897–940 (2002)
21. Mikheev, P., Okorokov, R., Sidorenko, G., Timofeeva, A.: Determination of energy costs of wind farms at all life cycle stages. Adv. Intell. Syst. Comput. **982**, 242–256 (2020)
22. Chekalin, A.V., Andreev, V.M., Ascheulov, Y.V., Chumakov, Y.S.: Aerodynamics for CPV: investigating the efficiency of using air fairings on solar trackers. In: AIP Conference Proceedings, vol. 2149, no. 1, p. 060002 (2019)
23. Semenov, A.S., Getsov, L.B.: Models of inelastic deformation and fracture of single-crystal nickel-base superalloys under thermal cyclic loading. In: ICF 2017 - 14th International Conference on Fracture, pp. 2378–379 (2017)
24. Kuznetsov, I.L., Salakhoutdinov, M.A., Gainetdinov, R.G.: Stand and test results of a 24 m span truss with elements of galvanized cold-formed profiles. News KSUAE **4**, 193–200 (2018)
25. Lalin, V., Galyamichev, A., Zdanchuk, E., Mutovkin, A., Dogru, S.: Wind loads on a high-rise building. Lect. Notes Civ. Eng. **70**, 551–562 (2020)
26. Solovev, S., Khrapunov, E.: Modeling of the mean wind loads on structures. Mag. Civ. Eng. **88**(4), 42–51 (2019)
27. Khrapunov, E., Solovev, S.: Advantages of modeling ABL properties to determine wind loads on structures. In: MATEC Web of Conferences, vol. 245 (2018)

Analysis of Distance Protection's Operation in Cases of Deep Saturation of Current Transformers

Anatoly Osintsev[1(✉)], Aleksandra Naumova[1], and Elena Gracheva[2]

[1] Novosibirsk State Technical University, K. Marksa Prospect, 20, 630073 Novosibirsk, Russia
osintsev@list.ru

[2] Kazan State Power Engineering University, Krasnoselskaya Street, 51, 420066 Kazan, Russia

Abstract. There is usually no information about permissible error of current transformers in modes succeed by large relative short-circuit current, at which microprocessor-based protections operate stably. By this reason, it is necessary to use data, defined for analog relays. It leads to value appreciation of a project because it is often essential to reduce current transformers' error in case of a short circuit fault. Therefore, it is necessary to define the value of current transformers' error, permitted for impedance relays. Conclusions of fundamental and applied sciences (mathematical analysis, theoretical foundations of electrical engineering), analytical methods of researching nonlinear circuits and digital signal processing were used. A simulation model was created for setting overall tests of the current trans-former (CT) system. It was a relay protection device that reflected all the important properties of studied objects and allowed an analysis of digital distance protection's operation at high levels of short-circuit currents. The factors influencing over digital distance protection's operation in case of deep saturation of CTs were revealed, and a certain algorithm for definition of the permissible CT errors was proposed. Stable operation of digital distance protection was observed in case of a fault nearby the place of current transformers' setting in all theoretically possible combinations of electrical system's power and length of a protected electric power transmission line. It is valid if electric load choice is carried with account for stable protection's operation in condition of a fault in the computational point and if voltage swell in secondary wirings is infeasible.

Keywords: Current transformers · Sigital distance protection's

1 Introduction

To date, microprocessor-based digital protections are widely used [1, 2]. Such protections have better characteristics compared with analog protection systems. The level of short-circuit currents increases due to electricity grid development and increasing of industrial capacity. It may become the reason of current transformers saturation in conditions of a fault nearby the place of current transformers' setting. It is important to

A. Sinitsyn (Ed.): SES 2020, LNCE 141, pp. 393–400, 2021.
https://doi.org/10.1007/978-3-030-67654-4_42

provide stable operation of relay protection in such modes [3–7]. The essence of the problem is that permissible error of current transformers, valid for analog impedance relays, is not justified in relation to digital protection, but is often used due to the lack of other data. Such method leads to increasing of protection's operation stability. By this reason, calculated secondary CT-load may seem excess, but in fact, it is permissible for digital protection if the calculation is carried as for analog relays. In such cases, there is no necessity to assume additional valuable measures to reduce current transformers' error.

2 Mathematical Model

To solve this problem, it is necessary to consider functioning of current transformers and measuring elements of the investigated devices in extreme modes offaults, when significant CT errors may occur, distorting secondary currents and, consequently, leading to a violation of relay protection stability. In this case, not only the current error is important, but also the angular error, which can also lead to unnecessary operation or failure in operation of protection.

Mathematical model of a part of an electrical network, which is necessary to study operation of digital distance e protection in case of a fault near CT setting, as well as all further considered objects, is created in MATLAB software (Fig. 1). Investigated model consists of a part of an electrical network consists of equivalent power system S with inductive reactance XS and generated voltage ES and an electric power transmission line L with inductive reactance XL.

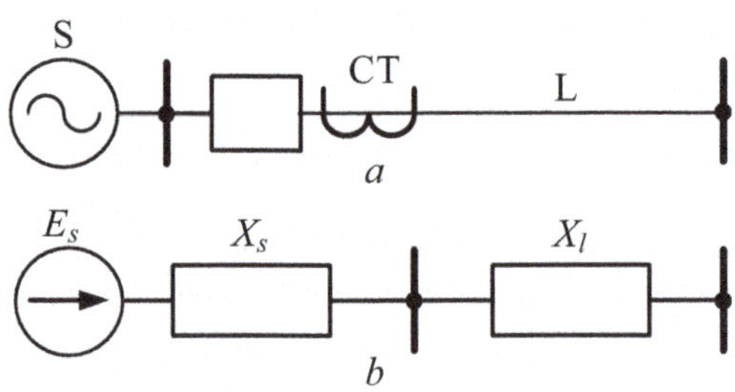

Fig. 1. Simulated part of an electrical network (a – primary electrical network, b – equivalent network).

To carry out a high-quality check of relay protection devices, it is necessary to obtain reliable information about shape of secondary current of a current transformer. The most difficult thing in modeling of a primary current transducer is mathematical description of the processes in its magnetic circuit, to which a large number of publications are devoted, for example [8–13]. Moreover, mathematical models of CT were developed as part of the creation of software systems for modeling electromagnetic transient processes in electric power systems in a number researches [14–16]. At the same time, mathematical description of the investigated object's operation should be as simple and clear as possible. In this regard, the following assumptions were made in the CT's model, which do not create large errors in the result:

- active and inductive resistance of the primary winding are equal to zero;
- the magnetic field inside the magnetic core is distributed regularly.

Description of the current transformer is compiled according to [17]. Saturation of the CT is taken into account by setting the magnetization curve in the form of an approximating function with selected coefficients [18].

The main parameters of current transformers (a core cross-sectional area SM, length of mean magnetic line lM, number of rounds of primary and secondary winding, the active resistance of the secondary winding) are determined from the passport data for a specific type of CT.

The voltage transformer model should be focused on the analysis of the processes in them in case of a fault.

Their impact on distance protection's operation is not significant. By this reason, it is enough to make scaling of primary signals' arrays, taking into account the CT ratio. The main purpose is to make an algorithm of distance protection's operation, precisely for it's measuring elements and logic section [19].

Parameters of the response characteristic are calculated automatically after entering of the line resistance value. Taking into account that nowadays manufacturers of relay protection devices provide a wide range of forms of distance protections response characteristic, we will consider only the most frequently used ones.

In the simplest case, it is a circle passing through zero of coordinate system. The diameter of the circle is equal to 85% of a protected line resistance. The inductive, active resistance of the power transmission line and the angle of maximum sensitivity are determined. A distant element with a circular characteristic passes through points Z1 and Z2, located at the ends of the diameter:

$$\begin{aligned} Z_1 &= (0;0); \\ Z_2 &= (0,85 \cdot R_L;\ 0,85 \cdot X_L); \\ R_L &= \frac{X_L}{tg\phi_L}. \end{aligned} \tag{1}$$

The middle of the interval connecting points Z1 and Z2 will be the center of the circle, which is the line of the response characteristic demarcation.

We will also consider generation of a response characteristic in the form of a polygon. Currently, microprocessor-based protections manufactured by different companies, including foreign ones, are used in Russian power systems. The most frequently used terminals for the protection of electrical networks with a voltage of 110 kV and above are manufactured by the company "Ekra" (Cheboksary) [20]. These terminals will be investigated further. Calculations of the operation settings, as well as the slope of the operation characteristic, were made according to the recommendations of the manufacturer. The measurement resistance vector Z(nT) will be in the area of operation of the digital distance protection with a polygonal characteristic if the following system of inequalities correspond to the studied case:

$$\left.\begin{array}{l}
X(nT) - \dfrac{X_6 - X_5}{R_6 - R_5} R(nT) + \dfrac{X_6 R_5 - X_5 R_6}{R_6 - R_5} \geq 0; \\
X(nT) - \dfrac{X_6 - X_5}{R_6 - R_5} R(nT) + \dfrac{X_6 R_5 - X_5 R_6}{R_6 - R_5} \geq 0; \\
X(nT) - \dfrac{X_6 - X_5}{R_6 - R_5} R(nT) + \dfrac{X_6 R_5 - X_5 R_6}{R_6 - R_5} \leq 0; \\
X(nT) - \dfrac{X_6 - X_5}{R_6 - R_5} R(nT) + \dfrac{X_6 R_5 - X_5 R_6}{R_6 - R_5} \leq 0; \\
X(nT) - \dfrac{X_6 - X_5}{R_6 - R_5} R(nT) + \dfrac{X_6 R_5 - X_5 R_6}{R_6 - R_5} \leq 0; \\
X(nT) - \dfrac{X_6 - X_5}{R_6 - R_5} R(nT) + \dfrac{X_6 R_5 - X_5 R_6}{R_6 - R_5} \geq 0.
\end{array}\right\} \quad (2)$$

where Rn and Xn are the coordinates of the response characteristic vertices, and R (nT) and X(nT) are, respectively, the coordinates of the measurement resistance on the plane of complex numbers. After obtainment of the results of calculation the arrays of the values of secondary currents and voltages at the place the protection installation an array of complex values of the measurement resistances is determined by the methods of digital filtration. This array of resistances is checked if it is in the area of operation of the simulated protection. It worth mention that when the voltage at the protection installation setting place has a certain value, that is close to zero, the distance protection model increases the measurement value of voltage by a certain amount. It is necessary to ensure the correct operation of the distant element in case of internal and external faultsnear the place of the relay protection setting due to the need to maintain a short-term or long-term polarizing voltage, which coincides or is close in phase with the measurement voltage, which decreases to zero at close short-circuits.

In this case, the uncertainty of operation and the lack of a clear direction of protection are excluded. To date there are several options for the formation of the polarizing quantity, including using voltages of different phases. However, for correct operation in case of close three-phase short-circuits, when all voltages drop to zero, it is necessary to use a memory block with voltage fixation of the pre-emergency (load) mode.

Operation of relay protection devices depends on several factors, not only on the level of error of the current transformer: it is important to know the ratio of the magnitude of the short-circuit current near the protection installation to the operating value. It means that in one case, stable operation of the protection with a higher error is possible, and it's failure with in case of a smaller error. Therefore, the worst case of the configuration of the electrical network, as well as the value of the secondary load resistance and the type of current transformer should be considered to determine the permissible level of error. A specific algorithm was developed for this task. A certain value of the load resistance is set. At the same time, the permissible resistance of the system according to the condition of overvoltage absence in case of a fault near the place of the protection's installation is determined. In addition, the minimum possible length of electrical power transmission line, at which a ten percent error of the current transformer is observed in case of a fault at the design point, is determined. The extremum, at which the measurement resistance formed at the output of the measuring element will be as close as possible to the line of the response characteristic demarcation, is defined using the method of sequential enumeration. The intervals between the end of the measurement resistance vector and each of the faces of the response characteristics in the form of a polygon are calculated, meanwhile it is defined if the resistance vector is in range of operation. Therefore, a certain value of the parameter will correspond to a specific load resistance, by which it can be defined whether the protection device has operated or not. If the response characteristic is circle-shaped, the distance from the measuring point to the center of the response characteristic should be determined. Such calculations should be made for a range of load resistance values in order to determine the "worst" value. The corresponding algorithm is created in MATLAB software.

3 Computational Experiments

A study of the operation of distance digital protections, namely the first stages, in CT saturation modes was carried out for this task. Two types of response characteristics are considered: a circle and a polygon. The results of the study are presented in Table 1.

Table 1. Analisis of digital distance protection's operation.

CT type	Z_n, Ω	X_s, Ω	L_L, km	Intervals between the end of the measurement resistance vector and the limit of the response characteristic, Ω		CT error, %
				Circle	Polygon	
TFM-110 (100/5)	1,5	4,86	65,76	2.1	0.46	86
TFM-110 (200/5)	1,8	2,86	36,85	1.1	0.29	81
TFM-110 (1200/5)	1,8	0,476	6,14	0.18	0.047	83

4 Results

If the load resistance of the CT is selected according to the conditions of infeasibility of a voltage swell in the secondary winding and 10% current error in case of fault at the computational point, then correct operation of the considered types of digital distance protection is observed with all possible configurations of the electrical network. This statement refers to current transformer applications TFM-110 with different transformation ratios.

At the same time, a comparison of the operation of distant elements with different response characteristics allows to assert that it is advisable to use a relay with a circular response characteristic as a high-speed line protection stage. In this case a certain margin of safety response is created (when using a polygon, the end measurement resistance vector will be located closer to the limit of the response characteristic, see Fig. 2).

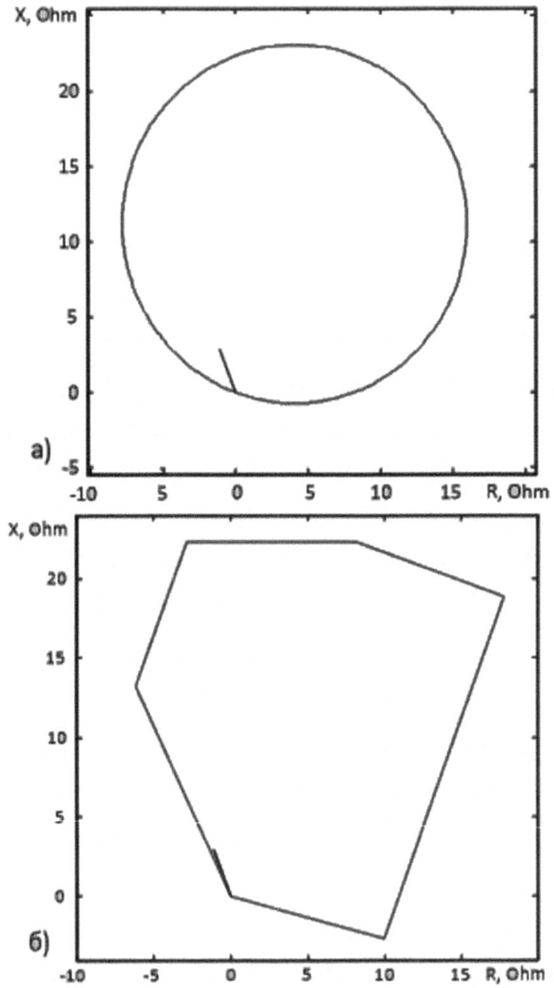

Fig. 2. Impedance protection's operation (the first line of the Table 1).

References

1. Shneerson, E.M.: Digital Relay Protection. Énergoatomizdat, Moscow (2007). (in Russian)
2. Svistunov, N.A.: Principles of construction systems of relay protection and automation based on microprocessor devices. Izvestiya Tula State Univ. Tech. Sci. **12–3**, 92–95 (2016)
3. Glazyrin, V.E., Frolova, E.I.: Analysis of the influence of transient processes in current transformers on remote elements' behavior using mathematical models. Electro Electr. Eng. Power Eng. Electr. Ind. **4**, 9–14 (2012)
4. Medeiros, R., Costa, F.: A wavelet-based transformer differential protection with differential current transformer saturation and cross-country fault detection. IEEE Trans. Power Delivery **22**(2), 789–799 (2018). https://doi.org/10.1109/TPWRD.2017.2764062
5. Kasztenny, B., Mazereeuw, J., Jones, K.: CT saturation in industrial applications – analysis and application guidelines [Electronic resource], GE Grid Solutions. https://store.gegridsolutions.com/faq/Documents/General/GET-8501.pdf. Accessed 29 Mar 2019
6. Kuzhekov, S.L., Degtyarev, A.A.: On the coordination of current transformers and relay protection and automation operation of electric-power systems in transient short-circuit states. Electr. Eng. **12**, 65–72 (2017)
7. Kuzhekov, S., Degtyarev, A., Vorobev, V.: Determination of time prior saturation of the current transformers in transient short-circuit mode. Power Stat. **1**, 42–47 (2017)
8. Drozdov, A.D., Platonov , V.V., et al. (eds.): Electric Circuits with Ferromagnetic Elements in Relay Protection, 2nd edn., Revised. and add, p. 256. Energoatomizdat, Moscow (1986)
9. Naumov, V.A., Shevtsov, V.M.: Mathematical models of current transformer in analysis of algorithms of differential protection. Elektr. Stat. **3**, 51–56 (2003)
10. Guerra, F.F., Mota, W.S.: Current transformer model. IEEE Trans. Power Delivery **22**(1), 187–194 (2007)
11. Annakkage, U.D., McLaren, P.G., Dirks, E., Jayasinghe, R.P., Parker, A.D.: A current transformer model based on the Jiles-Atherton theory of ferromagnetic hysteresis. IEEE Trans. Power Delivery **15**, 57–61 (2000)
12. Kezunovic, M.: Experimental evaluation of EMTP-based current transformer models for protective relay transient study. IEEE Trans. Power Delivery **9**(1), 405–413 (1994)
13. Litvinov, I.I., Osintsev, A.A., Glazirin, V.E.: Characteristic features of internal and external faults for use in differential protection. Advances in Engineering Research, vol. 133, pp. 425–432 (2017)
14. Glazyrin, V.E., Toropov, G.E.: Modeling of transient processes in groups of current transformers, vol. 3, no. 20, pp. 75–82. Sb. scientific. tr. NSTU, Publishing house of NSTU, Novosibirsk (2000)
15. Sop'yanik, V.: Computer calculation of transient and steady-state processes in current transformers and current circuits of the relay protection. Electric Stat. **2**, 48–52 (2004)
16. Prutik, A.F.: Modeling of relay protection and automation in the concept of all-mode modeling of electric power systems. In: Modern equipment and technologies: collection of articles. tr. XVI int. scientific-practical conf. students, graduate students and young scientists: in 3 volumes, t. 1, pp. 101–102. Publishing house of Vol. polytechnic un-ta, Tomsk (2010)
17. RD 153-34.0-35.301-2002. Instructions for checking current transformers used in relay protection and measurement circuits, Introduction 03.01.2003, p. 152. SPO ORGRES, Moscow (2003). (in Russian)
18. Korol, E.G.: Analysis of modeling methods of the magnetic characteristics of electromagnets for the magnetic field compensation in electrical equipment. Electr. Eng. Electromech. **2**, 31–34 (2007). (in Russian)

19. Osintsev, A.A., Frolova, E.I., Gozhenko, A.E.: Realization of distance protection operating characteristic. New in the Russian electric power industry: scientific and technical. Electron. Zhurn. **4**, 64–70 (2017)
20. Recommendations for settings calculation of backup protection of high voltage transmission lines based on terminals of NPP EKRA [Electronic resource], Recommendations for calculating settings, NPP EKRA LLC (2012). https://www.ekra.ru/engine/download.php?id=3738

Research of Wind Impact on Temporary Structures in the Presence of Obstacles in the Flow Path

Ilshat Galimullin(✉), Arthur Ippolitov, Gennady Shmelev, Lenar Khaidarov, and Ilya Cadyarov

Kazan State University of Architecture and Engineering, 1 Zelenaya Street, Kazan 420043, Russian Federation
alchimik96@mail.ru

Abstract. The article is devoted to the numerical study of the wind load on structures of various forms in the presence of obstacles in the path of the wind flow. The influence of shapes and sizes of obstructing structures on the wind load is also studied. Four main forms of temporary structures in the section were studied – "rectangle", "tent", "tribune", and "tower". When the distance between the obstacle and the studied object changes, the aerodynamic coefficient also changes. This influence is not taken into account in the regulatory documents such as SP "Loads and impacts". This regulatory document only considers various options for building density around the structure under construction. The schemes considered in the study are often used in various mass events. The results of the numerical study are relevant not only for temporary structures, but also for structures in principle. However, the heavier the structure, the smaller the share of its loads is the influence of wind.

Keywords: Structures · Wind load · Particle image velocimetry · Computational fluid dynamics · Obstacle · Wind-induced pressure

1 Introduction

Recently, various temporary structures have been used to accommodate spectators at cultural events [1, 2]. For such buildings and structures, the wind load plays a significant role [3–5]. Therefore, it is important to study the influence of barriers in front of a separate temporary structure on the loads on these structures [6, 7]. However, it is worth noting that the heavier the structure, the smaller the share of its loads is the influence of wind.

At the same time, both domestic and foreign regulatory documents provide a limited number of schemes of structures for calculating wind impacts. Alternative schemes are proposed to be considered using tests in specialized large-scale wind tunnels [8–11]. Unfortunately, such tests are very time-consuming, and for the most large-scale models they have to be performed in special laboratories, such experiments are near impossible for many scientists [12–15].

A. Sinitsyn (Ed.): SES 2020, LNCE 141, pp. 401–412, 2021.
https://doi.org/10.1007/978-3-030-67654-4_43

In modern construction, it is increasingly necessary to design buildings and structures of complex geometric configuration, requiring more precise determination of aerodynamic coefficients for the further application of wind loading. At the same time, the more complex the geometric configuration of a particular structure, the less suitable for use are standard schemes of structures from domestic and foreign normative documents. The existing regulatory documents do not include all aerodynamic factors and parameters, and therefore should be improved. Foreign regulatory documents describe the processes of aerodynamics of buildings and structures in more detail than domestic ones. They also recommend using load calculations based on experimental data and measurements for specific localities and locations in order to take into account the type of terrain, terrain and surrounding objects. The actual values of the aerodynamic coefficients may differ significantly from the values according to the SP "Loads and impacts" [16, 17].

In recent years, there has been an intensive development of computer systems for hydro-gas dynamics (CFD). Thus, there is an alternative to calculating temporary structures in specialized wind tunnels. Such methods of modeling wind impacts have sufficient accuracy and significantly reduce the time resources for calculating aerodynamic coefficients for temporary structures [18–23]. In computational hydro-gas dynamics, the most popular programs of the ANSYS complex (in particular, FLUENT and CFX), largely due to the ability to accurately model wind impacts and the visibility of the numerical modeling process.

The ANSYS FLUENT software package was selected for use in this numerical study as the most optimal for solving the set tasks. It is also important that the results obtained by this method can be further used for verification of wind tunnels and optimization of experiments carried out in them.

2 Materials and Methods

The effects of wind load on temporary structures are considered on the basis of a conditional rectangular object. Its dimensions are shown in Fig. 1.

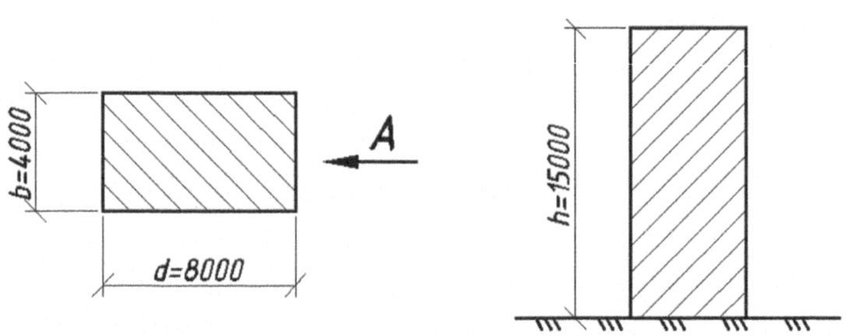

Fig. 1. General plan of the object under study.

The next stage involves analyzing the influence of wind flows on real temporary objects. There are several examples of such objects:

- Theatrical temporary construction of stands with a canopy;
- Tent;
- Sports facility of the tower type based on the real structure for holding the World Cup competitions in diving from extreme heights called “High diving” in Kazan.

General views of these types of structures are shown in Figs. 2, 3 and 4.

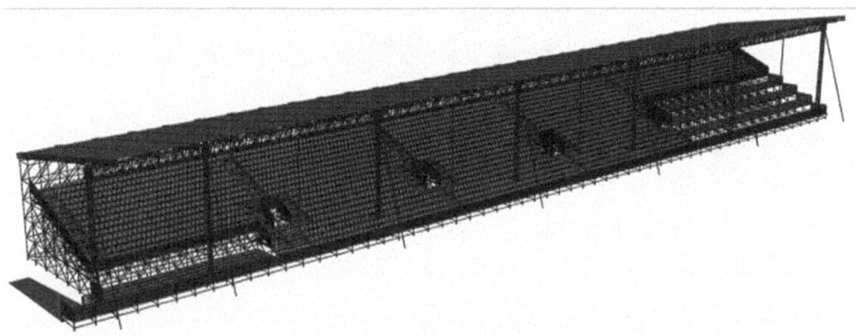

Fig. 2. General view of the structure of the stands with canopy.

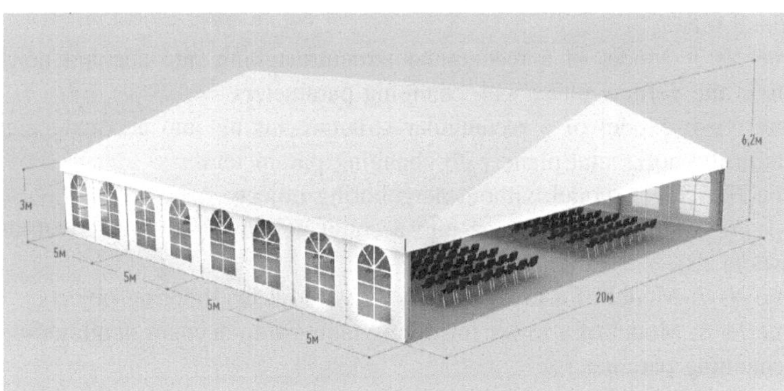

Fig. 3. General view of the tent.

Fig. 4. General view of the tower structure

The calculation is based on the following schemes:

Scheme № 1: Model of a rectangular structure without neighboring objects in the vertical plane.
Scheme № 2: Model of a rectangular structure without neighboring objects in the horizontal plane.
Scheme № 3: Model of a rectangular structure taking into account neighboring objects in the vertical plane with changing parameters.
Scheme № 4: Model of a rectangular structure taking into account neighboring objects in the horizontal plane with changing parameters.
Scheme № 5: Tent model without neighboring objects.
Scheme № 6: Tent model taking into account neighboring objects with changing parameters.
Scheme № 7: Model of a tower for diving without neighboring objects.
Scheme № 8: Model of a tower for diving taking into account neighboring objects with changing parameters.
Scheme № 9: Model of a stand with a canopy without neighboring objects.
Scheme № 10: Model of a stand with a canopy taking into account neighboring objects with changing parameters.

3 Results and Discussion

A detailed analysis of the rectangular structure model was performed, taking into account neighboring structures when changing the [z] parameter - the distance between objects. The results and corresponding diagrams are shown in the Figs. 5, 6, 7 and 8 below.

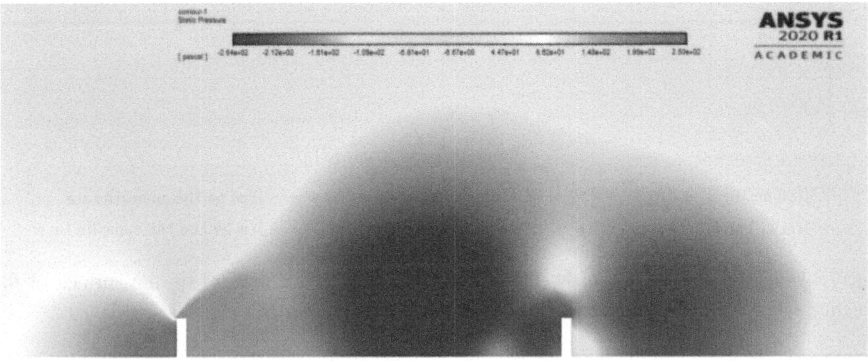

Fig. 5. The nature of the pressure distribution on the surface of the structure when changing the parameter [z]

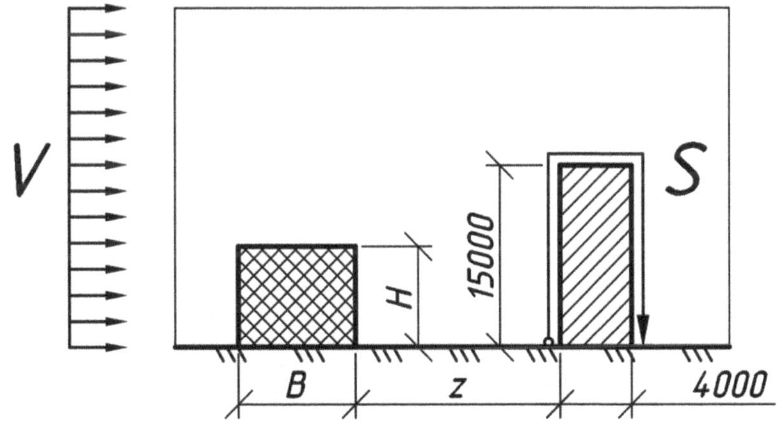

Fig. 6. The analyzed scheme

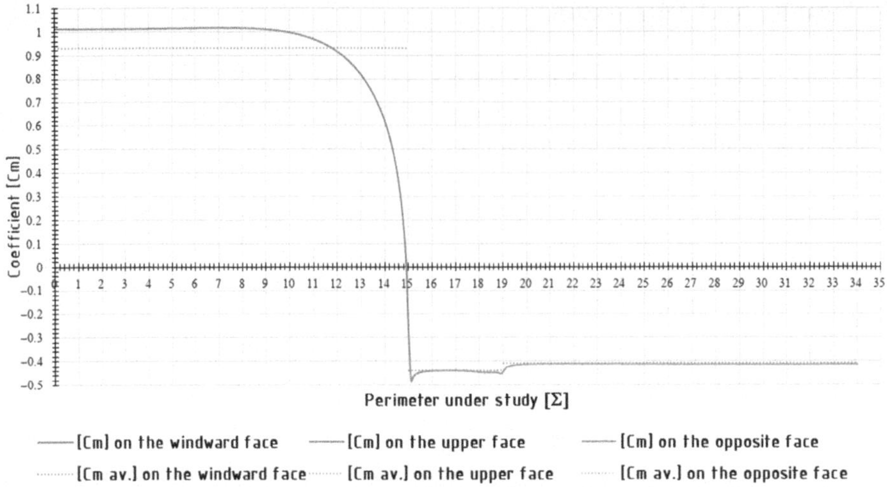

Fig. 7. Diagram of changes in the basic aerodynamic coefficient [Cm] on the surface of the structure when changing the parameter [z].

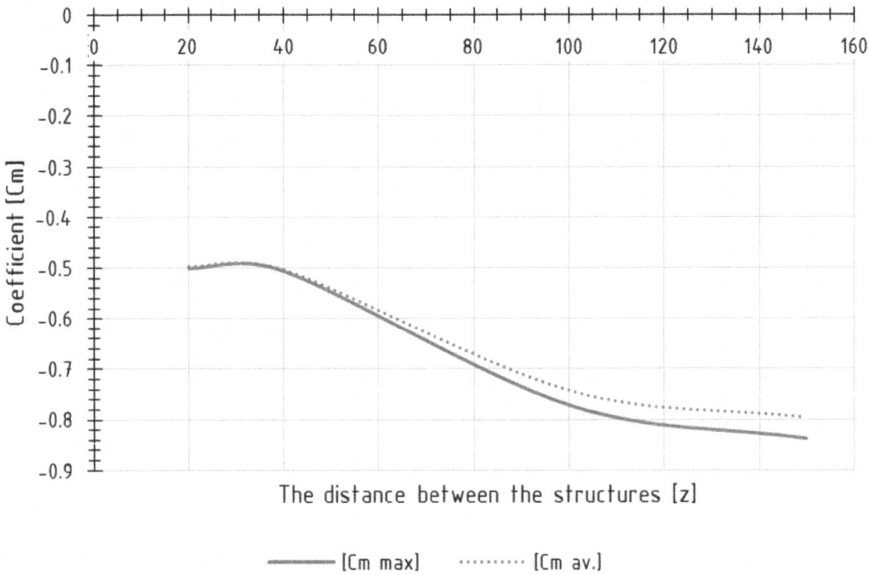

Fig. 8. Graph of the dependence of the basic aerodynamic coefficient [Cm] on the surface of the structure on the distance between objects [z].

The model of a rectangular structure was also analyzed, taking into account neighboring structures when changing the parameter [H]. The results and corresponding diagrams are shown in the Figs. 9, 10, 11 and 12 below.

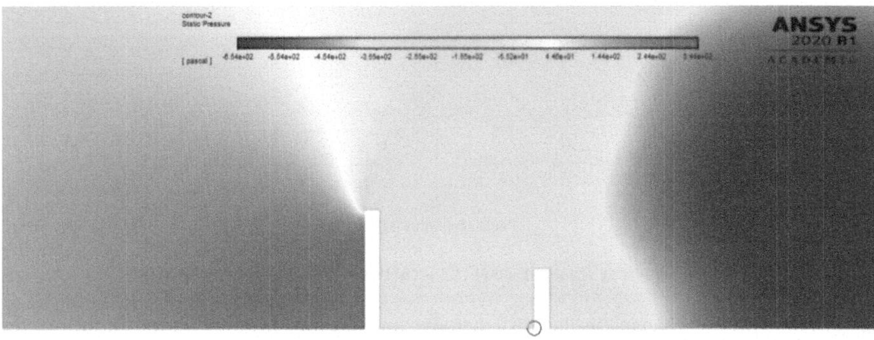

Fig. 9. The nature of the pressure distribution on the surface of the structure when changing the parameter [H]

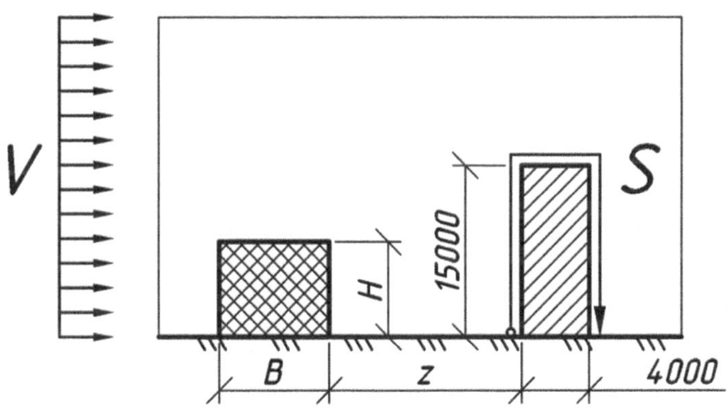

Fig. 10. The analyzed scheme.

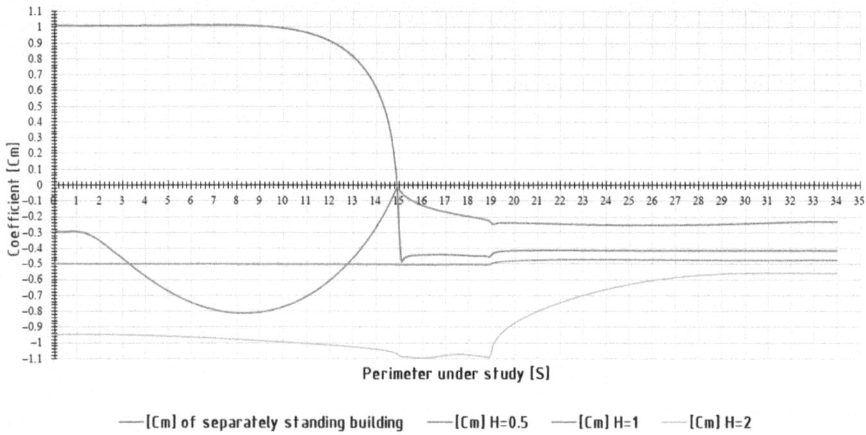

Fig. 11. Diagram of changes in the basic aerodynamic coefficient [Cm] on the surface of the structure when changing the parameter [H]

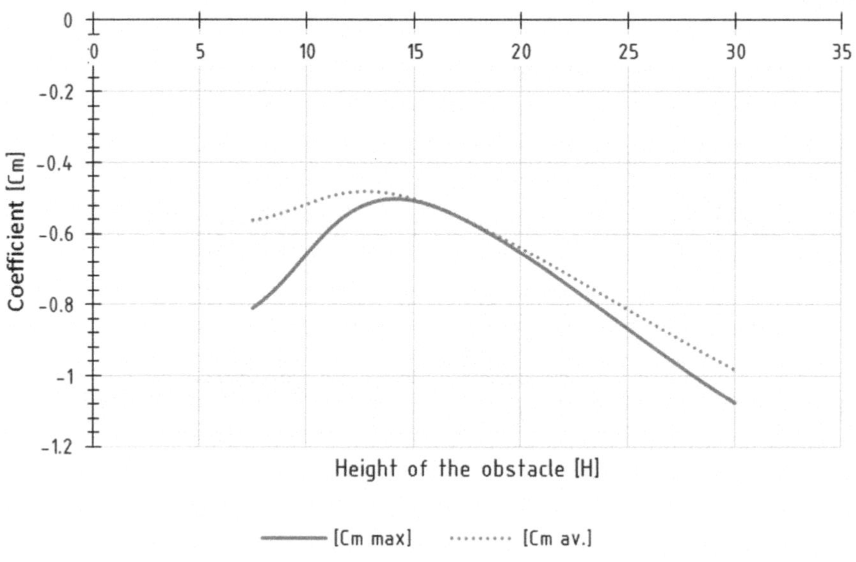

Fig. 12. Graph of the dependence of the basic aerodynamic coefficient [Cm] on the surface of the structure on the height of the obstructing object [H]

Finally, the model of a rectangular structure was analyzed, taking into account neighboring structures when changing the parameter [h] - height of the object under study. The results and corresponding diagrams are shown in the Figs. 13, 14, 15 and 16 below.

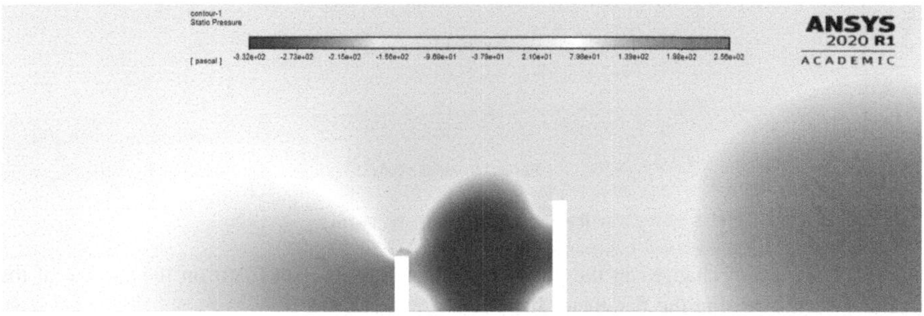

Fig. 13. The nature of the pressure distribution on the surface of the structure when changing the parameter [h].

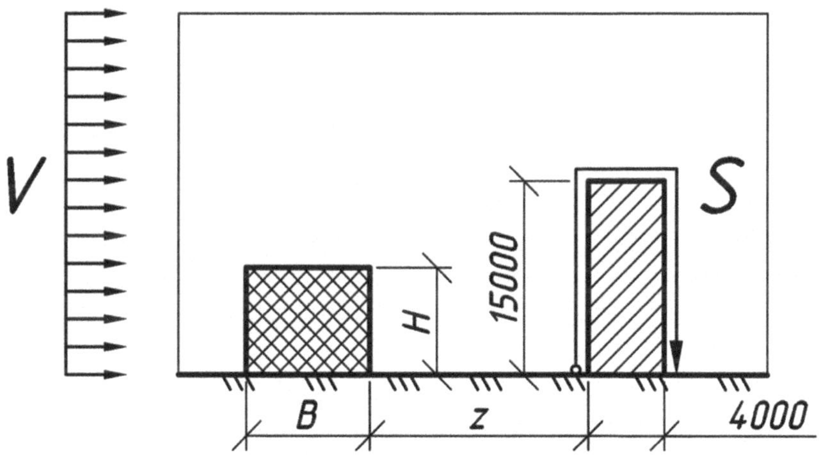

Fig. 14. The analyzed scheme.

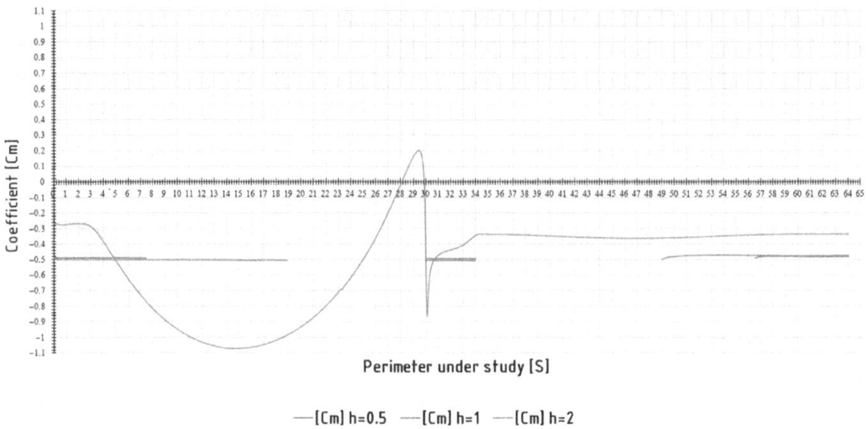

Fig. 15. Diagram of changes in the basic aerodynamic coefficient [Cm] on the surface of the structure when changing the parameter [h]

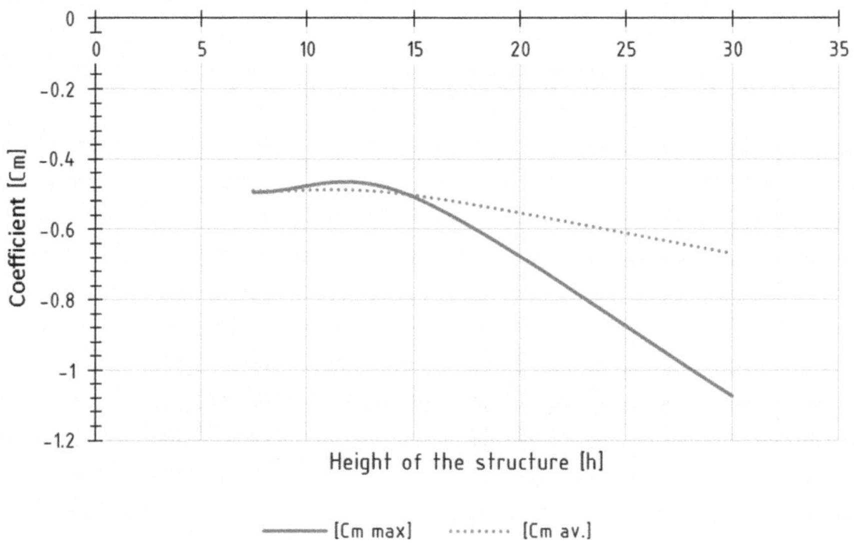

Fig. 16. Graph of the dependence of the basic aerodynamic coefficient [Cm] on the surface of the structure on the height of the object under study [h]

According to the same principle, a sequential analysis of the remaining schemes described in the previous section of the article was carried out.

4 Conclusion

Modern entertainment events often feature temporary structures of complex geometric shapes that differ from standard schemes of regulatory documents. For such structures, in particular, for stands with a canopy, it is important to accurately determine the aerodynamic coefficients that reflect the wind load.

Based on the results obtained in the course of numerical research, it is possible to draw the following main conclusions:

1. The ANSYS FLUENT was used for solving the set tasks of hydro gas dynamics.
2. A numerical study of the wind load on structures under various schemes of the presence of an obstacle in the path of the wind flow was performed.
3. It was found that the presence of an obstacle leads to a sharp decrease in the aerodynamic coefficients in comparison with both regulatory documents and calculations without obstacles.
4. It was found that changing the width of the obstructing structure practically does not affect the distribution of pressure on the surface of the object under study.
5. Increasing the distance between objects increases the impact of wind loads for both vertically oriented elements (walls) and inclined elements (tent ramps, stand's canopy).
6. It was also found that changing the height of obstructing structures affects buildings of different shapes in different ways. In general, increasing the height helps to reduce the overpressure behind the obstructing building, as evidenced by increasing negative values of coefficients. In addition, a decrease in the height of the obstacle below the object under study is accompanied by an extremely uneven distribution of pressure on the surface of the structure under study.
7. Results obtained by this method can be further used for verification of wind tunnels and optimization of experiments carried out in them.

References

1. Gordeev, V.N., Lantukh-Lyashenco, A.I., Pashinskiy, V.A., Perelmuter, A.B., Pichugin, S. F.: Loads and impacts on buildings and structures, pp. 195–222. IASW, Moscow (2007)
2. Overend, M., Zammit, K.: Wind loading on cladding and glazed façades. In: International Symposium on the Application of Architectural Glass, ISAAG, Nottingham, pp. 1–3 (2006)
3. Mendis, P., Ngo, T., Haritos, N., Hira, A., Samali, B., Cheung, J.: Wind loading on tall buildings. EJSE Special Issue Loading Struct. 41–42 (2007)
4. Ghosh, A.K.: International Workshop on Wind Engineering and Science, pp. 107–112. Elite Publishing House Pvt. Ltd., New Delhi (2004)
5. Yuan, X., Anumba, C.J., Parfitt, M.K.: Cyber-physical systems for temporary structure monitoring. Autom. Constr. **66**, 2–6 (2016)
6. Liu, C., He, L., Wu, Z., Yuan, J.: Experimental and numerical study on lateral stability of temporary structures. Arch. Civ. Mech. Eng. **18**, 1475–89 (2018)
7. Van Uffelen, G.M.: Wind-induced building interference: increase of wind loads on existing buildings after erection of new high-rises, vol. 5, pp. 1–3. EACWE, Florence (2009)

8. Hoffman, J., Johnson, C.: A new approach to computational turbulence modeling. Comput. Methods Appl. Mech. Eng. **195**(23–24), 2865–2870 (2006)
9. Hu, H., Yang, Z., Sarkar, P.: Dynamic wind loads and wake characteristics of a wind turbine model in an atmospheric boundary layer wind. Exp. Fluids **52**, 1277–1294 (2011)
10. Ahmad, S., Muzzammil, M., Zaheer, I.: Numerical prediction of wind loads on low buildings. Int. J. Eng. Sci. Technol. **3**, 59–60 (2011)
11. Crick, D., Grondin, G.Y.: Monitoring and Analysis of a Temporary Grandstand Structural Engineering Report No. 275, pp. 48–98. Department of Civil & Environmental Engineering of University of Alberta, Edmonton (2008)
12. Chandrangsu, T., Rasmussen, K.J.: Structural modelling of support scaffold systems. J. Constr. Steel Res. **67**, 866–875 (2011)
13. Sauca, A., Milchis, T., Gobesz, Z.: Wind loading on solar panels. Műszaki Tudományos Közlemények **10**, 73–78 (2019)
14 Killen, J.P., Letchford, C.W.: A parametric study of wind loads on grandstand roofs. Eng. Struct. **23**, 725–735 (2001)
15 Killen, J.P., Letchford, C.W.: Equivalent static wind loads for cantilevered grandstand roofs. Eng. Struct. **24**, 207–217 (2002)
16 Zhenggang, C., Ning, S., Yue, W., Shitao, P.: Gust response envelope approach to the equivalent static wind load for large-span grandstand roofs. J. Wind Eng. Ind. Aerodyn. **180**, 108–121 (2018)
17 Wang, F., Lam, K.M., Zu, G.B., Cheng, L.: Coherent structures and wind force generation of square-section building model. J. Wind Eng. Ind. Aerodyn. **188**, 177–193 (2019)
18 Zu, G., Lam, K.M.: Simultaneous measurement of wind velocity field and wind forces on a square tall building. Adv. Struct. Eng. **21**, 2241–2258 (2018)
19 Lam, K.M., To, A.P.: Generation of wind loads on a horizontal grandstand roof of large aspect ratio. J. Wind Eng. Ind. Aerodyn. **54–55**, 345–357 (1995)
20 Lam, K.M., Zhao, J.G.: Occurrence of peak lifting actions on a large horizontal cantilevered roof. J. Wind Eng. Ind. Aerodyn. **90**, 897–940 (2002)
21. Mikheev, P., Okorokov, R., Sidorenko, G., Timofeeva, A.: Determination of energy costs of wind farms at all life cycle stages. Advances in Intelligent Systems and Computing, vol. 982, pp. 242–256 (2020)
22. Chekalin, A.V., Andreev, V.M., Ascheulov, Y.V., Chumakov, Y.S.: Aerodynamics for CPV: investigating the efficiency of using air fairings on solar trackers. AIP Conf. Proc. **2149**, 060002–060011 (2019)
23. Semenov, A.S., Getsov, L.B.: Models of inelastic deformation and fracture of single-crystal nickel-base superalloys under thermal cyclic loading. In: ICF 2017 - 14th International Conference on Fracture, pp. 2378–2379 (2017)
24. Kuznetsov, I.L., Salakhoutdinov, M.A., Gainetdinov, R.G.: Stand and test results of a 24 m span truss with elements of galvanized cold-formed profiles. News KSUAE **4**, 193–200 (2018)

Mechanism of Adjusting the Cost of Consumed Electricity and the Improvement of Electricity Metering Devices Necessary for Its Implementation

Alexander Vinogradov[1], Maxim Borodin[2], Alina Vinogradova[1], Elena Zaugolnikova[3(✉)], Leonid Dolomanyuk[4], and Gennady Samarin[1]

[1] Federal Scientific Agroengineering Center VIM, 5 1st Institutsky Proezd, Moscow 109428, Russia
[2] Orel State Agrarian University, 69 General Rodina Street, Orel 302019, Russia
[3] NUST MISIS, Leninsky Prospect, 6, Moscow, Russia
vadimbolshev@gmail.com
[4] Kazan State Power Engineering University, 51 Krasnoselskaya Street, Kazan 420066, Russia

Abstract. The article shows the directions of improving electricity metering devices, including the providing of a mechanism for adjusting the cost of consumed electricity, depending on its quality, the source of distortion and the culprit of distortion. An example of a diagram of a metering device equipped with a function for adjusting the cost of electricity is given.

Keywords: Cost of consumed electricity · Electricity metering devices

1 Introduction

The most fully accepted methodology for assessing the reliability and quality of the supplied goods and services provided for the organization for the management of the unified national (all-Russian) electric grid and territorial grid organizations is set out in [1] and amendments to it [2]. However, the approach proposed in these documents is not devoid of drawbacks, the main of which is the form of obtaining the initial data - providing it according to the forms proposed by energy companies (at this stage, the information may be distorted).

Automatic data collection is not provided. In particular, the assessment of the quality of supplied electricity is based on an analysis of the number of incoming quality complaints, although not all consumers are able to assess it. Such mechanisms for stimulating energy companies to improve the reliability of power supply and quality of electricity, such as, for example, adjusting the cost of electricity depending on the number of interruptions in power supply and the quality of supplied electricity, are not given.

At the same time, improving the efficiency of power supply systems for consumers requires the development of not only technical devices to improve the reliability of

A. Sinitsyn (Ed.): SES 2020, LNCE 141, pp. 413–419, 2021.
https://doi.org/10.1007/978-3-030-67654-4_44

power supply, the quality of supplied electricity, reduce the time for technological connections [3], but also actions to introduce technical and economic mechanisms that stimulate the implementation of these developments. In turn, the proposed technical means should provide the possibility of implementing technical and economic incentive mechanisms to improve the quality of electricity and the reliability of power supply.

Analysis of existing incentive mechanisms of improving the quality of electricity. Scientists and industrialists have been developing technical and economic mechanisms to compensate for damage caused by low-quality electricity for a long time. So, in the article [4], the mechanisms of economic incentives to maintain the quality of electricity, proposed in the USSR [5, 6], are studied in sufficient details. These mechanisms consist in the payment of compensation payments to the person responsible for the distortion of the quality of electricity. The proposed amount of compensation is 25% of the cost of electricity with a quality that does not comply with regulatory documents. Additional price list No. 09-01-1980/11, issued by the USSR State Committee for Prices in 1984, indicates discounts on the electricity tariff for the consumer if the culprit of the electricity quality violation is the power supply company and, on the contrary, the premium to the tariff if the culprit is the consumer.

In [7], discounts/surcharges to the electricity fee are recommended to be set at 12% in case of violation of the electricity quality indicators (EQI). In addition, this work assumes the imposition of a one-time 10% fine of the cost of electricity in violation of the EQI.

These decisions are intended to stimulate an increase in the quality of electricity, but their implementation, in fact, is possible only if technical means are available to record violations of the quality of electricity. Currently, there are electricity metering devices that can monitor electricity quality indicators (EQIs), but they are not used to change the cost of electricity, and do not transmit information about the current quality of electricity to the network manager. There are also technical solutions for the execution of metering devices, involving the imposition of fines on the consumer and the energy supplying organization, given in [8, 9]. Nevertheless, these solutions have not yet found practical application, since a rather serious study of the legal issues of their use is required. The methods for determining the sources of distortion of the EQI are also imperfect, which can lead to incorrect imposition of penalties and further litigation between the consumer and the power supply organization. In addition, an acceptable method for adjusting the cost of electricity is required depending on its quality and source of distortion.

In [10] such a technique is proposed, and in [11] it is corrected taking into account GOST 32144–2013 [12]. Figure 1 shows a structural diagram of an electricity metering device has been developed that implements the specified method of adjusting the cost of electricity depending on its quality and the culprit of distortions [13].

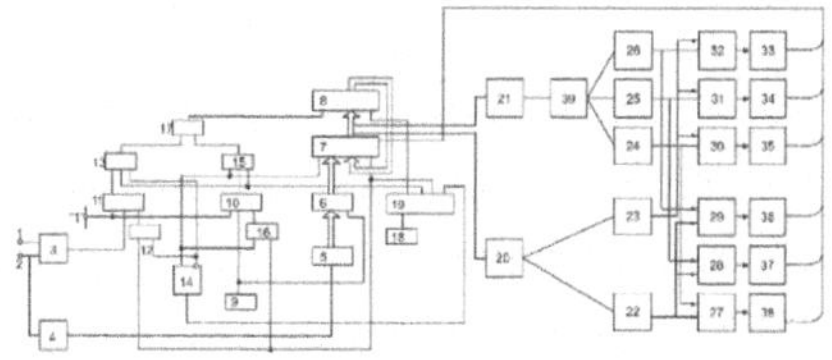

Fig. 1. Structural scheme of a metering device that implements the methodology for adjusting the cost of electricity depending on its quality.

The metering device contains current input voltage terminals 1 and 2, counter inputs 3, input of converter 4, information input of analog-to-digital converter 5 (ADC), information input of register 6, digital memory unit 7 (DMU), information input of counter 8 pulses (CP), sample pulse generator 9 (SPG), register write control input 6, clock input of the first D-flip-flop 10, information input of the second D-flip-flop 11, element And 12, element And 13, inverse output of SR-flip-flop 14, element And 15, element AND 16, element OR 17, pulse generator 18 (PG), distributor 19.

Pulse distributor 19 (PD), distortion source control unit BKII 20, distortion level control unit BKUI 21, block for fixing the fact of distortion of the power quality indicator by electric consumer EP 22, fixing unit the fact of distortion of the power quality indicator by the power system ES 23, blocks for monitoring the level of distortions 3ur. - Level 24, 2 - 25, 1 level. - 26, with elements I 27, I28, I29, elements I 30, I 31, I32, blocks that introduce a multiplying factor PK 33, PK 34, PK 35, or a reduction factor PK 36, PK 37, PK 38, time determination unit distortion of power quality indicators 39. The principle of operation of the meter is described in detail in [13, 14], therefore, this article is not considered. We only note that the device provides the possibility of storing and transferring data, as well as displaying information about the amount of consumed electricity, its cost, tariff, increasing (decreasing) coefficient, type and level of distortion, time of distortion of power quality indicators, data transmission, history accounting.

The technique implemented with the use of the metering device described above is not without its drawbacks, namely, that it involves the online adjustment of the electricity cost, which is not quite convenient when calculating the consumer for electricity and is rather difficult to implement.

2 Methodology

The proposed incentive mechanism to improve the quality of electricity and the reliability of power supply. New requirements for metering devices. Therefore, the following mechanism is proposed to stimulate consumers and the energy supplying organization to improve the quality of energy efficiency. It is assumed that the culprit for the distortion of the EE will be paid compensation for violations of the EE quality. But, unlike the methodology discussed above, compensation payments will be determined for a given billing period, for example, for a year or a month. The amount of the compensating payment is proposed to be determined taking into account the significance of the distortion of the PQE for a specific type of consumers, which is associated with the unequal damage from the distortion of the PQE for different consumers.

Compensation payment (Eq. 1) in case of power quality mismatch Kp(t1…t2) in the quality of electricity appears when the deviation of the parameters of the quality of electricity exceeds the values regulated by GOST [12] and includes the costs incurred by the energy supplying organization in the event that the distortion of the quality of electricity occurs through its fault.

The compensation payment can be determined:

$$K_p = K_{EQI} \cdot \frac{\sum (W_i \cdot \Phi)}{T_{stlmnt}} \cdot \left(\sum T_{total} - \sum T_{total.prms}\right) \tag{1}$$

K_{EQI} - coefficient characterizing the significance of the EQI distortion (in our case, the voltage deviation) for a given type of consumers or for the power supplying organization in the event that the EQI distortion occurred through the fault of the consumer, a dimensionless value; Wi is the volume of electricity consumption by the consumer for the intervals of the billing period, with the corresponding value of the electricity tariff, kW • h; Φi is the value of the electricity tariff, rubles/kWh; T_{stlmnt} - time of the settlement period, h (for example, year, 8760 h, or month, 672… 744 h); $\sum^{n}{}_{1}T_{total}$ - total time of power quality discrepancy for the billing period, h; $\sum^{n}{}_{1}T_{total.prms}$ - permissible total time of power quality discrepancy for the billing period, h.

In addition to the quality of electricity, an important factor in the efficiency of power supply is compliance with the requirements for reliability of power supply. Interruptions in power supply can be caused by various reasons, have different duration, but always harm consumers [15–17, 23], therefore, in case of exceeding the contractual values of interruptions in power supply, economic incentives should also be carried out, in this case, energy supplying organizations, to increase reliability. The implementation of the compensation mechanism in this case should be similar to the above, but in this case, requirements are imposed on the meter to control the reliability of the power supply.

Modern meters are equipped with the ability to fix time intervals when there was no voltage at the point of their connection. However, this data cannot be used as information on the duration of power outages, as it is not clear what caused the lack of voltage. To do this, it is required to equip them with the ability to monitor the voltage at a point up to the input switching device. Such solutions are considered in [18]. This will allow you to control the main indicator of the reliability of power supply - the time of interruptions in the supply of electricity and, if the permissible value of this time is exceeded, automatically determine the amount of compensation [15]. Compensation, as in the case of power quality, should be determined considering the significance of power outages for specific consumers. At the same time, to stimulate consumers to pay for electricity on time, metering should be carried out on a pay-and-use basis, as is done in mobile communications. That is, payment must be made in advance, a consumer account must be formed where he invests. In the absence of funds on the account, the supply of electricity is automatically terminated (or limited depending on the contractual conditions, consumer reliability category).

3 Results

Therefore, new additional requirements are formed for the electricity metering device:

- the ability to receive online data on tariff changes;
- fixing the volumes of electricity consumption at different tariff values;
- power quality control and fixation of time intervals with violations of the EQE;
- the possibility of investing in the counter of the EQE distortion significance factor;
- determination of the culprit of distortion;
- formation of the amount of compensation payments for violations of the EQE;
- monitoring the duration of power supply interruptions;
- the possibility of investing in the counter of the significance coefficient of interruptions in power supply;
- formation of the amount of compensation payments for violations of the reliability of power supply;
- payment for electricity on a pay-and-use basis.

The active development of technologies and means of data transmission, electricity metering, automation and intellectualization of electrical networks [19–22] suggests that the implementation of the proposed functionality of electricity meters is quite possible and necessary for the transition to the fourth technological mode of the economy.

4 Conclusion

Electricity metering in the context of the digitalization of the economy should be carried out using the capabilities of monitoring the quality of electricity and the reliability of electricity supply. In the event of a violation of the quality of electricity, the culprit of the distortion must make compensatory payments, which must be determined considering the significance of the distortion. Likewise, if the reliability of the power supply is violated, the damage incurred by the parties to the power supply agreement must be compensated. This creates new requirements for metering devices, which must ensure metering of electricity, considering the fulfillment of contractual conditions by the parties.

References

1. On approval of guidelines for calculating the level of reliability and quality of services supplied for the organization managing the unified national (All-Russian) electric grid and territorial grid organizations, Order of the Ministry of Energy of the Russian Federation 1256, 29 November 2016
2. On Amendments to the Methodological Guidelines for Calculating the Level of Reliability and Quality of Provided Services for an Organization for Management of the Unified National (All-Russian) Electric Grid and Territorial Grid Organizations, Order of the Ministry of Energy of the Russian Federation 544 (21 June 2017), approved by order of the Ministry of Energy of Russia 1256, 29 November 2016

3. Vinogradov, A.V.: Determination of the efficiency of power supply systems. Bull. NGIEI **7** (74), 26–35 (2017)
4. Ovseichuk, V.A.: Reliability and quality of power supply to consumers. Rationale for rationing. News of electrical engineering: information and reference publication, founder of CJSC. News Electrotech. **3**(81), 50–53 (2013)
5. On the approval of the rules for the use of electrical and thermal energy, Order of the USSR Ministry of Energy 310, 6 December 1981
6. On streamlining the system of economic (property) sanctions applied to enterprises, associations and organizations, Decree of the Government of the USSR 929, 30 July 1988
7. Optimization of reliability of electrical networks 110-35-10 (6) kV of JSC Lenenergo for the period up to 2010–2015 yrs. with the calculation of tariffs according to the RAB method, Report of ZAO PFC SKAF, Moscow (2009)
8. Skubilin, M.D., Pismenov, A.V., Mardamshin, Y.P.: Patent 2125269C1 Russian Federation, G01R21/133, G06F 17/00. Device for Electricity Measurement, applicant and patentee Taganrog State Radio Engineering University – 97114532/09, declared 28.081997, publ. 01.20.1999, Bul. 31, 9
9. Borodin, M.V., Vinogradov, A.V.: Patent 134338 Russian Federation, IPC G06F17/18 (2006.01). Metering device for electricity consumption and adjusting the cost of consumed electricity depending on its quality and source of distortion, applicant and patentee: Federal State Budgetary Educational Institution of Higher Professional Education Oryol State Agrarian University – 2013117164/08, declared 15.04.2013, publ. 10.11.2013, 14
10. Vinogradov, A., Borodin, M., Bolshev, V.,Makhiyanova, N., Hruntovich, N.: Improving the power quality of rural consumers by means of electricity cost adjustment. In: Renewable Energy and Power Supply Challenges for Rural Regions, pp. 312–341 (2019)
11. Borodin, M.V., Vinogradov, A.V.: Revision of the methodology for adjusting the cost of consumed electricity depending on its quality and the algorithm for its implementation in accordance with GOST for the quality of electricity. Bull. NGIEI **4**(83), 54–64 (2018)
12. GOST 32144-2013, Bibliographic record, Electric Energy, Electromagnetic compatibility of technical means. Electricity quality standards in general-purpose power supply systems, Interstate Council for Standardization, Metrology and Certification (MGS) [Electronic resource] (Moscow, Standartinform, 2014) 20. https://docs.cntd.ru/document/1200104301
13. Vinogradov, A.V., Borodin, M.V.: Patent No. 193102 Russian Federation, IPC G06F 17/18 (2019.10), Metering device for electricity consumption and adjustments and cost of consumed electricity depending on its quality, source of distortion and distortion time of power quality indicators, Applicant and patent holder: Federal State Budgetary Educational Institution of Higher Education Oryol State Agrarian University named after N.V. Parakhina, 2019114541/08, declared 13.05.2019, publ. 14.10.2019
14. Borodin, M.V., Vinogradov, A.V., Bolshev, V.E.: Accounting for electricity consumption and adjusting the cost of consumed electricity depending on its quality, source of distortion and time of distortion of power quality indicators. Innovations Agric. **3**(32), 330–339 (2019)
15. Vinogradov, A.V., Bolshev, V.E., Vinogradova, A.V., Borodin, M.V., Bukreev, A.V.: Technical and economic methods of increasing the efficiency of power supply systems for rural consumers. Bull. Agrarian Sci. Don **3**(47), 59–67 (2019)
16. Vinogradov, A., Bolshev, V., Vinogradova, A., Jasiński, M., Sikorski, T., Leonowicz, Z., Goňo, R., Jasińska, E.: Analysis of the power supply restoration time after failures in power transmission lines. Energies **13**, 2736 (2020)
17. Vinogradov, A., Vinograzdova, A., Bolshev, V.: Analysis of the quantity and causes of outages in LV/MV electrical grids. CSEE J. Power Energy Syst. **6**(3), 537–542 (2020)

18. Vinogradov, A., Bolshev, V., Vinogradova, A., Kudinova, T., Borodin, M., Selesneva, A., Sorokin, N.: A system for monitoring the number and duration of power outages and power quality in 0.38 kV electrical networks. In: Vasant, P., Zelinka, I., Weber, G.W. (eds.) Intelligent Computing & Optimization, ICO 2018, Advances in Intelligent Systems and Computing, vol. 866, pp. 1–10 (2019)
19. Bolshev, V.E., Vinogradov, A.V.: Review of foreign sources on the use of information networks in the infrastructure of smart networks smart grid. News High. Educ. Inst. Chernozem Reg. **55**(1), 8–18 (2019)
20. Kirkman, R.: Development in substation automation systems [Electronic resource]. https://www.ee.nsysu.edu.tw/isap2007CD/papers/f0083.pdf. Accessed 16 June 2020
21. Pereda, R., et al.: Substation automation experiences based on the IEC 61850 GOOSE messages [Electronic resource]. In: 23rd International Conference on Electricity Distribution, pp. 15–18 (2015). https://cired.net/publications/cired2015/papers/CIRED2015_0596_final.pdf. Accessed 18 June 2020
22. Bolshev, V.E., Vinogradov, A.V.: Promising communication technologies for the automation of power supply networks. Bull. Kazan State Energy Univ. 11, **2**(42), 65–82 (2019)
23. Gracheva, E., Alimova, A.: Calculating probability of faultless work of shop nets with the help of coefficients of ratio. In: 2019 International Russian Automation Conference. RusAutoCon **2019**, 8867727 (2019). https://doi.org/10.1109/RUSAUTOCON.2019.8867727

The Stability of the Thin-Walled Perforated Compressed Elements

Vladimir Rybakov[1(✉)], Quynh Ha[1], Anastasiia Gordeeva[2], and Karolina Urmanceva[3]

[1] Peter the Great St. Petersburg Polytechnic University, 29, Polytecknycheskaya Street, St. Petersburg 195251, Russia
fishermanoff@mail.ru

[2] Petrogradproekt, Ltd., 4, Novoroshchinskaya Street, St. Petersburg 196084, Russia

[3] Samarkand College of Light Industry and Economics, Bedil Street, 16, Samarkand 703018, Uzbekistan

Abstract. In this work, several variants of models of perforated thin-walled columns were implemented. The perforated column of 600 mm, 1600 mm and a wall thickness of 2,0 mm were selected and investigated on the central compression. The critical force acts as a required parameter, in which the columns loses its stability. The recommendations about the finite-elemental model engineering of the phenomenon of the loss of stability central compressed thin-walled elements have been developed. During this task, the calculation of perforated columns (and bar model) on compression with various variants of the boundary conditions were implemented. The results obtained provide a method of constructing a model of centrally compressed perforated profiles in SCAD Office. It also proposed to account for the geometric characteristics of model by a special program, for example using the program Tonus, where the geometric parameters of the real profile were assumed and entered, the thickness reduced by 25% assumed no perforated.

Keywords: Perforated thin-walled elements · Perforations · Compression · Stability · Finite-element method

1 Introduction

One of the most important criteria in today's world in construction engineering is not only the creation of quality and reliability of structures, but also evaluating the constructed speed. Therefore, in recent years the use of light gauge steel structions (LGSS) become increasingly popular. LGSS have several number of indisputable advantages such as quickly constructed, lightweight, assembly structures without the use of heavy machinery, save metal, low cost [1].

The basic characteristic of LGSS are cold-formed and light beams of thin galvanized plate. Thanks to its features, LGSS can solve the main problem of the reconstruction of kindergartens, and other old buildings reducing the load, destroying walls and foundations [1, 2].

A. Sinitsyn (Ed.): SES 2020, LNCE 141, pp. 420–432, 2021.
https://doi.org/10.1007/978-3-030-67654-4_45

The fundamentals of stability theory of bar system was established by Euler. The theory for calculating of thin bars was considered found by professor Timoshenko. The strength and stability of steel bent elements was studied Atavin, I.V., Heinilä, S., Björk, T. [3, 9]. Research thin-walled steel structures devoted by Danilov, A. [4] Heinisuo, M. [5], Garifullin, M. [6, 8], Nazmeeva, T.V. [7]. In [10] I-bars with a perforated wall on the calculated resistance and eccentric compression strength. In [11] were examined thin-walled rod elements of bent open section profiles dimensional stability. And many other quantitative experiments with perforated profiles could be found in [12–14].

One of commonly method using to model thin-walled structures is model them on the programs SCAD Office [15–21].

2 Method of Analysis

The goal of this work was to create recommendations for the finite element modeling of the phenomenon of loss of stability of centrally compressed thin-walled perforated columns. This paper deals with perforated profile with the length of 600 mm and 1600 mm and wall thickness of 2.0 mm, the construction of the profile shown in Fig. 1.

Fig. 1. Finite element model in SCAD Office.

2.1 Calculation of Perforated Thin-Walled Profile on Central Compression

During this task, the calculation of perforated columns on compression with various variants of the boundary conditions were implemented. The considered variants with the boundary conditions are shown below.

Variants with boundary conditions.

Variant 1.

All upper and lower nodes are superimposed on X and Y (Figs. 2 and 3).

Fig. 2. Consolidate upper nodes.

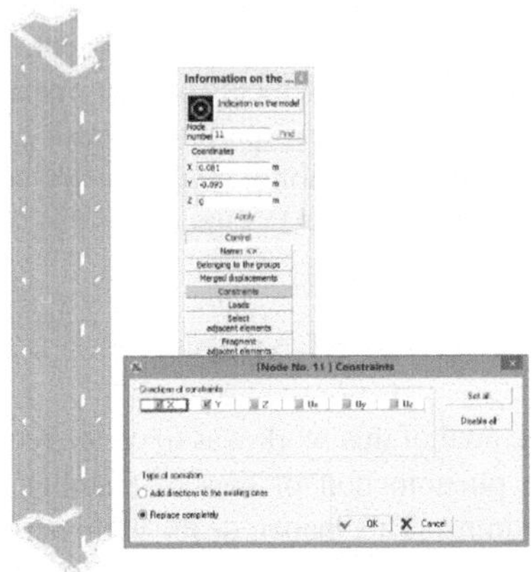

Fig. 3. Consolidate lower nodes.

On the central nodes are superimposed on Z. On the upper and lower nodes are considered the displacement on axis Z (Figs. 4 and 5).

Fig. 4. Consolidate central nodes.

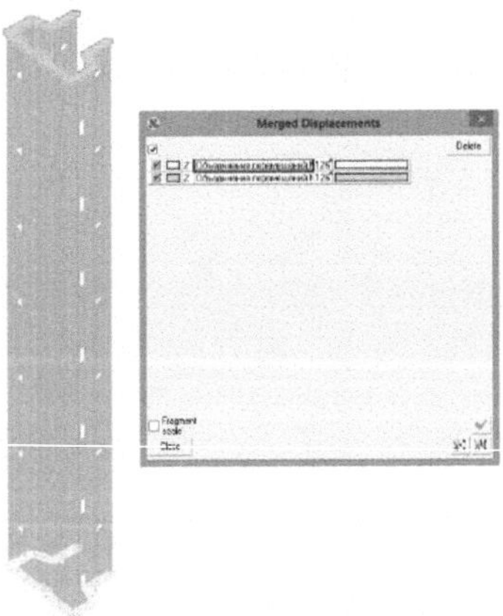

Fig. 5. Merged displacements.

Variant 2

All upper and lower nodes are superimposed on X and Y (Figs. 6 and 7).

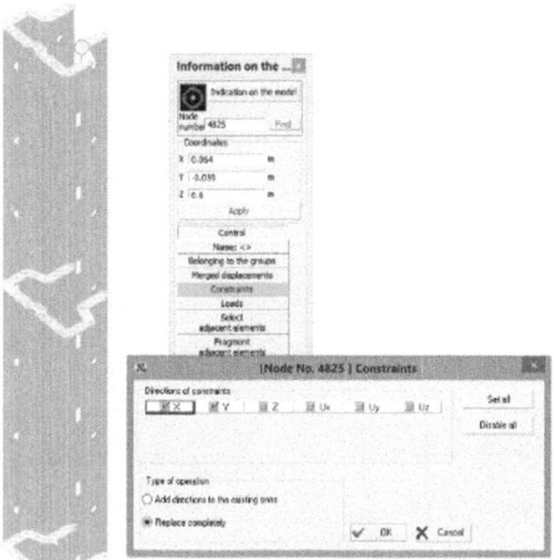

Fig. 6. Consolidate upper nodes.

Fig. 7. Consolidate lower nodes.

On the central bell nodes are superimposed on axis Z. On the upper and lower nodes are considered the displacement on axis Z (Figs. 8 and 9).

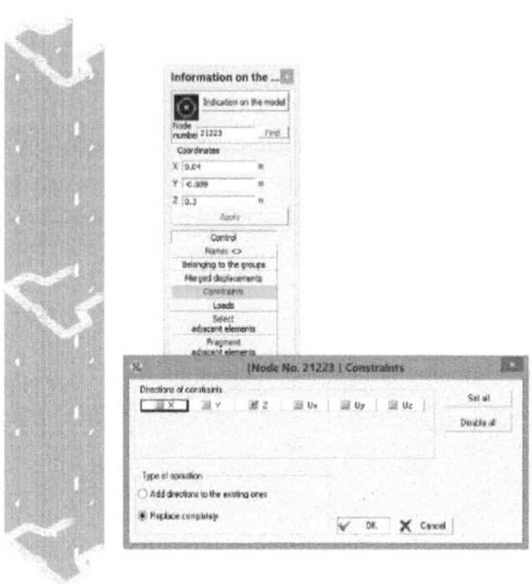

Fig. 8. Consolidate central bell nodes.

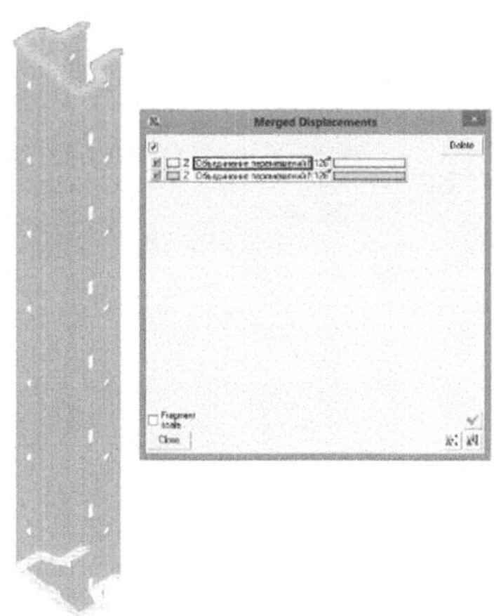

Fig. 9. Merged displacements.

Variant 3

All upper nodes are superimposed on X and Y, all lower nodes are superimposed on X, Y, Z (Figs. 10 and 11).

Fig. 10. Consolidate upper nodes

Fig. 11. Consolidate lower nodes.

On the upper and lower nodes are considered a displacement on axis Z (Fig. 12).

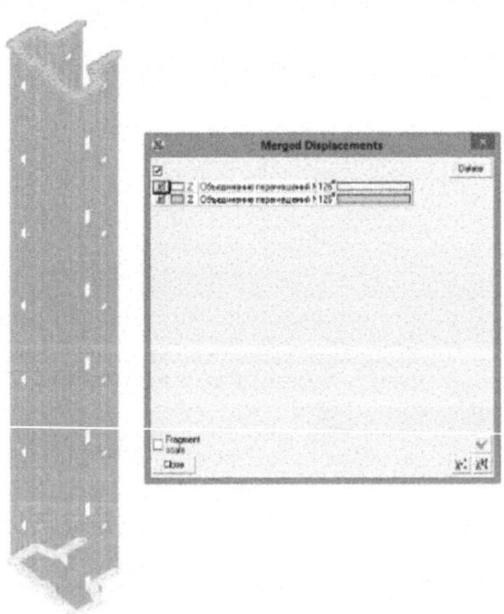

Fig. 12. Merged displacements.

Variant 4

All upper nodes are superimposed on X and Y, all lower nodes are superimposed on X, Y, Z (Figs. 13 and 14).

Fig. 13. Consolidate upper nodes.

Fig. 14. Consolidate lower nodes.

Calculation results of the stability with boundary conditions described above are shown in Table 1.

Table 1. The calculated value of critical forces.

Boundary conditions for	Length of columns, mm	Critical force (calculated), kN	Critical force (experience), kN	Critical force (Eucocode-3), kN
Variant 1	600	354.49216	147.5	136.001
Variant 2	600	354.49216		
Variant 3	600	354.714966		
Variant 4	600	286.065674		
Variant 5	1600	256.95318	134.3	117.204

From Table 1 can see that the calculated value of the critical force (calculation) exceeds the value obtained in experience of a few times. It is necessary to clarify the boundary conditions.

2.2 Clarification of Boundary Conditions

Variant 5

On the upper and lower nodes have established the finite stiffness Kx and Ky with 0,001 kN /mm. On the lower nodes are superimposed on axis Z (Figs. 15 and 16).

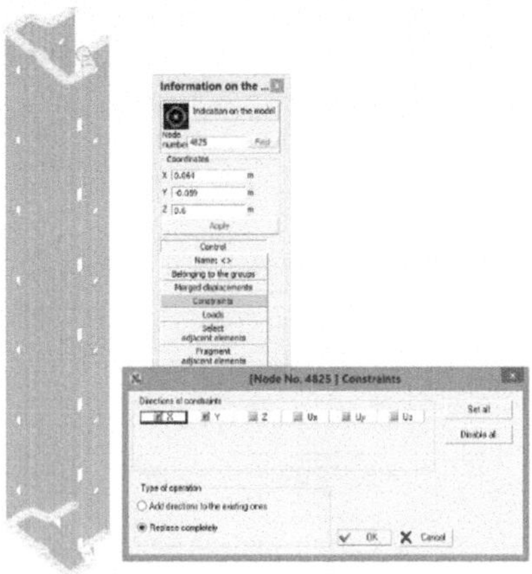

Fig. 15. Consolidate upper nodes.

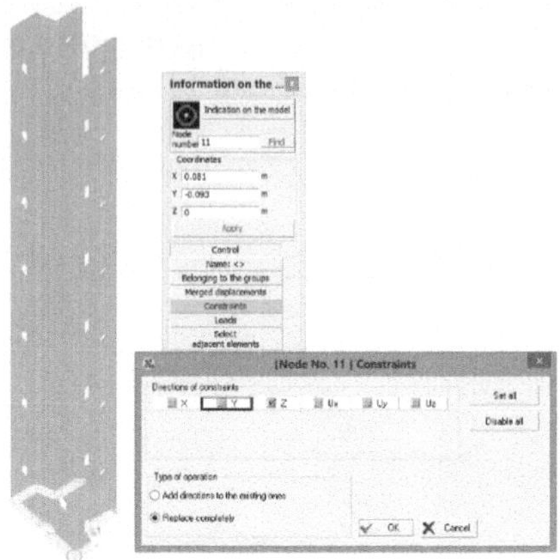

Fig. 16. Consolidate lower nodes.

On the upper and lower nodes are combined a displacement on axis Z (Fig. 17).

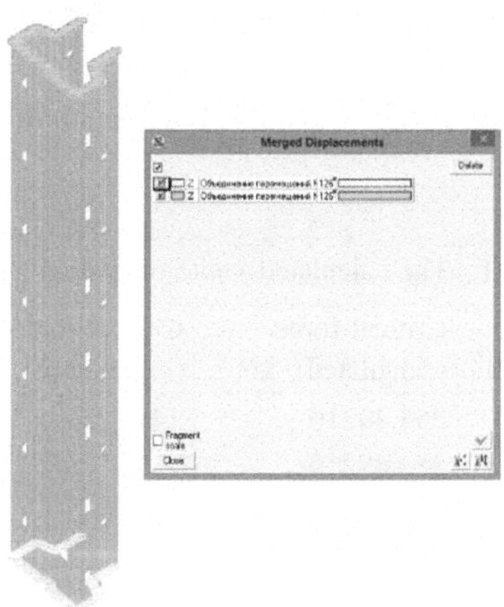

Fig. 17. Merged displacements.

The results of the calculation to the new boundary conditions are presented in Table 2.

Table 2. The calculated value of critical forces with new boundary conditions

Boundary conditions for	Length of columns, mm	Critical force (calculated), kN	Critical force (experience), kN	Critical force (Eucocode-3), kN
Variant 5	600	138.790894	147.5	136.001
	1600	133.190918	134.3	117.204

The values obtained for the critical force lies in the range between the critical force which calculated in accordance with Eurocode 3 and the critical force obtained experimentally, which gives an indication of the fact that the values obtained by numerical way is reliable.

2.3 Modeling Non-perforated Profiles

It was produced by finite element modeling in the shells of non-perforated profiles. This calculation was made in order to determine how non-perforated profile compressive strength compared to perforating. The results of the calculation are presented in Table 3.

Table 3. The calculated value of critical forces.

Boundary conditions for	Length of columns, mm	Critical force (calculated), kN	
		Perforated profiles	Non-perforated profiles
Variant 1	600	354.49216	436.157227
Variant 5	600	138.790894	332.8674321
Variant 1	1600	256.95318	322.247696
Variant 5	1600	133.190918	299.957275

A comparison of the results were obtained, it was found that the value of the critical force for unperforated profile exceeds the critical force for perforated profile almost 2.5 times.

2.4 Clarification Reduced the Profile of the Wall Thickness

The transition from a perforated profile to non-perforated was done by decreasing the thickness of the non-perforated wall profile. After numerous attempts, the required profile of the wall thickness has been found, that the value of the critical force non-perforated profile converges with little divergence from the critical force value perforated profile with wall thickness of 1.5 mm, regardless of the column 600 or 1600 mm. The value of critical force are shown in Table 4.

Table 4. The calculated value of critical forces.

Boundary conditions	Length of columns, mm	Critical force (calculated), kN		
		Perforated profiles	Non-perforated profiles	Non-perforated profiles with reduced wall thickness
Variant 1	600	354.49216	354.49216	436.157227
Variant 5	600	354.49216	138.790894	332.8674321

A comparison of the results were obtained, it was found that the value of the critical force for unperforated profile exceeds the critical force for perforated profile almost 2.5 times.

2.5 Modeling Non-perforated Profiles

The resulting value of the reduced wall thickness profile has been used in the calculation of the geometric characteristics of the profile. The geometric characteristics are necessary for the creation of a core model. Calculation were performed in Tonus. These geometric characteristics for the equivalent profile perforated with 2.0 mm wall thickness with non-perforated wall thickness of 1.5 mm is necessary to set the stiffness bars. The following boundary conditions were set for the calculation of bar models (Figs. 18 and 19):

Boundary conditions

Variant 6

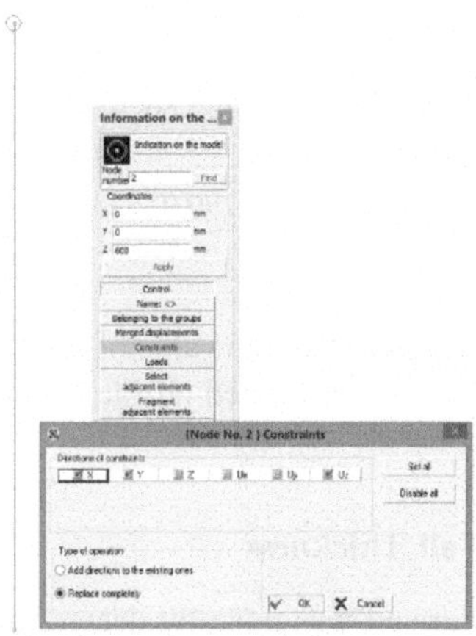

Fig. 18. Consolidate upper nodes

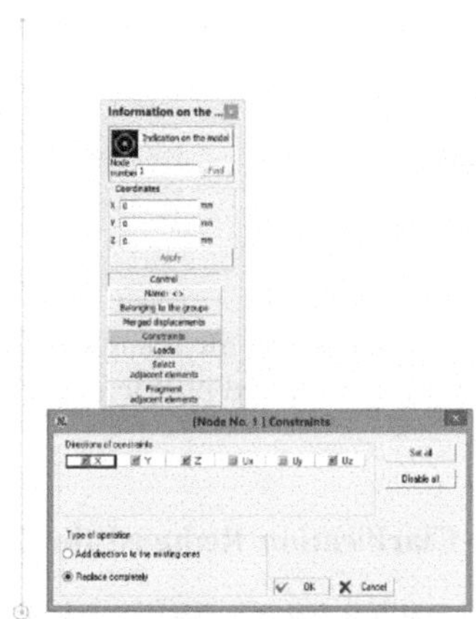

Fig. 19. Consolidate lower nodes.

The following results were obtained (Table 5):

Table 5. The calculated value of critical forces.

Boundary conditions	Length of columns, mm	Critical force (calculated), kN	
		Non-perforated profiles with reduced wall thickness	Bar model
90x1.5	600	137.424316	773.981094
90x1.5	1600	134.351807	108.102025

2.6 Clarification of Boundary Conditions for Bar Model Profile with Reduced Wall Thickness

Variant 7

See Figs. 20, 21 and 22.

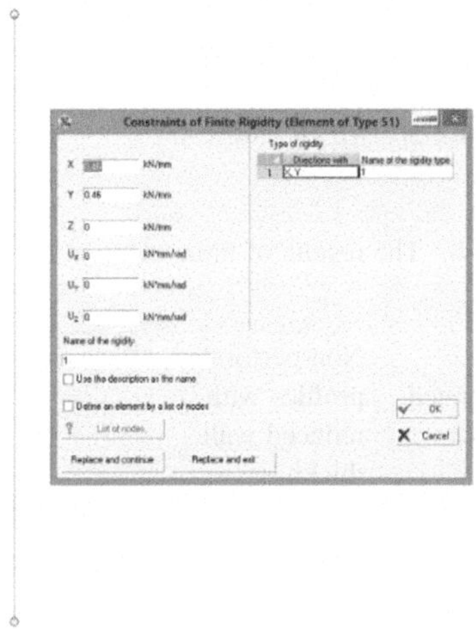

Fig. 20. Communicate finite stiffness Kx and Ky.

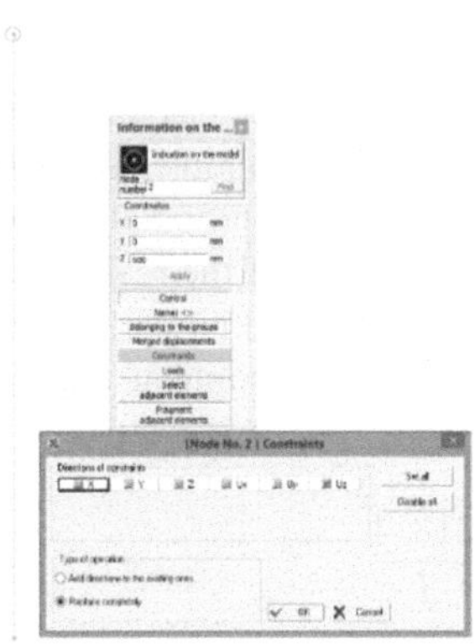

Fig. 21. Consolidate upper nodes.

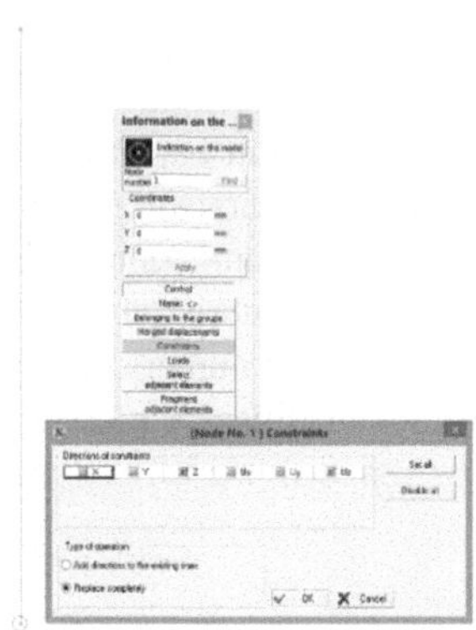

Fig. 22. Consolidate lower nodes.

3 Results

In this work, several variants of models of perforated thin-walled columns were implemented. The perforated column of 600 mm, 1600 mm and a wall thickness of 2,0 mm were selected and investigated on the central compression. The critical force acts as a required parameter, in which the columns loses its stability. The first stage was made of finite-element modeling perforated rack in the shells. Calculated model with various embodiments of the boundary conditions. In the specification of which it was made of the boundary conditions. Also it had been calculated on the stability non-perforated profile, which was carried out to further the replacement non-perforated perforated profile by reducing the wall thickness of the profile. Further, the design scheme was reduced from the shell to the bar. The geometric characteristics of the profile have been calculated taking into account the reduced thickness, which are used to calculate the core. It was also made to clarification of boundary conditions for bar model profile. The results are shown in Table 6.

Table 6. The results of finite element modeling

Critical force (calculation), kN						
Length of columns, mm	Perforated profile	Non-perforated profile	Non-perforated profiles with reduced wall thickness	Bar model	Experiment	Eurocode-3
600	138.79089	332.867432	137.424316	137.998581	147.5	136.001
1600	133.19091	299.957275	134.351807	108.102025	134.3	117.204

4 Conclusion

The results allow us to formulate the following recommendations to the central compressed perforated columns in general:

1) A method for calculation the stability of the columns could be calculated by using the finite element bar model in the software package, for example, SCAD Office.
2) A propose to characterize the geometric of model by a special program, for example using the program Tonus, where the geometric parameters of the real profile were assumed and entered, the thickness reduced by 25% assumed no perforated.
3) With the use of numerical experiments formula allows us to simulate the boundary conditions resiliently pliable bonds:

$$k = f(l) = \frac{3450}{l^2} + \frac{218,4}{l}$$

References

1. Brudka, Ya., Lubinski, M.: Legkie stalnye konstruktsii [Lightweight steel structures]. Stroyizdat, Moscow (1974). 331 p. (Rus)
2. Al Ali, M., et al.: Thin-walled cold-formed compressed steel members and the problem of initial imperfections. Procedia Eng. **40**, 8–13 (2012)

3. Heinilä, S., Björk, T., Marquis, G.: The influence of residual stresses on the fatigue strength of cold-formed structural tubes. ASTM Special Technical Publication 1508 STP: 200–215 (2009)
4. Danilov, A., Tusnina, O.: The joints of cold-formed purlins. J. Appl. Eng. Sci. **12**(2), 153–158 (2014)
5. Heinisuo, M., Lahdenmaa, J., Jokinen, T.: Experimental research on modular thin-walled steel structures. In: Proceedings of the 12th International Conference on Steel, Space and Composite Structures, pp. 85–97 (2014)
6. Garifullin, M., Bronzova, M., Sinelnikov, A., Vatin, N.: Buckling analysis of cold-formed c-shaped columns with new type of perforation. Eng. Struct. **126**, 63–68 (2016). https://doi.org/10.1201/b19249-13
7. Nazmeeva, T.V., Vatin, N.I.: A numerical investigations of notched C-profile compressed members with initial imperfections. Mag. Civ. Eng. **62**, 92–101 (2016). https://doi.org/10.5862/MCE.62.9
8. Garifullin, M., Sinelnikov, A., Bronzova, M., Kovacic, B., Kamnik, R.: Buckling behavior of cold-formed studs with thermal perforations. MATEC Web Conf. **73**, 1–11 (2016). https://doi.org/10.1051/matecconf/20167304011
9. Atavin, I.V., Melnikov, B.E., Semenov, A.S., Chernysheva, N.V., Yakovleva, E.L.: Influence of stiffness of node on stability and strength of thin-walled structure. Mag. Civ. Eng. **80**, 48–61 (2018). https://doi.org/10.18720/MCE.80.5
10. Kiymaz, G.: FE based mode interaction analysis of thin-walled steel box columns under axial compression. Thin-Walled Struct. **43**(7), 1051–1070 (2005). https://doi.org/10.1016/j.tws.2004.11.003
11. Casafont, M., Roure, F., Peki4Z, T.: Compression tests on uprights: checks for the effects of distortional buckling. In: 2nd ERF Workshop Tests on Racking Systems, Barcelona, Spain (2010)
12. Wilkinson, T.: Suitability of cold-formed rectangular hollow sections for use in plastic design, p. 184. The University of Sydney (2004)
13. Cortese, S.D.: Investigation of single span Z-section purlins supporting standing seam roof systems considering distortional buckling, p. 164. Virginia Polytechnic Institute and State University (2001)
14. Pham, C., Hancock, G.: Shear tests and design of cold-formed steel channels with central square holes. Thin-Walled Struct. **149** (2020). https://doi.org/10.1016/j.tws.2020.106650
15. Lalin, V.V., Rybakov, V.A., Morozov, S.A.: The finite elements research for calculation of thin-walled bar systems. Mag. Civ. Eng. (27), 53–73 (2012). https://doi.org/10.5862/MCE.27.7
16. Lalin, V., Rybakov, V., Sergey, A.: The finite elements for design of frame of thin-walled beams. In: Applied Mechanics and Materials, vol. 578–579, pp. 858–863 (2014).https://doi.org/10.4028/www.scientific.net/AMM.578-579.858
17. Rybakov, V., Sergey, A.: Mathematical analogy between non-uniform torsion and transverse bending 1 of thin-walled open section beams. In: Applied Mechanics and Materials, vol. 725–726, pp. 746–751 (2015). https://doi.org/10.4028/www.scientific.net/AMM.725-726.746. ISBN 1662-7482
18. Trubina, D., Pichugin, E., Rybakov, V., Garifullin, M., Sokolova, O.: Comparison of the bearing capacity of LST-profile depending on the thickness of its elements. In: Applied Mechanics and Materials, vol. 725–726, pp. 752–757 (2015). https://doi.org/10.4028/www.scientific.net/AMM.725-726.752
19. Maskaleva, V., Rybakov, V.: The analysis of convergence of finite element method for a foundation slab on the bilinear subgrade. In: Applied Mechanics and Materials, vol. 725–726, pp. 842–847 (2015). https://doi.org/10.4028/www.scientific.net/AMM.725-726.842

20. Trubina, D., Abdulaev, D., Pichugin, E., Rybokov, V.: Geometric nonlinearity of the thin-walled profile under transverse bending. In: Applied Mechanics and Materials, vol. 633–634, pp. 1133–1139 (2014). https://doi.org/10.4028/www.scientific.net/AMM.633-634.1133
21. Rybakov, V., Panteleev, A., Sharbabchev, G., Epshtein, E.: Snow-retaining system as a temporary decision for providing of the suitable temperature and humidity level of pitched roofs. In: Applied Mechanics and Materials, vol. 584–586, pp. 1876–1880 (2014). https://doi.org/10.4028/www.scientific.net/AMM.584-586.1876

Driving Change in the Brick Manufacturing Industry in India Through the RERA Regulatory Mechanism

Anthony De Sa[1(✉)], Satya N. Mandal[1], N. Sridharan[2], and Deepak Bajaj[1]

[1] RICS-SBE, Amity University, Noida, India
anthonydesa.ias@gmail.com
[2] School of Planning and Architecture, Bhopal, India

Abstract. The real estate sector in India is among the most affected by the Covid-19 pandemic. The two major issues that builders are confronted with in the post-Covid scenario are flight of migrant labour and shortage of raw-materials, particularly of bricks. This paper focuses on the technological, environmental and social issues that surround brick-production, impacting their supply to the real estate sector, and proposes a long-term solution which involves utilising the RERA regulatory mechanism. Mechanised units contribute less than 1% of the total brick production in India. The environmental hazards of brick manufacturing include air pollution due to the enormous use of coal, posing a hazard to the health of workers as well as to the surrounding environment, and also contribute to global warming by way of an enlarged carbon footprint. There seems to be a regulatory breakdown insofar as the brick manufacturing sector in India is concerned. Because of its direct stake in the sector, the regulatory mechanism established by the Real Estate (Regulation & Development) Act, 2016 could be entrusted with a monitoring role to drive technological change and safety, the legal basis for which already exists in the RERA Act. The paper proposes a minor amendment, which would catalyse labour reform as well as technological up-gradation in the brick making industry.

Keywords: Brick manufacturing industry · Regulatory mechanism · Real estate sector · RERA

1 Introduction

As the Indian economy, still in the process of "unlocking" in the aftermath of the Covid-19 pandemic, limps back to a desired normalcy, the sector that seems to be struggling the most is the real estate industry. Depending as it does on positive sentiment for bookings, and liquidity in the hands of buyers for purchases, the real estate sector is one of the hardest hit [1]. Even before the corona crisis made itself felt, real estate was experiencing acute distress (as evinced by the package that the Finance Ministry felt constrained to announce in November 2019, actually describing the sector as "ailing") [2]. "Recovery in a stalled real estate project is not like re-starting a shop,"

A. Sinitsyn (Ed.): SES 2020, LNCE 141, pp. 433–443, 2021.
https://doi.org/10.1007/978-3-030-67654-4_46

said the Member (Technical) of the Madhya Pradesh Real Estate Regulatory Authority in an interview. "It cannot be turned on and off at will. Once building activity is shut down, it takes a lot of special effort and time to get it re-started – something like a factory boiler [3]."

The President of the MP Chapter of the Confederation of Real Estate Developers Associations (CREDAI), Wasiq Hassan says that the two major issues that builders and promoters are accosted with in the post-Covid scenario are flight of migrant labour and shortage of raw-materials, particularly bricks. "While deliveries of steel and cement are gradually resuming, good quality bricks are difficult to come by because their production was dependent in large measure on migrant labour" [4].

While the challenges that confront the real estate sector are manifold and complex, this paper focuses on bricks – their availability and quality – and the environmental and social issues that surround brick-production, impacting their supply to the real estate sector, and proposes a long-term solution which involves, among other measures, regulation by the Real Estate Regulatory Authorities of the states. It is our contention that the legal basis for this already exists in the RERA Act, and with minor tweaking this could be made effective.

This paper reviews, very briefly, the literature relating to brick-kiln workers, based on empirical studies done in different states of India. The paper argues that while the problems of brick-kilns and their workers are indeed a serious issue of concern, and deserve to be urgently addressed, the manner in which they are addressed will have a significant impact on the real estate sector. The reason for this is that bricks are quite literally the building blocks of the construction industry. The response to dealing with the problems of brick-kiln workers has hitherto been largely focused on 'release' of bonded labourers, and their 'rehabilitation'. Rehabilitation has generally been seen as providing loans to start micro-enterprises at the village level. But skill-enhancement and technological innovation in the brick manufacturing industry itself is often ignored – probably because of the intrinsically exploitative nature of the ownership and management of these establishments.

It is our contention that there is room for a win-win situation in dealing with the problem, and ensuring adequate supply of quality bricks for the real estate sector, which is the highest employment generator in the country after agriculture [5], while simultaneously redressing the grievously abysmal conditions of working and living of brick-kiln workers.

The recent exodus of migrant labour due to the effects of the lockdown of the economy as part of the official strategy to combat the Covid pandemic, and the chaos and tremendous human suffering that ensued, showed that unfortunately the existing statutory framework for the welfare of migrant workers did not prove adequate to the challenge. However, if, because of its direct stake in the sector, the regulatory mechanism established by the Real Estate (Regulation & Development) Act, 2016 could be entrusted with a role in keeping with the mandate outlined in the objectives of the RERA Act, it could catalyse labour reform as well as technological up-gradation in the brick making industry, thus positively enhancing the impact of the RERA Act on the real estate sector in India.

2 Implications for Real Estate

Government of India announced that it intended to achieve its 'Housing for All' target two years ahead of schedule with the sanctioning of 11.2 million houses under various schemes by March 2020 [6]. Since government schemes constitute only a segment of the entire house-construction industry, the normal construction activity in the private sector – both organised and unorganised (self-built houses being a common practice in the rural areas) – would be in addition to this. Thus the demand for bricks would rise proportionally. It is clear that a shortage of quality bricks could prove a bottle-neck in the achievement of these ambitious housing targets.

Brick-making in India is a labour-intensive micro-scale industry, manufacturing taking place in a large number of coal-fired kilns spread all over the country [7]. It is the brick that links the kiln-worker to the house-owner, though in between them is a chain of activities, interspersed with middlemen and practically no physical value addition, except that of transportation, selling, buying and using. At one end of this spectrum is the kiln-worker, who ironically is generally houseless, and at the other end is the house-owner, who himself may range from among the highest income group to the low (but not quite the lowest, since he is among the just 65% of Indians who own a pucca house). (For calculation see Table 1).

Table 1. Percentage of Indians owning *pucca* houses.

Total Households*	246.74 million
Houses with grass of thatch roofs*	37.13 million
Houses with polythene or plastic sheets for roofs*	15.57 million
Percentage of Indians who own houses**	86.6%
Thus percentage of Indians owning *pucca* houses: 86.6 − [{246.74 − (37.13 + 15.57)}100/246.74]	65.2%

*Table HH-2A of the Census of India, 2011. www.censusindia.gov.in [27]
** Houses – Household Amenities and Assets, 2011 [28]
Assumption: Houses roofed with hand-made country tiles or asbestos or GI sheets are included as *pucca*
Source: Authors

Maithel, Wanjule et al. and HZZK [7–9] estimate brick production to be between 240 and 250 billion pieces per year. The basis for this estimation does not seem to be based on an empirical survey, but is apparently based on their estimates of the number of kiln-workers, which in turn is based on the estimate of brick-kilns functioning in the country. The problem is that the estimates for kilns vary considerably from 1,00,000 [8], to 1,40,000 [7], to 1,50,000 [10], right up to 1,80,000 [9]. The estimated number of workers also varies from 10 million [7, 10], to 12 million [11], up to 15 million [8].

Irrespective of the inexactitude of these estimates, it does not detract from the fact that bricks are the most basic in-put for all construction activity – even more fundamental than cement and steel. John & Ateeq point out that the demand for bricks and the growth of the brick industry have a direct relationship with the growth of urbanization and urban development in terms of civil infrastructure and domestic construction [12].

The impact on the real estate sector is not limited to the quantity of supply, but also to the quality of the bricks. The methods of producing and firing these bricks described in most reports seem to indicate that little has changed since medieval times. Mechanised and semi-mechanised units contribute less than 1% of the total production [7]. Apparently the availability of cheap labour acts as a disincentive for technological upgradation. It seems that it is not just the labour that is captive, but also the market. Up to 80% of all bricks used in India are traditional clay kiln-fired [13]. The standardization in dimensions (20 cms × 10 cms × 5 cms) are imprecise in the production lots, and quality control measures relating to them are purchaser-dependent, i.e. they would be either accepted or rejected in the market place, rather than controlled at the production stage.

3 Issues Relating to Labour in Brick Production

In view of the fact that, as already stated, the brick manufacturing industry is labour intensive, it is natural that both quantity and quality of bricks would be dependent on the skill level and productivity of the workers. Productivity in turn is dependent on the working conditions and general well-being of workers [14].

Five empirical studies were reviewed, chosen because of their geographical spread. Though of uneven quality, they give one a broad-brush picture of the situation in the country as a whole. The most recent one by Kazi and Bote relates to Maharashtra in the west [15]; two studies were done in 2016 – one by Das in Assam in the east [16], and another by Jadi & Durgam in Andhra Pradesh in the south [17]; two studies relating to Punjab in the north were reviewed, one by Kainth done in 2009 [18], and another by John & Ateeq in 1998 [12]. Though over two decades old, this latter was considered to view the issues relating to brick-kilns over time. A study done on brick-kiln workers in Rajasthan, representing the central region, by the Aajeevika Bureau, supported by the Overseas Development Institute of London, is by far the best, in methodology as well as analysis.

A major theme that recurs in these studies is that a significant proportion of the labour working in kilns is migrant, in some cases upwards of 50% [14, 16–18]. The migrant labourers are not all from outside the state though – two thirds of them are from other parts of the same state, and one-third from outside. But this varies from state to state. Some states like Punjab and Maharashtra have a greater percentage of out-of-state workers, while states with higher poverty levels get their workers from closer by.

The reasons that constrain workers to migrate for such jobs can be seen as 'push' and 'pull' factors [14, 16, 18]. 'Push' factors are those that compel poor people to leave home in search of work – poverty, unemployment and indebtedness. 'Pull' factors are those that attract them to specific areas – the promise of employment and higher wages. Both these sets of factors are clearly inter-twined; there is only a conceptual difference between them. In practice it is very difficult to pinpoint the exact motivations of a particular worker.

The Aajeevika study found that while 60% were motivated by lack of opportunities, 20% followed friends and relatives and 15% were directly recruited by middlemen. It sums this up by saying that "migration is a livelihood strategy" [14].

The Aajeevika statement, however, gives the impression that migration is always a choice; whereas the reality is not so. One shortcoming in this analysis is that it looks as though these are mutually exclusive motivations. But this is obviously not so. For example, lack of opportunities would underlie a hundred per cent of the cases. And then the other two options (following friends and relatives, or being directly recruited by middlemen) should be seen as the means by which this search for opportunities was realized. If this is done, it would emerge that 57% followed friends and relatives and 43% were directly recruited by middlemen. The Andhra Pradesh study shows that often middlemen deliberately ensnare poverty stricken villagers into a debt trap by advancing them money knowing full well that they would be unable to repay, and then obliging them into offering themselves as labourers in distantly located kilns [17]. In fact, Pal quoting Madan Mishra of the recently founded National Struggle Committee for Brick Kiln Workers, says that "the root cause for this practice of modern slavery is primarily the recruitment process" [11].

70% of the labourers migrate with their families, and 75% of them are below the age of 40 [16]. The exact figures vary from state to state. But the trend is uniform – the majority of migrants are accompanied by their families, and the majority of them are in the younger age group. However, the more disturbing finding is that even little children participate in the work, principally because there are often no schools in the vicinity.

Expectedly, the vast majority of labourers are either illiterate or possess only a very basic education. The Assam study showed that while 62% were illiterate, 22% had studied up to Class III, and only one person among those surveyed had studied up to class VII [16].

The literature is unanimous in listing the abysmal working and living conditions of brick-kiln workers. The workers labour for upto16 h a day "without shade or shelter" [14]. The tasks are repetitive and hard. The work is accident-prone, replete with occupational-hazards, including the constant inhalation of sand and dust. [14]. The living conditions "are almost sub-human" – to quote Dr. Vamsi Vakulabharanam of the University of Hyderabad, in Wainwright [10] – there in no proper arrangement for water, leave alone drinking water, no electricity, no sanitation, no separate kitchen [16]. The locations are often remote with no health or medical facilities or dispensary located nearby even though the labourers suffer from several ailments, including those that are work-related [17]. The remoteness of the locations and the almost captive nature of their clientele induce even the grocery shops to over-charge heftily [14].

Though the violence and abuse applies to all workers, those who are migrant feel the difficulties more acutely, since they have little recourse to relief so far away from home. Even after taking up jobs in these distant kilns, more than 93% of the labourers had no savings whatsoever [16]. The mate (supervisor) lends money, often without a written record [14], ensuring that they remained in a perpetual debt-trap – thus making them virtually bonded.

Wainwright, quoting Andrew Brady of Union Solidarity International (USI), a UK-based civil society organization, terms it "modern-day slavery", involving "horrific abuses of minimum wage rates and health and safety regulations" [10].

4 Environmental and Technology-Related Issues

Brick firing is an energy intensive process. Coal, wood and agricultural wastes are used as fuels. One of the major environmental hazards of brick manufacturing is air pollution due to the enormous use of coal – at least 35 million tons a year is consumed by kilns [8]. Brick-kilns are the fifth largest consumer of coal in the country, spewing out huge quantities of particulate matter and green house gases [13]. Combustion of coal and other biomass fuels in brick kilns results in the emissions of suspended particulate matter (SPM), black carbon (BC), sulphur dioxide (SO2), oxides of nitrogen (NOx), and carbon monoxide (CO), posing a hazard to the health of workers as well as to the surrounding environment, and also contribute to global warming [7].

Erosion of top-soil due to indiscriminate and unscientific mining of clay, is another major problem. A judgement of the MP High Court dated 8th March, 2018 in PIL 2592/2006, Bakir Ali v. State of Madhya Pradesh & Others directs the authorities to take appropriate steps to control mining of clay due to ecological sensitivity [19].

The Patrika reported in May 2019 that the Ministry of Environment, Forests & Climate Change had decided to shut down all traditional brick-kilns in the country within one year [20]. More than a year has elapsed since then, and we are no where nearer this goal. Of course, Down To Earth questions this very objective as being one that is impossible to achieve. While strongly advocating maximization of utilisation of fly-ash and reducing consumption of clay and air pollution from coal-fired brick kilns, it does a back-of-the-envelope calculation to show that at the very most 110 tons of fly-ash is available for producing non-fired bricks which are produced from a mixture of fly ash, sand, lime, gypsum and cement. Even if cent per cent of this available fly-ash is used for this purpose – itself an impractical expectation – it would meet just one-third of the demand for bricks in the country. "This draft notification, therefore, is a classic case of a top-down law that cannot be implemented simply because it would shut down the construction industry," says the article [13].

Information relating to the various types of technology employed by kilns in India is given in Table 2. It will be seen that the bulk of kilns use the FCBTK (Fixed Chimney Bull's Trench Kiln) technology or open clamp technology, both of which are energy-inefficient. To add to it, often the lower grade of bituminous coal is purchased by kiln owners, resulting in higher air emissions. The improved technology kilns account for less than 5% of the total, while substantially mechanized kilns comprise less than 1% of the total [7].

Table 2. Brick kiln technologies currently prevalent in India.

Kiln type	Regional spread	Contribution to brick production
Fixed Chimney Bull's Trench Kiln (FCBTK)	Indo-Gangetic plains (North and East) and clusters in South and West India	70%
Clamps	Central, West and South India	25%
Zig-zag fired kilns, Down-draught kilns, Hoffman Kilns	West Bengal, and a few clusters in North India	3–4%
Vertical Shaft Brick Kilns (VSBK)	Central and East India	>1%

Source: Based on Report of SAARC Energy Centre [7]

5 Regulatory Breakdown

The combination of social, environmental, and technological challenges that beset the brick-making industry in the country inevitably and inexorably have a negative impact on the economy of the construction sector as a whole, and the real estate sector in particular, which is the user of this product.

It is not as though there is a lack of legislation to deal with the problems of the kiln-workers. There is the Bonded Labour System (Abolition) Act (1976), the Child Labour (Prohibition and Regulation) Act (1986), the Inter-State Migrant Workmen Act (1979), and a host of other legislation relating to minimum wages, workmen's compensation, and equal remuneration. There is also the Factories Act (1948). The registers required to be maintained for the purpose are described in detail in the Legal Guide developed by Anti-Slavery International [21].

The traumatic fall-out of the recent exodus of migrant workers due to the lockdown ordered in an attempt to control the spread of the Covid-19 pandemic, is a glaring example of the neglect of the regulation envisioned in the Inter-State Migrant Workmen Act. The Act in itself is a fine piece of legislation. It applies to every establishment in which five or more inter-state migrant labourers are employed, including the contractors who supply such labour. The establishments are required to be registered with specified registering authorities, and the contractors required to obtain licences from the home state as well as the host state of the workmen. Pass books are to be issued to each workman and registers maintained. Displacement allowance and journey allowance is required to be paid, in addition to wages. Basic amenities, including medical facilities and protective clothing are required to be provided to the workers. All this is to be checked periodically by Inspectors appointed for the purpose [22]. Had these provisions been followed in letter and spirit, when the pandemic struck the repatriation of migrant workers could have been accomplished in a far more orderly fashion than the chaotic reality, and the acute human misery it engendered could have been minimised. The fact that this did not happen is testimony to the shoddy implementation of the Act.

Insofar as kilns are concerned, we have seen that approximately half the labourers are migrant. Periodic inspections relating to basic living amenities and medical facilities would automatically benefit even the non-migrant component of the labour. This

means that in spite of provisions existing, these are not followed. If nothing else, this points to regulatory breakdown.

There is also the Environment Protection Act (1986) to deal with the pollution generated by brick-kilns. The mechanisms established by the Act are required to promote the adoption of improved technology to mitigate such pollution [23]. If such technology is actually adopted, it would have the spin-off benefit of increasing productivity and product quality.

A number of technological options are available, including for retro-fitting. These include adoption of efficient kiln technologies, promotion of internal fuel usage and mechanized coal stoking and burner systems – technologies that have been successfully adopted in China (which produces more than half the world's bricks) and elsewhere in South-East Asia [7]. It is therefore all the more confounding that in spite of these relatively simple and inexpensive technologies being available for so long, they are not being adopted in India at the pace that they should have been.

Unfortunately there seems to be lack of focus on the brick-kiln industry, because when seen in the total context of all migrant workers, kiln-workers form only a small percentage of them, and when seen in the context of the entire industrial scenario, kilns are only a small part of national industry. Thus, ironically, in spite of the construction sector employing 51 million people directly and 45 million people indirectly, and being the second largest contributor to national employment [5], and contributing over 275 billion rupees to the country's GDP [24] comprising over 8% of the GDP [25], a critical raw material in-put relating to it, i.e. bricks production, is neglected.

6 Using the RERA Regulatory Mechanism

The efforts of civil society organisations and solutions suggested for ameliorating the problems of kiln-workers are generally focused on the workers rather than on the industry. With the exception of two studies (the Aajeevika Bureau one and the SAARC report), the rest advocate rehabilitation of the labourers through skilling in other trades, and perhaps providing loans under social welfare schemes so that they could set up their own micro-enterprises far removed both physically and metaphorically from brick-production. This is perhaps because the entire brick-kiln industry is seen as intrinsically exploitative.

Sudhir Katiyar of the National Struggle Committee for Brick Kiln Workers says, "When it comes to the implementation of the laws one can see total lawlessness. Our focus is not just to rehabilitate people but to ensure systemic changes in the industry," [11]. And herein lies the key. Systemic changes that include regulatory enhancement and technological thrust are needed if the social problems of the workers, the environmental problems of the units and the in-put requirements of the user industries are to be resolved in the interests of the national economy.

An unreformed brick-kiln sector dependent on archaic technology and medieval level of skills inevitably reflects adversely on the quality of bricks. The existing mechanisms do not seem to be working at optimum level not because they are inadequate in themselves, but because of a lack of close and specific monitoring.

Since the brick-making industry is of critical importance to the real estate sector, it is our contention that the framework established for the regulation and development of the real estate sector could be an effective instrument to achieve this objective.

The Real Estate (Regulation & Development) Act of 2016 established a Real Estate Regulatory Authority (RERA) in each state. Unique among regulatory authorities, RERA is charged not only with the regulation of the sector, but also with its development [26]. Specific sections talk about RERA's mandate to promote the real estate sector; Sect. 32(i) equips RERA with the right (and duty) to advise the state government on issues relating to the development of the real estate. Section 33(1) actually makes it incumbent on state governments to make a reference to RERA on a proposed law or policy or a change in an existing one that may affect the real estate sector before promulgating it. Though the government is not bound by this advice, the very fact that it has to seek it is significant. Section 33(3) requires RERA to "take suitable measures for the promotion of advocacy, creating awareness and imparting training" about laws and policies relating to the real estate sector.

Particularly relevant in this context is Sect. 32(e) which mandates RERA to take "measures to encourage construction of environmentally sustainable and affordable housing, promoting standardization and use of appropriate construction materials, fixtures, fittings and construction techniques."

Section 34(h) provides for state governments entrusting RERA with performing any other functions related to real estate. If mandated to do so, RERA would be the ideal body to monitor whether, for example, Inspectors and registering authorities appointed under the Inter-State Migrant Workmen Act are doing their duty and complying with the relevant provisions. Similarly, the RERA Act requires that all real estate projects be registered with the Authority before they can be marketed. The Act equips the Authority to impose any condition necessary for the better implementation of the provisions. RERA checks the various permissions required to be taken by a real estate developer before a project is registered; one of these is the environmental clearance. So again, if mandated to do so, RERA could direct the Pollution Control Board to make it a condition that real estate developers purchase bricks only from kilns that comply with technological up-gradation conditions that the PCB may have imposed on brick-kilns while licensing them.

Since the mandate to RERA to promote and develop the real estate sector already exists in principle, all that remains to be done is introducing an explicit provision for this in the Act. Section 37 of the RERA Act empowers RERA to issue directions to any promoter (which term includes builders and developers), consumer (allottee or property buyer) and real estate agent (brokers). This section would have to be amended to include all stake-holders in the real estate sector, more particularly any government department or agency. This in fact, would be the only minor amendment required in the RERA Act in order to enable the real estate regulatory framework to monitor the implementation of the various social and environmental legislations insofar as brick-kilns are concerned.

It is not our contention that all problems would be immediately solved at a stroke with this amendment; but it is indeed felt that the Real Estate Regulatory Authorities of various states, due to their intimate stake in the construction industry, are eminently suited to play an effective role in monitoring the implementation of social welfare and environmental provisions relating to brick-kilns and kiln-workers and driving technological change in the brick manufacturing sector.

References

1. PTI (Press Trust of India): Corona virus: CREDAI says Property Sales, Construction Affected, The New Indian Express, March 2020. https://www.newindianexpress.com/business/2020/mar/19/coronavirus-credai-says-property-sales-construction-affected-seeks-government-help-to-tide-over-c-2118820.html
2. India Today: Government to infuse Rs 25,000 crore to revive stalled housing projects, November 2019. https://www.indiatoday.in/business/story/how-indians-are-spending-during-coronavirus-pandemic-1708804-2020-08-07
3. Kapaley, A.: Member (Technical), Madhya Pradesh Real Estate Regulatory Authority. Personal interview, July 2020
4. Hassan, W.: President CREDAI, MP Chapter. Personal interview, July 2020
5. India Today: Here's why Indian construction industry requires more skills in this fast-paced era, February 2020. https://www.indiatoday.in/education-today/jobs-and-careers/story/here-s-why-indian-construction-industry-needs-more-skills-in-this-fast-paced-era-1645985-2020-02-13#
6. Ramnani, V.: Housing for All target of 1.12 crore houses, August 2019. https://www.moneycontrol.com/news/business/real-estate/housing-for-all-target-of-rs-1-12-crore-houses-to-be-achieved-two-years-in-advance-hardeep-singh-puri-4354771.html
7. Maithel, S.: Evaluating Energy Conservation Potential of Brick Production in India, SAARC Energy Centre, Islamabad, Pakistan (2013). https://www.saarcenergy.org/wp-content/uploads/2016/02/India-Report-on-Brick.pdf
8. Wanjule, P.B., Chandanshiv, S.B., Aswale, S.: Brick making in India – history. Int. J. Mark. Financ. Serv. Manage. Res. **4**(11), 11–16 (2015)
9. HZZK: The Brick Market, Habla Zig-Zag Kilns, Australia (2017). https://www.hablakilns.com/the-brick-industry/the-brick-market/
10. Wainwright, O.: Blood Bricks: How India's urban boom is built on slave labour, The Guardian, London, January 2014. https://www.theguardian.com/artanddesign/architecture-design-blog/2014/jan/08/blood-bricks-india-urbanisation-human-rights-slave-labour
11. Pal, S.: Time to Free Brick Kiln Workers from Modern Slavery, Newsclick, March 2019. https://www.newsclick.in/time-free-brick-kiln-workers-modern-slavery
12. John, J., Ateeq, N.: Migrant Labour in the Brick Kilns of Punjab, Centre for Education and Communication (CEC Working Paper), New Delhi, September 1998. www.cec-india.org
13. Bhushan, C.: No Bricks in the Wall, Down To Earth, May 2019. https://www.downtoearth.org.in/blog/air/no-bricks-in-the-wall-64510
14. Bureau, A.: Udaipur and Overseas Development Institute, London: Brick-By-Brick, A Study on the Situation of Migrant Brick Kiln Workers from Kotra, March 2008. http://www.aajeevika.org/assets/pdfs/Brick%20by%20Brick%20-%20An%20Occupational%20Profile%20of%20Brick%20Kiln%20Migrant%20Workers.pdf
15. Kazi, R., Bote, M.: A cross-sectional study to find the working conditions of brick kiln workers. Indian J. Appl. Res. **9**(7), 14–16 (2019)
16. Das, M.: Factors affecting migration of workers in brick industry with special reference to Barpeta district of Assam. Int. J. Humanit. Soc. Stud. **4**(2), 29–32 (2016)
17. Jadi, R., Durgam, S.: A report on migrant labour in the brick kilns industry at Quthbullapur Mandal, Rangareddy district. Int. J. Humanit. Soc. Sci. Res. **2**(6), 21–24 (2016)

18. Kainth, G.S.: Push and Pull Factors of Migration: A case study of brick kiln industry of Punjab state. Asia-Pacific J. Soc. Sci. **1**(1), 82–116 (2009)
19. Indian Kanoon – Law Search Engine: Judgement of MP High Court regarding illegal mining of clay, March 2018. https://indiankanoon.org/doc/15904325/
20. Rathod, V.: Traditional Brick-kilns to be shut down within one year. Hindi news report in Patrikaon, May 2019. https://www.patrika.com/neemuch-news/conventional-brick-kilns-will-be-closed-in-a-year-4529561/
21. Somasekarappa, G. (undated): Legal Guide – Identification, Rescue, Rehabilitation and Prosecution of Bonded Labour in the Brick Kiln Industry. Anti-Slavery International, London. https://www.antislavery.org/wp-content/uploads/2019/12/Legal-Guide-bonded-labour-India-brick-kiln.pdf
22. Inter-state Migrant Workmen (Regulation of Employment and Conditions of Service) Act, 1979. https://clc.gov.in/clc/acts-rules/inter-state-migrant-workmen
23. Environment (Protection) Act, 1986. https://www.indiacode.nic.in/bitstream/123456789/4316/1/ep_act_1986.pdf
24. Keelery, S.: Contribution of construction to India's GDP 2016–2019, July 2020. https://www.statista.com/statistics/1069818/india-gdp-contribution-by-construction/#statisticContainer
25. Dixit, S., Pandey, A.K., Mandal, S.N., Bansal, S.: A study of enabling factors affecting construction productivity: Indian scenario. Int. J. Civ. Eng. Technol. **8**(6), 741–758 (2017)
26. Real Estate (Regulation and Development) Act, 2016. http://mohua.gov.in/cms/TheRealEstateAct2016.php
27. Census of India: Office of the Registrar General & Census Commissioner, India, New Delhi. Table HH-1 and Table HH-2A (2011). www.censusindia.gov.in
28. Registrar General & Census Commissioner, India: Houses - Household Amenities and Assets, 2011, New Delhi (2012). https://censusindia.gov.in/2011census/hlo/Data_sheet/India/HLO_Indicators.pdf

Farm Advisory Services for Farmers Using SWAT and APEX Model

Rajesh Kumar Rai[1], Ashvani Kumar Gosain[2], Priyanka Singh[3(✉)], and Saurav Dixit[4]

[1] Department of Agricultural Engineering, Udai Pratap College, Varanasi 221 002, Uttar Pradesh, India
[2] Department of Civil Engineering, Indian Institute of Technology Delhi (IITD), Hauz Khas, New Delhi 110016, India
[3] Department of Civil Engineering, Amity School of Engineering and Technology, Amity University, Noida, Uttar Pradesh, India
priyanka24978@gmail.com
[4] RICS, School of Built Environment, Amity University, Noida, Uttar Pradesh, India

Abstract. This study aims to simulate hydrological processes at farm and basin scale in the Manjira sub-basin of the Godavari river basin in India. In addition, the study also aims to simulate crop yields in the region at three scales: the scale of the sub-basin, the scale of a watershed and the scale of a farm within the sub-basin. The simulations have been performed using SWAT and APEX model. The study focuses on farm advisory services to be provided at farm level to the farmers based on agro-hydrological model simulation results. This study pave the way for farmers, researchers and advisors to work in a collaborative manner to enhance the farm produce and manage the farm resources in optimum way.

Keywords: SWAT and APEX model · Farm · Agro-hydrological model

1 Introduction

The model system are now a days deployed in agriculture. It represents the model's ability to simulate the real condition of the farm agriculture which help in reaching an appropriate decision making that is named here as farm advisory services. Furthermore, the adaptation options and farm advisory are proposed for the study area based on the model results.

Climatic impact on agricultural production is a serious concern, as it is directly linked to food security and poverty alleviation system of the country. Evidences of the changes in temperature and rainfall from their normal values have been found on scientific basis. These climatic factors have direct impact on agriculture and any change in them are bound to have significant impact on crop yield. Thus, the issue of climate influenced impact on agriculture becomes a serious concern (Rai and Gosain 2018).

Due to the changing nature of weather parameters, the farmers have no surety for potential crop yield and investment return. Favourable weather has always been a matter of chance for both irrigated and rain-fed farmers. In addition, agriculture has to

A. Sinitsyn (Ed.): SES 2020, LNCE 141, pp. 444–458, 2021.
https://doi.org/10.1007/978-3-030-67654-4_47

face many continuously evolving challenges such as climate change, environmental concerns, market globalisation, food safety and quality, and population growth (Kaur et al. 2012; Pathak et al. 2014). Keeping this point in view, the requirement of a modelling system is genuinely felt that can work in a holistic manner to solve all the farm related problems in a comprehensive way. Simultaneously, such a system should also assimilate local agricultural information and human interventions, and should allow farm resources management in attaining potential crop yield.

In the prospect of model system some crucial indicators were needed to assess and monitor the Paddy growth in response to the prevailing weather for practical advisory services to be provided at farm level. Because, the system was intended to be developed for farm level agricultural operations and their anticipated potential users might be small scale farmers, it was intended to use models that are able to handle such requirements and ground realities.

A model system is not a clairvoyant subject rather it's overt to the scientific community based on real-time observations. The APEX model with geographic information system framework provides a user friendly approach to simulate crop yield and farm management details, and assess the climate change impact on crop yield. Once validated, it is ready to respond to local climatic conditions to simulate agricultural crop growth at farm level in real manner. The user friendly computer simulation system with geographic information system framework with SWAT and APEX, including helping tools such as VizSWAT visualisation and analysis tool, as well as model customisation facility, and database management system has provided number of ways to achieve desired results with desired accuracy. Thus, it has been termed as an intelligent agricultural system that works on "systemic" (Perini and Susi 2004) approach i.e., problem is analysed in terms of all the knowledge and the data it depends on.

The coupling is not uncommon and has been already in use under various fields of study. Coupling of the models includes a set of models, data, and assessment methods that together forms the basis for constructing an appropriate modelling system and are responsible for consuming and producing information as part of the system (Whelan et al. 2014; Laniak et al. 2013). Models coupling various processes into a unified framework are seen as useful tools to help analyse the alternatives (Kelly et al. 2013). Therefore, the aforesaid system has become the rational reason for use of coupled modelling approach for farm management under present and climate change scenario using agro-hydro-climatic indicators.

Modelling of an agricultural system is an agro-hydrological model system that assist farmers in farm resource management in response to the prevailing climate. The system is intended to predict the crop yield well in advance and helps to take appropriate action to avoid crop failures due to inappropriate anticipated weather. It also allow farmers to manage farm resources, agricultural operations, as well as time of agricultural operations (since it affect crop growth and reproduction). The model system has capability to assess the spatial and temporal weather impact on crop and it helps in enhancing the crop water productivity and chemical use efficiency. It helps in understanding the health of the agricultural system so that the sensitive parameters can be identified and necessary control measures can be taken to avoid significant losses.

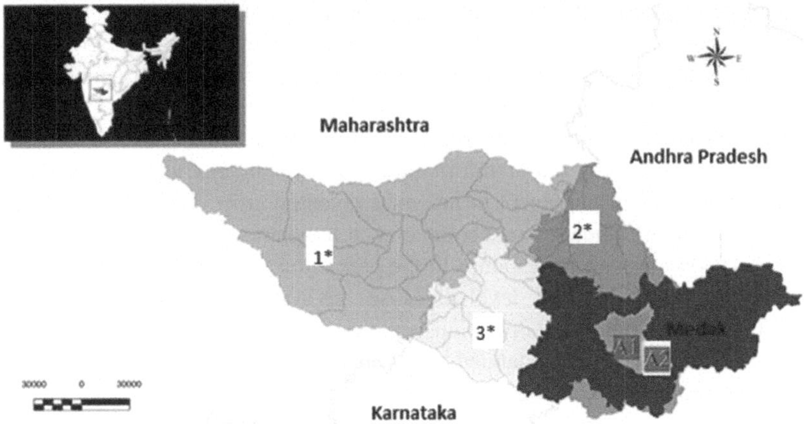

Fig. 1. Area modelled using SWAPP approach shown in VizSWAT interface. Figure shows Manjira Sub-basin, Medak District, and village farm (A2). A1 is the watershed area modelled in SWAPP and A2 is the village farm modelled in APEX. Inset shows the study area on geographical map of the India. 1*, 2* and 3* are the portions of Maharashtra, Andhra Pradesh and Karnataka States respectively contributing to Manjira Sub-basin.

Indicators that have direct impact on Paddy have been identified to manage the farm resources to get potential crop yield in response to the prevailing weather. The indicators are categorised as agricultural, hydrological, and climatic indicators (agro-hydro-climatic indicators). Agricultural advisory based on these indicators has been proposed to alleviate the weather impact on Paddy yield at farm level.

In this study the SWAT and APEX model has been used to provide an advisory at farm level using the model results.

2 Materials and Methods

The working detail and model setup has been given below.

2.1 Brief Detail of Crop Grown in the Study Area

Paddy (Oryzasativa L. ssp. Indica) seed variety is "Cotton Dora Sannalu" (MTU 1010; a Paddy variety developed by Andhra Pradesh Rice Research Institute & Regional Agriculture Research Station Maruteru).The average crop duration is 110–125 days after transplantation. Expected Paddy yield is 4.0–4.5 t ha^{-1}. Plant height vary from 0.80–1.40 m. Water use efficiency for Paddy is 0.37 kg m^{-3} for Rabi 2009–10; 0.43 kg m^{-3} for Rabi 2010–11; and 0.52 kg m^{-3} for Kharif 2011. Irrigation is done by pumping water released from Singur Dam.

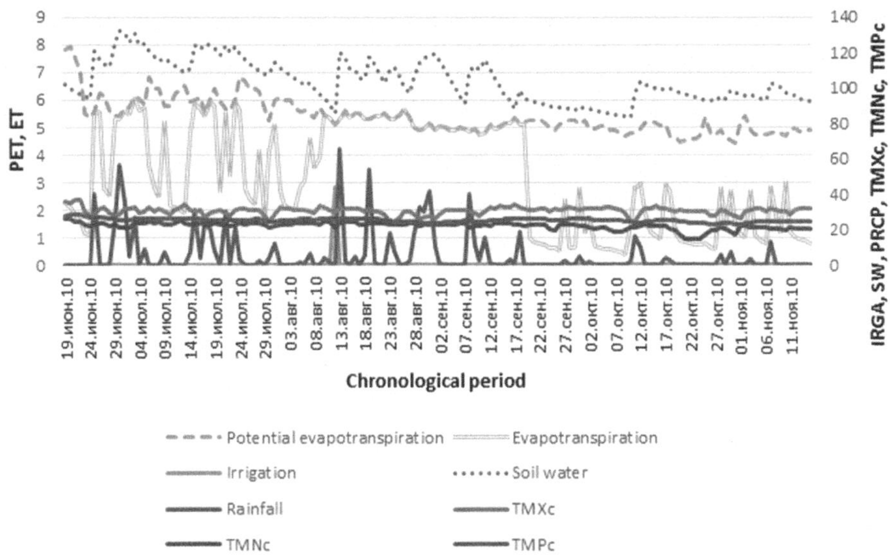

Fig. 2. Farm level hydrological parameters for Kharif Season (Primary axis: potential evapotranspiration (PET mm), and evapotranspiration (ET mm). Secondary axis: rainfall (PRCP mm), irrigation (IRGA mm), soil water (SW mm). TMXc, TMNc, and TMPc are monthly mean estimates of daily maximum, minimum, and average temperatures (°C) for the mean altitude of the catchment).

2.2 Paddy Yield Simulation at Village Farm Level

During the experimental period, different soil, plant, and water measurements were carried out to assess indicators. Agricultural management practices that were adopted at farm were input in APEX to simulate Paddy yield for the farm of 26.5 ha. Simulations were performed for the year 2010. The study area can be easily seen in Fig. 1.

2.2.1 Kharif Season

Major agricultural events that have been simulated are; sowing (19-June), transplanting (21-July), and harvesting (15-Nov) for the year 2010. Total 118 days after transplantation has been used for the crop growth simulation. Simulated and observed Paddy crop water productivity has been given in Table 1. The farmer's field under observation has been simulated. The bio-physical parameters of the farm have been given in Table 2. Hydrological parameters such as rainfall, potential evapotranspiration, crop evapotranspiration, irrigation, and soil water have been shown in Fig. 2.

The model applies irrigation, when soil water become lower than the crop water requirement (automatic irrigation). Observed and simulated Paddy yield is 4.20 and 4.24 t ha^{-1}, respectively. Rainfall has been found to be sufficient, so the irrigation water has been applied only one time (Fig. 2) when the soil water has reached critical level. Observed water use efficiency has been found to be better in Kharif Season (Fig. 2) in comparison to Rabi Season (Fig. 3). The rainfall is usually simulated in the model on basis of grid data input to the model system. Therefore, the simulated values and the observed values are different in the Table 1. Crop evapotranspiration values can be calculated by many calculator available using meteorological data Deelstra et al. (2012).

Table 1. Calculation of crop water productivity and comparison of observed and simulated data of Paddy for the year 2010 (Kharif Season).

Crop water productivity parameters	OBS[1]	SIM[2]
Paddy yield [t ha^{-1}]	4.20	4.24
Irrigation applied [mm]	375.0	57.4
Number of irrigations [mm]	8	1
Average irrigation [mm]	374.9	57.4
Crop water productivity [kg m^{-3}]	0.41	0.70
Water saved [mm] = 317.5 (observed irrigation – simulated irrigation)		

OBS[1] = Observed values; SIM[2] = Simulated values.

Table 2. Bio-physical parameter comparison of observed data recorded from field and simulated results.

Bio-physical parameters	OBS[1]	SIM[2]
Crop evapotranspiration (mm)	465	375.5
Rainfall (mm)	627	545.4
Mean air temperature (°C)	25.3	24.9
Paddy yield (t ha^{-1})	4.2	4.24

OBS[1] = Observed values; SIM[2] = Simulated values.

2.2.2 Rabi Season

The critical cropping dates used for simulation are: sowing (21-Dec), transplanting (18-Jan), and harvesting (30-Apr) for the year 2010. Total 103 days after transplantation has been used for the crop growth as was observed from the local farmers. Paddy crop water productivity has been given in Table 1.The major bio-physical parameters simulated in the model is given in Table 2. Hydrological parameters has been shown in Fig. 3.

It can be seen that the potential evapotranspiration increases as the days gets hotter in summer, so model has calculated the crop evapotranspiration and has applied irrigation accordingly when the soil water has reached a critical level. Observed and simulated Paddy yield is 5.20 and 5.22 t ha^{-1}, respectively. It can be observed that there is negligible rainfall during Rabi Season (Fig. 3), and so the irrigation frequency is much more in comparison to Kharif Season. Irrigation has been applied on the basis of crop evapotranspiration requirements scheduled in the model when reached a minimum level so the irrigation trend has been shown as triangles in Fig. 3. Soil Water shown in the Fig. 3 as SW is an average status over the day.

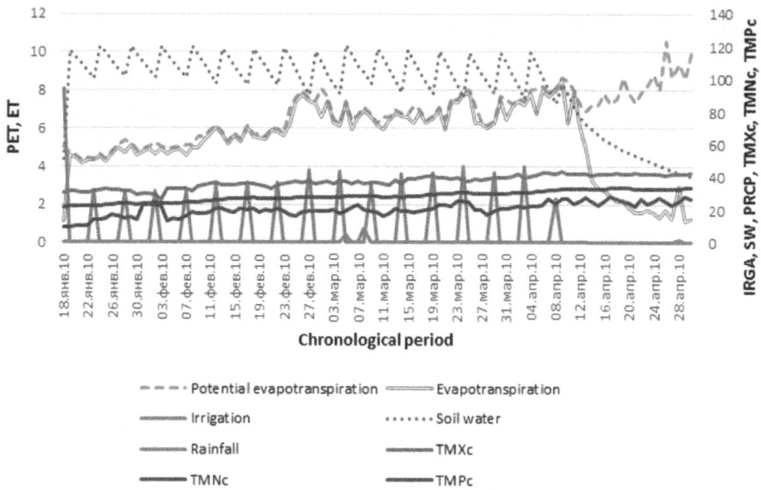

Fig. 3. Farm level hydrological parameters for Rabi Season (Primary axis: potential evapotranspiration (PET mm), and evapotranspiration (ET mm). Secondary axis: rainfall (PRCP mm), irrigation (IRGA mm), soil water (SW mm). TMXc, TMNc, and TMPc are monthly mean estimates of daily maximum, minimum, and average temperatures (°C) for the mean altitude of the catchment).

Table 3. Calculation of crop water productivity and comparison of observed and simulated data of Paddy for the year 2010 (Rabi Season).

Crop water productivity parameters	OBS[1]	SIM[2]
Paddy yield [t ha^{-1}]	5.20	5.22
Irrigation applied [mm]	1426.0	699.1
Number of irrigations [mm]	19	17
Average irrigation [mm]	75.05	41.10
Crop water productivity [kg m^{-3}]	0.35	0.76
Water saved [mm] = 726.8 (observed irrigation – simulated irrigation)		

OBS[1] = Observed values; SIM[2] = Simulated values.

Table 4. Biophysical parameter comparison of observed data recorded from field and simulated results.

Biophysical parameters	OBS[1]	SIM[2]
Crop evapotranspiration (mm)	605	563.6
Rainfall (mm)	19	19.2
Mean air temperature (°C)	29.4	27.0
Paddy yield (t ha^{-1})	5.2	5.22

OBS[1] = Observed values; SIM[2] = Simulated values.

3 Results and Discussion

3.1 Agro-Hydro-Climatic Indicators

Indicators that are practically easy to obtain have been categorised under agricultural, hydrological, and climatic indicators. Possible adaptations tested at farm level by using agricultural (bio-physical) indicators are shift in planting dates (transplantation age), length of growing season, changes in amount and timing of irrigation (irrigation scheduling), and plant spacing (density and population). Hydrological indicators are crop evapotranspiration (consumptive use), and soil water content. Climatic parameters that are most important from crop growth and agricultural point of view are temperature and rainfall.

The evaluation of these indicators for Paddy yield has been given in Table 5. The developed farm advisory in response to the indicators have been verified with other sources available in scientific literature to confidently disseminate the agricultural advisory services suggested to farmers on simulation basis using agro-environmental based-model.

Table 5. Indicators and their impact on crop yield.

IND[1]	Change in parameter	Change in average Paddy yield
TRA[2]	5 days earlier (fixed harvest date)	0.15% (↓)
	5 days later (fixed harvest date)	3.95% (↓)
LEN[3]	5 days shift earlier (constant length of growing season)	0.15% (↓)
	5 days shift later (constant length of growing season)	4.02% (↓)
SPA[4]	200 plants m^{-2}	0.60% (↓)
	250 plants m^{-2}	1.31% (↑)
TMP[5]	1% (↑) in Tmin and Tmax	2.28% (↓)
	1% (↓) in Tmin and Tmax	1.71% (↑)

IND[1] = Indicators; TRA[2] = Shift in planting dates (transplantation age); LEN[3] = Length of growing season (duration of crop growth); SPA[4] = Plant spacing (density and population); TMP[5] = Temperature.

3.1.1 Agricultural (Bio-Physical) Indicators

3.1.1.1 Transplantation

In general, the plant sowing begins as the plant available water in the top soil is available within a specific range. The lower limit and upper limit of water indicates that soil conditions are too dry and wet for sowing (Janssen et al. 2009; Therond et al. 2011).

Proper planting time, planting density, irrigation, and chemical inputs are essential for crop growth and it has been found that one month time after sowing is sufficient for transplantation for Paddy (Gadgil and Kumar 2006).

In the year 2010, for Kharif Season, 20-July has been found to be critical for transplantation (determined by model simulation) for the study area at farm level. Transplanting before or after critical date results in decreasing Paddy yield. The model results show that delaying four days (→) after critical date results in 3% decrease (↓) in Paddy yield, while transplanting five days before (←) the critical date results in 0.15% (↓) decrease in Paddy yield (Table 5).

3.1.1.2 Length of Growing Season of Paddy Crop

Generally, 118 ± 3 days after transplantation is the growing season for Kharif Season and 103 ± 3 days after transplantation for Rabi Season that has been found to be suitable for Paddy. Proper heat unit or number of degree-days (Thornthwaite and Mather 1955) should be sufficiently available to the crop so that it can attain sufficient biomass required for the potential crop yield. These heat units should be available in addition to sufficient fertiliser and irrigation water thus minimising all the crop stresses such as temperature, plant water (excess (flood) and deficit (drought)), root aeration, and fertiliser (nitrogen and phosphorous) stress (Williams et al. 2012).

Model simulation shows that decreasing (↓) the length of growing season (with critical transplant date fixed and varying the harvest date) decreases (↓) the crop yield, while increase (↑) in crop growth duration have no significant impact on crop yield after attaining the potential yield and remains constant thereafter.

Model simulation results show that five days shift (←) in length of growing season before critical date (keeping length of growing season constant that is 118 days for Kharif Season) has resulted in decrease of crop yield by 0.15% (↓) and shifting (→) after critical date has resulted in decrease of crop yield by 4.02% (↓) (Table 5).

3.1.1.3 Plant Spacing and Population

APEX does not simulate tillering (Williams et al. 2012). Therefore, required plant population is imperative to input to the model. Plant spacing is a key parameter in leaf area development, which depends on radiation capture and radiation capture is related with heat unit and thus, crop growth and yield is affected.

Precision farming is the modern concept where land grading is done to achieve the desired land slope to facilitate the water application to the crop and the water drainage from the farm. It uses the laser technique to achieve the land grading. Thereafter, proper spacing between the plant-to-plant and row-to-row spacing is maintain throughout the field to alleviate the inter-competition between plant for nutrient, water and light. Precision farming is modern concept in Paddy and it maintains the desired plant spacing and population density (Rowshon et al. 2006). By maintaining the plant spacing and population density the radiation capture and thus the heat unit of crop can be maintained as per the biomass growth requirement. APEX model works on heat unit (HU) concept to model the crop growth and crop yield.

Optimum plant population yield maximum (Baloch et al. 2006; Oghalo 2011). Proper plant spacing that is between row-to-row and plant-to-plant and population density (plants m^{-2}) should be maintained throughout the farm.

Decreasing (↓) the plant population (critical value determined by simulation) could result in low crop yield and increasing (↑) plant population have shown increase in crop yield, but in principle increase in plant population should have no significant change in yield after critical value which lies in between 200–250 plants m^{-2} (Table 5), because it is biologically not possible that a crop will keep on adding biomass. Population density of 159–304 plants m^{-2} yield maximum (Baloch et al. 2002).

3.1.1.4 Irrigation Scheduling

In Kharif Season sufficient rainfall has been observed, and this is the reason that the number of irrigations applied is very less in comparison to the Rabi Season that has experienced minimal rainfall (Figs. 2 and 3, Tables 1, 2, 3 and 4).

From water use efficiency point of view over-irrigation is avoided, on the contrary even deficit irrigation is practised with care. Regulated deficit irrigation is modern concept that can be used with sprinkler and drip irrigation systems with precision farming.

Model apply water to the farm according to plant water needs (Arnold et al. 2013; Williams et al. 2012) and irrigation and rainfall is balanced to manage crop water requirements. Time, frequency, and quantity of irrigation is very important to keep soil water maintained at the required level. Frequency and amount should be maintained according to crop evapotranspiration value and surplus water is to be provided to fulfil the losses such as drainage, runoff, and seepage. Irrigation is applied, when the soil water reaches a critical level in absence of rainfall.

3.1.2 Hydrological Indicators

The two hydrological indicators are crop evapotranspiration and soil water content.

3.1.2.1 Crop Evapotranspiration

Crop evapotranspiration affects the hydrological cycle of the agricultural area and plays a pivotal role in determination of crop yield. Evapotranspiration is balanced by rainfall and irrigation water. Rainfall and irrigation (frequency and amount) both values should fulfil the consumptive water use and should be sufficient to provide water needed for metabolic processes. The amount of water used in metabolic process is very less in comparison to evapotranspiration, but its presence cannot be ignored, as it is the water amount that contributes to biomass growth.

Simulation results show that the average Paddy evapotranspiration (ET) for Kharif Season is 3.18 mm day^{-1} and for Rabi Season is 5.36 mm day^{-1} that means ETKharif < ETRabi. Seasonal values for Kharif is 465 mm and for Rabi is 605 mm. The reason behind this lies in the fact that the mean air temperature for Kharif is 25.3 °C and for Rabi is 29.4 °C and the ET values are plenary governed by the temperature and plant stomatal conductance parameters.

3.1.2.2 Soil Water Content

Soil water content depends on rainfall, surface runoff, farm drainage, evapotranspiration, and water inflow-outflow (Park et al. 2013). Below certain critical value, it can harm crop growth and in extreme cases it may result in even plant death. On the contrary, its excess leads to flooding and that is harmful too as crop requires proper root aeration.

Simulation results shows that the average soil water on calculated on daily basis for Kharif Season is 103.91 mm and for Rabi Season is 96.02 mm for the year 2010.

3.1.3 Climatic Parameters

The two climatic parameters are surface air temperature and rainfall.

3.1.3.1 Surface Air Temperature

Simulation results show that the average surface air temperature (TEMP) for Kharif Season is 25.27 and for Rabi Season is 28.38 °C day^{-1} that means TEMPKharif < TEMPRabi.

Temperature affects almost every component of hydrological cycle. Temperature affects soil water content and thus irrigation water demand (Fig. 2 and 3). With the rise (↑) in temperature the evaporative demand increases (↑).

Crop productivity is very sensitive to the cardinal temperatures such as minimum (TMN) and maximum (TMX) temperature and threshold temperature or average base temperature (TBSC) (TBSC depends on crop type; for Paddy TBSC is 10 °C). With 1% (↑) increase in TMX there was 1.17% (↓) decrease in crop yield and by decreasing TMX by 5% (↓) there was increase in crop yield by 4.27% (↑) (when threshold values are not taken into account).

With the increase of TMN by 1% (↑) there was decrease of 0.93% (↓) in yield and a decrease in TMN by 1% (↓) resulted in increase of 0.70% (↑) in crop yield (when threshold values are not taken into account).

It has been found by simulation that with 1% (↑) in TMN and TMX there was decrease in crop yield by 2.28% (↓), while decrease in TMN and TMX by 1% (↓) shows increase in crop yield by 1.71% (↑).

3.1.3.2 Rainfall

Rainfall and irrigation water go hand-in-hand in compensating crop water requirement (Fig. 2 and 3). Decrease in rainfall amount has to be compensated by irrigation water.

Qualitative and quantitative measures of rainfall (number of storm, storm pattern, timing, duration, and intensity) affects most to the agricultural regions. If rainfall is uniform and fulfil the evaporative demand, then it is considered to be suitable for the crop growth. Rainfall directly affects the irrigation demand.

Use of modern irrigation technology has been suggested in case of less rainfall to improve water use efficiency in the study area. A decision tree mentioned in Turral et al. (2011) should be made according to the field demand and situation of the local agriculture at farm level. The decision tree could play an important role in managing farm which deals with wide range of available options to be adopted in the area. Both water drainage and water conservation plays an important role in excess and scarce water condition respectively.

For the study area, in year 2010, the average rainfall for Kharif Season is 5.63 and for Rabi Season is 0.16 mm day^{-1} (Tables 1, 2, 3 and 4).

3.2 Agricultural Advisory Services

The agriculture extension services have invariably found that farmers usually lack information regarding agricultural innovations and modern technologies, though they are well versed in conventional and traditional methods. To combat the present farming challenges they need to adopt modern techniques. There has been always a gap between researchers and farmers, though many Krishi Vigyan Kendra (Farm Science Centre) and such allied institutes/centres are obliged to bridge this gap in India.

It is known that farming has always been a laborious job in India, especially for small scale farmers. The only option to make farming smarter and not harder is to adopt modern techniques supported by computer modelling. Modelling approach in agriculture provides an opportunity to farmers in resource management and decision making. Realistic model approach has been presented to best manage the farm resources to obtain potential crop yield. It provides an opportunity to control the major constraints such as chemical and irrigation water needs according to the rainfall and crop evapotranspiration, along with other resource availability. Among the advisory services, the options such as irrigation scheduling have been showcased in Fig. 2 and 3, and have been described in Table 5.

3.3 Crop Yield Anticipation Before Harvest

This model cum agricultural system (highly developed model system such as SWAT and APEX which can be made up to the level by calibration and validation where it can sense the local level agro-meteorological dataset and when provided with the futuristic weather dataset it could model the crop yield in realistic manner) can be used to anticipate future crop yield at farm level to avoid crop failures and to gain potential crop yield, along with farm resource management. The major constraint in anticipation of crop yield before harvest is the availability of futuristic weather data. The Government/scientific agencies are providing weather forecast of a few days only such as India Meteorological Department for seven days, while other long-term forecast are questionable and less reliable. To simulate crop growth, season-long (at least four months) weather forecast is needed. The only option left is the use of tools that generate weather data for the cropping season based on statistical methods.

Weather generators can be used discreetly to generate future weather data using experience, government forecast, and observed normal data. There is no harm in using programme tools to generate data based on observed weather statistics for about next three-four months, as every model is based on some mathematical and statistical sciences only, but it should be used deliberately based on personal experience and knowledge of the local weather. One such tool is WXGN (standalone weather generator; a combination of WGEN (Weather Generator) and CLIGEN (CLImate GENerator)) that has been used in EPIC (Environmental Policy Integrated Climate) and APEX (Sharpley and Williams 1990).

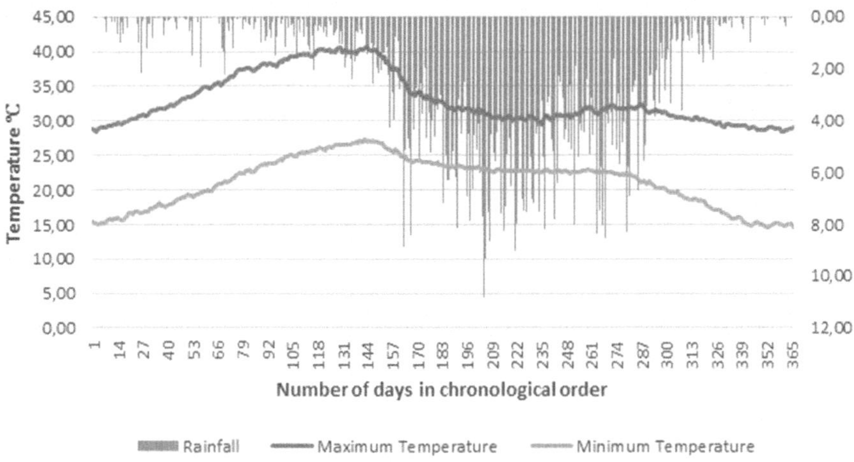

Fig. 4. Daily normal weather data prepared from IMD data for the period 1969–2007.

The success of using model cum agricultural system for crop yield anticipation before harvest lies only in getting actual data and detailed information at farm level, along with weather forecast data to simulate future crop yield. Using proper statistical techniques, one can generate season-long data from observed normal data statistics. Such daily observed data has been used for the village (Fig. 4). Using normal weather data and other farm details the Paddy yield obtained for Kharif and Rabi Season is 4.29 and 5.40 t ha^{-1}, respectively.

One can use forecast data from India Meteorological Department and normal weather data for the rest of future period, and can predict the crop yield before harvest. It not only helps in farm resource management, but also help in anticipation of crop failures prior to harvest. The best thing that could be attributed to this technique is that once model is setup for a particular location, the farm advisors only need to manage future weather and farm resources to predict the future crop yield.

4 Conclusions

It has been found that temperature and rainfall is changing from their normal values in Indian context. It is the spatial and temporal variation of the weather parameters that has great impact on crop. The model system has provided a way that can manage farm resources and predict the crop yield. The system also admonish the farmers well in time about crop failure and anticipate the incurring crop loss. Effect of frost and hot spell can be anticipated well in time. Advisory for best management practices and optimum crop production can be provided at farm or at field level using model system for sustainable agricultural system. As the future of agriculture is serious business, such tools can provide indicators that can help in managing crop growth and farm resources in attaining maximum crop yield with cost-effective farm inputs. In this chapter various aspects related to the farm advisory services has been explained and in this way the second objective to develop a model system for agricultural advisory services based on agro-hydro-climatic indicators that are easy to obtain and representative of the spatial and temporal variations in constituent parameters has been achieved.

The specific (not exclusive) conclusions from this chapter has been summarized as below:

1. Once the model set-up has been done, the crop water productivity parameters in terms of Paddy yield [t ha^{-1}], irrigation applied [mm], number of irrigations [mm], average irrigation [mm], and crop water productivity [kg m^{-3}] were determined for the Kharif and Rabi season which were found to be as 4.24, 57.4, 1, 57.4, and 0.70, respectively for Kharif season and 5.22, 699.1, 17, 41.10, and 0.76, respectively for Rabi season for the simulation data which are in good agreement of the observed data obtained from the observations at farm.
2. According to the model simulation we could have saved 317.5 mm of water in Kharif season and 726.8 mm in Rabi season by exercising model obtained irrigation schedule.
3. The observed and simulated bio-physical values for the model were obtained as crop evapotranspiration (mm) 465 and 375.5 and the Paddy yield (t ha^{-1}) 4.2 and 4.24 for Kharif season and crop evapotranspiration (mm) 605 and 563.6, and the Paddy yield (t ha^{-1}) 5.2 and 5.22 for Rabi season.
4. The shift in planting dates have shown a great impact in crop yield. When the dates were fixed at five days earlier (fixed harvest date) the decline in crop yield were found by 0.15% and when transplantations were done at five days later (fixed harvest date) then also the decline of 3.95% has been obtained. This shows that right amount of heat unit accumulation is necessary for biomass growth that resulted to in terms of crop yield.
5. Duration of crop growth has also good impact on crop yield. When the duration were shifted earlier by five days only (constant length of growing season) then it resulted to a decline of 0.15% on the contrary at shift of five days later (constant length of growing season) it resulted to more decline by 4.02%. This could also be attributed to heat unit concept as explained earlier.
6. Plant spacing has also shown good impact on crop yield. When the plant spacing were kept at 200 plants m^{-2} then a decline of 0.60% were obtained but a sharp increase in crop yield were obtained at plant spacing of 250 plants m^{-2} that resulted to increase crop yield percent by 1.31%.
7. We have almost no control over weather temperature but the model shows that optimum temperature are imperative for the potential crop growth. A 1% increase in Tmin and Tmax resulted to 2.28% decline in crop yield while 1% decrease in Tmin and Tmax resulted to sharp increase in crop yield by 1.71%.

Acknowledgements. Authors are thankful to associated project partners. The research project was fully sponsored by CLIMAWATER, a collaborative project between Indian Institute of Technology Delhi (IITD), New Delhi, India, Bioforsk, Norway; and International Water Management Institute (IWMI), Hyderabad, India.

References

Arnold, J.G., Kiniry, J.R., Srinivasan, R., Williams, J.R., Haney, E.B., Neitsch, S.L.: SWAT2012 input/output file documentation, Texas Water Resource Institute TR—439 (2013). https://swat.tamu.edu/media/69296/SWAT-IO-Documentation-2012.pdf

Baloch, A.W., Soomro, A.M., Javed, M.A., Ahmad, M., Bughio, H.R., Bughio, M.S.: Optimum plant density for high yield in rice (Oryza sativa L.). Asian J. Plant Sci. **1**, 114–116 (2002)

Baloch, M.S., Awan, I.U., Hassan, G.: Growth and yield of rice as affected by transplanting dates and seedlings per hill under high temperature of Dera Ismail Khan. Pak. J. Zhejiang Univ. Sci. B. **7**(7), 572–579 (2006)

Deelstra, J., Kakumanu, K.R., Rai, R.K., Nagothu, U.S.: Water use efficiency testing and improvement in the Godavari River Basin. [India]. In: Nagothu, U.S., Gosain, A.K., Palanisami, K. (eds.) Water and Climate Change: An Integrated Approach to Address Adaptation Challenges, pp. 105–122. Mac. Pub. India Ltd, New Delhi (2012)

Gadgil, S., Kumar, K.R.: The Asian monsoon, agriculture and economy. In: Wang, B. (ed.) The Asian Monsoon. Springer (2006). https://meteo.edu.vn/DATA/Books/The%20Asian%20Monsoon/

Janssen, S., Oomen, R., Hengsdijk, H., van Ittersum, M.K.: Agricultural management module. Production Enterprise Generator, Production Technique Generator, Simple Management Translator and Technical Coefficient Generator. SEAMLESS project deliverable (PD 3.3.3). SEAMLESS Integrated Project EU FP6 (contract nr. 010036), Wageningen (2009). https://www.seamlessassociation.org, Wageningen. https://ageconsearch.umn.edu/bitstream/57932/2/Report_44_PD3.3.3.3.pdf

Kaur, H., Jalota, S.K., Kanwar, R., Vashisht, B.B.: Climate change impacts on yield, evapotranspiration and nitrogen uptake in irrigated maize (Zea mays)–Wheat (Triticum aestivum) cropping system: a simulation analysis. Ind. J Agri. Sci. **82**(3), 213–219 (2012)

Kelly, R.A., Jakeman, A.J., Barreteau, O., Borsuk, M.E., ElSawah, S., Hamilton, S.H., Henriksen, H.J., Kuikka, S., Maier, H.R., Rizzoli, A.E., van Delden, H., Voinov, A.A.: Selecting among five common modelling approaches for integrated environmental assessment and management. Environ. Modell. Softw. **47**, 159–181 (2013)

Laniak, G.F., Olchin, G., Goodall, J., Voinov, A., Hill, M., Glynn, P., Whelan, G., Geller, G., Quinn, N., Blind, M., Peckham, S., Reaney, S., Gaber, N., Kennedy, R., Hughes, A.: Integrated environmental modelling: a vision and roadmap for future. Environ. Modell. Softw. **39**, 3–23 (2013)

Oghalo, S.O.: Effect of population density on the performance of upland rice (Oryza Sativa) in a forest-savannah transition Zone. Int. J. Sustain. Agri. **3**(2), 44–48 (2011)

Park, C., Lee, J., Koo, M.: Development of a fully-distributed daily hydrologic feedback model addressing vegetation, land cover, and soil water dynamics (VELAS). J. Hydrol. **493**, 43–56 (2013)

Pathak, H., Pramanik, P., Khanna, M., Kumar, A.: Climate change and water availability in Indian agriculture: impacts and adaptation. Ind. J Agri. Sci. **84**(6), 671–679 (2014)

Perini, A., Susi, A.: Developing a decision support system for integrated production in agriculture. Environ. Modell. Softw. **19**, 821–829 (2004)

Rai, R.K., Gosain, A.K.: Farm level agro-hydrological simulation using the SWAT model for Manjira sub-basin, India. New Agric. **29**(2) 2018. Accepted Paper

Rowshon, M.K., Amin, M.S.M., Hassan, S.M.H., Shariff, A.R.M., Lee, T.S.: New performance indicators for rice-based irrigation systems. Paddy Water Environ. **4**(2), 71–79 (2006)

Sharpley, A.N., Williams, J.R.: EPIC: Erosion/Productivity Impact Calculator 1. Model Doc., Tech. Bul. 1768. U.S. Deptt. Agric. (USDA) (1990). https://organicroots.nal.usda.gov/download/CAT10698097/PDF

Therond, O., Hengsdijk, H., Casellas, E., Wallach, D., Adam, M., Belhouchette, H., Oomen, R., Russell, G., Ewert, F., Bergez, J., Janssen, S., Wery, J., van Ittersum, M.K.: Using a cropping system model at regional scale: low-data approaches for crop management information and model calibration. Agric. Ecosyst. Environ. **142**, 85–94 (2011)

Thornthwaite, C.W., Mather, J.R.: The Water Balance. Pub. Climatology. Drexel Institute of Technology, vol. 8, no. 1, p. 104 (1955)

Turral, H., Burke, J., Faure, J.M.: Climate Change, Water and Food Security. Food and Agriculture Organisation of the United Nations (FAO), Rome, Italy (2011). https://www.fao.org/docrep/014/i2096e/i2096e.pdf

Whelan, G., Kim, K., Pelton, M.A., Castleton, K.J., Laniak, G.F., Wolfe, K., Parmar, R., Babendreier, J., Galvin, M.: Design of a component-based integrated environmental modelling framework. Environ. Modell. Softw. **55**, 1–24 (2014)

Williams, J.R., Izaurralde, R.C., Steglich, E.M.: Agricultural Policy/Environmental eXtender Model. Theo. Doc. Ver.0806. Texas A&M Univ. (TAMU) Texas, USA (2012). https://apex.tamu.edu/media/72000/the-apex0806-theoretical-documentation.pdf

Conditional-Constant Component in the Total Consumption of an Energy Resource and Its Influence on the Energy Efficiency of Industrial Consumers

Nadezhda Hruntovich[1(✉)], Aleksey Kapansky[1], Svetlana Jhukovets[1], Y. Shenets[2], Mikhail Malashanka[3], and Elena Gracheva[4]

[1] Sukhoi State Technical University of Gomel, Prospect Octiabria, 48, 246746 Gomel, Republic of Belarus
gruntovich@tut.by

[2] Public Corporation "Gazprom Transgaz Belarus", Nekrasova 9, Minsk, Republic of Belarus

[3] Energy Efficiency Department of the State Committee for Standardization of the Republic of Belarus, Svobody sq. 17, Minsk 220030, Republic of Belarus

[4] Kazan State Power Engineering University, Str. Krasnoselskaya, 51, 420066 Kazan, Russia

Abstract. For industrial consumers with a simple dependence (or close to it) of energy consumption on the volume of output, the possibility of assessing the conditional- constant component of costs using the one-factor model "consumed energy resource-volume of output" is considered. The article highlights the concept of general-plant conditional-constant energy consumption and technological one, which is associated with the heating and maintenance of technological equipment in a working condition. For natural gas used in both the technology of cement production and glass production, the technological conditional-constant component is determined by the cost of energy to maintain the furnaces in a working condition. Using the example of a modern enterprise for the production of nitrogen fertilizers, an assessment of the conditional-constant component of electricity consumption for various industries was made in conditions of changing the production program. For a group of 88 industrial electricity consumers and 32 industrial consumers using either fuel (natural gas) or thermal energy in the technological process, the weight of the conditional-constant component of the consumption was investigated according to the production load. It is shown that industries with a higher weight of the conventional-constant component in the total energy consumption also have a greater horizontal regulation capacity for energy efficiency, i.e. the ability to change the total and specific energy consumption while changing production volumes.

Keywords: Energy efficiency · Industrial consumers

A. Sinitsyn (Ed.): SES 2020, LNCE 141, pp. 459–470, 2021.
https://doi.org/10.1007/978-3-030-67654-4_48

1 Introduction

The importance of the conditional-constant component of the energy resource costs (the part of the energy resource consumption independent of the production load) is recognized by both industrial power engineers and energy auditors. In publications devoted to issues of energy efficiency management (EEF) of modern industries, it is shown that the presence of a conditional-constant component in the consumption of energy resources determines [1–5]:

- regulatory capacity for energy efficiency by changing the volume of production (horizontal regulation);
- the current state of energy efficiency;
- the efficiency of the implemented energy saving measures.

The latter is especially important, since it is an indisputable fact that with the introduction of seemingly effective technologies, the effectiveness of the measures being introduced is significantly lower than the forecast values or may not be achieved at all [6, 7].

However, the question concerning the assessment and separation of the share of the conditional-constant component of costs from the total volume of fuel and energy resources consumption is now being raised. There are two ways to solve this problem. The first is the development of an annual energy balance of an industrial consumer and the separation of a conditionally constant component of energy consumption from its structure. This is the most laborious way, as it involves the collection of a large amount of data on the FER receivers and information on their operating modes. The second way is based on the development of regression models linking energy consumption and the volume of production, which is important for enterprises operating with variable technological load [8–13].

2 Research Results

For most industrial consumers of fuel and energy resources, the technological subsystem can be represented by two main subsystems: raw materials and technological equipment. Since the qualitative and quantitative indicators of technological equipment are relatively stable over relatively short periods of time (a year or two), and the raw materials which were used also have relatively stable characteristics, then the model describing the "energy – technology" interconnection will represent the dependence of energy consumption only on volumes of products, that means to be simple. The presence of a conditional-constant component of the consumption of energy resources in the total consumption of fuel and energy resources determines the hyperbolic nature of the dependence of the specific costs of fuel and energy resources on production volumes (Fig. 1) and has a well-known form for enterprises with a simple (or close to it) interconnection between energy and technology [14, 15]:

a) for fuel:

$$b_{sp.fuel} = b_{sp.tech} + \frac{B_{cond}}{P}, \quad (1)$$

b) for electrical energy:

$$w_{sp.el} = w_{sp.techn} + \frac{W_{cond}}{P}, \quad (2)$$

where B_{cond}, W_{cond} – a conditional-constant component of fuel and electrical energy consumption, respectively, independent of the volume of output; $b_{sp.tech}$, $w_{sp.techn}$ – specific technological energy consumption per unit output; P – the volume of production.

The conditional-constant component of energy consumption largely depends on the type of energy resource. For electrical energy, the conditional-constant component of the consumption includes the energy consumption for the operation of lighting, ventilation, auxiliary production needs (production associated with the repair of equipment, storage facilities, administrative and utility rooms). For industries that have furnace equipment in the technological process, it is also possible to single out the technological conditionally constant component of electricity, which is associated with heating and maintaining the technological equipment in working order. For fuel, for example, natural gas, which is used in the technology of both cement production and glass production, the conditional-constant component is also determined by the cost of energy to maintain the furnaces in working order. That is, in the conditional-constant component of the energy resource consumption, the technological conditional-constant component is clearly distinguished. First of all the conditional-constant component is determined by the seasonal factor, as well as the quantitative and qualitative characteristics of the equipment that determine it [16–18].

Table 1. Annual models of unit costs of EE production "K-1".

Title	Model of total power consumption, W, kWh	Model of specific power consumption, $w_{sp.el}$, kWh/t	Specific technological power consumption, $w_{sp.techn}$, kWh/t	Conditional-constant power consumption, W_{cond}, kWh
Workshop power consumption model	$144.8 \cdot P + 103008.9$	$144.8 + \frac{103008.9}{P}$	144.8	103008.9
Plant-wide power consumption model	$180.35 \cdot P + 181887.9$	$180.35 + \frac{181887.9}{P}$	180.3	181887.9

Table 2. Evaluation of the EE structure in the workshop and its general production consumption for the production of "K-1".

Output volume, *P*, t/day	Electricity consumption, *W*, kWh	The share of technological energy consumption W_{techn} in the total consumption *W*, %	Share of conditional-constant power consumption W_{cond} in total consumption *W*, %
General workshop electricity consumption			
70	113144.9	9.8	91.0
90	116040.9	12.7	88.8
110	118936.9	15.5	86.6
140	123280.9	19.7	83.6
170	127624.9	23.9	80.7
200	131968.9	28.1	78.1
Plant-wide power consumption			
70	194508.9	6.9	93.5
90	198114.9	8.9	91.8
110	201720.9	10.9	90.2
140	207129.9	13.9	87.8
170	212538.9	16.9	85.6
200	217947.9	19.8	83.5

For consumers of fuel and energy resources with a significant weight of a conditional-constant component in the total volume of fuel and energy, neglection of the factor "change in the volume of output" causes gross errors in assessing energy efficiency from the implemented energy saving measures and predicting it in conditions of changes in the production program [19, 20].

On the example of an enterprise for the production of nitrogen fertilizers (production of "K-1" and "K-2"), an assessment of the conditional-constant component of electricity consumption was carried out using the dependence models Wspl.el = f(P).

For the production of "K-1" in accordance with the generated database on daily production rates and general production electricity consumption of the workshop, annual models of the dependence of specific electricity consumption on the volume of production were built (formula 2).

Figure 1 demonstrates the actual values of specific general production and shop specific electricity consumption from the volume of production. The parameters of K-1 production models are presented in Table 1. As we can see from Table 1, the transition from the workshop electricity consumption to the general production consumption entails an increase in the conditional-constant component by 76.6%.

Table 2 presents the results of assessing the share of the conditional-constant component in the total power consumption and the ratio of the technological consumption of EE to the conditional-constant consumption of EE under conditions of change in the production load.

The data (Table 2) demonstrates that in the total general and shop production power consumption, the share of the conditional-constant component Wcond is not constant when the consumer operates under the conditions of a changing production program. In the workshop power consumption, the weight of the conditional-constant component of power consumption reaches 91% with a minimum volume of production and decreases to 78.1% with maximum productivity. In general production power consumption the share of the conditionally constant component of electricity consumption at a minimum capacity of 70 t/day reaches 93.5%, and at a maximum capacity of 200 t/day – 83.5%.

The weight of the technological component of energy consumption also changes depending on the performance. For shop "K-1" the ratio of shop Wtechn to Wcond is 9.8% with a minimum capacity of 70 t/day and 28.1% with a maximum capacity of 200 t/day. In general, for the production of "K-1" the considered ratio of Wtechn to Wcond decreases: for a minimum productivity of 70 tons/day - 6.9%, for a maximum productivity of 200 tons/day – 19.8%.

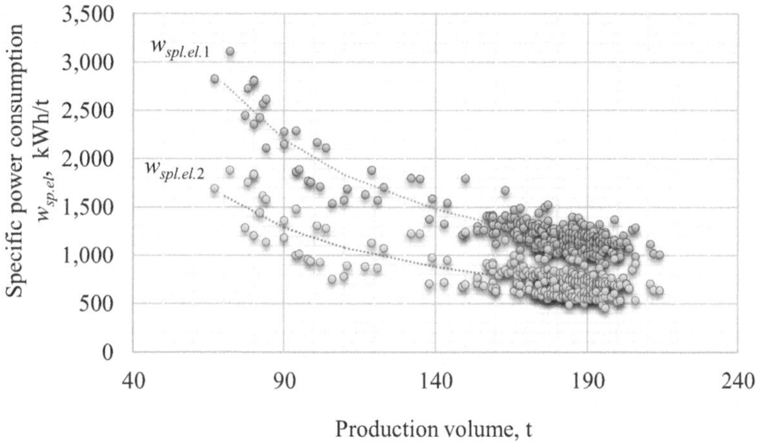

Fig. 1. Dependence of general production ($w_{spl.el.1}$) and workshop ($w_{spl.el.2}$) specific electricity consumption on the output of production of "K-1".

To improve energy efficiency, a new K-2 production facility was put into operation. Figure 2 shows the actual values of specific energy consumption for "K-1" and "K-2".

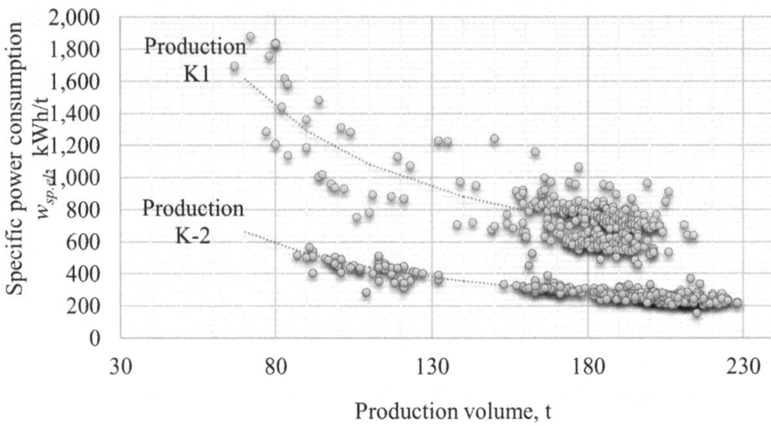

Fig. 2. Comparison of energy efficiency of production facilities “K-1” and “K-2”.

As we can see from Fig. 2, the implementation of the new production “K-2” allowed not only to increase the daily productivity from 200 t/day to 230 t/day, but also to significantly reduce the unit electricity consumption in the workshop. The ratio of the share of the conditionally constant component in the workshop power consumption and the ratio of the technological and conditionally constant component is presented in Table 3.

Introduction “K-2” into service allowed increasing the EEF by an average of 60%. At the same time, it should be noted that the ratio of the technological and general workshop component of electricity consumption for a production volume of 70 tons/day decreased from 9.8% to 6.6%, and for a maximum productivity of 200 tons/day it decreased by 9.3%. The share of the conditionally constant technological component in the general shop power consumption increased from 91% to 93.8% with a productivity of 70 tons/day, and from 78.1% to 84.2% with a daily productivity of 200 tons/day.

From the analysis of the data obtained, it follows that at the moment the production of “K-2” has reached the limit values for energy efficiency. There is no energy efficiency reserve in the technology. If we return to Fig. 3, it can be seen that for the productivity range Pday from 70 to 200 tons for the K-1 technology, the variation in the specific power consumption ranged from 1616.4 kWh/t to 659.8 kWh/t. At the same time, for the K-2 technology, with the same productivity variation, the specific consumption changes from 663.7 to 258.9 kWh/t. The ratio of the change in the specific power consumption Δwsp.k1 to the change in production volumes ΔPK1 in the first case is: (1616.4–659.8)/(200–70) = 956.6/130 = 7.36 kWh. In the second case, the ratio will be (663.7–258.9)/(200–70) = 404.8/130 = 3.11 kWh.

The authors carried out studies to assess the weight of the conditional-constant component of the consumption of fuel and energy resources in its total consumption for enterprises, individual industries of various industries on the basis of a formed database of daily, monthly, quarterly and annual regimes. Table 4 represents a fragment of the developed models for various industries.

Table 3. Comparison of the technological and conditional-constant components of electricity for the production of K1 and K2.

Production volume, P, t/day	Production «K1»				Production «K2»				Energy efficiency change, %
	Specific power consumption, $w_{sp.el}$, kWh/t	Power consumption, W. kWh	Part of the technological in the total power consumption, %	Part of constant in total power consumption, %	Specific power consumption, $w_{sp.el}$, kWh/t	Power consumption, W. kWh	Part of the technological in the total power consumption, %	Part of constant in total power consumption, %	
70	1616.4	46 457	6.6	93.8	663.7	113 145	9.8	91.0	−58.9
90	1289.3	47 277	8.5	92.2	525.3	116 041	12.7	88.8	−59.3
110	1081.2	48 097	10.3	90.6	437.2	118 937	15.5	86.6	−59.6
140	880.6	49 327	13.2	88.4	352.3	123 281	19.7	83.6	−60.0
170	750.7	50 557	16.0	86.2	297.4	127 625	23.9	80.7	−60.4
200	659.8	51 787	18.8	84.2	258.9	131 969	28.1	78.1	−60.8

Table 4. Parameters of models of specific expenditures of fuel and energy resources of industries in various industries.

Kind of product	Type of mathematical model	Units of specific energy consumption	Notes
Production «K-1»	$w_{sp}(P) = 144{,}8 + 103008{,}9/P$	kWh/t	Average annual model of specific electricity consumption
Liquid nitrogen fertilizers	$w_{sp}(P) = 73{,}81 + 29459{,}7/P$	kWh/t	
Methanol production	$w_{sp}(P) = 187{,}2 + 388930{,}3/P$	kWh/t	
Carbamide	$w_{sp}(P) = 674 + 11434{,}8/P$	Mcal/t	Average annual model of specific heat consumption
Liquid nitrogen fertilizers	$w_{sp}(P) = 101{,}2 + 87581{,}1/P$	Mcal/t	
Methanol production	$w_{sp}(P) = 101{,}2 + 87581{,}1/P$	Mcal/t	
Sheet glass	$w_{sp}(P) = 0{,}023 + 106{,}2/P$	Ton of fuel equivalent/t	Average annual model of specific gas consumption
Tires	$w_{sp}(P) = 5214 + 2196933/P$	Mcal/ conditional bus	Average monthly model of specific heat energy consumption
Alcohol production	$w_{sp}(P) = 4{,}86 + 2050{,}9/P$	Mcal/ (0,1·liter)	
Engineering products	$w_{sp}(P) = 0{,}055 + 147{,}51/P$	kWh/Kg	
Release of plastic products	$w_{sp}(P) = 1{,}04 + 16851/P$	kWh/Kg	Average monthly model of specific electricity consumption
Engineering products	Среднемесячная ЭЭ $w_{sp}(P) = 205{,}87 + 74461/P$	kWh/conditional unit	
Common consumption goods	$w_{sp}(P) = 9{,}7 + 42059/P$	Gcal/conditional unit	

Figure 3 shows the results of ranking the ratio of the weight of the conditional-constant electricity consumption in the total electricity consumption for the studied sample, consisting of 88 industries. The Wcond/W ratio is in a wide range from 2.9% (alcohol production) to 96.6% (mechanical engineering products).

Figure 4 provides a quantitative estimate of the share of constant electricity consumption in total electricity consumption. It was found that for 42% of enterprises in the sample, the share of a conditional-constant component in the total energy consumption is over 50%. The proportion of permanent electricity consumption can vary greatly depending on the production load. Its weight with a minimum volume of production can increase significantly (Fig. 3).

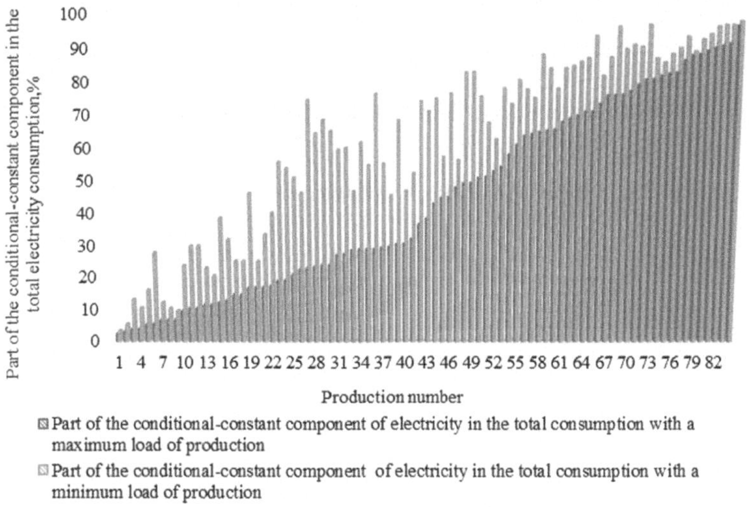

Fig. 3. The results of ranking the conditional-constant consumption to the total consumption of heat energy for the studied industrial enterprises.

For the studied group of enterprises, a 3-fold (and more) increase in the weight of the conditional-constant component of power consumption with a decrease in the volume of production was established at 6 industrial enterprises, 22 enterprises have 2 to 3 times, from 1.5 to 2 times - 19 enterprises, below 1.5 times - 38 enterprises.

For industries with high growth rates of the share of a conditional-constant from the total volume of consumed energy, with a decrease in the volume of output, the specific consumption of energy for output will rapidly grow. The highest multiplicity of the ratio of conditionally constant with the minimum and maximum volume of production was noted for industries with a conditionally constant share of 10 to 60% with the maximum volume of production.

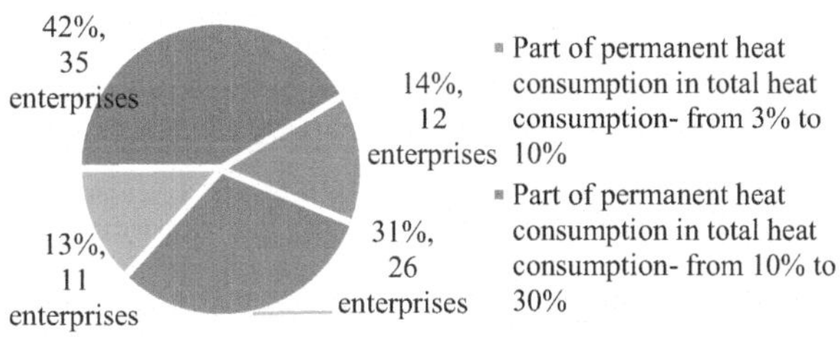

Fig. 4. Quantitative evaluation of the share of constant electricity consumption in total electricity consumption.

Figure 5 shows the results of the Wcond/W ranking for 32 enterprises using either fuel (natural gas) or thermal energy in the technological process. In contrast to electricity, the developed one-factor models of the dependence of the specific consumption of fuel or heat energy on the volume of output relate directly to technology. Structurally, the costs of heat energy for burning and ventilation of buildings and structures, as well as for hot water supply are allocated separately.

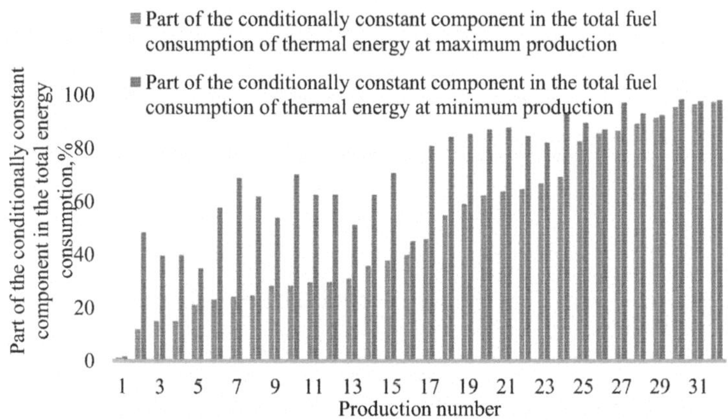

Fig. 5. The results of ranking the conditional-constant consumption to the total consumption of heat energy for the studied industrial enterprises.

Therefore, the obtained models of specific fuel or thermal energy consumption are analyzed for the allocation of a technological conditionally constant component, which can further influence the formation of the resulting energy efficiency when implementing energy saving measures. As can be seen from Fig. 5 and Fig. 6, 29 out of 32 industries have a high share of the technological conditional-constant component in the total technological energy consumption: from 20 to 98%.

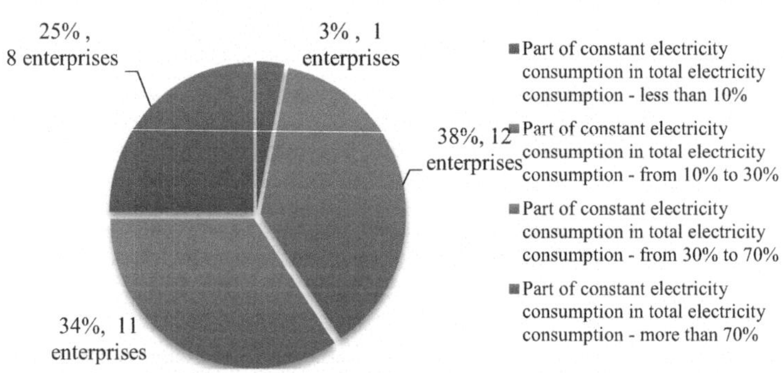

Fig. 6. Quantitative evaluation of the share of constant electricity consumption in total electricity consumption.

This group includes industries: tires, methanol, alcohol, sheet glass, ceramic products, etc. A necessary criterion for energy efficiency of such industries is maintaining a high technological load. In such conditions, the influence of the conditional-constant technological component on the formation of the resulting EEF is reduced to the maximum. For 24 out of 32 industries, with a decrease in the volume of production, the weight of the technological conditional-constant component of the fuel and energy resources consumption increases from 1.1 to 4.1 times, compared to the maximum production load.

3 Conclusions

For electricity, high values of the share of constant electricity consumption may indicate that the manufacture has not reached the design capacitance and is heavily overloaded with machinery that is used quite inefficiently. Such manufactures should be deeply analyzed by energy auditors.

Manufactures that have a higher weight of the conditional-constant component in the total energy consumption also have a greater horizontal regulation capacity for energy efficiency, i.e. the ability of both individual structural elements and the entire technical system as a whole to change the specific and total costs of energy resources by changing the volume of production.

References

1. Melentiev, L.: Selected Works, Methodology of System Researches in Power Engineering, p. 289. Nauka, Moscow (1995)
2. Voropay, N.: System Research in the Energy Sector: Retrospective of Scientific Directions of the SEI – ISEM, p. 686. Nauka, Novosibirsk (2010)
3. Ilkevich, N.: Multilevel Modeling of the Development of Gas Supply Systems, p. 217. Nauka, Novosibirsk (2014)
4. Ilkevich, N., Dzyubina, T., Kalinina, Z., Okuneva, S.: Simulation of the Equilibrium of the Flows of the Cost of Supply and Demand on Natural Gas, Pipeline Systems of Power Engineering: Methodical and Applied Problems of Mathematical Modelling, pp. 318–331. Nauka, Novosibirsk (2015)
5. Peregudov, F.: Introduction to System Analysis, p. 367. Vysshaja shkola, Moscow (1989)
6. Shevchenko, M.: Development of private models for monitoring the regional gas supply system. Technol. Audit Prod. Reserves **6**(2(26)), 40–46 (2015)
7. Galustov, G., Brovchenko, S., Meleshkin, S.: Mathematical Modeling and Forecasting in Technical Systems: Educational, p. 30. Publishing House of TTI JUFU, Taganrog (2008)
8. Hruntovich, N.V., Kapanski, A.A., Baczynski, D., Vagapov, G.V., Fedorov, O.V.: Optimization of a variable frequency drive pump working on a water tower. In: E3S Web of Conferences, vol. 124, p. 05060 (2019)
9. Fedorov, O.V.: Expeditious forecasting of power consumption. In: 2017 International Conference on Industrial Engineering. Applications and Manufacturing (ICIEAM). IEEE (2017)

10. Gracheva, E.I., Naumov, O.V.: Estimation of power losses in electric devices of the electrotechnical complex. In: International Conference on Industrial Engineering. Applications and Manufacturing (ICIEAM), 6 (2019)
11. Shpiganovich, A.A., Pushnitsa, K.A., Churkina, E.V., Fedorov, O.V.: Features of the operation of power supply systems of ferrous metallurgy enterprises. Ferrous Metals **5**, 56–61 (2017)
12. Gracheva, E.I., Alimova, A.N.: Calculating methods and comparative analysis of losses of active and electric energy in low voltage devices. In: International Ural Conference on Electrical Power Engineering (UralCon), pp. 361–367 (2019)
13. Gruntovich, N., Moroz, D., Shalonik, V.: About necessity of study the structure and regularities of the functioning of regional gas supply systems. Energeticheskaja strategija **1**, 42–46 (2017)
14. Moroz, D., Kapanskij, A.: Development of models of gas consumption by structural elements of the regional gas supply system. In: In the Collection: Energy Saving and Efficiency in Technical Systems, Materials of the IV International Scientific and Technical Conference of Students, Young Scientists and Specialists, pp. 416–418, Tambov, State Technical University (2017)
15. Moroz, D.R., Hruntovich, N.V.: Regional gas supply system considered from the standpoint of system analysis and regularities of its functioning. Energetika, Proc. CIS Higher Educ. Inst. Power Eng. Assoc. **61**(4), 359–371 (2018). https://doi.org/10.21122/1029-7448-2018-61-4-359-371. (in Russian)
16. Kantjukov, R., Sukharev, M., Meshalkin, V., Gimranov, R., Popov, A., Ryzhenkov, I.: Forecast gas consumption – the basis for making rational decisions on the structure and technological parameters in the design and reconstruction of the gas supply system. Electron. Sci. J. Oil Gas Bus. **1**, 201–221 (2015)
17. Lee, H.: Willis, Spatial Electric Load Forecasting. CRC Press, Boca Raton (2002)
18. Fildes, R.: The evaluation of extrapolative forecasting methods. Int. J. Forecast. **8**(1), 81–98 (1992)
19. Scott, A.J., Fildes, R.: Correspondence on the selection of error measures for comparisons among forecasting methods. J. Forecast. **14**(1), 67–71 (1995)
20. Kapanski, A., Hruntovich, N., Bakhur, S., Markaryants, L., Dolomanyak, L.: Optimize the cost of paying for electricity in the water supply system by using accumulating tanks. In: E3S Web of Conferences, vol. 178, p. 01065 (2020)

Retraction Note to: Thermal Conductivity of Geopolymer Concrete with Different Types of Aggregate

Oleg Krotov, Pavel Gromyko, Marina Gravit, Svetlana Belyaeva, Shuhrat Sultanov, Yuriy Lazarev, Marina Petrochenko, Viktor Velichkin, and Marcelo de Andrade Romero

Retraction Note to:
Chapter "Thermal Conductivity of Geopolymer Concrete with Different Types of Aggregate" in: A. Sinitsyn (Ed.): *Sustainable Energy Systems: Innovative Perspectives*, LNCE 141, https://doi.org/10.1007/978-3-030-67654-4_22

The Editor has retracted this chapter because it has been previously published by the same author group [1]. This chapter is therefore redundant.

Oleg Krotov, Pavel Gromyko, Marina Gravit, Svetlana Belyaeva and Shuhrat Sultanov agree to this retraction; Yuriy Lazarev, Marina Petrochenko, Viktor Velichkin and Marcelo de Andrade Romero have not responded to any correspondence from the editor/publisher about this retraction.

[1] Oleg Krotov et al 2021 IOP Conf. Ser.: Mater. Sci. Eng. 1030 012018 https://doi.org/10.1088/1757-899X/1030/1/012018

The retracted version of this chapter can be found at https://doi.org/10.1007/978-3-030-67654-4_22

A. Sinitsyn (Ed.): SES 2020, LNCE 141, p. C1, 2022.
https://doi.org/10.1007/978-3-030-67654-4_49

Author Index

A. Sinitsyn (Ed.): SES 2020, LNCE 141, pp. 471–473, 2021.
https://doi.org/10.1007/978-3-030-67654-4

The manufacturer's authorised representative in the EU is Springer Nature Customer Service Centre GmbH, Europaplatz 3, 69115 Heidelberg, Germany. If you have any concerns regarding our products, please contact ProductSafety@springernature.com

Printed and bound by CPI Group (UK) Ltd, Croydon, CR0 4YY

30/09/2026

02225650-0016